Thermal Structures for Aerospace Applications

Earl A. Thornton
University of Virginia
Charlottesville, Virginia

EDUCATION SERIES
J. S. Przemieniecki
Series Editor-in-Chief
Air Force Institute of Technology
Wright-Patterson Air Force Base, Ohio

Published by
American Institute of Aeronautics and Astronautics, Inc.
1801 Alexander Bell Drive, Reston, Virginia 20191-4344

American Institute of Aeronautics and Astronautics, Inc., Reston, Virginia

Library of Congress Cataloging-in-Publication Data

Thornton, Earl A. (Earl Arthur). 1936–
Thermal structures for aerospace applications/Earl A. Thornton.
p. cm. – (AIAA education series)
Includes bibliographical references and index.
1. Space vehicles – Thermodynamics. 2. Airframes – Design and construction. 3. Thermal stresses. 4. Aerothermodynamics.
I. Title. II. Series.
TL800, T48 1996 629.47–dc20 96-30497
ISBN 1-56347-190-6

Texts Published in the AIAA Education Series

Thermal Structures for Aerospace Applications
Earl A. Thornton, 1996
Structural Loads Analysis for Commercial Transport Aircraft: Theory and Practice
Ted L. Lomax, 1996
Spacecraft Propulsion
Charles D. Brown, 1996
Helicopter Flight Dynamics: The Theory and Application of Flying Qualities and Simulation Modeling
Gareth Padfield, 1996
Flying Qualities and Flight Testing of the Airplane
Darrol Stinton, 1996
Flight Performance of Aircraft
S. K. Ojha, 1995
Operations Research Analysis in Quality Test and Evaluation
Donald L. Giadrosich, 1995
Radar and Laser Cross Section Engineering
David C. Jenn, 1995
Introduction to the Control of Dynamic Systems
Frederick O. Smetana, 1994
Tailless Aircraft in Theory and Practice
Karl Nickel and Michael Wohlfahrt, 1994
Mathematical Methods in Defense Analyses Second Edition
J. S. Przemieniecki, 1994
Hypersonic Aerothermodynamics
John J. Bertin, 1994
Hypersonic Airbreathing Propulsion
William H. Heiser and David T. Pratt, 1994
Practical Intake Aerodynamic Design
E. L. Goldsmith and J. Seddon, Editors, 1993
Acquisition of Defense Systems
J. S. Przemieniecki, Editor, 1993
Dynamics of Atmospheric Re-Entry
Frank J. Regan and Satya M. Anandakrishnan, 1993
Introduction to Dynamics and Control of Flexible Structures
John L. Junkins and Youdan Kim, 1993
Spacecraft Mission Design
Charles D. Brown, 1992
Rotary Wing Structural Dynamics and Aeroelasticity
Richard L. Bielawa, 1992
Aircraft Design: A Conceptual Approach Second Edition
Daniel P. Raymer, 1992
Optimization of Observation and Control Processes
Veniamin V. Malyshev, Mihkail N. Krasilshikov, and Valeri I. Karlov, 1992
Nonlinear Analysis of Shell Structures
Anthony N. Palazotto and Scott T. Dennis, 1992
Orbital Mechanics
Vladimir A. Chobotov, 1991
Critical Technologies for National Defense
Air Force Institute of Technology, 1991

(Continued on next page)

Texts Published in the AIAA Education Series (Continued)

Defense Analyses Software
J. S. Przemieniecki, 1991
Inlets for Supersonic Missiles
John J. Mahoney, 1991
Space Vehicle Design
Michael D. Griffin and James R. French, 1991
Introduction to Mathematical Methods in Defense Analyses
J. S. Przemieniecki, 1990
Basic Helicopter Aerodynamics
J. Seddon, 1990
Aircraft Propulsion Systems Technology and Design
Gordon C. Oates, Editor, 1989
Boundary Layers
A. D. Young, 1989
Aircraft Design: A Conceptual Approach
Daniel P. Raymer, 1989
Gust Loads on Aircraft: Concepts and Applications
Frederic M. Hoblit, 1988
Aircraft Landing Gear Design: Principles and Practices
Norman S. Currey, 1988
Mechanical Reliability: Theory, Models and Applications
B. S. Dhillon, 1988
Re-Entry Aerodynamics
Wilbur L. Hankey, 1988
Aerothermodynamics of Gas Turbine and Rocket Propulsion,
Revised and Enlarged
Gordon C. Oates, 1988
Advanced Classical Thermodynamics
George Emanuel, 1988
Radar Electronic Warfare
August Golden Jr., 1988
An Introduction to the Mathematics and Methods of Astrodynamics
Richard H. Battin, 1987
Aircraft Engine Design
Jack D. Mattingly, William H. Heiser, and Daniel H. Daley, 1987
Gasdynamics: Theory and Applications
George Emanuel, 1986
Composite Materials for Aircraft Structures
Brian C. Hoskins and Alan A. Baker, Editors, 1986
Intake Aerodynamics
J. Seddon and E. L. Goldsmith, 1985
Fundamentals of Aircraft Combat Survivability Analysis and Design Robert E. Ball, 1985
Aerothermodynamics of Aircraft Engine Components
Gordon C. Oates, Editor, 1985
Aerothermodynamics of Gas Turbine and Rocket Propulsion
Gordon C. Oates, 1984
Re-Entry Vehicle Dynamics
Frank J. Regan, 1984

Published by
American Institute of Aeronautics and Astronautics, Inc., Reston, Virginia

Foreword

Aircraft structures designed for supersonic flight are subjected to aerodynamic heating caused by the air in the boundary layer being progressively slowed down. This process generates heat and consequently all external surfaces on the aircraft are heated. This in turn leads to nonuniform transient temperatures that produce thermal stresses and deformations. Likewise, spacecraft structures are subjected to aerodynamic heating during launch and re-entry phases of their operations, in addition to intensive solar radiation while operating in space.

The introduction of thermal effects into the theory of elasticity and structural analysis dates back to the first half of the last century when the French scientist J. M. C. Duhamel presented his papers on the generalization of the fundamental theorems of elasticity to include thermal strains and stresses caused by nonuniform temperature distribution in an elastic body. Since that time, technical literature covered the subject very extensively, but only very few textbooks appeared on the analysis and design of thermal structures. This recent addition to the AIAA Education Series discusses specifically the thermal behavior of aerospace structures. *Thermal Structures for Aerospace Applications* by Professor Earl A. Thornton is an authoritative source on the subject of design and analysis of aerospace structures in a thermal environment. This new text allows the reader to develop a good understanding of the physical behavior of thermal structures and to gain an appreciation for the role of classical thermal and stress analyses for thermal structures. In addition, the text discusses modern computational methods (e.g., finite element thermal-structural analysis) which provide a better insight into the realistic behavior of structural components or major structural assemblies subjected to thermal loading.

The text provides not only the structural analysis for thermal structures but also the accompanying thermal analysis, comprising aerodynamic heating and heat transfer from and within the structure itself. Also, many specialized problems with thermal aerospace structures are discussed, such as: thermally induced vibrations, thermal buckling, and thermoviscoplasticity.

The AIAA Education Series embraces a broad spectrum of theory and application of different disciplines in aeronautics and astronautics, including aerospace design practice. The series also includes texts on defense science, engineering, and technology. It serves as both teaching texts for students and reference materials for practicing engineers and scientists. *Thermal Structures for Aerospace Applications* will be a valuable text for graduate level courses and aerospace design teams. It complements several other texts on aircraft and spacecraft design previously published in this series.

J. S. Przemieniecki
Editor-in-Chief
AIAA Education Series

This page intentionally left blank

Table of Contents

This page intentionally left blank

Preface

As aircraft flight speeds have increased and space missions have mandated complex structures, the need for deeper understanding of aerospace thermal-structural behavior has grown. At the same time, the development of computer technology has provided engineers and researchers with the capability for performing sophisticated analyses under realistic flight conditions. The analyses are usually based on engineering principles implemented in complex computer programs that have evolved to address the details required for modern flight vehicle design.

Because a knowledge of fundamentals is required before complex computer analyses can be performed effectively, our purpose is to study basic problems. We delve into aerospace thermal-structural behavior so that readers may develop an understanding of the basic physical behavior of thermal structures, gain an appreciation for the role of classical engineering thermal and stress analyses, and apply computational methods to provide insight into realistic behavior.

The first two chapters of the book provide an overview of thermal-structural problems encountered in high-speed flight and orbiting space structures. Chapter 1 presents a review of basic aspects of high-speed flight, beginning with a historical description of the evolution of thermal structures from World War II through the National Aerospace Plane research of the 1980s. The basic concepts of aerodynamic heating are described, and aerothermal load effects are highlighted. Chapter 1 concludes with a brief description of thermal-structural design of high-speed vehicles. Chapter 2 provides a similar review of basic aspects of thermal structures for orbiting spacecraft but uses a slightly different approach. It begins by using a microwave radiometer spacecraft to illustrate some complexities of orbiting space structures. Then environmental heating is described, and fundamental features of thermal and structural behavior are introduced.

After these first two chapters have acquainted us with basic problems encountered in aerospace thermal structures, the next three chapters lay theoretical foundations for thermal-structural analysis by introducing the subjects of continuum mechanics, heat transfer, and thermoelasticity. Even though the chapters are designed to lay theoretical foundations, a practical approach is taken. The intent is to strive for clarity and understanding of basic principles rather than mathematical sophistication. Chapter 3 presents basic concepts and equations from continuum mechanics for stress, strain, the conservation laws, and constitutive behavior. Chapter 4 describes basic concepts, equations, and solutions for heat transfer in aerospace structures. It is a blend of classical and modern material. The classical material includes formulations and solutions for a number of boundary-value problems; the modern material introduces computational methods and provides several illustrations of temperature distributions in modern aerospace structures. Chapter 4 is a brief description of classical thermoelasticity. Again, the emphasis rests on presenting basic formulations, fundamental equations, and solutions for a few classical boundary-value problems. The chapters on heat transfer and thermoelasticity bridge the gap between the basicconcepts from continuum mechanics and practical thermal-stress analysis presented in the two chapters that follow.

Chapters 6 and 7 consider thermal-stress analysis of rods, beams, and plates. Here the emphasis is on basic engineering theory and formulations for classical boundary/initial-value problems. Solutions for a number of examples illustrate thermal-stress behavior of each idealization. These engineering formulations of thermal structures set the stage for the finite element analyses in the next chapter.

Chapter 8 presents a self-contained development of finite element analyses for thermal and structural problems. The method of weighted residuals is used to derive the finite element equations so that no prior knowledge of variational principles is assumed. Finite element formulations, solution algorithms, and numerical results are given for both heat transfer and thermal stress problems. Several examples illustrate the approach and demonstrate steady and transient thermal-structural responses of aerospace structures. The chapter provides the essential finite element background for the final chapters of the book.

Each of the three concluding chapters discusses a specialized thermal structures topic important for advanced aerospace applications: thermally induced vibrations, thermal buckling, and thermoviscoplasticity. Chapter 9 addresses thermally induced vibrations, beginning with a historical review from early theoretical work to more recent thermally induced vibration problems encountered on spacecraft appendages. Later, analyses of uncoupled and coupled flexural vibrations, as well as thermal flutter, are presented. Chapter 9 concludes with an analysis of thermally induced flexural vibrations of the Hubble Space Telescope solar arrays. Chapter 10 describes thermal buckling of beam columns, plates, and shells. Detailed thermal-buckling analyses of several beam-column and plate problems are presented, and shell buckling is introduced for a circular cylindrical shell. The book concludes with a discussion of the relatively new subject of thermoviscoplasticity. Unified constitutive models are illustrated by the Bodner–Partom model. A finite element formulation and a solution algorithm applicable to several constitutive models are developed in detail. Applications of finite element analyses are presented for convectively cooled hypersonic structures.

Throughout the book, numerous references are given to provide readers with detailed background and supplementary material. In addition, several chapters contain bibliographies on specialized subjects beyond the scope of the book. The subjects include supersonic transports, hyperbolic heat conduction, temperatures in aircraft structures, and effects of temperature on structural vibrations.

Interest in aerospace thermal structures over the last 40 years has been cyclical, depending on national priorities. Not many engineering schools offer courses in thermal structures routinely, and engineers often acquire basic principles of thermal structures without classroom instruction. For both classroom and self-instruction, this book serves as an introductory text that presents thermal structures in a clear, easy-to-understand manner. It is aimed primarily at aerospace and mechanical engineers, as well as students majoring in engineering mechanics and engineering science. The emphasis is on establishing a clear understanding of thermal structures to facilitate research and to solve practical, real-world problems. The engineering science and mathematics prerequisite is an advanced undergraduate background typical of U.S. academic programs. Numerous problems at the end of each chapter reinforce concepts in the text and introduce complementary material.

For a graduate course at the University of Virginia, I have used Chapters 1 and 2 as introductory material, lecturing on some topics and assigning the remainder as outside reading. A significant portion of the course focuses on Chapters 3–5, followed by selected topics from Chapters 6–11. These latter topics usually include lectures on beams and plates, but other topics depend on student interests and ongoing research. Although there is too much material to cover in one semester, students should be able to read the omitted portions of the text on their own.

While a graduate student at the University of Illinois in 1961, I first encountered *Theory of Thermal Stresses* by Professors Bruno A. Boley and Jerome H. Weiner. Their book was to influence me and a generation of thermal-stress analysts most significantly. For their contribution I am most appreciative and, if portions of this book bear any similarity to their classic text, it can only be because I have used their book so much as a text and reference.

I take this opportunity to thank my colleagues and students for their assistance in the preparation of *Thermal Structures for Aerospace Applications*. The origins of the book date from the 1970s, when I was teaching at Old Dominion University in Norfolk, Virginia, and conducting research in aerospace structures at the NASA Langley Research Center. Several Old Dominion students contributed to my research at that time, but two graduate students from that period deserve special mention for research efforts that contributed to this book. To Dr. Pramote Dechaumphai, now at Chulalongkorn University in Thailand, and Dr. Kumar Tamma at the University of Minnesota, I express most sincere appreciation for their efforts during that period. Over the last few years, students at the University of Virginia made important contributions both in research and manuscript preparation. Their names appear as co-authors with mine on several recent papers that are cited in the book. They have also used portions of the book as a text in a graduate course at the University of Virginia. Many homework and examination problems that they "debugged" appear as problems in this book. For their help, energy, and enthusiasm, I am very grateful. Paul Hernan, Rob Marino, and Joe Blandino drew a substantial number of new illustrations for the text and problems. Several other students, such as Yool Kim, David Gulick, Peter Chung, Jim Kolenski, and Rick Foster, performed numerical computations. Marshall Coyle, Rory McLeod, and Mark Rowley contributed experimental results.

Over the years, I have benefited greatly from research interactions at NASA Langley. Without that association, this book would not have happened. Many good friends (both retired and active) and former and current students deserve recognition, but I will not attempt to list them because of the possibility of unintentionally omitting someone. They all deserve my heartfelt thanks.

During the evolution of the book, I had several discussions with Dean John S. Przemieniecki of the Air Force Institute of Technology, who is the editor of this AIAA series. Not only were his editorial comments very valuable, but he provided several excellent papers on thermal stresses from his earlier work in the aerospace industry of Great Britain. For his suggestions, comments, encouragement, and patience, I am most grateful.

At the University of Virginia, Paulette Hughes performed numerous activities in the preparation of the manuscript. She obtained a great many references and performed a number of other tasks with skill, patience, and good humor. Thanks, too, to Karen Harper for her help in preparing the manuscript. Finally, the University of Virginia deserves special recognition. Through the University's Academic Enhancement Program, significant resources were invested in thermal-structures research, both in student support and development of the Thermal Structures Laboratory. The author is most pleased to acknowledge the university's significant contribution to the development of this book.

Earl A. Thornton
July 1996

1
High-Speed Flight

1.1 Introduction

The design of structures for winged flight vehicles that fly through the Earth's atmosphere, either to and from space or in sustained flight, poses severe challenges to structural designers. A major component of the challenge is the selection of materials and design structures that can withstand the aerothermal loads of high-speed flight. Aerothermal loads exerted on the external surfaces of the flight vehicle consist of pressure, skin friction (shearing stress), and aerodynamic heating (heat flux). Pressure and skin friction have important roles in aerodynamic lift and drag, but aerodynamic heating is the predominant structural load. Aerodynamic heating is extremely important because induced elevated temperatures can affect the structural behavior in several detrimental ways. First of all, elevated temperatures degrade a material's ability to withstand loads because elastic properties such as Young's modulus are significantly reduced. Moreover, allowable stresses are reduced, and time-dependent material behavior such as creep come into play. In addition, of course, thermal stresses are introduced because of restrained local or global thermal expansions or contractions. Such stresses increase deformation, change buckling loads, and alter flutter behavior. The advent of the National Aerospace Plane offered structural engineers new challenges for the design of thermal structures for high-speed flight.

1.2 Evolution of Thermal Structures

1.2.1 From World War II to Sputnik

The need to understand aerothermal loads and the design of thermal structures have their origins in the late 1940s. In World War II, airplane speeds had become high enough for compressibility phenomena to have a significant role in performance. Transonic phenomena were not well understood and, over a period of years, the phrase *sound barrier* came into use. The need for a transonic research airplane was recognized during the war and, in 1944, the design development of the Bell X-1 program was initiated.[1] The X-1 was enormously successful, and the flight of Captain Charles E. Yeager on October 14, 1947, proved beyond doubt that manned aircraft could fly faster than the speed of sound. An advanced version of the aircraft, the X-1B shown in Fig. 1.1 flew several research missions for NACA to study aerodynamic heating effects. The original X-1 aircraft, as well as the advanced version, used aluminum construction throughout. Measured skin temperatures are shown for a NACA research mission flown in January 1957 at Mach 1.94. Note that skin temperatures are low, less than 200°F. Thereafter, supersonic flight speeds increased rapidly, and the need to consider aerodynamic heating became evident. The difficulties presented by high temperatures accompanying flight at supersonic speeds became known as the *thermal barrier*. For about 10 years, the thermal barrier caused concern that large structural weight increases would be required to keep material temperatures within allowable values.

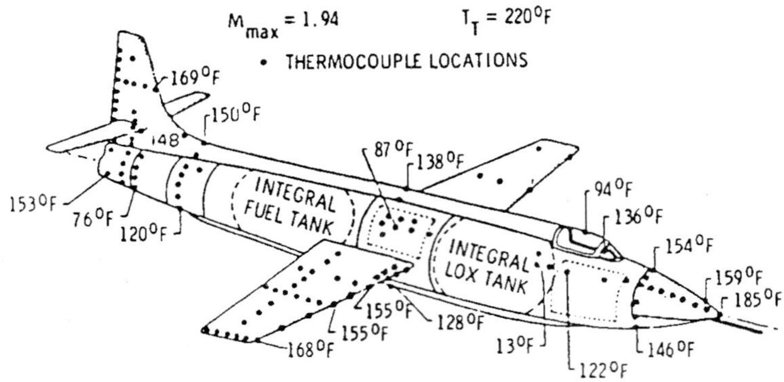

Fig. 1.1 Maximum measured temperatures on X-1B airplane, Mach 1.94, 1957 (Ref. 12).

Subsequently, researchers[2–3] found that these concerns did not materialize because the problems were overcome through research and development of effective thermal structures.

After the first supersonic flight, research and development of high-speed aircraft intensified. A contract for the design, development, and construction of two X-2 swept-wing, supersonic research aircraft was awarded to the Bell Aircraft Corporation in 1947. The X-2 was the first aircraft structure designed for aerodynamic heating.[4] Until the X-2, speeds had not been high enough for the structure to be affected adversely by aerodynamic heating. For increased strength at elevated temperatures, the fuselage was constructed from K-Monel, and aerodynamic skins used stainless steel. A drawing shown in Fig. 1.2 illustrates the X-2 structure. The X-2 became the first research aircraft to explore the thermal barrier with speeds above Mach 2.5. On September 27, 1956, the X-2 achieved its maximum speed of Mach 3.2, but the aircraft went out of control and the pilot, Captain Milburn G. Apt, was killed.

During this period, theoretical and experimental studies were in progress to develop the technologies needed for high-speed aircraft. Considerable research was under way in compressible flow and thermal structures and materials. Excellent descriptions of these efforts are available through survey articles and collections of

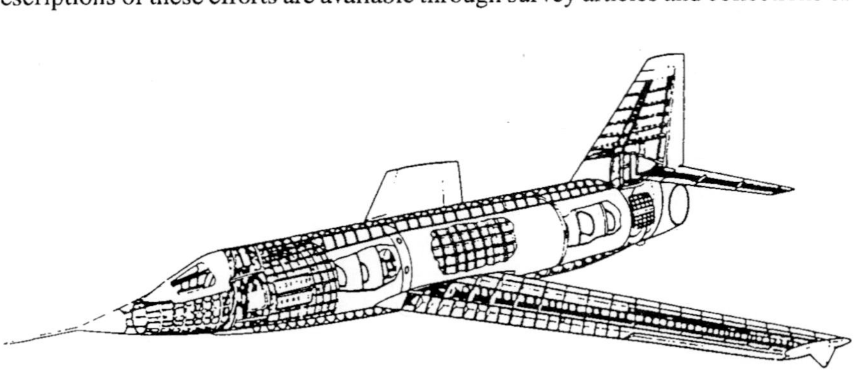

Fig. 1.2 X-2, K-Monel internal structure and stainless steel skin, Mach 3.2, 1956 (Ref. 4).

papers in books describing thermal-structures conferences. The November 1955 *Applied Mechanics Review* article by Professor N. J. Hoff[5] of Polytechnic Institute of Brooklyn (now Polytechnic University) describes high-temperature effects in aircraft structures, with emphasis on the effects of temperature on buckling and creep. In December 1955, Professor R. L. Bisplinghoff presented the Nineteenth Wright Brothers Lecture to the Institute of Aeronautical Sciences in Washington, D.C. The paper,[6] including discussion by several prominent researchers, was published in April 1956. Professor Bisplinghoff's paper gives a comprehensive review of structural considerations for high-speed flight and an excellent account of design and analysis practices. A book published by the Advisory Group for Aeronautical Research and Development (AGARD) of the North Atlantic Treaty Organization (NATO) in 1958[7] describes effects of high temperatures on aircraft structures caused by aerodynamic heating.

The next major flight program that stimulated thermal-structural research was the X-15. The X-15 had complex origins, including the prewar and postwar work of German scientists Eugene Sanger and Irene Bredt who, in 1944, outlined a hypersonic, rocket-propelled aircraft. The evolution of their ideas, which contributed to the development of the X-15, is described in the award-winning paper by Richard P. Hallion.[8] Further descriptions of the X-15 program are given by Hallion[9] and by NASA Langley research scientist John V. Becker.[10]

On June 24, 1952, the prestigious NACA Committee on Aeronautics charged the agency with studying problems of manned and unmanned flight at altitudes between 12 and 50 miles and speeds of Mach 4 to Mach 10. By 1954, the NACA Langley Laboratory (now NASA's Langley Research Center) had formed a hypersonic study team with John Becker as the chairman. According to Becker,[10] the unprecedented problems of aerodynamic heating and high-temperature structures appeared so formidable that they were viewed as "barriers" to hypersonic flight. Nevertheless, the NASA group evolved a baseline design that closely resembled the ultimate X-15 configuration. Three X-15 aircraft were built by North American Rockwell.

A thick-skinned, heat-sink approach was adopted to suit the short-duration missions of the X-15. A typical research mission lasted 10–12 min.[11] Surfaces exposed to aerodynamic heating were made of Inconel X, a nickel alloy. Internal structures not subject to high temperatures were made of titanium. Skin temperatures were designed for a maximum of 1200°F. Figure 1.3 shows maximum temperatures experienced in an X-15 flight.

On October 4, 1957, the Soviet Union orbited *Sputnik I*, the world's first artificial satellite. This event changed the nation's priorities for high-speed, high-altitude

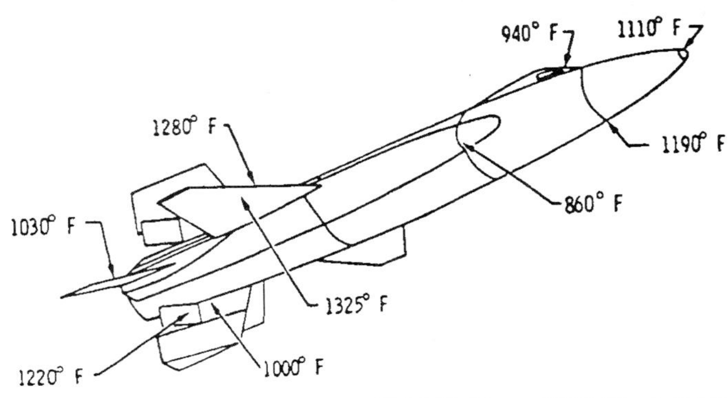

Fig. 1.3 Measured temperatures on an X-15 flight, ca. 1965, Mach 5.0 (Ref. 11).

flight, making the X-15 program vital to America's national prestige. The X-15 program served the nation well, accomplishing 199 missions between 1959 and 1968. The X-15 was the first and, to date, the only manned vehicle capable of flying atmospheric missions at Mach 5 for altitudes of 100,000 ft or higher. It made many contributions to the understanding of hypersonic flight, including the design of thermal structures.

From World War II to *Sputnik*, the nation's research programs were focused on high-speed flight of aircraft. In thermal structures, the NACA Langley laboratory made significant contributions. Many of the Langley research efforts were led by Richard R. Heldenfels. After his retirement, he described research conducted at Langley from 1948 to 1958 on structural problems caused by aerodynamic heating.[12]

1.2.2 After Sputnik

After the success of *Sputnik*, the U.S. high-speed flight research program was broadened to include a major emphasis on manned space flight. The National Aeronautics and Space Act of 1958 created NASA, and NACA became NASA on October 1, 1958.

The expanded scope of the research is reflected in the 1960 paper by Heldenfels,[3] which describes proposed re-entry structures as well as space vehicles and space structures. The proceedings[13] of a conference held in Cambridge, Massachusetts, on July 25, 1961, contain papers describing research on thermal structures for manned and unmanned re-entry vehicles. The conference was concerned primarily with thermal protection systems for lifting vehicles. Two types of thermal protection systems are described: *cool-structure* and *hot-structure* approaches. In the cool-structure approach, the load-bearing structure is insulated from high temperatures by an external heat shield. In the hot-structure approach, the load-bearing structure operates at nearly skin temperatures.

A NASA–University Conference on the Science and Technology of Space Explorations was held in Chicago, Illinois, on November 1–3, 1962. The conference proceedings describe research on structure for launch vehicles, winged aerospace vehicles, and planetary entry vehicles. The paper by Mathauser[14] gives a good account of Langley research on thermal-structural problems for winged vehicles. Types of winged vehicles under consideration by NASA include a research airplane (the X-15), a re-entry glider, and a large hypersonic aircraft. The re-entry glider is representative of a X-20 Dyna-Soar vehicle that was to be launched with a ground-based booster. The hypersonic aircraft was to possess horizontal takeoff capability and be able to sustain hypersonic flight. All three vehicles utilize radiation-cooled hot structures. Radiation equilibrium temperatures [see Eq. (1.13)] estimated for a re-entry glider are shown in Fig. 1.4.

The re-entry glider and hypersonic aircraft of the Mathauser paper are hypothetical vehicles used in NASA's fundamental research studies. However, the Dyna-Soar project sponsored by the U.S. Air Force did lead to the final design of the Boeing X-20. The events leading to the Dyna-Soar program, its evolution, and subsequent demise are described by Miller[4] and Hallion.[8] The Boeing development program began with a contract award in November 1959. The X-20 was designed to provide a piloted, maneuverable vehicle for conducting experiments in the hypersonic and orbital flight regime. However, the X-20 program was canceled in December 1963 before the first vehicle was completed. The Dyna-Soar program accelerated progress in several technologies that ultimately were applicable to the

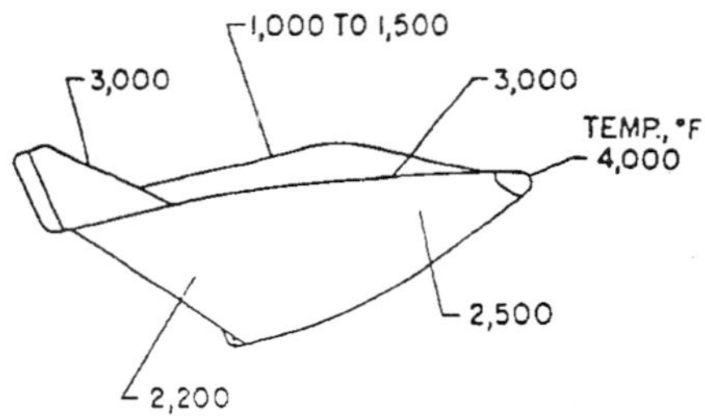

Fig. 1.4 Radiation equilibrium temperatures for a re-entry glider computed by NASA in 1962 (Ref. 14).

Space Shuttle. A review of the Dyna-Soar winged spacecraft technology program appears in the 1961 paper by Yoler.[15] The structural design utilized a René 41 nickel superalloy primary structure, a columbium alloy heat shield, a graphite and zirconia nose cap, and molybdenum alloy leading edges.

Concurrent with these activities, the United States' manned space flight program accelerated rapidly. The effort began with the Mercury program, including the Alan B. Shepard Jr. suborbital flight on May 5, 1961, and America's first orbital flight by John H. Glenn Jr. on February 20, 1962. The effort continued with the Gemini program and the first two-man flight by John W. Young and Virgil I. Grissom on March 23, 1965. The Gemini program achieved the first rendezvous and docking in space and the first American "space walk." The Apollo program began with the October 11–22, 1968, flight by Walter M. Shirra Jr., Donn F. Eisele, and R. Walter Cunningham. The program reached its zenith with the historic flight of Apollo 11 by Neil A. Armstrong, Edwin E. Aldrin Jr., and Michael Collins and the first lunar landing on July 20, 1969. The lunar program concluded with *Apollo 17* making the sixth and last lunar landing in December 1972. America's first Earth-orbiting space station, Skylab, was launched atop a Saturn V booster on May 14, 1973. Three 3-man crews visited the space station, with the last mission returning to Earth in February 1974. The rendezvous and docking of an Apollo spacecraft with a Russian Soyuz craft in Earth orbit on July 18, 1975, closed out the Apollo program.

All of the manned spacecraft missions used blunt-body re-entry vehicles and ablative heat shields to dissipate aerodynamic heating. These ballistic or semiballistic vehicles had small lift-to-drag ratios, permitting only limited maneuverability.

The lack of a follow-on winged research flight vehicle to succeed the X-15 forced thermal-structures research in the 1960s into a period of more fundamental, rather than applied, research. The 1962 paper by Mathauser[14] indicates that research had begun at Langley on hypersonic structures. A 1966 paper by Heldenfels[16] discusses in more detail structural prospects for hypersonic vehicles. Figure 1.5 shows isotherms on a hypersonic vehicle assumed to be cruising at Mach 8 at 88,000 ft. Temperatures shown are radiation equilibrium temperatures expected during a typical flight of one to two hours' duration. The paper discusses thermal-structural designs for fuselage tanks for liquid hydrogen, wings, and engine structures. Hydrogen fuel-cooled structures for engines and passive hot structures of high-temperature materials for airframes are described. The proceedings of a conference held at Langley in November 1971 contain several papers describing basic hypersonic vehicle research; a paper by Anderson and Kelly[17] reviews the technology base for propulsion structures, primary structures, and liquid hydrogen tanks. A paper by Shore[18] reviews research on convectively cooled structures in the 1960s and 1970s.

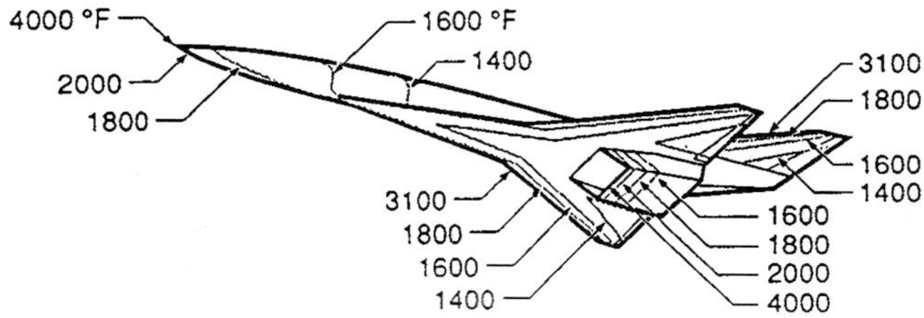

Fig. 1.5 Equilibrium surface temperatures for a NASA hypersonic vehicle concept for sustained flight at Mach 8 at 88,000 ft, 1966 (Ref. 16).

1.2.3 Space Shuttle

The Space Shuttle resulted from a perceived need in the 1960s for a logistical spacecraft to support orbital space stations. After the lunar landing in 1969, however, NASA recognized that funds would not be available to support both the Shuttle and a space station. Justification for the Shuttle shifted from space station support to its use as a substitute for expendable launch systems. With strong support from the Department of Defense, the preliminary analysis phase began in February 1969 with contracts to Lockheed, General Dynamics, McDonnell Douglas, and North American-Rockwell. Rockwell eventually became the prime contractor, and construction of the actual Shuttle Orbiter began in June 1974. Rockwell completed this vehicle, the *Enterprise*, in September 1976. Difficulties with the main engines and thermal protection system delayed the second Shuttle, the *Columbia*, until 1981. Piloted by astronauts John W. Young and Robert L. Crippen, *Columbia* completed the Shuttle's first orbital flight on April 14, 1981. Further details of its history are given by Hallion.[8]

The Orbiter basically has a conventional skin-stringer aluminum aircraft structure. The design of the thermal protection system (TPS) required that structural temperatures be kept at less than 350°F. The thermal protection is composed of two types of reusable surface insulation (RSI) tiles. The RSI tiles covering the Orbiter are made of coated silica fiber. The two types differ only in surface coating to provide protection for different temperature environments. The low-temperature insulation (LRSI) consists of 8-in.-square silica tiles and covers the top of the vehicle, where temperatures are less than 1200°F. The high-temperature insulation (HRSI) consists of 6-in.-square tiles that cover the bottom and some leading edges of the Orbiter, where temperatures are below 2300°F. Reinforced carbon–carbon (RCC) is used for the nose cap and wing leading edges, where temperatures are above 2300°F. Flexible, reusable surface insulation (FRSI) is used at locations where temperatures are less than 700°F. Orbiter isotherms for a normal flight are shown in Fig. 1.6. Further details of the Orbiter are given in Ref. 19. The launch of the first Shuttle flight was delayed by problems with the TPS; an article by Langley researchers Paul A. Cooper and Paul F. Holloway[20] explains the problems and their resolution.

1.2.4 National Aerospace Plane (X-30)

In his State of the Union address in February 1986, President Ronald Reagan made the initial announcement of research on development of the National Aerospace Plane (NASP). The November 1986 article by Robert L. Williams[21] defines the program objective: "to develop, and then demonstrate in an experimental

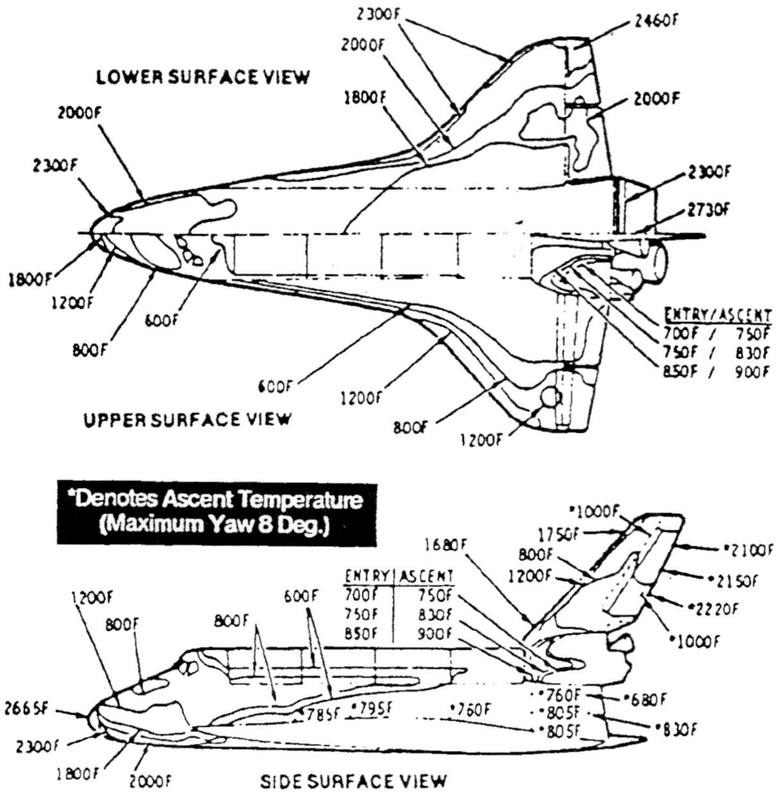

Fig. 1.6 Space Shuttle expected surface temperatures, 1976 (Ref. 19).

flight vehicle (the X-30), the requisite technologies to permit the nation to develop both military and civil vehicles capable of operating at sustained hypersonic speeds within the atmosphere and/or operating as space launch vehicles for delivering payloads into orbit." Under joint sponsorship by NASA and the Department of Defense (DOD), technology development began in 1986. By 1988, McDonnell Douglas, General Dynamics, and Rockwell had won separate contracts to proceed with development of the airframe. Pratt and Whitney and Rocketdyne had won contracts for propulsion system development. A consortium of these contractors participated in a 30-month materials and structures augmentation project.[22]

The competitive nature of the development program and security issues have limited publications describing thermal structures for the X-30. However, at the 30th AIAA Structures, Structural Dynamics, and Materials Conference, David A. Ellis of the McDonnell Douglas Corporation presented an overview paper[23] describing the concepts under consideration as well as the types of trade studies being conducted. Figure 1.7 shows representative surface temperatures for a typical NASP hypersonic condition. Candidate primary structure design concepts include hot structures, actively cooled structures, and thermally protected structures. The design of the center fuselage is particularly challenging because the external skin is subjected to aerodynamic heating whereas the internal structure is exposed to cryogenic temperature due to the hydrogen fuel.

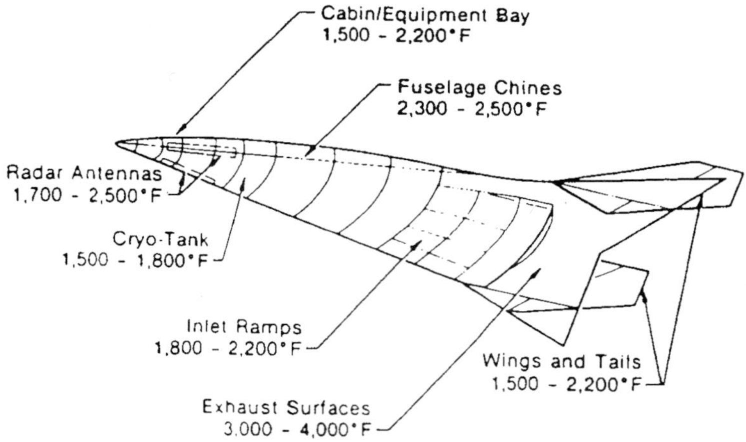

Fig. 1.7 Representative surface temperatures for the NASP, 1989 (Ref. 23).

1.3 Aerodynamic Heating

The prediction of aerodynamic heating has challenged thermal analysts since before World War II. The problem involves solving the equations that describe the conservation of mass, momentum, and energy for fluid flow. For altitudes less than 300,000 ft, the atmosphere may be assumed to be a continuum, and the set of conservation equations are known as the Navier–Stokes equations. For two-dimensional flow, the Navier–Stokes equations governing viscous, compressible flow are a set of four nonlinear, partial differential equations with mixed hyperbolic, parabolic, and elliptic behavior. A solution of these equations with appropriate boundary conditions and constitutive equations for the fluid provides distributions for the density ρ, the velocity components u and v, the pressure P, and the temperature T throughout the fluid. At the fluid/aerodynamic surface interface, the aerodynamic heating is computed from Fourier's law using the fluid thermal conductivity and the derivative of temperature normal to the wall.

Prior to the digital computer, there was no hope of solving the complete Navier–Stokes equations. Instead, analysts devoted substantial effort to obtaining approximate solutions, and with the help of experimental data, they developed engineering relations to predict aerodynamic heating as well as skin friction. The engineering relations typically are based on boundary-layer or other approximations. The relations are restricted by a number of assumptions, and the application of computers to the problem of aerodynamic heating has contributed significantly to a better understanding of the phenomenon and improved reliability of predictions.

There is substantial literature describing aerodynamic heating, and the esoteric details will be left for thermal analysts. Because of the interdisciplinary nature of thermal-structural problems for high-speed vehicles, however, some fundamental aspects are presented here. Early approaches for the prediction of aerodynamic heating are described in the 1956 paper by Van Driest[24] and in the 1960 text by Robert W. Truitt.[25] A further discussion appears in the 1989 text by John D. Anderson Jr.[26]

The preceding section mentioned that aerodynamic heating effects become significant at Mach numbers above 2.5. For Mach numbers up to about 5, the flow is considered supersonic, and above Mach 5, the flow is termed *hypersonic.* The

NASP will experience aerodynamic heating through the supersonic and hypersonic regimes, with a peak Mach number of about 25. As the Mach number increases above 5, new physical phenomena become of progressively greater importance, making hypersonic flow more complex than supersonic flow. These phenomena include 1) fluid dynamics effects that limit the validity of boundary layer approximations and 2) high-temperature effects that introduce chemical reactions. In addition, at altitudes above 300,000 ft, low-density effects restrict the application of continuum models.

1.3.1 Classical Boundary-Layer Predictions

The concept of the boundary layer was introduced by Prandtl in 1903 for incompressible flows. The extension of the concept to compressible flows began in the 1930s, and analytical work continued until the mid-1950s. Thereafter, most research turned to the development of computational methods for solving aerodynamic heating problems.

The boundary-layer equations are obtained from the Navier–Stokes equations by an order-of-magnitude analysis for the thin fluid layer next to the body, where viscous effects dominate. Outside the boundary layer, the fluid is assumed inviscid, and the flow is described by the Euler equations. For steady, two-dimensional laminar flow, the boundary-layer equations are

$$\begin{gathered}\frac{\partial(\rho u)}{\partial x}+\frac{\partial(\rho v)}{\partial y}=0\\ \frac{\partial u}{\partial x}+\rho v\frac{\partial v}{\partial y}=\frac{\partial}{\partial y}\left(\mu\frac{\partial u}{\partial y}\right)\\ \frac{\partial P}{\partial y}=0\\ \rho u\frac{\partial(c_pT)}{\partial x}+\rho v\frac{\partial(c_pT)}{\partial y}=\frac{\partial}{\partial y}\left(k\frac{\partial T}{\partial y}\right)+u\frac{\partial P_e}{\partial x}+\mu\left(\frac{\partial u}{\partial y}\right)^2\end{gathered}\tag{1.1}$$

In the preceding equation, P_e denotes the pressure at the edge of the boundary layer, ρ is density, and T is temperature. The fluid specific heat is c_p, the thermal conductivity k, and the viscosity μ. In general, these fluid properties are temperature-dependent. The boundary-layer equations are nonlinear but exhibit parabolic behavior. The perfect-gas law is used to relate pressure, density, and temperature. For further details, including boundary conditions, see Ref. 26.

For some simple but important cases, analytical solutions to the boundary-layer equations have been obtained. The approach involves making changes of independent and dependent variables to yield simpler, nonlinear, nondimensional equations. These transformations lead to the concept of self-similar solutions for problems such as a flat plate. Typically, the nondimensional equations can be solved numerically for various values of the Mach number M, the Reynolds number Re, and the Prandtl number Pr.

The solution for the flow over a flat plate obtained by Van Driest in 1952[27] demonstrates typical features. The problem consists of a flat plate with specified uniform temperature T_w (Fig. 1.8). The inviscid flow outside the boundary layer has constant velocity u_∞ and temperature T_∞. Within the boundary layer, the viscosity varies with temperature according to Sutherland's law,[26] and the Prandtl

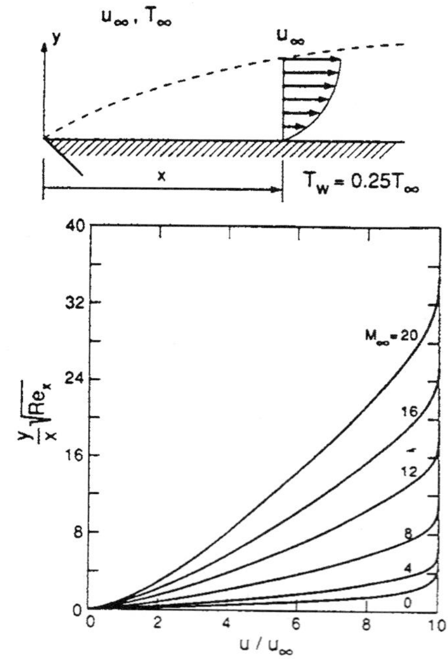

a) Velocity profiles

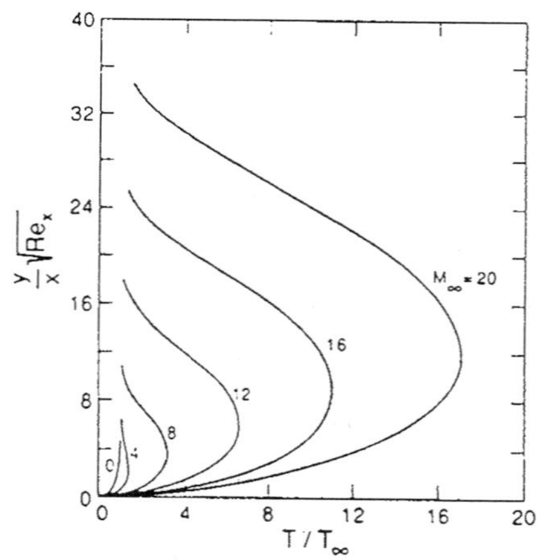

b) Temperature profiles

Fig. 1.8 Compressible boundary-layer flow, 1952 (Ref. 27).

number $Pr = \mu c_p/k$ is assumed constant ($Pr = 0.7$). The ratio of the wall to freestream temperature is taken as $T_w/T_\infty = 0.25$. Figure 1.8a shows typical u velocity profiles, and Fig. 1.8b shows typical temperature profiles. In these figures, the Reynolds number $Re_x = \rho_\infty u_\infty x/\mu_\infty$. The flat-plate boundary-layer solution illustrates results representative of high-speed flows. The velocity profiles show that the boundary-layer thickness increases rapidly with increasing Mach number.

In fact, Anderson[26] shows that a hypersonic boundary-layer thickness increases approximately as Mach number squared. One consequence of a large hypersonic boundary-layer thickness is that the viscous boundary layer alters the outer inviscid flow, limiting the accuracy of boundary-layer predictions. This behavior is called *viscous interaction.*

The temperature profiles show the fluid behavior that is of great importance to the structure. The important point is that the peak temperature within the boundary layer is higher than the freestream temperature. Moreover, this peak temperature increases rapidly with increasing Mach number. The high fluid temperature within the boundary layer is due to viscous dissipation. Viscous dissipation is the process in which the large kinetic energy of the high-speed flow is converted to thermal energy by boundary-layer shearing stresses. The aerodynamic heating rate is proportional, by Fourier's law, to the slope of the temperature profile at the wall ($y = 0$). Figure 1.8b shows that fluid temperatures within the boundary layer rapidly increase with increasing Mach number. This is the basic reason for the need for special thermal structures for high-speed flight.

Classical, self-similar boundary solutions have also been derived for the stagnation region of a blunt body. An important result of this analysis was to show for hypersonic flows that the stagnation heating rate varies inversely with the square root of the blunt-body radius. Thus, to reduce aerodynamic heating, the vehicle nose and leading-edge regions need to be as blunt as possible.

1.3.2 Convective Boundary Condition

For determining structural temperatures, aerodynamic heating is often represented as a convective boundary condition. Boundary-layer analysis of compressible flow shows that the heat transfer between the boundary layer and the wall can be expressed as

$$q_w = h(T_{\text{aw}} - T_w) \tag{1.2}$$

where q_w is the local heating rate (e.g., Btu/ft^2-s) at the wall, h is the convection coefficient, T_{aw} the adiabatic wall temperature, and T_w the wall temperature. The adiabatic wall temperature is the temperature the fluid attains for no heat transfer between the fluid and wall. Temperature profiles illustrating the heat-transfer possibilities for high-speed flow are shown in Fig. 1.9. The figure illustrates that, for wall temperatures less than T_{aw}, the fluid flow heats the wall; when the temperature at the wall is T_{aw}, there is zero heat transfer; and, for wall temperatures greater than T_{aw}, the wall heats the fluid. Equation (1.2) shows that aerodynamic heating is directly proportional to the difference between the adiabatic wall temperature and the actual wall temperature. The adiabatic wall temperature is always higher than the freestream temperature T_∞ and "drives" the heat transfer.

The heat-transfer coefficient is often written in nondimensional form as

$$C_H = \frac{h}{\rho_e u_e c_p} \tag{1.3}$$

or, using Eq. (1.2), as

$$C_H = \frac{q_w}{\rho_e u_e c_p (T_{\text{aw}} - T_w)} \tag{1.4}$$

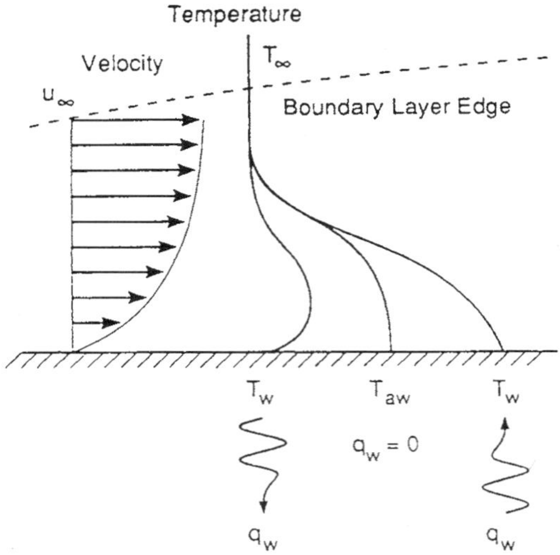

Fig. 1.9 Velocity and temperature profiles in high-speed flow.

where C_H is the Stanton number. Another alternative is to define the Stanton number in terms of fluid enthalpies, that is,

$$C_H = \frac{q_w}{\rho_e u_e (H_{\text{aw}} - H_w)} \tag{1.5}$$

where H_{aw} is the adiabatic wall enthalpy and H_w the wall enthalpy. In this approach, enthalpy replaces temperature as an unknown in the flow analysis. For a calorically perfect gas (constant specific heats), $h = c_p T$; otherwise, $\mathrm{d}h = c_p(T)\mathrm{d}T$.

Thus, to use a convective boundary condition, it is necessary to know the Stanton number and the adiabatic wall temperature. Both, in general, are local quantities; that is, they vary with position along the wall. The classical flat-plate boundary-layer solution will serve as an illustration. Extensive studies of the flat-plate problem[24] show that the adiabatic wall temperature can be calculated with good accuracy by

$$T_{\text{aw}} = T_\infty \left[1 + r \frac{\gamma - 1}{2} M_\infty^2 \right] \tag{1.6}$$

where γ is the ratio of fluid specific heats $\gamma = c_p / c_v$ and r is the recovery factor. Analytical and experimental results show that the recovery factor can be computed as

$$r = P_r^{1/2} \qquad \text{(Laminar flow)}$$
$$r = P_r^{1/3} \qquad \text{(Turbulent flow)}$$

The boundary-layer analysis for the flat-plate flow shows that the Stanton number has the form

$$C_H = \frac{f(M_e, P_r, \delta, T_w, T_e)}{\sqrt{Re_x}} \tag{1.7}$$

where f denotes a functional relationship. Eckert,[28] in 1956, showed that, by using the concept of a reference temperature, a simple formula for C_H could be developed from the corresponding result for low-speed incompressible flow. This approach, called the *reference temperature method*, computes the Stanton number from

$$C_H = \frac{0.332}{\sqrt{Re_x^*}}(P_r^*)^{-2/3} \qquad \text{(Laminar)} \tag{1.8a}$$

$$C_H = \frac{0.185}{(\log_{10} Re_x^*)^{2.584}}(P_r^*)^{-2/3} \qquad \text{(Turbulent)} \tag{1.8b}$$

where Re_x^* and Pr^* are evaluated at a reference temperature T^*; that is,

$$Re_x^* = \frac{\rho^* u_e x}{\mu^*} \qquad P_r^* = \frac{\mu^* c_p^*}{k^*} \tag{1.9}$$

where P^*, c_p^*, and k^* are evaluated at a reference temperature T^*. The reference temperature is computed as

$$T^* = T_e[1 + 0.032M_e^2 + 0.58(T_w/T_e - 1)] \tag{1.10}$$

where, as before, the subscript e denotes quantities at the edge of the boundary layer. Further details, including expressions for skin friction and equations for turbulent flow, are given in the Eckert paper,[28] the text by Anderson,[26] and other convective heat-transfer texts.

1.3.3 Limitations of Classical Boundary-Layer Theory

Classical boundary-layer theory provides basic insight to the understanding of the fundamentals of aerodynamic heating. The boundary-layer solution, in some cases, yields very practical engineering results that are frequently used. But boundary-layer theory has limitations that deserve mentioning.

Classical self-similar solutions are limited to a few problems with simple geometries (e.g., the flat plate) and simple boundary conditions (e.g., constant wall temperature). In the age of the computer, however, this limitation is not very serious because the boundary-layer equations, Eqs. (1.1), are solved numerically. In fact, since the equations are parabolic, marching methods may be used, and very efficient computer programs have been developed. When combined with inviscid flow programs to determine variables at the edge of the boundary layer, very effective analytic procedures have been developed. Some of these methods are described in a 1987 paper by Fred R. DeJarnette et al.,[29] which reviews approximate methods for aerodynamic heating analysis.

There are several flow situations to which the boundary-layer approximation of uncoupling the inviscid and viscous analysis does not apply. Two such situations are those in which viscous interaction or flow separation occurs. As mentioned earlier, viscous interaction refers to flows in which the viscous boundary layer interacts with the inviscid flow. Such a situation may occur for hypersonic flows with relatively low Reynolds numbers. Boundary-layer thicknesses vary inversely with the square root of the Reynolds numbers, and so low Reynolds number flows have thick boundary layers. Flow separation refers to the case in which the flow turns away from the wall into itself, introducing a local recirculation region next to the wall. Flow separation does not occur when the pressure decreases in the flow direction. However, when the pressure increases in the flow direction (called an *adverse pressure gradient*), flow separation occurs, and boundary-layer theory does not apply. Flow separation may occur in several problems of importance in high-speed flight. X-15 experience provides two such examples: 1) flow over protuberances and 2) shock/boundary-layer interactions. In both examples, an adverse pressure gradient occurs and flow separation develops. A consequence of flow separation in both instances is to increase significantly the local aerodynamic heating. For these problems, boundary-layer theory is not applicable, and the phenomenon is governed by the Navier–Stokes equations.

1.3.4 Navier–Stokes Solutions

One of the major advances in computational mechanics in the last four decades is the development of the capability of solving the Navier–Stokes equations computationally. The survey article by Dwoyer et al.[30] describes the capability of computational fluid dynamics (CFD) for hypersonic aircraft. The Navier–Stokes equations, written in conservation form, are

$$\frac{\partial}{\partial t}\{U\} + \frac{\partial}{\partial x}\{E\} + \frac{\partial}{\partial y}\{F\} = 0 \tag{1.11}$$

where $\{U\}$ is a vector of conservation variables and $\{E\}$ and $\{F\}$ are vectors of the flux components in the x and y directions, respectively. These vectors are

$$\begin{aligned}
\{U\}^T &= [\rho \quad \rho u \quad \rho v \quad \rho\mathcal{E}] \\
\{E\}^T &= [\rho u \quad \rho u^2 + P \quad \rho uv \quad \rho u\mathcal{E} + Pu] \\
&\quad - [0 \quad \sigma_x \quad \tau_{xy} \quad u\sigma_x + v\tau_{xy} - q_x] \\
\{F\}^T &= [\rho v \quad \rho uv \quad \rho v^2 + P \quad \rho v\mathcal{E} + Pv] \\
&\quad - [0 \quad \tau_{xy} \quad \sigma_y \quad u\tau_{xy} + v\sigma_y - q_y]
\end{aligned} \tag{1.12}$$

where $\mathcal{E}$ is the total energy. Each of the flux vectors contains two vectors of components representing inviscid and viscous flux contributions. The conservation equations are supplemented by a thermodynamic equation of state relating pressure, temperature, and density (typically, the perfect-gas law). In the viscous flux components, the stresses σ_x, σ_y, and τ_{xy} are related to the velocity gradients, assuming Stokes' hypothesis. The heat fluxes q_x and q_y are related to temperature gradients by Fourier's law. For air, the temperature-dependent viscosity is computed from Sutherland's law, and the thermal conductivity is computed assuming a constant Prandtl number of 0.72. Considerable success has been achieved in numerical solutions for the two-dimensional Navier–Stokes equations using supercomputers. Numerical solution schemes include finite difference, finite volume and, more recently, finite element methods. Three-dimensional flows remain a challenge although a viscous solution[31] around an X-24c lifting body has been obtained. Such solutions, even with the largest supercomputers, are not capable of resolving three-dimensional flow details, particularly aerodynamic heating rates. An important problem for which Navier–Stokes solutions provided valuable insight was shock interaction phenomenon for the NASP.

In 1967, NASA conducted a series of X-15 flights with a dummy hypersonic ramjet engine mounted on a pylon under the rear of the fuselage. On the third

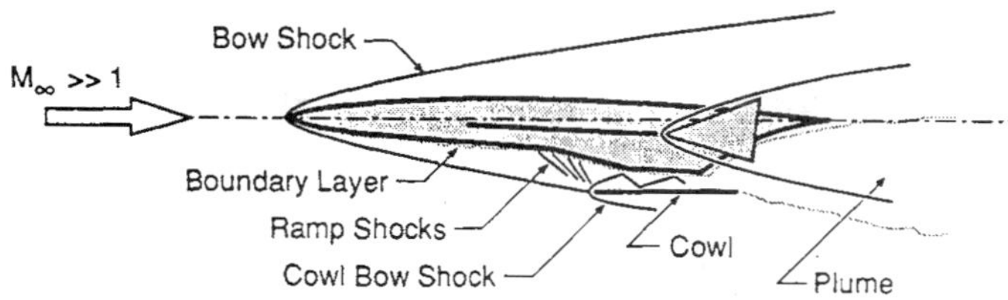

Fig. 1.10 Overall flowfield for the NASP (Ref. 34).

flight with the dummy engine on October 3, 1967, the X-15 reached a maximum Mach number of 6.7 at an altitude of 99,000 ft. During the flight,[32] severe pylon structural damage was experienced as a result of complex shock impingement and interference effects on local aerodynamic heating. Since then, shock interference heating has been recognized as a critical problem for high-speed vehicles because extreme pressure and heat-transfer rates can occur in highly localized regions where the interference pattern impinges on the surface.

Shock interference heating was an important consideration in the design of the cowl leading edge of the engine structure of the NASP. The problem was strong motivation for studies of shock interference heating on leading edges. Figure 1.10 shows the overall flowfield for the NASP. The interaction between the aircraft bow shock and the cowl bow shock can cause severe local aerothermal loads on the cowl leading edge. Shock interference heating on leading edges has been studied experimentally by Wieting and Holden[33] and computationally by Klopfer and co-workers[34–35] as well as Thareja et al.[36] Dechaumphai et al.[37] studied the flow–thermal-structural behavior of leading edges. Figure 1.11 shows the problem statement for a Navier–Stokes solution and a comparison of experimental and predicted heating-rate distributions. The comparative surface heating rates (Fig. 11b) show very good agreement, but the paper points out that there was some difference in predictions of the undisturbed heating rate q_0 that normalizes the curves. Note that the results show that the interference heating rate is almost 15 times the undisturbed level. Peak interference heating rates for NASP flight conditions and geometries can be as high as 70,000 Btu/ft^2-s. Such high local heating presents a formidable challenge to the structural designer. To put this value in perspective, a typical 1-kW portable hair dryer produces about 30 Btu/ft^2-s; solar heating is 0.12 Btu/ft^2-s.

A discussion of aerothermodynamics of transatmospheric vehicles from the design perspective appears in the papers of Tauber and Menees[38] and Tauber and Adelman.[39]

1.4 Aerothermal Load Effects on Flight Structures

In his classic 1956 paper, Professor R. L. Bisplinghoff[6] identified the basic structural and aeroelastic considerations for high-speed flight. From a structural perspective, the considerations include 1) deterioration of mechanical properties at elevated temperatures, 2) thermal stresses introduced by temperature gradients, 3) modification of stiffness and vibration properties, and 4) aeroelastic instabilities. These considerations remain important today. The general problem consists of determining the behavior of the flight structure under aerothermal loads, primarily aerodynamic pressures and heating. The behavior of a flight structure can be classified according to whether the response is quasistatic or dynamic. In

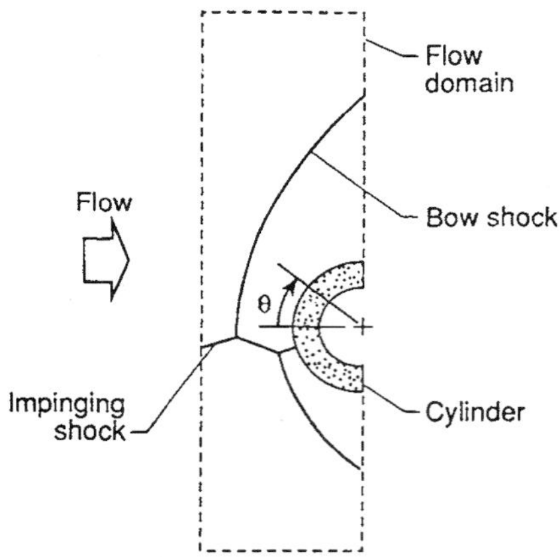

a) Shock interaction

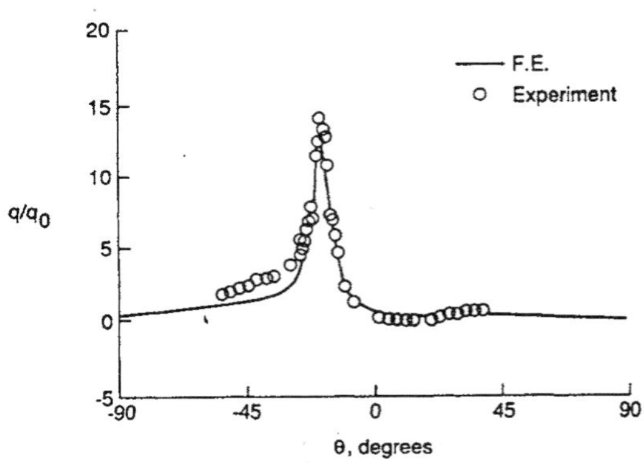

b) Comparative surface heating rate distribution

Fig. 1.11 Shock interaction problem for cylinder (Ref. 37).

a quasistatic response, inertia forces are negligible, and the structure responds to the aerothermal loads slowly. In a dynamic response, inertia forces have a significant role, and the structure responds with vibratory motions. The determination of the quasistatic response of structures has traditionally been called *thermal stress analysis*. The determination of the structural dynamic response, considering the interaction of the deforming structure and the aerodynamic pressure in the presence of aerodynamic heating, is called *aerothermoelasticity*. An area of recent concern is the response of structures to acoustic loads. The determination of the dynamic response of a structure to random fluctuating pressures in the presence of aerodynamic heating may be called *aerothermoacoustics*.

1.4.1 Quasistatic Interactions

To determine the quasistatic response, a logical approach is to separate the aerothermal-structural problem into distinct uncoupled problems by assuming weak coupling between the external aerodynamic flow and the structural response. This assumption is permissible when structural deformations are too small to alter the external flow. During X-15 flights,[9,10] several unexpected thermal problems were encountered as a result of intense local aerodynamic heating stemming from quasistatic interactions. Windshield damage occurred when thermal buckling of the retainer frame caused intense local heating in the glass. These problems were identified and solved during flight tests. According to Ref. 10, however, "the really important lesson here is that what are minor and unimportant features of a subsonic or supersonic aircraft must be dealt with as prime design problems in a hypersonic airplane."

A more recent example[40] of a quasistatic flow–structural interaction is thermal protection systems tested in the Mach 7 eight-foot high-temperature tunnel at the NASA Langley Research Center. The tests show that panels "bowed up" into the flow to produce heating rates that are up to 1.5 times greater than flat-plate predictions. Thornton and Dechaumphai[41] used a finite element approach to study coupled flow, thermal, and structural behavior of aerodynamically heated panels. For the Mach 6.6 conditions studied, panel deformations introduce shocks, expansions, and recirculation regions in the flow. The effect of convex panel deformation is to increase local heating rates on windward surfaces and decrease them on leeward surfaces.

In the overview paper[23] on the design of an airframe structure for NASP, the interaction between the external flow and the thermoelastic deflections of a movable wing is described. The paper notes that consideration of the interaction resulted in the prediction of lower deflections and stresses, which translated into reduced structural weight.

The recent analyses of quasistatic flow–structural interactions illustrate some of the important effects of this behavior on flight structures. Because of their multidisciplinary nature, however, the analyses are difficult, expensive, and additionally, have not been validated by experimental data.

1.4.2 Dynamic Interactions

In a 1963 paper, I. E. Garrick[42] of NASA Langley surveyed developments in aerothermoelasticity. The classical aeroelastic triangle representing interactions among the fields of aerodynamics, elasticity, and inertia was extended to include thermal effects. A new figure, the aerothermoelastic tetrahedron, illustrated the interdisciplinary aspects of aerothermoelasticity. The paper also discusses aeroelastic consideration of the X-15, effects of transient heating on vibration frequencies, and panel flutter. Early in the X-15 flight program, the pilot reported a rumbling noise at high dynamic pressures. This turned out to be panel flutter of large areas of the skin of the side fairings and tail. The problem was strong motivation for NASA studies in the 1960s and 1970s of panel flutter at elevated temperatures. An assessment[43] of flutter model testing relating to the NASP provides an excellent summary of flutter literature for supersonic and hypersonic flight.

The response of aerodynamically heated structures to acoustic loads is an important consideration because of acoustic fatigue. Reference 44 notes that, as demonstrated by operational experience for a variety of aircraft, intense acoustic pressures can cause fatigue failures of lightweight structures. Preliminary estimates of acoustic loadings for the NASP indicate pressure levels well into the range in which acoustic fatigue failures have occurred in the past. Thermal effects are important because 1) acoustic data for high-temperature structures and materials are scarce, 2) thermal prestrain and buckling can affect strain levels significantly, and 3) temperature can change material properties and the fatigue lifetime as expressed in the stress vs number of fatigue cycles (S–N) diagram. Efforts to develop computational methods for aerothermoacoustics are described in the dissertation by Locke[45] and the paper by Locke and Mei.[46]

1.5 Design of Thermal Structures

The design of thermal structures is a complex process that involves consideration of the flight vehicle trajectory, aerothermal loads, and thermal structural concepts and materials. A near-optimum design involves tradeoffs among these and other factors. The design of thermal structures is too complex to be discussed in detail, but fundamentals will be cited.

1.5.1 Flight Regime

To determine aerothermal loads and heat transfer to the vehicle, the flight trajectory must be determined. The trajectory is determined by mission requirements. The basic equations of flight mechanics, hypersonic aerodynamics, and re-entry heating appear in the text by Wilbur L. Hankey.[47] Design studies[48] for hypervelocity vehicles and for the NASP[23] describe relationships between mission requirements and thermal-structural design concepts.

A parameter used in the design of thermal structures is radiation equilibrium temperature, that is, the upper level that the surface of a structure can reach. An energy balance at the surface of a structure at radiation equilibrium states that the aerodynamic heat flux given by Eq. (1.2) is equal to the heat flux emitted by radiation, or

$$q_w = h(T_{\text{aw}} - T_r) = \sigma \epsilon T_r^4 \tag{1.13}$$

where T_r is the radiation equilibrium temperature, ϵ the surface emissivity, and σ the Stefan–Boltzmann constant. Equation (1.13) states that all incident aerodynamic heat flux is emitted by radiation; none of the incident flux is conducted into the structure. Actual surface temperatures may be lower than radiation equilibrium temperatures because of conduction heat transfer into the structure.

Two key factors of the flight regime for thermal structural design are radiation equilibrium temperatures and exposure times. Radiation equilibrium temperatures for the X-15, the Space Shuttle, and the NASP can be compared in Figs. 1.3, 1.6, and 1.7, respectively. The exposure time for the NASP vis-à-vis other flight vehicles is shown in Fig. 1.12. The important point is that the design of hypersonic cruise vehicles like the NASP differs from previous designs because of much longer flight durations. A third important factor of hypersonic flight is use of liquid hydrogen fuel. Among other considerations, cryogenic hydrogen introduces significant structural temperature gradients. Both maximum temperatures and gradients influence structural design concepts and material selection.

1.5.2 Thermal Protection Systems

Hypersonic flight vehicles require thermal protection systems to withstand sustained aerothermal loads. The thermal protection system, the supporting airframe, and the engine structure are examples of thermal structures. The design of a thermal protection system is based on the principle that the energy transmitted by the hot boundary-layer flow must be absorbed or rejected by the thermal protection system. Figure 1.12 identifies thermal protection systems used by hypersonic flight vehicles. For relatively short missions, the Mercury and Apollo spacecraft absorbed the thermal energy through ablative heat shields. The Space Shuttle absorbs re-entry heating and thermally insulates the airframe by a very effective, but fragile, tile system. For relatively short missions, the X-15 absorbed the aerodynamic heating by using a skin and an airframe of high-temperature metallic

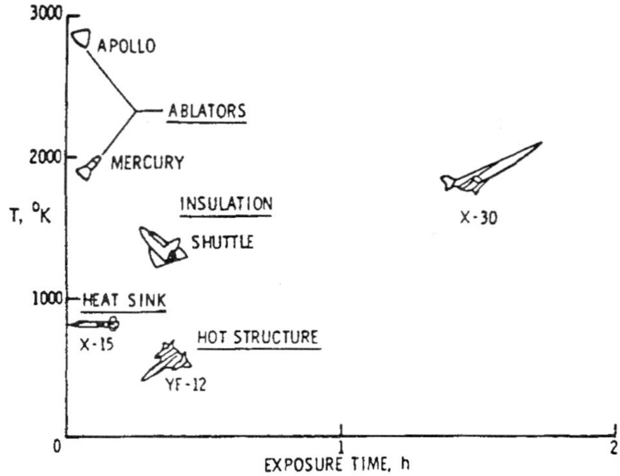

Fig. 1.12 Flight regimes for hypersonic vehicles (after Ref. 17).

materials. For long-duration but lower-temperature flights, the YF-12 (or SR-71) used a high-temperature titanium structure with high emissivity surface coating at radiation equilibrium to reject the aerodynamic heating. For sustained, very high temperature flights, a combination of several thermal protection concepts is likely to be used. Candidate thermal protection systems for the NASP include hot structures, insulated structures, and convectively cooled structures. Convectively cooled airframe structures are relatively new, with no previous flight experience. Design and experimental studies[18] conducted over the last 20 years indicate that these systems will be effective in absorbing thermal energy in regions of intense local heating, such as shock interference heating on engine structures.

1.5.3 Materials

The evolution of thermal structures for high-speed flight vehicles has placed ever-increasing demands on material performance. Structural designs typically require high-stiffness, thin-gauge materials that can be fabricated into complex, built-up structures. Thermal structural designs typically require high-strength, low-density materials that retain their desirable properties at elevated temperatures. For the NASP there was a substantial research effort to develop new aerospace materials to meet these challenges. References 49–56 describe progress in material development for thermal structures.

The performance of the NASP depended on the development of new, lightweight materials that can perform at temperatures higher than today's materials can stand. Today's high-temperature materials, such as nickel-based alloys, cannot be used because structural weight limits mandate lighter-weight materials. The structural weight limits require the use of thin-gauge, low-density alloys and composites. Figure 1.13 compares specific strength and stiffness of several advanced materials. Below 2000°F, the structural designer has several alternatives but, above 2000°F, choices are limited.

In addition to good thermal-structural performance, a material must perform well in its chemical service environment. The chemical service environment normally encountered in hot structures is an oxidizing environment. The effects of

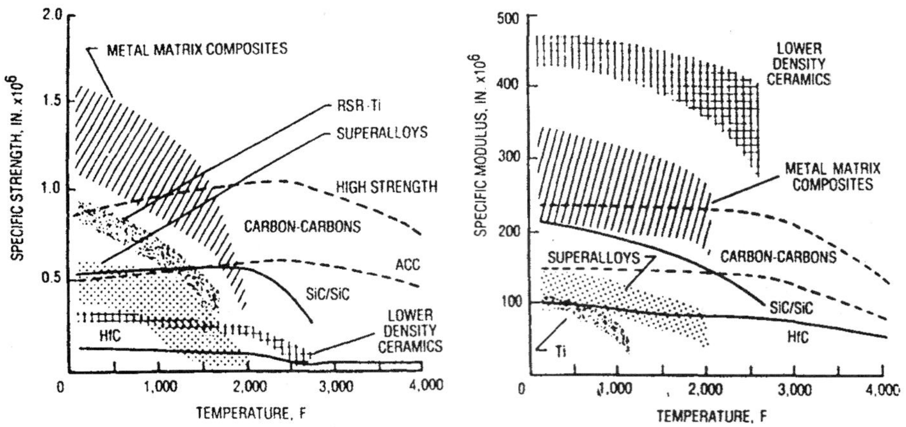

Fig. 1.13 Specific strength and stiffness of advanced materials (Ref. 52).

oxidation are addressed by the use of oxidation barrier coatings. When the chemical environment includes hydrogen, however, such as is encountered in NASP engine structures, there are significant unresolved problems. Little experience exists on the effects of hydrogen on the thermal-structural behavior of most aerospace structural materials, but evidence indicates that exposure to hydrogen at elevated temperature seriously degrades material performance. Hydrogen environment effects on advanced alloys and composites are described by Howard G. Nelson.[57] The hydrogen compatibility problem is an emerging challenge for the development of the next generation of high-temperature materials.

1.6 Closing Remarks

This chapter traces the evolution of thermal structures from the early days of supersonic flight to the research and development for the National Aerospace Plane. Fundamental concepts of aerodynamic heating, aerothermal load effects on flight structures, and design of thermal structures are reviewed. Major advances in technology have occurred that provide the foundations for the design of thermal structures for flight vehicles in the twenty-first century.

References

[1]Hallion, R. P., *Supersonic Flight, The Story of the Bell X-1 and Douglas D-558*, Macmillan, New York, 1972.

[2]Hoff, N. J., "The Thermal Barrier-Structures," *Transactions of the American Society of Mechanical Engineers*, Vol. 77, No. 5, 1955, pp. 759–763.

[3]Heldenfels, R. R., "Frontiers of Flight Structural Design," *Aeronautics and Astronautics*, Proceedings of the Durand Centennial Conference held at Stanford University, Aug. 5–8, 1959, edited by N. J. Hoff and W. G. Vincenti, Pergamon, New York, 1960.

[4]Miller, Jay, *The X-Planes, X-1 to X-29*, Specialty Press, Marine on St. Croix, MN, 1983.

[5]Hoff, N. J., "High Temperature Effects in Aircraft Structures," *Applied Mechanics Reviews*, Vol. 8, No. 11, 1955, pp. 453–456.

[6]Bisplinghoff, R. L., "Some Structural and Aeroelastic Considerations of High Speed Flight," The Nineteenth Wright Brothers Lecture, *Journal of the Aeronautical Sciences*, Vol. 23, No. 4, 1956, pp. 289–329.

[7]Hoff, N. J. (ed.), *High Temperature Effects in Aircraft Structures*, published for AGARD, NATO, Pergamon, New York, 1958.

[8]Hallion, R. P., "The Path to the Space Shuttle: The Evolution of Lifting Reentry Technology," *Journal of the British Interplanetary Society*, Vol. 36, No. 12, 1983, pp. 523–541.

[9]Hallion, R. P., "X-15: Highest and Fastest of Them All," *Flight International*, Dec. 23, 1978, pp. 2256–2262.

[10]Becker, J. V., "The X-15 Program in Retrospect," *Raumfahrtforschung*, March–April 1969, pp. 45–53.

[11]Stillwell, W. H., "X-15 Research Results with a Selected Bibliography," NASA SP-60, 1965.

[12]Heldenfels, R. R., "Historical Perspectives on Thermostructural Research at the NACA Langley Aeronautical Laboratory from 1948 to 1985," NASA TM 83266, Feb. 1982.

[13]Glaser, P. E. (ed.), *Aerodynamically Heated Structures*, Prentice–Hall, Englewood Cliffs, NJ, 1962 (proceedings of a conference sponsored by Air Force Office of Scientific Research and Arthur D. Little, Inc.).

[14]Mathauser, E. E., "Research, Design Considerations, and Technological Problems of Structures for Winged Aerospace Vehicles," NASA SP-28, Dec. 1962.

[15]Yoler, Y., "Dyna-Soar—A Review of the Technology," *Aerospace Engineering*, Vol. 20, No. 8, 1961, pp. 8–9, 56–67.

[16]Heldenfels, R. R., "Structural Prospects for Hypersonic Air Vehicles," *Aerospace Proceedings 1966*, Vol. 1, edited by J. Bradbrooke, J. Bruce, and R. R. Dexter, Spartan Books, New York, 1967, pp. 561–583.

[17]Anderson, M. S., and Kelly, H. N., "Structures Technology for Hypersonic Vehicles," *Vehicle Technology for Civil Aviation, The Seventies and Beyond*, NASA SP-292, Nov. 1971, pp. 179–192.

[18]Shore, C. P., "Review of Convectively Cooled Structures for Hypersonic Flight," NASA TM-87740, May 1986.

[19]Staff of Johnson Space Center, *Space Shuttle*, NASA SP-407, 1976.

[20]Cooper, P. A., and Holloway, P. F., "The Shuttle Tile Story," *Astronautics and Astronautics*, Vol. 19, Jan. 1981, pp. 24–36.

[21]Williams, R. L., "National Aerospace Plane: Technology for America's Future," *Aerospace America*, Nov. 1986, pp. 18–22.

[22]Newman, N. D., and Pinckert, R. E., "Materials for the NASP," *Aerospace America*, Vol. 27, May 1989, pp. 24–25.

[23]Ellis, D. A., "Overview-Design of an Efficient Lightweight Airframe Structure for the National Aerospace Plane," AIAA Paper 89-1406CP, April 1989; see also *Proceedings of 10th Structures, Structural Dynamics, and Materials Conference*, AIAA, Washington, DC, 1989, pp. 2157–2163.

[24]Van Driest, E. R., "The Problem of Aerodynamic Heating," *Aeronautical Engineering Review*, Vol. 15, No. 10, 1956, pp. 26–41.

[25]Truitt, R. W., *Fundamentals of Aerodynamic Heating*, Ronald Press, New York, 1960.

[26]Anderson, J. D. Jr., *Hypersonic and High Temperature Gas Dynamics*, McGraw–Hill, New York, 1989.

[27]Van Driest, E. R., "Investigation of Laminar Boundary Layer in Compressible Fluids Using the Crocco Method," NACA TN 2597, Jan. 1952.

[28]Eckert, E. R. G., "Engineering Relations for Heat Transfer and Friction in High-Velocity Laminar and Turbulent Boundary-Layer Flow Over Surfaces with Constant Pressure and Temperature," *Transactions of the ASME*, Vol. 78, No. 6, 1956, pp. 1273–1283.

[29]DeJarnette, F. R., Hamilton, H. H., Weilmuenster, K. J., and Cheatwood, F. M., "A Review of Some Approximate Methods Used in Aerodynamic Heating Analysis," *Journal of Thermophysics,* Vol. 1, No. 1, 1987, pp. 5–12.

[30]Dwoyer, D. L., Kutler, P., and Povinelli, L. A., "Retooling CFD for Hypersonic Aircraft," *Aerospace America,* Vol. 25, Oct. 1987, pp. 32–41.

[31]Shang, J. S., and Scherr, S. J., "Navier–Stokes Solution of the Flow Field Around a Complete Aircraft," AIAA Paper 85-1509, July 1985.

[32]Watts, J. D., "Flight Experience with Shock Impingement and Interference Heating on the X-15-2 Research Airplane," NASA TM-X-1669, Oct. 1968.

[33]Wieting, A. R., and Holden, M. S., "Experimental Study of Shock Wave Interference Heating on a Cylindrical Leading Edge at Mach 6 and 8," AIAA Paper 87-1511, June 1987.

[34]Klopfer, G. H., and Yee, H. C., "Viscous Hypersonic Shock-on-Shock Interaction on Blunt Cowl Lips," AIAA Paper 88-0233, Jan. 1988.

[35]Klopfer, G. H., Yee, H. C., and Kutler, P., "Numerical Study of Unsteady Viscous Hypersonic Blunt Body Flows with an Impinging Shock," NASA TM-100096, April 1988.

[36]Thareja, R. R., Stewart, J. R., Hassan, O., Morgan, K., and Peraire, J., "A Point Implicit Unstructured Grid Solver for the Euler and Navier–Stokes Equations," AIAA Paper 88-0036, Jan. 1988.

[37]Dechaumphai, P., Thornton, E. A., and Wieting, A. R., "Flow-Thermal-Structural Study of Aerodynamically Heated Leading Edges," *Journal of Spacecraft and Rockets,* Vol. 26, No. 4, 1989, pp. 201–209.

[38]Tauber, M. E., and Menees, G. P., "Aerothermodynamics of Transatmospheric Vehicles," *Journal of Aircraft,* Vol. 24, No. 9, 1987, pp. 594–602.

[39]Tauber, M. E., and Adelman, H. G., "Thermal Environment of Transatmospheric Vehicles," *Journal of Aircraft,* Vol. 25, No. 4, 1988, pp. 355–363.

[40]Shideler, J. L., Webb, G. L., and Pittman, C. M., "Verification Tests of Durable Thermal Protection System Concepts," *Journal of Spacecraft and Rockets,* Vol. 22, No. 6, 1985, pp. 598–604.

[41]Thornton, E. A., and Dechaumphai, P., "Coupled Flow, Thermal, and Structural Analyses of Aerodynamically Heated Panels," *Journal of Aircraft,* Vol. 25, No. 11, 1988, pp. 1052–1059.

[42]Garrick, I. E., "A Survey of Aerothermoelasticity," *Aerospace Engineering,* Vol. 22, No. 1, 1963, pp. 140–147.

[43]Reed, III, W. H., Hanson, P. H., and Alford, W. J. Jr., "Assessment of Flutter Model Testing Relating to the National Aerospace Plane," NASP Contractor Rept. 1002, Dynamic Engineering Inc., July 1987.

[44]Mixson, J. S., and Roussos, L. A., "Acoustic Fatigue: Overview of Activities at NASA Langley," NASA TM 89143, April 1987.

[45]Locke, J. E., "A Finite Element Formulation for the Large, Deflection Random Response to Thermally Buckled Structures," Ph.D. Dissertation, Old Dominion Univ., Norfolk, VA 1988.

[46]Locke, J. E., and Mei, C., "A Finite Element Formulation for the Large Deflection Random Response of Thermally Buckled Beams," *AIAA Journal,* Vol. 28, No. 12, 1990, pp. 2125–2131.

[47]Hankey, W. L., *Re-Entry Aerodynamics,* AIAA Education Series, AIAA, Washington, DC, 1988.

[48]Shih, P. K., Prunty, J., and Mueller, R. N., "Thermostructural Concepts for Hypervelocity Vehicles," AIAA Paper 88-2295, April 1988; see also related article, "Structures for Hypervelocity Flight," *Aerospace America,* Vol. 27, May 1989, pp. 28–31.

[49]Pritchett, T. R., "Advanced Technology Aluminum Materials for Aerospace Applications," *Light Metal Age,* Vol. 44, Nos. 7–8, 1986, pp. 10–14.

[50]Steinberg, M. A., "Materials for Aerospace," *Scientific American,* Vol. 255, No. 4, 1986, pp. 66–72.

[51]Robinson, J. J., "Getting the Space Plane Off the Ground," *Journal of Metals,* Vol. 39, No. 7, 1987, pp. 8–9.

[52]Jackson, L. R., Dixon, S. C., Tenney, D. R., Carter, A. L., and Stephens, J. R., "Hypersonic Structures and Materials: A Progress Report," *Aerospace America,* Vol. 25, Oct. 1987, pp. 24–30.

[53]Cannon, P., "Aerospace Materials," *Journal of Metals,* Vol. 40, No. 5, 1988, pp. 10–14.

[54]Rainer, R. A., and Ekvall, J. C., "Elevated-Temperature Alloys for Aircraft Structure," *Journal of Metals,* Vol. 40, No. 5, 1988, pp. 16–18.

[55]Destegani, J. D., "Advances in Inter-Metallics," *Advanced Materials and Processes,* Vol. 135, No. 2, 1989, pp. 37–41.

[56]Ronald, T. M. F., "Advanced Materials to Fly High in NASP," *Advanced Materials and Processes,* Vol. 135, No. 5, 1989, pp. 29–37.

[57]Nelson, H. G., "Hydrogen Environment Effects on Advanced Alloys and Composites in Aerospace Structures," *Thermal Structures and Materials for High Speed Flight,* edited by Earl A. Thornton, Vol. 140, Progress in Astronautics and Aeronautics, AIAA, Washington, DC, 1992, pp. 383–399.

Problems

1.1. Write a short paper describing some of the thermal-structural design aspects of the airframe for the Concorde supersonic transport. The paper should be typed, double-spaced, and three to five pages in length, excluding references. Suggested content should include the following:

(a) A definition of flight path (altitude, Mach number, etc.) for a typical trans-Atlantic flight.

(b) A definition and the properties of the airframe material.

(c) A brief discussion of the airframe construction, including a large foldout showing cutaway views of the airframe.

(d) A discussion of some of the basic thermal-structural design issues.

(e) A brief description of laboratory thermal-structural tests.

(f) Any other thermal-structural aspects that you feel are relevant or interesting.

1.2. In an experiment at sea level, air is flowing over a flat plate at 500 mph. The freestream temperature is 60°F, and the pressure is 14.7 psia. Assuming laminar flow, compute at 1 ft behind the leading edge:

(a) The adiabatic wall temperature

(b) The convection coefficient

1.3. Consider the thin, diamond-shaped wing shown in Fig. 4.19 that is flying steadily at Mach 2.0 in the atmosphere at an altitude of 50,000 ft. For simplicity, assume that the angle of attack is zero. Assume a perfect gas. Perform all calculations, and present your results in SI units.

(a) Using basic equations of inviscid, compressible, supersonic flow, compute the flow conditions behind the oblique shock at the leading edge and those condi-

Table P1.4 Mission profile

Time, min	Altitude, ft	Mach No.
0	0	0
25	60,000	2.4
265	70,000	2.4
290	0	0

tions behind the expansion of the apex of the diamond-shaped airfoil. For each flow region, compute the Mach number and the flow velocity, as well as the temperature, pressure, and density of the air.

(b) Using the Eckert reference temperature method (Sec. 1.3.2), compute distributions of convective heat-transfer coefficients along the top surface of the wing for laminar and turbulent flows. Compare the distributions of the laminar and turbulent convection coefficients in graphs of h vs x along the wing surface.

1.4. Use the Eckert reference temperature method to provide engineering estimates for heating rates for a wing of a supersonic transport. The mission profile is given by Table P1.4. Between these data points, quantities are assumed to vary linearly. In the following calculations, use properties for the Standard Atmosphere.

To estimate the aerodynamic heating rate, the wing is assumed to be a flat plate at zero angle of attack. Neglect any shock waves. Compute the aerodynamic heating rate at points $x = 12$ in. and $x = 60$ in. behind the leading edge of the plate. Assume that plate T_w is at radiation equilibrium temperature throughout the mission profile. For the two points on the plate, compute and plot the following quantities as a function of time for the mission profile:

(a) The plate radiation equilibrium temperature $T_w = T_r$ (°F).
(b) The convection coefficient h (Btu/ft^2-s-°F).
(c) The adiabatic wall temperature T_{aw} (°F).
(d) The heating rate q_w (Btu/ft^2-s).

For each quantity, show histories for 1) laminar flow and 2) turbulent flow.

Selected Bibliography on Supersonic Transports

For convenience of reference, a number of publications are listed that describe the development of supersonic transports. The publications are listed in chronological order.

British–French Concorde

Satre, P., "Design for Mach 2.2," *Flight International*, Vol. 85, April 1964, pp. 649–651.

Harrison, N. F. G., "The Concorde," *Flight International*, March 1965, pp. 317–328.

Russell, A. E., "Concorde Conception," *Flight International*, Dec. 12, 1965, pp. 950–953.

N'Guyen, V. P., and Ripley, E. L., "Design Philosophy and Fatigue Testing of Concorde," *Proceedings of 5th ICAF Symposium*, Melbourne, Australia, May 1967, pp. 403–437.

Anon., "BAC SUD-Aviation Concorde," *Flight International,* Vol. 91, May 1967, pp. 799–816.

Morisset, J., "Concorde Flight Testing," *Flight International,* Dec. 1967, pp. 987–988.

Harpur, N. F., "Concorde Structural Development," *Journal of Aircraft,* Vol. 5, No. 2, 1968, pp. 176–183.

Harpur, N. F., "Structural Development of the Concorde," *Aircraft Engineering,* March 1968, pp. 18–30.

Rettie, I. H., and Lewis, W. G. E., "Design and Development of an Air Intake for a Supersonic Transport Aircraft," *Journal of Aircraft,* Vol. 5, No. 6, 1968, pp. 513–521.

Blackall, T. E., *Concorde: The Story, the Facts and the Figures,* G. T. Foulis, Henley-on-Thames, England, UK, 1969.

Davis, J., *The Concorde Affair from Drawing Board to Actuality,* Regnery, Chicago, 1969.

Satre, P., "Supersonic Air Transport—True Problems and Misconceptions," *Journal of Aircraft,* Vol. 7, No. 1, 1970, pp. 3–12.

Burgess, E. H., "The Concorde Flight Test Programme," *Aeronautical Journal,* Vol. 74, No. 716, 1970, pp. 636–642.

Anon., "Concorde NC," *Flight International,* Vol. 99, Jan. 1971, pp. 63–68.

Taylor, J. (ed.), "Concorde," *Jane's All the World Aircraft,* Haymarket Publishing Group, London, 1971, pp. 120–122.

Morgan, M., "A New Shape in the Sky," *Aeronautical Journal,* Vol. 76, 1972, pp. 1–18.

Wilson, A., *The Concorde Fiasco,* Penguin, Harmondsworth, England, UK, 1973.

Chevalier, R., "Anglo-French Collaboration—Concorde and Supersonic Transport," *Aeronautical Journal,* Vol. 78, No. 758/759, 1974, pp. 61–64.

Chevalier, R., "Concorde and Safety: Design, Testing and Certification," *Aeronautical Journal,* Vol. 78, No. 762, 1974, pp. 277–283.

Costello, J., and Hughes, T., *The Concorde Conspiracy,* Scribner, New York, 1976.

Knight, G., *Concorde: The Inside Story,* Stein and Day, New York, 1976.

Calvert, B., *Flying Concorde,* Fontana Paperbacks, Chaucer Press, Bungay, England, UK, 1981.

Leyman, C. S., "A Review of the Technical Development of Concorde," *Progress in Aerospace Science,* Vol. 23, 1986, pp. 185–238.

Lefer, H., "Air France Concorde: A Valuable Symbol," *Air Transport World,* No. 1, 1986, pp. 46–49.

Woolsey, J., "Aérospatiale Proposes Second Generation Concorde," *Air Transport World,* No. 3, 1986, pp. 36–37.

Orlebar, C., *The Concorde Story,* Temple, Twickenham, England, UK, 1986.

Shifrin, C. A., "Britain and France Begin Concorde Follow-On Study," *Aviation Week & Space Technology,* May 14, 1990, pp. 24–26.

U.S. Supersonic Transport (SST)

Borger, J. G., and Hibbs, W. F., "What's an Economical SST?," *Aeronautics and Astronautics,* Vol. 2, No. 9, 1964, pp. 16–24.

Anon., "Evolution of the Lockheed Supersonic Transport," *Lockheed Horizons,* Issue No. 1, Spring 1965, pp. 5–19.

Anon., "Mach Three Technology, Lockheed's Supersonic Transport: Part 1—The Background; Aerodynamics; Propulsion; Airframe Structure," *Flight International,* Jan. 20, 1966.

Anon., "Mach Three Technology, Lockheed's Supersonic Transport: Part 2—Thermal

Management, Fuel, Flying Control, Hydraulic and Electrical Systems," *Flight International,* Jan. 27, 1966.

Heppe, R. R., and Englebry, C. R., "Supersonic Transport Design Evolution," *Lockheed Horizons,* Issue No. 5, Second Quarter, 1966, pp. 5–33.

Shapiro, N., and Edwards, S. S., "Noise and the SST," *Lockheed Horizons,* Issue No. 5, Second Quarter, 1966, pp. 54–67.

Thompson, J., and Parnell, J., "Sonic Boom and the SST," *Lockheed Horizons,* Issue No. 5, Second Quarter, 1966, pp. 68–77.

Donaldson, D. R., and Kenworthy, R. J., "Fatigue Design and Test Program for the American SST," *Proceedings of 5th ICAF Symposium,* Melbourne, Australia, May 1967, pp. 437–476.

Dwiggins, D., *The SST: Here It Comes, Ready or Not,* Doubleday, New York, 1968.

Herron, E. A., *Cobra in the Sky: The Supersonic Transport,* Macmillan, New York 1968.

Corning, G., *Supersonic and Subsonic Airplane Design,* 3rd ed., Braun–Brumfield, Ann Arbor, MI, 1970; see Appendices II and III for Supersonic Transport design data.

Swan, W. C., "Design Evolution of the Boeing 2707-300 Supersonic Transport—Part I: Configuration Development, Aerodynamics, Propulsion and Structures," *AGARD Conference Proceedings 147,* Oct. 1973.

Cooper, P. A., and Heldenfels, R. R., "The NASA Structures and Materials Program for Supersonic Cruise Aircraft," *Astronautics and Aeronautics,* Vol. 14, May 1976, pp. 26–37.

Horwitch, M., *Clipped Wings—The American SST Conflict,* MIT Press, Cambridge, MA, 1982.

McLean, F. E., *Supersonic Cruise Technology,* NASA SP-472, 1985.

Soviet SST

Moon, H., *Soviet SST: The Technopolitics of the Tupolev-144,* Orion Books, New York, 1989.

U.S. High-Speed Civil Transport (HSCT)

Mattill, J. I., "A New Launch Toward an SST," *Technology Review,* July 1985, pp. 75–76.

Rowberg, R. E., Hancock, K., and Hill, C. T., "Commercial High Speed Aircraft Opportunities and Issues," Congressional Research Service, Library of Congress 89-163SPR, March 1989.

Staff, "Study of High-Speed Civil Transports," NASA CR 4235, Douglas Aircraft Company, Dec. 1989.

Rosen, R., and Williams, L. J., "The Rebirth of Supersonic Transport," *Technology Review,* Feb.–March 1993, pp. 23–29.

2 Orbiting Space Structures

2.1 Introduction

Orbiting space structures present structural analysts with significant challenges in analysis and design.[1–3] The large sizes of many structures make them distinct from past Earth satellites. Deployable antenna concepts up to 100 m and erectable structures several hundred meters in diameter have been under consideration. To gain low weight and high stiffness, the structures may employ open, latticework construction. Designs include pretensioned, cable-stiffened structures and space trusses. To achieve low weight and high stiffness and to minimize thermal distortions, the designs make extensive use of advanced composite materials.

Often, an important design constraint lies in controlling structural deformations within close tolerances. Operational requirements for a 100-m antenna design can limit surface distortions to a few millimeters. Such stringent operational requirements have focused attention on analysts' capabilities for predicting small deformations with high accuracy. An important factor emphasizing the need for effective methods of analysis is that the size of the proposed structure will prohibit the ground testing customary for past spacecraft. The control of large space structures' deformations within small tolerances requires careful consideration of several effects on the structural response. Thermally induced forces are only one of several on-orbit loads that must be considered, including pretensioning, orbital positioning thrusts, gravity gradient forces, and atmospheric drag.

The determination of thermally induced forces is a complex, interdisciplinary problem requiring consideration of orbital mechanics, heat transfer, and structural mechanics. Typical heat load, thermal, and structural analyses of space structures will be discussed by using a design of a future spacecraft to illustrate specific modeling and analysis characteristics. Then heating, thermal, and structural analyses will be discussed.

2.2 Microwave Radiometer Spacecraft

The design details of a microwave radiometer illustrate a few of the complexities of orbiting space structures. Details of an early microwave radiometer concept are described by Garrett,[4] and a more recent concept is discussed by Wahls et al.[5] The more recent low-frequency microwave radiometer concept (LFMR) is described here.

2.2.1 Microwave Radiometry

Mission to Planet Earth is a NASA program to monitor and study the Earth's hydrologic, biogeochemical, and climate cycles on a global scale. The type of observations to be made include various surface, atmospheric, and oceanic changes that occur in the global Earth system because of environmental and man-made conditions. The complement of spacecraft needed to support the desired observations

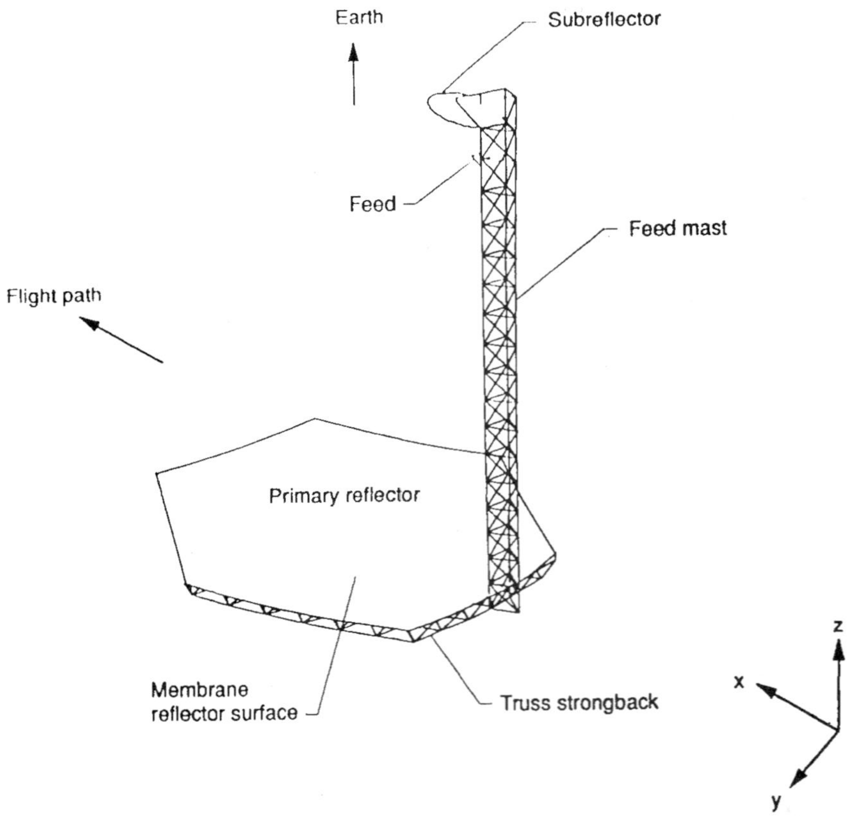

Fig. 2.1 Low-frequency microwave radiometer (LFMR), Ref. 5.

includes low-inclination low-Earth-orbit (LEO), polar LEO, and geosynchronous-Earth-orbit (GEO) platforms, each carrying a payload of instruments and sensors. The LFMR (Fig. 2.1) is a large antenna concept aboard a geostationary platform. The LFMR operates at frequencies of 6–37 GHz for the measurements it will make, which include monitoring precipitation and ocean surface temperatures, surface currents, and wind patterns.

2.2.2 LFMR Structure

The primary reflector is a 15-m-diam offset-fed paraboloid. It consists of a tetrahedral truss structure (referred to as a *strongback*) supporting a membrane reflector surface. The membrane, made of 0.5-mil-thick aluminized Kapton, is attached to the strongback at each of the strongback top-surface joints. The tetrahedral truss provides a stiff, stable support for the membrane surface to minimize the distortion of the overall reflector shape. The 12-bay truss is composed of graphite–epoxy tube members with aluminum alloy joints and end fittings. Typical tube members have an outside diameter of 2.2 cm and a length of 1.18 m. A structural-dynamics finite element model of the truss consists of 235 nodes, 954 beam elements, and 362 lumped masses. The feed mast is a deployable, retractable triangular truss constructed of graphite–epoxy tubes with aluminum alloy hinges and fittings. An equivalent structural model of 18 beam elements represents the feed mast.

The electromagnetic performance of the antenna can be affected adversely by numerous factors. Among these is the geometric displacement of the various structural components of the antenna from their optimum positions. For example, the surface roughness of a reflector antenna is defined as the root mean square (rms) of the reflector surface distortions relative to the best-fit paraboloid through the distorted surface locations. For the LFMR, the allowable rms error is 8.11×10^{-5} m, or 3.2 mils. Thus, structural components such as the primary reflector and feed mast must be capable of maintaining their shapes within design allowables. Design of the LFMR includes the determination of the structural-dynamics response induced by onboard disturbances such as mechanical scanning operations by the subreflector. Thermal-structural deformations must also be considered.

2.3 Environmental Heating

An orbiting space structure may be heated by environmental and onboard heat sources. The sun and Earth are the primary environmental sources. Onboard heating can come from many sources, such as power systems, electronic equipment, and heat-rejection systems. Given that characteristics of onboard heat sources are strongly dependent on the specific design, only heating from environmental sources will be considered here.

2.3.1 Heating Rates

Environmental heating rates—the sum of solar flux, Earth-emitted radiation, and Earth-reflected radiation (albedo)—depend strongly on altitude and orientation of structural members. Heating rates on members of a structure may vary significantly with member orientation and vary strongly with time during the orbit. Fundamental concepts of spacecraft orbital heating are described in Refs. 6–9.

A surface in Earth orbit is heated by radiation from both the sun and the Earth (Fig. 2.2). The heating rate per unit area on a surface is a function of the magnitude of the incident heat flux, the orientation of the surface with respect to the incident heat flux, and the absorbtivity of the surface. Because surface absorbtivity is a function of wavelength, it is necessary to distinguish between surface absorbtivity for solar radiation a_s and Earth radiation, a_e.

The solar flux in Earth orbit is approximately 1350 W/m^2. The solar heat received by a surface is given by

$$q_s = 1350 a_s \cos \psi \qquad (\mathrm{W/m^2}) \tag{2.1}$$

where ψ is the angle between the solar-flux vector and the surface normal.

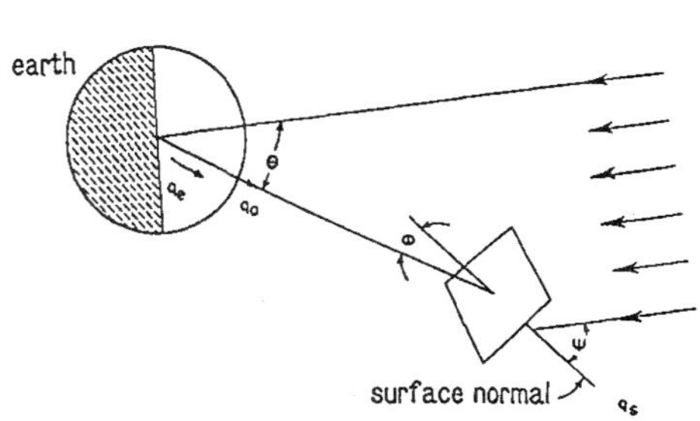

Fig. 2.2 Orbital heating sources and geometry.

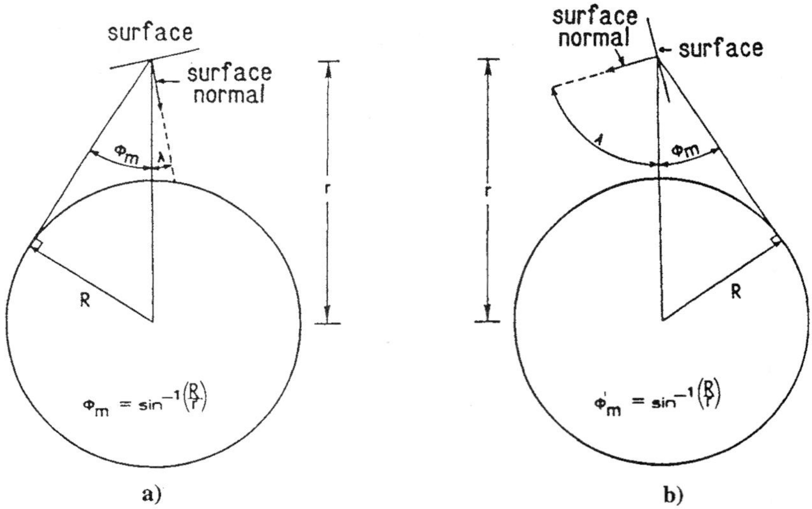

Fig. 2.3 Earth heating geometry for view factor calculations.

The radiation emitted by the Earth can be approximated by assuming the Earth to be a blackbody radiating at $T_e = 289$ K. The Earth-emitted radiation absorbed by a surface is given by

$$q_e = \sigma T_e^4 a_e F \tag{2.2}$$

The view factor F (also called *shape factor* or *configuration factor*) gives the fraction of the total radiant energy leaving the Earth that arrives at the surface. Cunningham[10] shows that a surface orbiting the Earth can be oriented such that either one side is heated or both sides are heated. If only one side is heated (Fig. 2.3a),

$$\lambda + \phi_m \leq \pi/2 \tag{2.3}$$

and the view factor is given by

$$F = \cos\lambda / H^2 \tag{2.4}$$

where λ is the angle between the surface normal and the heat flux, and $H = r/R$. The radius of the Earth is R, and the distance from the center of the Earth to the heated surface is r. The angle ϕ_m is defined in Fig. 2.3. If both sides of the surface are heated (Fig. 2.3b),

$$\pi/2 - \phi_m < \lambda \leq \pi/2 + \phi_m \tag{2.5}$$

and a separate view factor must be calculated for each side. The total view factor for the surface is then the sum of the two. The view factors for each side are given by

$$F = \frac{2}{\pi}\left[\frac{\pi}{4} - \frac{\sin^{-1}}{2}\left[\frac{(H^2-1)^{1/2}}{H\sin\lambda}\right] + \frac{1}{2H^2}\left\{\cos\lambda\cos^{-1}\left[-(H^2-1)^{1/2}\cot\lambda\right] - (H^2-1)^{1/2}\left[1 - H^2\cos^2\lambda\right]^{1/2}\right\}\right] \tag{2.6}$$

Table 2.1 Typical orbital heating rates

Orbit	Heating rates, W/m²		
	Solar	Earth emitted	Albedo
Low Earth	1240	310	380
Geosynchronous	1240	8	10

Solar albedo, Earth-reflected heating, depends on the solar albedo factor AF. The solar albedo factor is defined as the fraction of solar radiation striking the Earth that is reflected back into space. It varies both seasonally and geographically, but a good approximation can be obtained by using $AF = 0.36$. Albedo heating is given by

$$q_a = 1350(AF)a_s F \cos\theta \qquad (\mathrm{W/m^2}) \tag{2.7}$$

Albedo heating occurs over only the daylight side of the Earth, and it varies as the cosine of the reflection angle θ (Fig. 2.2). Modest[11] has shown that, whereas the view factor is a complicated function of both altitude and attitude, a good approximation can be obtained by using the view factors described by Eqs. (2.3–2.6). The wavelength of albedo radiation is similar to that of solar radiation, and so the absorbtivity a_s is used.

The total heating of an orbiting surface is thus the sum of solar, Earth-emitted, and albedo heating:

$$q = q_s + q_e + q_a \tag{2.8}$$

where q_s, q_e, and q_a are defined in Eqs. (2.1), (2.2), and (2.7), respectively. Table 2.1 gives typical heating rates for a low Earth and geosynchronous orbit.

The Earth's shadow has two regions: 1) the umbra, which is totally shadowed from the sun, and 2) the penumbra, which is only partially shaded (Fig. 2.4). Representative orbital data are tabulated in Table 2.2 for a geosynchronous Earth orbit (GEO) and for a low Earth orbit (LEO). Table 2.2 gives the transit times of a structure through the umbra and penumbra in GEO and LEO. Because the transit time through the penumbra is negligible compared with that through the umbra, the penumbra can be disregarded and the structure considered to have entered the umbra directly.

During shadow transit, incident heating is greatly reduced since the structure receives only Earth-emitted heating. The sharp reduction in heating during shadow transit has a pronounced effect on the thermal and structural behavior.

Table 2.2 Typical orbit data

Orbit	Radius, km	Period, s	Transit time, s	
			Umbra	Penumbra
Low Earth	6,878	5,675	2,144	8
Geosynchronous	42,253	86,400	4,167	128

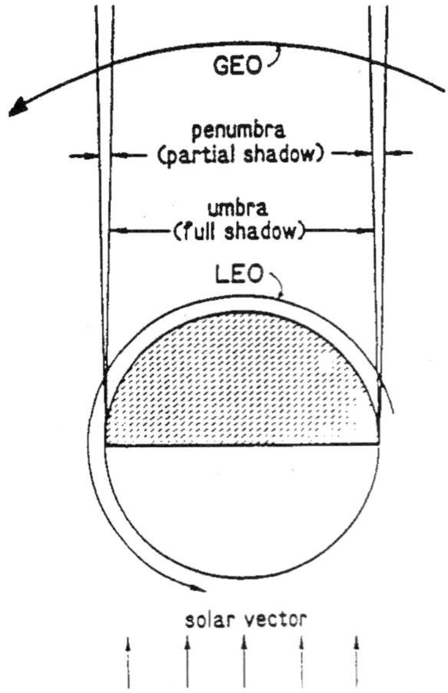

Fig. 2.4 Orbit and Earth shadow definition.

Kreith[6] shows that a satellite will enter the shadow of the Earth (Fig. 2.5) if $\tau > \cos^{-1}(R/r)$, where τ is the angle between the plane of the orbit and the terminator (the line marking the beginning of the Earth's shadow). The angle β is the angular position of the satellite in a circular orbit, with $\beta = 0$ at the satellite noon position, the point in the orbit closest to the sun. Then the value of β at which the satellite enters the Earth's shadow is given by

$$\beta_{\text{in}} = \sin^{-1}\left[\frac{\sin(\cos^{-1} R/r)}{\sin \tau}\right] + \frac{\pi}{2} \tag{2.9}$$

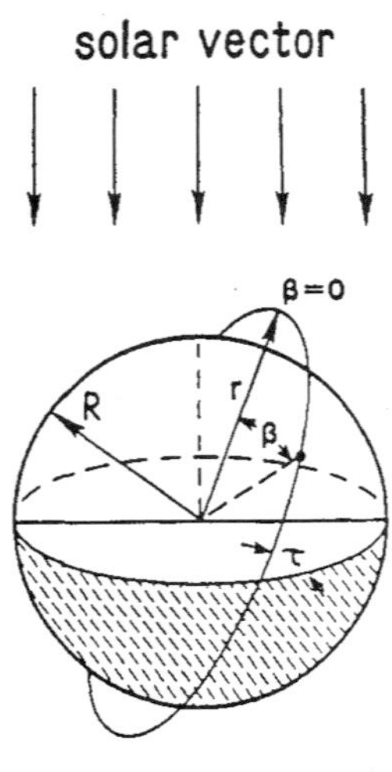

Fig. 2.5 Earth-orbit geometry.

The time spent in the shadow is given by

$$t_s = \frac{\pi r}{V_s}\left\{1 - \frac{2}{\pi}\left[\sin^{-1}\left(\frac{\sin\{\cos^{-1} R/r\}}{\sin \tau}\right)\right]\right\} \tag{2.10}$$

The time for the entire orbit is given by

$$t = \frac{2\pi r}{V_s} \tag{2.11}$$

where V_s is the circular satellite velocity given by

$$V_s = (gR^2/r)^{1/2} \tag{2.12}$$

Hence, Eqs. (2.9–2.12) can be used to find the time of shadow entry,

$$t_{\text{in}} = t\beta_{\text{in}}/2\pi \tag{2.13}$$

and the time of shadow exit,

$$t_{\text{out}} = t_{\text{in}} + t_s \tag{2.14}$$

This section describes the heating of a single surface in Earth orbit. For a structure with several flat surfaces, the heating is computed for each surface, taking into account the surface orientation. For other surfaces such as tubular members, the net incident flux can be computed by projecting the illuminated area on a flat surface normal to the incident flux vector.

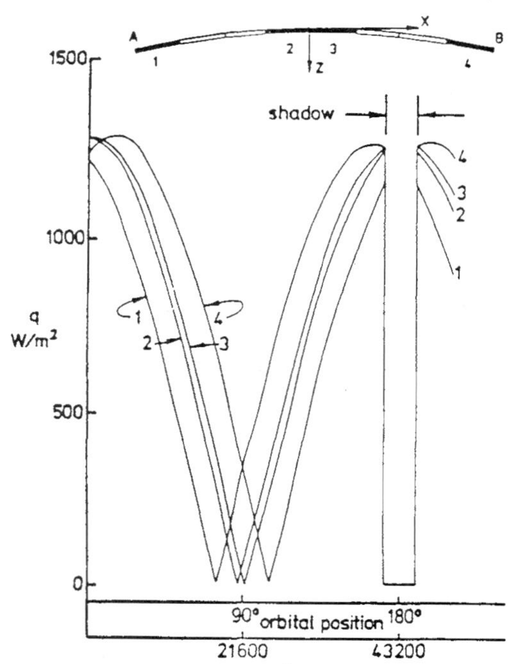

Fig. 2.6 **Heating histories for parabolic truss in GEO.**

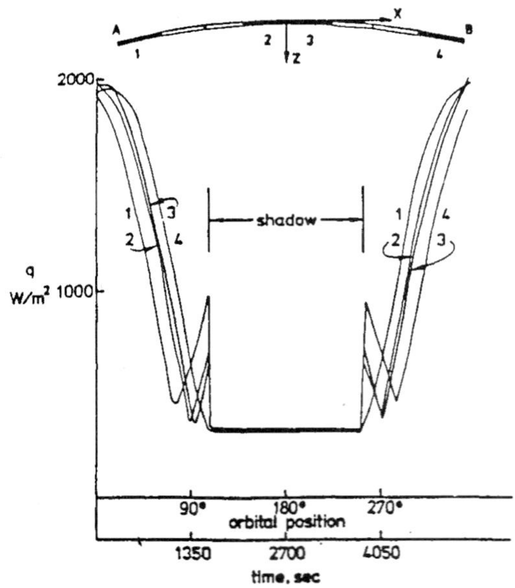

Fig. 2.7 Heating histories for parabolic truss in LEO.

In some structures, there may be several members with a common orientation, so that these members experience the same heating throughout the orbit. A heating analysis of a curved structure, such as the spherical membrane reflector surface (MRS) tetrahedral truss (Fig. 2.1), must consider all members individually because there are no similarities in orientation. Mahaney and Strode[12] discuss environmental heating on an Earth-facing parabolic tetrahedral truss in an ecliptic plane orbit. Two results (Figs. 2.6 and 2.7) may be used to illustrate details of structural heating rates. Figure 2.6 presents heating histories of four members on the main diagonal of a parabolic truss in GEO. Figure 2.7 presents the heating histories of the same four members in LEO. There are two major differences: 1) The magnitude of the heating in LEO is greater than that in GEO because the LEO is closer to the Earth and 2) shadow transit covers about 41% of the orbit in LEO vs only about 4.8% in GEO.

Onboard heating may be more important than environmental heating for some structures. Thermal management of waste heat may have a more significant effect than environmental heating because of higher levels of onboard heating and because heating distributions from onboard heat sources are more localized.

2.3.2 Shadowing Effects

For large space structures such as the MRS system (Fig. 2.1), there are three types of shadowing effects: 1) shadowing of the structure by opaque surfaces such as the solar panels, 2) partial shadowing by large semitransparent surfaces such as antenna mesh, and 3) shadowing of structural members by other slender structural members. Detailed consideration of any of these shadowing effects is difficult because of the complex geometry involved and the time-dependent character of each problem.

Partial shadowing of large space structures by antenna meshes is discussed in Refs. 4 and 13. These analysts use a mesh transmissivity that depends on mesh characteristics (e.g., openings per inch) and the angle of incidence between the heating vector and the mesh normal. Chambers et al.[13] show that the mesh can cause complete shadowing of the structure at points during an orbit and that a detailed consideration of mesh shadowing is critical for defining structural heating.

Consideration of shadowing by slender structural members has been customarily omitted for latticework structures such as trusses. This assumption has been questioned,[14] and particularly for nearly planar Earth-facing structures. For these structures, significant shadowing can occur when the solar vector is nearly tangent to the orbital path. Mahaney and Thornton[15] describe the effects of shadowing on the thermal-structural behavior of a single member and a section from the center column of a gravity gradient space station. For the trusses studied, the study shows that member self-shadowing significantly alters the thermal-structural response and should be included for high-precision predictions. The inclusion of shadowing complicates and increases the costs of analyses appreciably.

2.4 Thermal Behavior

A typical member of a space structure experiences conduction heat transfer combined with radiation heat exchanges from nearby structural members and the other spacecraft components. The discussion here is limited to the unique thermal analysis problems of large latticework structures, such as the MRS shown in Fig. 2.1. In Chapter 4, heat transfer in structures will be discussed in more detail.

2.4.1 Heat Transfer

In an open latticework structure, radiation exchanges between members can be neglected in comparison to incident heating and emitted radiation. Chambers et al.[13] show that the view factor between two nearby members in an open structure is typically less than 0.001, indicating that almost all the radiant energy emitted by a member is lost to space. In general, temperatures of structural members vary along the lengths and around perimeters of the members. Most truss members are hollow rods (tubes), and internal radiation may be significant. Thermal analysis of a large truss taking all these details into account is impractical because of the prohibitively large number of unknowns required. Consequently, a number of simplifying assumptions are customarily made to permit thermal analysis at acceptable costs.

Structural-member temperature distributions depend strongly on material and surface properties. In general, material and surface properties are temperature-dependent and vary throughout the orbit. The thermal conductivity of metallic materials (aluminum or metal matrix composites) is much higher than the thermal conductivity of a resin matrix (graphite–epoxy) composite material. Temperature distributions in members made from these materials differ significantly, and these differences have an important effect on the required analytical models. For instance, the temperature variation around the perimeter of an aluminum tube (ratio of outside diameter to wall thickness = 25) subject to solar heating in LEO is negligible. However, the temperature variation around the perimeter of a graphite–epoxy tube (ratio of outside diameter to wall thickness = 100) is significant.

In addition to the temperature variation around the tube perimeter, truss members' temperatures may vary along their length as a result of axial conduction. A member's axial temperature variation depends on the combination of conduction and radiation heat transfer. In Refs. 16–18, there are studies of typical truss-member temperature distributions using the finite element method. Temperature distributions of aluminum and graphite–epoxy composite members differ significantly.

Aluminum members have a more nonuniform axial temperature distribution than composite material members because of conduction through the joints of the truss. Temperature distributions along composite members are so nearly uniform that, if joints are neglected, it is valid to assume that a single truss member is isothermal. Thus, for trusses with composite members, each member's mean temperature can be computed separately. Temperatures in a truss with composite members can be computed more easily than in a truss with aluminum members, where each member's spatial temperature distribution must be computed.

The isothermal concept is widely used for preliminary design analysis of large space structures such as the MRS system. Chambers et al.[13] account for heat conduction around the tube perimeter; graphs of correction factors for cylindrical and square tubes are given vs a characteristic tube parameter. Chambers and co-workers also describe an approximate method of computing the diametrical temperature difference due to perimeter heat conduction and interior radiation heat transfer. Calculation of these temperature gradients is important in determining bending deflections that result in a reduction of the structural-member allowable column buckling load. Temperature distributions in the cross section of laminated composite tubes may also be important for the long-term fatigue behavior of the composite.

2.4.2 Thermal Response

As the preceding section indicates, the space structure's material significantly influences the heat-transfer behavior and the thermal response. Certainly, for all-composite material structures, an isothermal-member model is an excellent approximation for average member temperatures and leads to uncoupled thermal analyses. For a metallic structure or a nonhomogeneous structure with both metallic (e.g., joints) and composite members, a more general analysis is required to determine member spatial-temperature distributions as a function of time.

Transient temperature histories for four members of a graphite–epoxy tetrahedral truss are shown in Figs. 2.8 and 2.9. The temperature histories shown

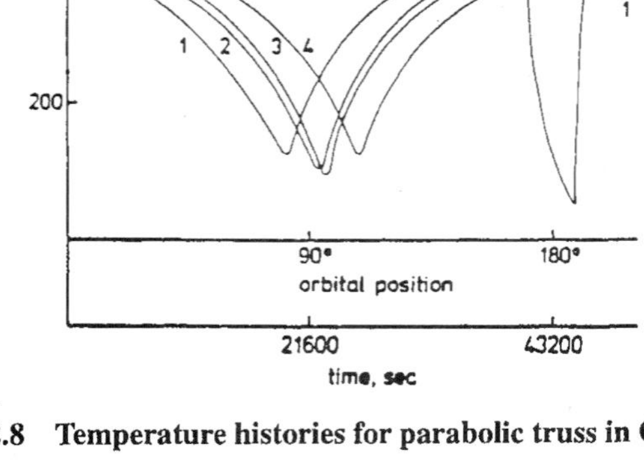

Fig. 2.8 Temperature histories for parabolic truss in GEO.

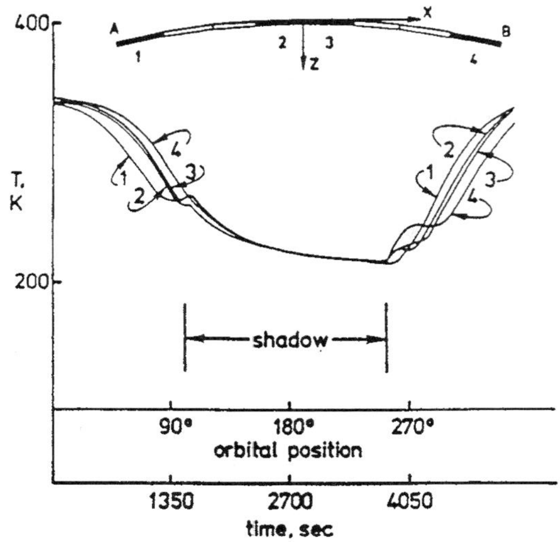

Fig. 2.9 Temperature histories for parabolic truss in LEO.

in these figures were computed using the isothermal member concept, and they correspond to the heating histories for the same members shown in Figs. 2.6 and 2.7. The temperature histories are similar to the heating histories, with two important differences. First, in LEO, the truss is never at radiation equilibrium. Radiation equilibrium is not achieved because the shorter orbital period causes the heating rates to change much faster. Second, the much greater heating of the Earth increases the magnitude of the temperatures but also moderates temperature excursions.

2.5 Structural Behavior

For analyses of the thermal-structural responses of orbiting large space structures, the temperature distribution is regarded as a known function of time. This approach assumes that structural deformations are small enough to allow absorbed heating and temperature distributions to remain unaltered by the structural deformations. In uncoupled thermal-structural analyses, structural deformations and stresses are determined as the final step in a sequential computation of heat loads, thermal response, and structural response. In some problems, however, structural deformations may be large enough to alter heat loads and temperature distributions, requiring a coupled thermal-structural analysis. The following discussion first considers uncoupled thermal-structural behavior.

2.5.1 Structural Models

The thermal-structural response of a complex space structure like the MRS system shown in Fig. 2.1 is usually computed from a finite element model. The complete structure is represented as an assembly of elements chosen to represent the structural characteristics of the beams, cables, and truss members. The thermal environment affects the structural analysis through 1) material properties that may be temperature-dependent and 2) equivalent thermal forces and moments that depend on integration of the temperature distributions over member volumes. Two types of structural models are used: 1) discrete models, in which finite elements represent structural members in detail, and 2) continuum models, which replace a lattice-type structure with repetitive geometry by an equivalent elastic continuum, such as a beam or a plate.

The discrete model approach gives the best representation of the space structure's mass, stiffness, and equivalent thermal loads. With the exception of pretensioned cables and membranes, major structural components, such as trusses and beams, exhibit linear behavior and can be analyzed effectively using linear, small-deflection finite element programs. Pretensioned cables and membranes, however, are known to have significant nonlinear force-deflection characteristics and may experience large deflections and, consequently, warrant nonlinear analysis.[19–20]

Continuum models[21,22] have been developed as a practical solution method for overall response and preliminary design studies. The continuum models are shear flexible beams and plates. Equivalent elastic constants are developed for the continuum models in terms of the material properties and geometry of the original lattice structure. For overall isothermal structural behavior, such as vibration and buckling, continuum models give excellent agreement in numerical examples for beamlike and platelike lattice grids.

2.5.2 Material Properties

Advanced composite materials are widely used in large space structures because of their high stiffness and low coefficient of thermal expansion (CTE). Reliable material data are a prerequisite to accurate thermal-structural response predictions. The determination of these data for advanced composites in the space environment is difficult and is the subject of current research. The major environmental parameters in the space environment are low pressure (high vacuum), ultraviolet radiation, ionizing radiation, and thermal cycling. References 23–28 describe these environmental effects and present advances in the measurement of the CTE and the effects of the space environment. The CTE of a composite member depends on the laminate lay-up and the material properties of individual laminae. Johnson et al.[24] present a compilation of CTE data for composite materials and a procedure for computation of lamina properties by lamination theory. The stability of structural material properties in a long-term space environment is not well understood. Experiments described in Ref. 23 indicate that significant changes occur in the CTE because of the combined effects of vacuum and thermal cycling.

2.5.3 Quasistatic Response

Most thermal-structural analyses of large space structures neglect dynamics effects to use a quasistatic response analysis and perform a sequence of static analyses at selected points in the transient thermal response. To illustrate characteristics of the structural response of a large space truss, typical deformations of a 43-m-diam parabolic tetrahedral truss (Fig. 2.10) are shown in Figs. 2.11 and 2.12. The transient thermal-structural response was computed quasistatically[12] for the 109-node, 420-element truss in GEO using the temperature histories shown in Fig. 2.8.

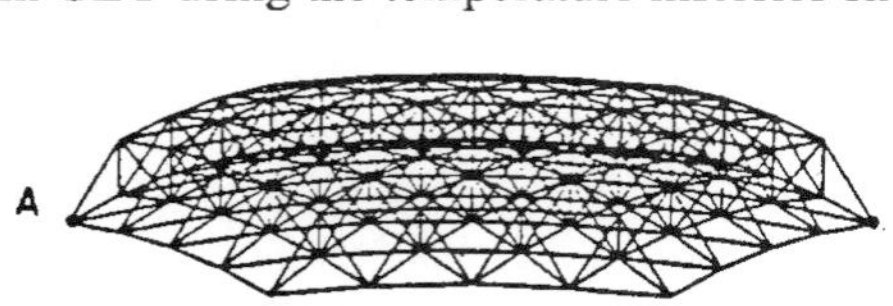

Fig. 2.10 Four-ring parabolic tetrahedral truss.

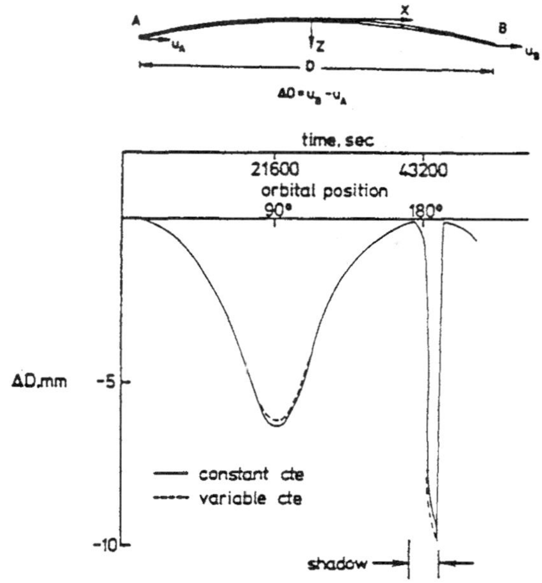

Fig. 2.11 Change in diameter of parabolic truss in GEO.

The structural response may be characterized by the change in diameter of the truss, by a shear deformation between the faces, and by out-of-plane distortions of the faces. Figure 2.11 shows the change in diameter of the diagonal of the parabolic truss in GEO. The change in diameter follows the temperature histories in Fig. 2.8. Figure 2.12 presents the shear deformation of the parabolic truss. The shearing deformation is computed from the displacements u_1 and u_2 of two joints and the two faces of the truss shear with respect to each other because of changes in the length of the core members. As the faces of the parabolic truss shear, they experience an out-of-plane distortion. The maximum normal nodal displacement is about 1 mm and the maximum root mean square (rms) out-of-plane distortion 0.6 mm. In contrast, a flat truss exhibits no out-of-plane surface distortion. As the structure orbits, the deformation of the core members causes the faces to shear with respect to each other, but the faces remain perfectly flat. These results are based on temperatures computed by neglecting member-to-member shadowing effects.

2.5.4 Thermally Induced Vibrations

Thermally induced vibrations of beams and plates were studied by classical analytical methods in the 1950s.[29] The studies showed that structural inertia assumes importance only in exceptional cases, and that for most analyses, inertia effects may be disregarded to permit quasistatic thermal-structural analyses. Very thin beams and flat plates were the exceptions. For these structural components, thermal shock introduced structural vibrations with amplitudes up to twice the corresponding quasistatic deflection.

Thermally induced vibrations of space structures became known during the flight of the OGO-IV spacecraft in the 1960s. On that flight, an 18-m (60-ft) experiment boom sustained a solar-induced large-amplitude oscillation that severely compromised spacecraft performance. A detailed study of the problem[30–31] showed that a

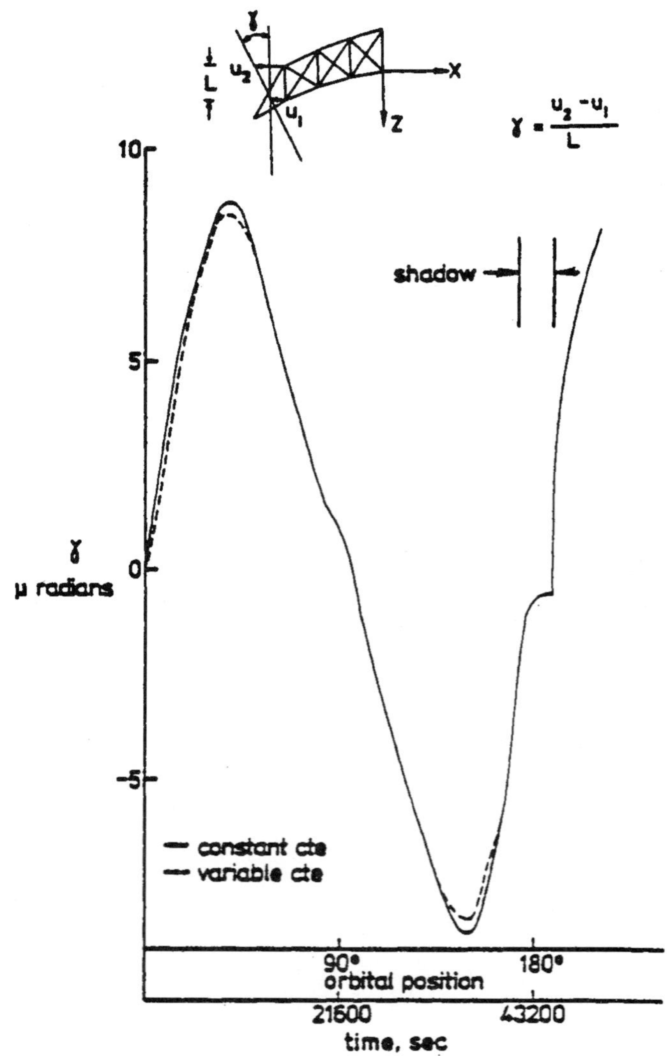

Fig. 2.12 Shear deformation of parabolic truss in GEO.

coupled thermal-structural analysis predicted thermally induced torsional-flexural vibrations consistent with the observed phenomenon. The study showed that the thermally induced vibrations could be eliminated by increasing the boom torsional rigidity. Thermally induced oscillations are described in detail in Chapter 9.

2.6 Closing Remarks

This chapter describes some of the unique features of thermal-structural problems for orbiting structures. Equations needed for the prediction of environmental heating rates are presented. An overview is given of the unique features of thermal and structural analyses for orbiting structures. In subsequent chapters, further details of thermal and structural analyses will be developed.

References

[1]Card, M. F., and Boyer, W. J., "Large Space Structures—Fantasies and Facts," AIAA Paper 80-0674, May 1980.

[2]Thornton, E. A., and Paul, D. B., "Thermal-Structural Analysis of Large Space Structures—An Assessment of Recent Advances," *Journal of Spacecraft and Rockets*, Vol. 22, No. 4, 1985, pp. 385–393.

[3]Malla, R. B., Nash, W. A., and Lardner, T. J., "Thermal Effects on Large Space Structures," *Journal of Aerospace Engineering*, Vol. 1, No. 3, 1988, pp. 171–192.

[4]Garrett, L. B., "Interactive Design and Analysis of Future Large Spacecraft Concepts," *Spacecraft Radiative Transfer and Temperature Control*, edited by T. E. Horton, Progress in Astronautics and Aeronautics, Vol. 83, AIAA, New York, 1982, pp. 345–376.

[5]Wahls, D. M., Farmer, J. T., and Sleight, D. W., "On-Orbit Structural Dynamic Performance of a 15-Meter Microwave Radiometer Antenna," NASA TP-3041, Dec. 1990.

[6]Kreith, F., *Radiation Heat Transfer for Spacecraft and Solar Power Plant Design*, International Textbook, Scranton, PA, 1962, pp. 95–96.

[7]Wertz, J. R., and Larson, W. J., *Space Mission Analysis and Design*, Kluwer, Dordrecht, The Netherlands, 1991, pp. 370–386.

[8]Fortescue, P. W., and Stark, J. P. W., *Spacecraft Systems Engineering*, Wiley, New York, 1991, pp. 275–302.

[9]Griffin, M. D., and French, J. R., *Space Vehicle Design*, Education Series, AIAA, Washington, DC, 1991, pp. 371–394.

[10]Cunningham, F. G., "Power Input to a Small Flat Plate from a Diffusely Radiating Sphere with Applications to Earth Satellites," NASA TN D-710, Aug. 1961.

[11]Modest, M. F., "Solar Flux Incident on an Orbiting Surface After Reflection from a Planet," *AIAA Journal*, Vol. 18, No. 6, 1980, pp. 727–730.

[12]Mahaney, J., and Strode, K. B., "Fundamental Studies of Thermal Structural Effects on Orbiting Trusses," AIAA Paper 82-0650, May 1982.

[13]Chambers, B. C., Jensen, C. L., and Coyner, J. V., "An Accurate and Efficient Method for Thermal–Thermo-Elastic Performance Analysis of Large Space Structures," AIAA Paper 81–1178, June 1981.

[14]O'Neill, R. F., and Zich, J. L., "Space Structure Heating: A Numerical Procedure for Analysis of Shadowed Space Heating of Sparse Structures," *Spacecraft Radiative Transfer and Temperature Control*, edited by T. E. Horton, Progress in Astronautics and Aeronautics, Vol. 83, AIAA, New York, 1982, pp. 377–395.

[15]Mahaney, J., and Thornton, E. A., "Self-Shadowing Effects on the Thermal-Structural Response of Orbiting Structures," *Journal of Spacecraft and Rockets*, Vol. 24, No. 4, 1987, pp. 342–348.

[16]Mahaney, J., Thornton, E. A., and Dechaumphai, P., "Integrated Thermal-Structural Analysis of Large Space Structures," NASA CP-2216, *Symposium on Computational Aspects of Heat Transfer in Structures*, NASA Langley Research Center, Nov. 1981, pp. 179–198.

[17]Thornton, E. A., Mahaney, J., and Dechaumphai, P., "Finite Element Thermal-Structural Modeling of Orbiting Truss Structures," NASA CP-2215, Part 1, *Large Space Systems Technology—1981*, Third Annual Technical Review, NASA Langley Research Center, Nov. 1981, pp. 93–108.

[18]Thornton, E. A., Dechaumphai, P., and Wieting, A. R., "Integrated Finite Element Thermal-Structural Analysis with Radiation Heat Transfer," AIAA Paper 82-0703, May 1982.

[19]Bushnell, D., "Prediction of Loads on Antenna Ribs Due to Mesh Deployment," *Journal of Spacecraft and Rockets*, Vol. 17, July–Aug. 1980, pp. 290–302.

[20]Thornton, E. A., Dechaumphai, P., and Pandey, A. K., "Finite Element Thermal-Structural Analyses of a Cable-Stiffened Orbiting Antenna," *Journal of Spacecraft and Rockets*, Vol. 23, No. 6, 1986, pp. 620–624.

[21]Noor, A. K., Anderson, M. S., and Greene, W. H., "Continuum Models for Beam- and Plate-like Lattice Structures," *AIAA Journal*, Vol. 16, Dec. 1978, pp. 1219–1228.

[22]Nayfeh, A. H., and Hefzy, M. S., "Continuum Modeling of the Mechanical and Thermal Behavior of Discrete Large Structures," *AIAA Journal*, Vol. 19, June 1981, pp. 766–773.

[23]Tennyson, R. C., "Composite Materials in a Space Environment," AIAA Paper 80-0678, May 1980.

[24]Johnson, R. H., Kural, M. H., and Mackey, G. B., "Thermal Expansion Properties of Composite Materials," NASA CR-165632, Lockheed Missiles and Space Company, Sunnyvale, CA, July 1981.

[25]Tenney, D. R., and Bowles, D. E., "Space Radiation Effects on Dimensional Stability of Composites," *Proceedings of Fourth European Symposium on Spacecraft Materials in Space Environment* (Toulouse, France, Sept. 1988), Capedues Editions, 1989, pp. 133–144.

[26]Herakovich, C. T., and Hyer, M. W., "Damage-Induced Property Changes in Composites Subjected to Cyclic Thermal Loading," *Engineering Fracture Mechanics*, Vol. 25, Nos. 5/6, 1986, pp. 779–791.

[27]McComb, H. G., Jr., Murrow, H. N., and Card, M. F., "Structures and Materials Technology for Hypersonic Aerospacecraft," NASA TM 102583, NASA Langley Research Center, Jan. 1990.

[28]Milkovich, S. M., Herakovich, C. T., and Sykes, G., "Space Radiation Effects on the Thermo-Mechanical Behavior of Graphite-Epoxy Composites," *Journal of Composite Materials*, Vol. 20, Nov. 1986, pp. 579–593.

[29]Boley, B. A., and Weiner, J. H., *Theory of Thermal Stresses*, Wiley, New York, 1960, pp. 339–345.

[30]Frisch, H. P., "Thermally Induced Vibrations of Long Thin-Walled Cylinders of Open Section," *Journal of Spacecraft and Rockets*, Vol. 7, Aug. 1970, pp. 897–905.

[31]Frisch, H. P., "Thermally Induced Response of Flexible Structures: A Method for Analysis," *Journal of Guidance and Control*, Vol. 3, Jan.–Feb. 1980, pp. 92–94.

Problems

2.1. Tethers have become of interest for a variety of space missions. Write a short paper discussing potential thermal-structural problems that tethers may encounter in space missions. Describe:

(a) Some applications of tethers in space.
(b) How tethers are made, including materials and construction.
(c) Possible temperature distributions.
(d) Potential quasistatic and/or dynamic thermal-structural problems.

2.2. A long tether extends radially from a platform that is in a circular low Earth orbit. The lower length of the tether, say, for $0 \leq x \leq a$, extends into the upper atmosphere and is aerodynamically heated. For the remaining portion of the tether, for $a \leq x \leq L$, aerodynamic heating is negligible. The entire length of the tether receives the usual three forms of radiant heating and experiences surface radiation energy losses to deep space. We wish to determine the temperature distribution $T(x, t)$ of the tether.

Formulate an approach for determining $T(x, t)$. State the basic assumptions, and write appropriate fundamental equations and boundary conditions. Briefly discuss a computational approach for determining the tether temperature distribution for a typical Earth orbit.

This page intentionally left blank

3
Concepts from Continuum Mechanics

3.1 Introduction

Continuum mechanics unifies basic principles that describe the behavior of solids, liquids, and gases. Historically, the basic principles that underlie subject areas such as elasticity, plasticity, and fluid mechanics were developed independently over a long period of time. In this century, the notions of stress, deformation, and flow were recognized as common to both solids and fluids. The unification of these notions with the basic laws of physics as applied to an idealized deformable medium has become the fundamental approach of modern engineering mechanics. Continuum mechanics in some texts may be presented in a highly rigorous manner. Our presentation will be relatively unsophisticated, with a view toward establishing fundamental concepts applicable to thermal-structural problems.

This chapter begins the development of the fundamental principles that will be applied to thermal-structural problems in succeeding chapters. The basic concepts from continuum mechanics are presented to serve as a foundation for the detailed analyses of thermal-structural behavior that will follow. For readers who have studied continuum mechanics previously, the chapter serves as a review and introduces notation used in later chapters. For readers without such a background, the chapter is self-contained, although briefer and more specific than developments of continuum mechanics presented in popular texts.

3.2 Continuum

In continuum mechanics,[1–4] the molecular structure of a material is ignored by assuming the medium is continuous everywhere. Thus, if a small volume of a continuum is shrunk to a point, the volume will always contain material. Continuum mechanics, as applied to structures, deals with macroscopic variables such as displacements, stresses, strains, and temperature. These variables exist at every point and vary smoothly from point to point within the continuum. The displacements, stresses, and strains are tensors. This means that, at a point, these quantities have several components, and the magnitudes of the components vary with the orientation of a coordinate system located at the point. Temperature, however, is a scalar quantity that is independent of coordinate system orientation. If a continuum problem is steady, these quantities vary only with position; in a Cartesian system, a point is located at the position given by (x_1, x_2, x_3). If a problem is unsteady, the quantities vary with time t as well.

3.3 Stress Tensor

On a small rectangular parallelepiped at a point within a continuum, the stress state is characterized by nine stress components (Fig. 3.1a). The components are represented by σ_{ij}, where i and j take on the values 1, 2, 3. Three of the stress components, $\sigma_{11}, \sigma_{22}, \sigma_{33}$, are normal stresses, and the remaining six, $\sigma_{12}, \sigma_{13}, \sigma_{21}$,

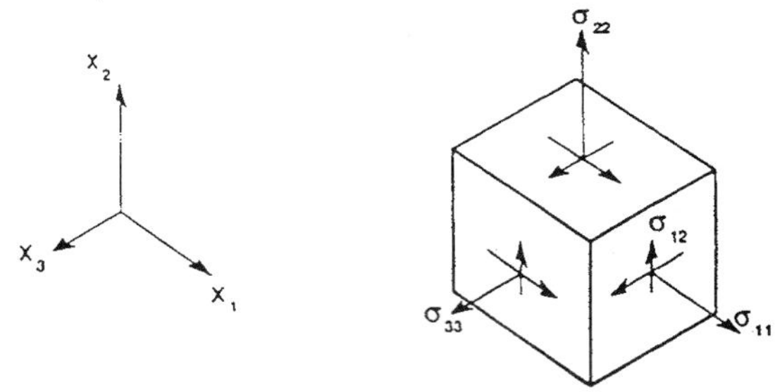

a) Cartesian stress components

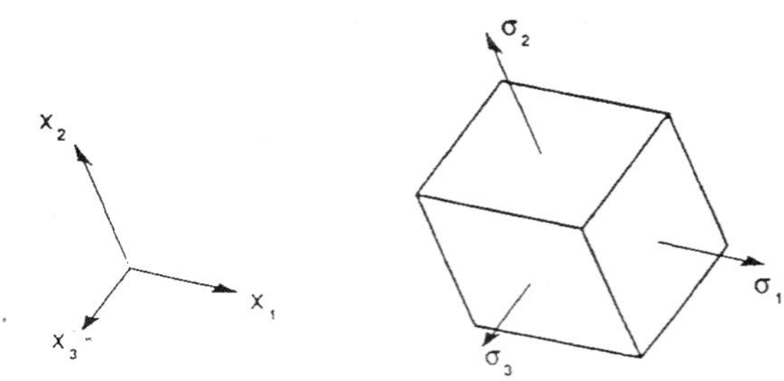

b) Principal stress components

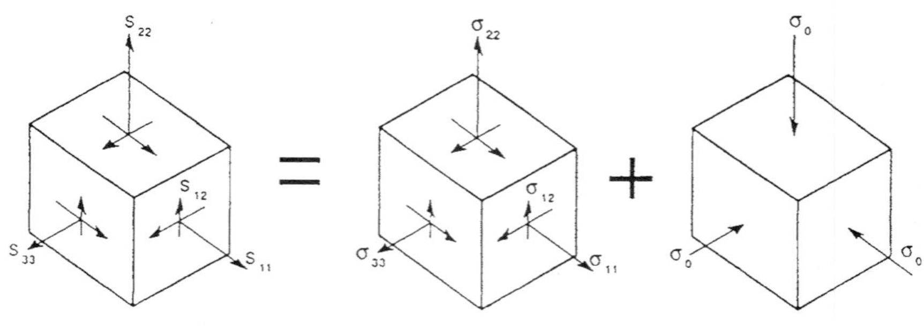

c) Deviatoric stresses

Fig. 3.1 Stress at a point.

$\sigma_{23}, \sigma_{31}, \sigma_{32}$, are shearing stresses. The magnitudes of the stress components may vary with position and time; that is, $\sigma_{ij} = \sigma_{ij}(x_1, x_2, x_3, t)$. In addition, magnitudes of the stress components depend on the orientation of a local coordinate system. At each point, an orientation can be found that produces extreme values for the normal stresses and causes shearing stresses to be zero. This orientation denotes principal axes, and the corresponding normal stresses are the principal stresses, $\sigma_1, \sigma_2, \sigma_3$ (Fig. 3.1b). The values of the principal stresses are given by the three roots of the cubic algebraic equation,

$$-\sigma^3 + I_1\sigma^2 - I_2\sigma + I_3 = 0 \tag{3.1}$$

where I_1, I_2, I_3 are given by

$$
\begin{gathered}
I_1 = \sigma_{11} + \sigma_{22} + \sigma_{33} \\
I_2 = \begin{vmatrix} \sigma_{11} & \sigma_{12} \\ \sigma_{21} & \sigma_{22} \end{vmatrix} + \begin{vmatrix} \sigma_{11} & \sigma_{13} \\ \sigma_{31} & \sigma_{33} \end{vmatrix} + \begin{vmatrix} \sigma_{22} & \sigma_{23} \\ \sigma_{32} & \sigma_{33} \end{vmatrix} \\
I_3 = \begin{vmatrix} \sigma_{11} & \sigma_{12} & \sigma_{13} \\ \sigma_{21} & \sigma_{22} & \sigma_{23} \\ \sigma_{31} & \sigma_{32} & \sigma_{33} \end{vmatrix}
\end{gathered}
\tag{3.2}
$$

The vertical bars in the equations for I_2 and I_3 denote determinants. The quantities I_1, I_2, I_3 are called *stress invariants* because they have the same value independently of the local coordinate system orientation. Thus, the invariants may also be expressed in terms of the principal stresses,

$$
\begin{gathered}
I_1 = \sigma_1 + \sigma_2 + \sigma_3 \\
I_2 = \sigma_1\sigma_2 + \sigma_1\sigma_3 + \sigma_2\sigma_3 \\
I_3 = \sigma_1\sigma_2\sigma_3
\end{gathered}
\tag{3.3}
$$

Experimental studies have shown that plastic deformation of metallic materials is independent of hydrostatic pressure. To characterize the stress components that induce plasticity, alternative stress components known as the *deviatoric stresses* are defined by subtracting uniform hydrostatic normal stresses from the original stress tensor. The hydrostatic stress is defined by

$$\sigma_0 = \tfrac{1}{3}(\sigma_{11} + \sigma_{22} + \sigma_{33}) = \tfrac{1}{3}\sigma_{ii} \tag{3.4}$$

and the deviatoric stresses (Fig. 3.1c) are given by

$$S_{ij} = \sigma_{ij} - \sigma_0\delta_{ij} \tag{3.5}$$

where δ_{ij}, is the Kronecker delta, $\delta_{ij} = 0$ if $i \neq j$, and $\delta_{ij} = 1$ if $i = j$. In Eq. (3.4), the summation convention is illustrated; that is, a repeated subscript means that the terms are summed for $i = 1, 2, 3$. Principal deviatoric stresses are roots of

$$-S^3 + J_2 S + J_3 = 0 \tag{3.6}$$

where the deviatoric stress invariants are given by

$$J_2 = \tfrac{1}{2}S_{ij}S_{ji} \qquad J_3 = \tfrac{1}{3}S_{ij}S_{jk}S_{ki} \tag{3.7}$$

and the summation convention is employed. The first invariant, J_1, is identically zero, so that a term with S^2 does not appear in Eq. (3.6). The second deviatoric stress invariant, J_2, is often used in classical plasticity theory. The von Mises yield criterion of plasticity[3] states that yielding of material at a point occurs when the stress state reaches a level to cause $J_2 = \sigma_y^2/3$, where σ_y is the material yield stress determined from a uniaxial tension test. The equation given by the von Mises yield criterion defines a yield surface. At a material point, if the principal stresses lie within the yield surface, elastic behavior occurs. If the principal stresses fall on the yield surface, plastic deformation occurs.

3.4 Strain Tensor

In response to surface forces (sometimes called *surface tractions*), body forces, and temperature changes, a rectangular parallelepiped of material within a continuum will translate, rotate, and deform. The relative motions of the edges of the parallelepiped are measures of the deformation of the material. At a point, nine components of the strain tensor ε_{ij} characterize the relative deformations. Three of the strain components, ε_{11}, ε_{22}, ε_{33}, are normal strains along the coordinate axes. The normal strains measure extensions or contractions of line elements. The remaining six strain components, ε_{12}, ε_{13}, ε_{21}, ε_{23}, ε_{31}, ε_{32}, are shearing strains. The shearing strains measure relative angle changes between the coordinate axes. As with the stress components, magnitudes of the strain components at a point depend on the orientation of the local coordinate axes. At each point, an orientation of the coordinate axes can be found that produces principal strains. The principal strains at a point can be computed from an equation similar to Eq. (3.1), and strain invariants are defined by equations similar to Eq. (3.2).

As the continuum undergoes deformation, each point experiences a displacement that is represented by the Cartesian components $u_i, i = 1, 2, 3$. As with the stress and strain tensors, the displacement components u_i are functions of position and time, $u_i = u_i(x_1, x_2, x_3, t)$. The strain components ε_{ij} at a point are related to the displacement components by

$$\varepsilon_{ij} = \frac{1}{2}\left[\frac{\partial u_i}{\partial x_j} + \frac{\partial u_j}{\partial x_i} + \frac{\partial u_k}{\partial x_i}\frac{\partial u_k}{\partial x_j}\right] \tag{3.8}$$

where the summation convention is used. A typical normal strain written out is

$$\varepsilon_{11} = \frac{\partial u_1}{\partial x_1} + \frac{1}{2}\left[\left(\frac{\partial u_1}{\partial x_1}\right)^2 + \left(\frac{\partial u_2}{\partial x_1}\right)^2 + \left(\frac{\partial u_3}{\partial x_1}\right)^2\right] \tag{3.9}$$

These equations are the strain-displacement relations. The equations represent a nonlinear relationship between the displacement gradients and strains. Often, displacement gradients are assumed small and the last term in Eq. (3.8), which involves products and powers of displacement gradients, is neglected in comparison to the first two terms. The result is the linear strain-displacement relations,

$$\varepsilon_{ij} = \frac{1}{2}\left[\frac{\partial u_i}{\partial x_j} + \frac{\partial u_j}{\partial x_i}\right] \tag{3.10}$$

Equation (3.8) or (3.10) shows that the strain tensor is symmetric, that is, $\varepsilon_{ij} = \varepsilon_{ji}$. Symmetry of the tensor means that there are only six independent strain components. Under severe conditions, structures may experience large deformations, and the nonlinear strain-displacement relations may be required. In these circumstances, the definitions of the stress components are reinterpreted relative to the undeformed and deformed configurations of the structure.[1]

3.5 Conservation Equations

For general continuum formulations for solids, the conservation laws that must be considered are conservation of mass, linear momentum, conservation of angular momentum, and conservation of energy. In our discussion, density is assumed constant, and conservation of mass is satisfied identically.

3.5.1 Linear Momentum

We consider now the conservation of linear momentum for the continuum. Figure 3.2 shows a continuum contained within a volume V and having a surface ∂V. Within the continuum, each infinitesimal element of volume $\mathrm{d}V$ experiences body forces per unit mass denoted by f_i, $i = 1, 2, 3$. On the surface of the continuum, each infinitesimal element of surface area $\mathrm{d}A$ experiences surface tractions, forces per unit area, denoted by T_i, $i = 1, 2, 3$. Conservation of linear momentum requires that the sum of these forces integrated over the volume and surface area of the continuum be zero for each coordinate direction. Thus,

$$\Sigma F_i = 0$$

yields

$$\int_V \rho f_i \, \mathrm{d}V + \int_{\partial V} T_i \, \mathrm{d}A = 0 \qquad i = 1, 2, 3 \tag{3.11}$$

where ρ denotes the constant material density. At each point on the surface of the continuum, the surface tractions are related to the components of the stress tensor by

$$T_i = \sigma_{ij} n_j \tag{3.12}$$

where n_j, $j = 1, 2, 3$, are the components of a unit vector $\hat{n}$ normal to the surface. Substituting Eq. (3.12) in Eq. (3.11) yields

$$\int_V \rho f_i \, \mathrm{d}V + \int_{\partial V} \sigma_{ij} n_j \, \mathrm{d}A = 0 \tag{3.13}$$

At this point, we wish to convert the surface integral in Eq. (3.13) to a volume integral. The Gauss theorem from vector calculus provides the general relationship

$$\int_{\partial V} g_i n_i \, \mathrm{d}A = \int_V \frac{\partial g_i}{\partial x_i} \, \mathrm{d}V \tag{3.14}$$

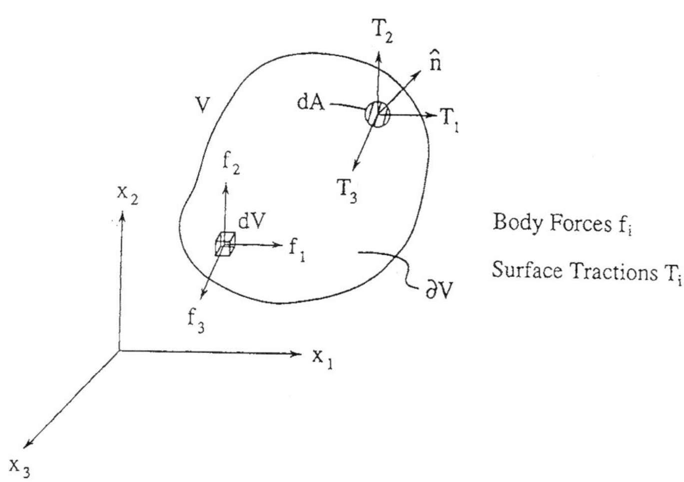

Fig. 3.2 Continuum with body forces and surface tractions.

where g_i is an arbitrary function, $g_i = g_i(x_1, x_2, x_3)$. Using Eq. (3.14), we may then replace the second term in Eq. (3.14) by a volume integral, with the gradient of the stress tensor as the integrand. After this transformation, we may write Eq. (3.13) as

$$\int_V \left[\frac{\partial \sigma_{ij}}{\partial x_j} + \rho f_i\right] \mathrm{d}V = 0 \tag{3.15}$$

Because the volume V is arbitrary we argue that, if Eq. (3.15) is to be satisfied for any arbitrary volume, we must require that the integrand be zero. Thus,

$$\frac{\partial \sigma_{ij}}{\partial x_j} + \rho f_i = 0 \qquad i = 1, 2, 3 \tag{3.16}$$

Equation (3.16) represents the three conservation equations for linear momentum. If we wish to write the three equations of motion, we use D'Alembert's principle from dynamics. D'Alembert's principle allows us to include the mass times acceleration of each point within the continuum as reversed effective (inertia) forces. This approach permits us to write the body force in Eq. (3.16) as two parts,

$$\rho f_i = -\frac{\partial^2 u_i}{\partial t^2} + B_i$$

The first term on the right-hand side, $-\rho \partial^2 u_i / \partial t^2$, is the inertia force, where the second time derivative of a displacement u_i denotes an acceleration component. The second term on the right-hand side, B_i, denotes a body force per unit volume. With this substitution, Eq. (3.16) yields the three equations of motion for a continuum,

$$\frac{\partial \sigma_{ij}}{\partial x_j} + B_i = \rho \frac{\partial^2 u_i}{\partial t^2} \qquad i = 1, 2, 3 \tag{3.17}$$

These partial differential equations must be satisfied at every point within the volume V. They are solved subject to appropriate boundary conditions on the surface ∂V and appropriate initial conditions. If the accelerations on the right-hand side are zero, the equations represent equilibrium of the continuum. Written out, the equations of motion are

$$\begin{aligned}
\frac{\partial \sigma_{11}}{\partial x_1} + \frac{\partial \sigma_{12}}{\partial x_2} + \frac{\partial \sigma_{13}}{\partial x_3} + B_1 &= \rho \frac{\partial^2 u_1}{\partial t^2} \\
\frac{\partial \sigma_{21}}{\partial x_1} + \frac{\partial \sigma_{22}}{\partial x_2} + \frac{\partial \sigma_{23}}{\partial x_3} + B_2 &= \rho \frac{\partial^2 u_2}{\partial t^2} \\
\frac{\partial \sigma_{31}}{\partial x_1} + \frac{\partial \sigma_{32}}{\partial x_2} + \frac{\partial \sigma_{33}}{\partial x_3} + B_3 &= \rho \frac{\partial^2 u_3}{\partial t^2}
\end{aligned} \tag{3.18}$$

Conservation of angular momentum applied to the infinitesimal element of volume shows that the stress tensor is symmetric, that is, $\sigma_{ij} = \sigma_{ji}$. Symmetry of the stress tensor means that there are only six independent stress components.

3.5.2 Energy

Conservation of energy for a continuum considers all energy forms possible for both thermal and mechanical behavior. Thermal-energy forms include energy sources within the continuum, energy transferred across the surface by heat transfer, and thermal energy stored within the material by the material's thermal capacity. Mechanical forms of energy include the work performed by the stresses, body forces, and surface tractions, as well as the kinetic energy due to the motion of the material. Thus, in the most general form, the conservation of energy equation is a complicated relationship representing the continuum's thermal and mechanical behavior.

That the conservation of energy contains both thermal and mechanical forms of energy is consistent with our experience. We intuitively expect for solid materials that there is a relationship between the thermal-energy state as expressed by temperature and the material's mechanical state as expressed by stress, strain, and deformation. We expect that, if the thermal-energy state is changed by the addition or removal of thermal energy, the mechanical state will change also. With some reflection, we may realize that the converse is true as well. For, if we change the mechanical state, there will be an accompanying change in the thermal-energy state of the material. One example in our experience reinforces this concept. If we inelastically deform a material by stretching it in a uniaxial tension test, we observe an increase in temperature. This increase in temperature represents a conversion of mechanical to thermal energy. In the uniaxial tension test, work is performed on the material by the testing machine as the applied forces move through the deformations caused by the inelastic material behavior. Within the material, the deformation process converts the work into an increase in thermal energy, which we sense as an increase in temperature.

Our intuitive insight and experience thus lead us to conclude that there is a coupling between the thermal- and mechanical-energy states of a solid material. In thermal-structural mechanics, the thermal–mechanical coupling is characterized as either one-way coupling or two-way coupling. One-way coupling denotes the situation in which a change in the thermal-energy state causes a change in the mechanical state. To be specific, a change in temperature causes stress and/or strain with deformation. In one-way coupling, however, the converse is *assumed* not to occur. In other words, changes in the mechanical state do not cause a change in the thermal state; that is, a change in stress or a deformation is assumed not to cause a change in temperature. In contrast, two-way coupling denotes the general situation in which the thermal- and mechanical-energy states are coupled so that changes in the thermal state alter the mechanical state and vice versa. Two-way coupling is the situation represented by the general conservation of energy equation of continuum mechanics.

We describe one-way and two-way coupling because there are two forms of the conservation of energy for thermal structures. The most widely used conservation of energy equation for solid materials, the equation used, for example, in heat-transfer texts, models one-way coupling. The general conservation equation that appears in continuum mechanics texts models two-way coupling. By neglecting appropriate terms in the general conservation of energy equation, the simpler equation corresponding to one-way coupling is obtained. In this book, we derive the conservation of energy for one-way coupling.

We consider now conservation of energy for a continuum, assuming one-way thermal–mechanical coupling. Figure 3.3 shows a continuum contained within

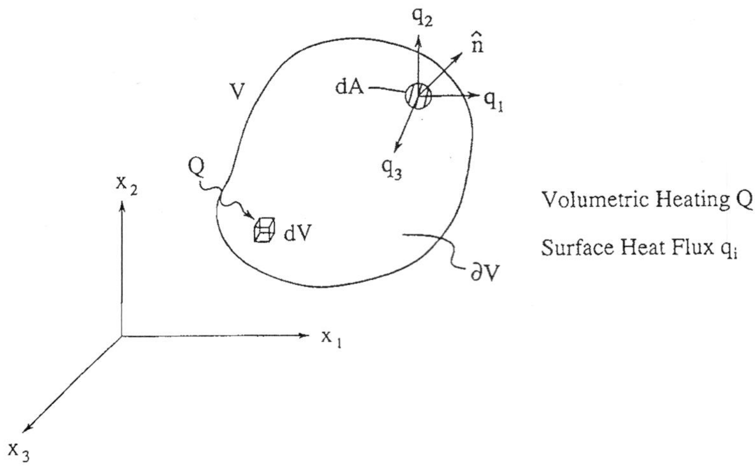

Fig. 3.3 Continuum with volumetric heating and surface heat flux.

a volume V and having a surface ∂V. Within the continuum, each infinitesimal element of volume $\mathrm{d}V$ experiences volumetric heating at a rate denoted by $Q = Q(x_1, x_2, x_3, t)$. On the surface of the continuum, each infinitesimal element of surface area $\mathrm{d}A$ experiences heat transfer per unit area at a rate denoted by a heat-flux vector $\vec{q}$, whose components are q_i, $i = 1, 2, 3$. Note that $q_i = q_i(x_i, x_2, x_3, t)$.

We use the conservation of energy equation in the form

$$\dot{E}_{\text{in}} + \dot{E}_{\text{gen}} = \dot{E}_{\text{sto}} + \dot{E}_{\text{out}} \tag{3.19}$$

where $\dot{E}_{\text{in}}$ denotes the rate of energy entering the control volume, $\dot{E}_{\text{gen}}$ the rate of energy generated within the control volume, $\dot{E}_{\text{sto}}$ the rate of energy stored within the control volume, and $\dot{E}_{\text{out}}$ the rate of energy leaving the control volume. In SI units, each term in Eq. (3.19) has the unit of joule per second (J/s), which is a watt (W). Thus, the conservation of energy equation (3.19) represents in rate form the balance of energy for the control volume.

Consider now the application of Eq. (3.19) to the continuum shown in Fig. 3.3. No energy is entering the control volume, so that

$$\dot{E}_{\text{in}} = 0$$

But energy is generated within the continuum by the volumetric heating rate Q. Integrating over the control volume produces

$$\dot{E}_{\text{gen}} = \int_V Q \mathrm{d}V$$

The equation that we write for the energy stored within the continuum depends on our assumption of one- or two-way coupling. By assuming one-way coupling, we neglect energy stored within the continuum by deformation. We assume that the internal energy u per unit volume depends only on the temperature of the material and the material's intrinsic ability to store internal energy by its thermal capacity. Thus, for one-way coupling, we assume, in the conservation of energy equation, that $u = u(T)$, where T denotes temperature as a function of position and time, i.e., $T = T(x_1, x_2, x_3, t)$. Then, for a temperature change ΔT at a point, we may write

$$\Delta u = \rho c(T)\Delta T$$

where $c(T)$ is the material's specific heat. Note that the specific heat $c = c(T)$ denotes that the specific heat is temperature-dependent. In rate form, the energy stored per unit volume is

$$\frac{\partial u}{\partial t} = \rho c(T)\frac{\partial T}{\partial t}$$

or, for the complete control volume, the rate of energy stored is

$$\dot{E}_{\text{sto}} = \int_V \rho c(T)\frac{\partial T}{\partial t}\mathrm{d}V$$

The final term in Eq. (3.19) is the energy leaving the control volume, which we obtain by integrating the normal component of the heat-flux vector over the surface ∂V. We obtain the normal component of the heat-flux by performing the dot product of the heat-flux vector with a unit-normal vector to the surface. Then, integrating over the surface, we may write

$$\dot{E}_{\text{out}} = \int_{\partial V} q_i n_i \mathrm{d}A$$

Then we substitute the four rates of energy into Eq. (3.19) to write the total balance of energy as

$$\int_V Q\mathrm{d}V = \int_V \rho c(T)\frac{\partial T}{\partial t}\mathrm{d}V + \int_{\partial V} q_i n_i \mathrm{d}V$$

Continuing, we use the Gauss theorem, Eq. (3.14), to write the surface integral as a volume integral. After this transformation, we may write the result as a single volume integral,

$$\int_V \left[\rho c(T)\frac{\partial T}{\partial t} + \frac{\partial q_i}{\partial x_i} - Q\right]\mathrm{d}V = 0$$

and, because the volume V is arbitrary, the integrand must vanish. Thus, the final result is

$$\rho c(T)\frac{\partial T}{\partial t} + \frac{\partial q_i}{\partial x_i} - Q = 0 \tag{3.20}$$

Equation (3.20) represents a single scalar equation for conservation of energy for a continuum, assuming one-way thermal–mechanical coupling. Written out, the equation is

$$\rho c(T)\frac{\partial T}{\partial t} + \frac{\partial q_1}{\partial x_1} + \frac{\partial q_2}{\partial x_2} + \frac{\partial q_3}{\partial x_3} - Q = 0 \tag{3.21}$$

We recognize that, with temperature-dependent specific heat, the energy equation is a nonlinear partial differential equation. A structure's specific heat is often assumed constant.

Several texts[1–3] derive the conservation of energy equation for two-way thermal–mechanical coupling. In general, the conservation of energy equation has the form

$$\rho\frac{\partial u}{\partial t} + \frac{\partial q_i}{\partial x_i} - \sigma_{ij}\frac{\partial \varepsilon_{ij}}{\partial t} - Q = 0 \tag{3.22}$$

where now the internal energy depends on the state of strain and temperature, $u = u(\varepsilon_{ij}, T)$. The third term on the right-hand side of Eq. (3.22), the summation of the products of stress components and strain-rate components, represents the rate of work done by the internal stresses. When we make the assumptions of one-way coupling that 1) the internal energy is a function only of temperature and 2) the surface tractions and body forces do not perform work, then Eq. (3.22) reduces to the form established in Eq. (3.21). An important practical difference between the two conservation of energy equations is that, for one-way coupling, the energy equation is independent of the structure's state of stress, strain, and deformation. Thus, we may solve Eq. (3.21) independently for the structure's temperature distribution. In contrast, Eq. (3.22) must be solved simultaneously with the structure's equations of motion because it depends on the stresses.

Under special circumstances, significant conversion of mechanical to thermal energy may occur. For instance, during metal-forming operations with rapid, inelastic deformations, material temperatures increase. This type of extreme behavior is highly unusual in aerospace thermal-structural applications. More frequently in aerospace applications, significant thermal energy is supplied by external sources such as aerodynamic heating. Then the assumption of one-way coupling may be justified on the basis that the internal thermal energy created by mechanical–thermal conversion is small compared to the externally supplied energy. For readers interested in pursuing two-way coupling further, a review paper, Ref. 16, discusses two-way coupling in detail.

3.6 Boundary and Initial Conditions

The equations describing the motion, typically Eq. (3.18), are solved subject to boundary conditions specifying either the displacement components or surface tractions on all external surfaces of the solid. Specifying displacements means that external supports either prohibit motion in one or more directions or enforce motion according to a prescribed history. Specifying surface tractions means that surfaces are subject to known forces per unit surface area (e.g., pressures) or that surfaces are completely free to deform—the surface tractions are zero. The mechanical initial conditions specify initial values of the structure's displacements and velocities. In addition, if the structure has been subject to a previous load history that induced permanent deformation, initial values of stresses and strains are required.

The energy equation (3.20) is solved subject to boundary conditions specifying either the temperature or normal heat flux on all external surfaces of the solid. Specifying temperature requires that an external means be available to control the surface temperature. Specifying the normal surface heat flux implies that the heat flux is related to a known heat-transfer mechanism at the surface, or that heat transfer across a surface is zero. Possible heat-transfer mechanisms at the surface include 1) a known incident heat-flux history, 2) convection heat transfer due to fluid motion next to the surface, or 3) emitted or incident radiation heat transfer. The mathematical expressions specifying convection and radiation boundary conditions involve the solid's unknown surface temperature. Thermal initial conditions specify initial values of the solid's temperature.

3.7 Constitutive Relations

By *constitutive relations* we mean the equations that characterize the behavior of the continuum's material. These are the equations that depend on a material's properties, which are determined experimentally. We have in mind two sets of material constitutive relations. The first set characterizes the structural behavior, and the second set characterizes the thermal behavior. The first set, we might say, is the stress–strain relations; the second set is the heat-flux–temperature relations.

For a long time, Hooke's law, the constitutive equation for a linear, elastic solid, has served as the basis for linear, elastic structural analysis. In more recent times, constitutive relations for inelastic behavior have been developed. These new constitutive models provide analysts with the capability of simulating complex inelastic structural response that heretofore has not been possible.

3.7.1 Linear Elastic Behavior

For linear, elastic behavior, the stress components are related to the strain components by generalized Hooke's law.[6–9] In indicial notation, this constitutive relation may be written for a homogeneous, isotropic material as

$$\sigma_{ij} = \lambda\,\delta_{ij}\varepsilon_{kk} + 2G\varepsilon_{ij} - (3\lambda + 2G)\delta_{ij}\alpha(T - T_{\text{ref}}) \tag{3.23}$$

where λ and G are material parameters known as the *Lamé constants*. In classical elasticity texts,[6–9] the second Lamé constant is denoted by μ instead of G. The Lamé constants are related to more familiar engineering constants by

$$\lambda = \frac{\upsilon E}{(1+\upsilon)(1-2\upsilon)} \qquad G = \frac{E}{2(1+\upsilon)} \tag{3.24}$$

where E is the modulus of elasticity, and υ is Poisson's ratio. In Eq. (3.23), α is the coefficient of thermal expansion, and T_{ref} is the reference temperature for zero thermal stress. For small temperature changes, E, υ, α are constant, but for large temperature changes, the elastic properties are temperature-dependent. Then, for temperature variations throughout a solid, the properties vary from point to point.

3.7.2 Elastic and Inelastic Behavior

For small stresses, strains, and small temperature changes, the behavior of structural members is elastic, and the boundary/initial-value problem can be formulated using Eq. (3.23). For metals, strains within the elastic limit are typically less than about 0.4%; small temperature changes are less than, say, 200°F. Analysis of an elastic continuum with small temperature changes is called *thermoelasticity*. As stress and strain levels rise and/or large temperature changes occur, the material behavior may become inelastic. For slow rates of loading under isothermal conditions, inelastic material behavior is described by the theory of *plasticity*. The classical theory of plasticity provides a theoretical description of the stress–strain behavior for an elastic–plastic response. The theory is based on the hypothesis that plastic strains occur after the stress state at a point reaches a critical level. As previously mentioned, the distinction between elastic and plastic behavior at a point is made according to a yield criterion. After yielding occurs at a point, plastic strains are induced and permanent deformation occurs.

Plastic strains, under most circumstances, depend on the loading history. A classical relation of plasticity is the Prandtl–Reuss flow rule.[10–13] This rule relates

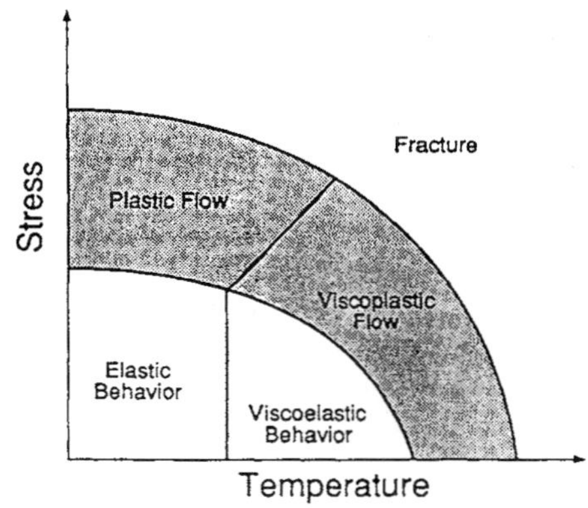

Fig. 3.4 Schematic representation of ductile material behavior.

plastic strains to the deviatoric stresses,

$$\frac{\partial \varepsilon_{ij}^{p}}{\partial t} = S_{ij}\frac{d\lambda}{dt} \tag{3.25}$$

where the superscript p denotes plasticity, and λ is a scalar factor (not to be confused with the Lamé constant) that varies during deformation. The concept of a yield criterion, together with the flow rule, provides a basis for the formulation of boundary-value problems in plasticity. Plasticity theory rests on the assumption that material behavior is independent of time since load rates are assumed small.

Under rapid loading or at high temperature levels, material behavior becomes time-dependent. A classical illustration of time-dependent inelastic behavior is creep. If a metallic bar is subjected to a constant tensile force with the stress below the yield point, a final elongation will be reached that is predicted by elastic theory. If the temperature of the bar is then raised to a high level, the elongation of the bar will increase continuously with time. Such time-dependent deformation under constant load is known as *creep*. Creep behavior is important in practical applications such as jet engines, where turbine blades at high temperature are subjected to centrifugal forces. Material scientists characterize material behavior using deformation mechanism diagrams.[14] A schematic of such a diagram (Fig. 3.4) for a ductile material illustrates the notions discussed in this section.

Under high rates of loading, with the material at elevated temperature, structures will experience inelastic behavior that includes rate-dependent plastic (viscoplastic) deformations. Most metals exhibit viscoplastic behavior at temperatures above 40% of their melting temperature. Analysis of structural problems that involve thermoviscoplastic effects requires constitutive models far more complex than those used in elasticity and plasticity. Constitutive models that include both plasticity and creep in a single set of relations are called *unified constitutive equations.*

3.7.3 Unified Constitutive Equations

During the last 20 years, unified constitutive equations have been under development by several investigators.[15] The models are based on microphysical behavior of materials, guided by phenomenological considerations, and employ concepts of continuum mechanics. Various investigators have developed different models for viscoplastic behavior, so that there are currently several sets of unified constitutive equations. Typically, the equations use the concept of internal-state variables to represent the evolution of the material behavior. The equations also involve a number of material parameters (some of which are temperature-dependent) that must be determined from experiments. Typical experiments include constant strain-rate uniaxial tests, creep tests and, for some models, constant strain-rate cyclic tests. At this time, there is no single constitutive model that can predict all aspects of inelastic material behavior. Moreover, comprehensive material-parameter data are not yet available.

3.7.4 Bodner–Partom Constitutive Model

One of the best-known constitutive models is the one introduced in 1968 by S. R. Bodner. This model assumes small strains (less than 5%). To consider the time-dependent character of the behavior, we write the equations in rate form. The total strain-rate tensor is $\dot{\varepsilon}_{ij}$, where the dot denotes differentiation with respect to time. The total strain-rate tensor is separated into elastic and inelastic components:

$$\dot{\varepsilon}_{ij} = \dot{\varepsilon}^e_{ij} + \dot{\varepsilon}^p_{ij} \tag{3.26}$$

where the elastic component is related to stresses by the time derivative of Hooke's law, Eq. (3.23). The inelastic strain-rate component is derived by Bodner and Partom, beginning with the Prandtl–Reuss flow rule, Eq. (3.25). The result is

$$\dot{\varepsilon}^p_{ij} = \left(S_{ij}/\sqrt{J^2}\right) D_0 \exp\left[-\tfrac{1}{2}\left(Z^2/3J_2\right)^2\right] \tag{3.27}$$

where D_0 is a material parameter (the limiting strain rate in shear), n a temperature-dependent material parameter, and Z a load-history–dependent parameter called the *internal-state variable* that represents the hardness of the material for resistance to plastic flow. The evolution equation for the state variable for isotropic hardening is

$$\dot{Z} = m_1[Z_1 - Z]\dot{W}_I - A_1 Z_1[(Z - Z_2)/Z_1]^{r_1} \tag{3.28}$$

with the initial condition $Z(0) = Z_0$. In the first term, Z_1 is the limiting value of Z, and m_1 is the hardening rate; the inelastic work rate is

$$\dot{W}_I = \sigma_{ij}\dot{\varepsilon}^p_{ij} \tag{3.29}$$

which is taken as a measure of hardening; Z_2 is the minimum value of Z at a given temperature; A_1 and r_1 are temperature-dependent material parameters. Equation (3.28) for the state variable is an ordinary differential equation with time as the independent variable.

In contrast to classical elasticity and plasticity, the Bodner–Partom unified constitutive model [see Eq. (3.26)] assumes that, for all load levels, there is inelastic strain. In addition, the model does not employ a yield criterion. In applications, the inelastic strain component is small in comparison to the elastic component at low load levels and becomes significant only when inelastic phenomena become prominent.

3.7.5 Heat Flux–Temperature Relations

The classical, widely used equation that relates the heat-flux components q_i to the temperature distribution $T(x_1, x_2, x_3, t)$ is Fourier's law. Fourier's law states that the heat flux is linearly proportional to the continuum's temperature gradients. For a general anisotropic material, it is commonly represented as

$$q_i = -k_{ij}\frac{\partial T}{\partial x_j} \qquad i = 1, 2, 3 \tag{3.30}$$

where k_{ij} is the material's thermal-conductivity tensor. Continuum texts show that the conductivity tensor is symmetric, i.e., $k_{ij} = k_{ji}$. The minus sign assures that the heat flow occurs from a higher to a lower temperature. In other words, a negative temperature gradient corresponds to a positive heat flow. Written out, Eq. (3.28) represents three scalar equations:

$$\begin{aligned} q_1 &= -\left(k_{11}\frac{\partial T}{\partial x_1} + k_{12}\frac{\partial T}{\partial x_2} + k_{13}\frac{\partial T}{\partial x_3}\right) \\ q_2 &= -\left(k_{21}\frac{\partial T}{\partial x_1} + k_{22}\frac{\partial T}{\partial x_2} + k_{23}\frac{\partial T}{\partial x_3}\right) \\ q_3 &= -\left(k_{31}\frac{\partial T}{\partial x_1} + k_{32}\frac{\partial T}{\partial x_2} + k_{33}\frac{\partial T}{\partial x_3}\right) \end{aligned} \tag{3.31}$$

For an anisotropic material, the off-diagonal terms in the thermal-conductivity tensor couple a heat flux in one direction with temperature gradients in all three directions. Such behavior might occur in a laminated, fibrous composite material. For an isotropic material, such as a typical metal, there is only a single material property the thermal conductivity k. Then Eq. (3.28) simplifies to

$$q_i = -k\frac{\partial T}{\partial x_i} \qquad i = 1, 2, 3 \tag{3.32}$$

or

$$q_1 = -k\frac{\partial T}{\partial x_1} \qquad q_2 = -k\frac{\partial T}{\partial x_2} \qquad q_3 = -k\frac{\partial T}{\partial x_3} \tag{3.33}$$

In the isotropic case, the heat-flux component in each coordinate direction depends only on the temperature gradient in that direction.

In general, over a broad temperature range, a material's thermal-conductivity properties vary with temperature. But, for small temperature changes, thermal conductivities are often assumed constant.

The conduction of heat in solids as modeled by Fourier's law predicts that thermal disturbances are propagated with infinite speed. Although the infinite propagation speed is not exact for most solids, the propagation of thermal energy occurs with such high velocity that Fourier's law is sufficiently accurate for aerospace applications. Recent studies show that the propagation of thermal energy at finite wave speeds may be important in problems involving thermal pulse times of extremely small duration or for cryogenic fluids. Heat propagation at finite wave speeds is sometimes called "second sound." At the end of the chapter, there is a bibliography on hyperbolic heat conduction.

3.8 Closing Remarks

This chapter introduces basic concepts of continuum mechanics that will be used in following chapters as we develop thermal and structural analyses for heated aerospace structures. The treatment of continuum mechanics here is brief, with an emphasis on topics relevant to thermal structures. For a broader, deeper understanding, readers should refer to continuum mechanics, elasticity, and plasticity texts.

References

[1]Fung, Y. C., *Foundations of Solid Mechanics*, Prentice–Hall, Englewood Cliffs, NJ, 1965.

[2]Malvern, L. E., *Introduction to the Mechanics of a Continuous Medium*, Prentice–Hall, Englewood Cliffs, NJ, 1969.

[3]Spencer, A. J. M., *Continuum Mechanics*, Longman Scientific and Technical, London, 1988.

[4]Frederick, D., and Chang, T. S., *Continuum Mechanics*, Allyn and Bacon, Boston, 1965.

[5]Krieth, F., *Principles of Heat Transfer*, 3rd ed., Intext, Scranton, PA, 1973.

[6]Timoshenko, S., and Goodier, J. N., *Theory of Elasticity*, 3rd ed., McGraw–Hill, New York, 1970.

[7]Love, A. E. H., *A Treatise on the Mathematical Theory of Elasticity*, Dover, New York, 1944.

[8]Sokolnikoff, I. S., *Mathematical Theory of Elasticity*, 2nd ed., McGraw–Hill, New York, 1956.

[9]Sechler, E. E., *Elasticity in Engineering*, Wiley, New York, 1952.

[10]Hill, R., *The Mathematical Theory of Plasticity*, Oxford Univ. Press, Oxford, England, UK, 1950.

[11]Phillips, A., *Introduction to Plasticity*, Ronald, New York, 1956.

[12]Prager, W., *An Introduction to Plasticity*, Addison–Wesley, Reading, MA, 1959.

[13]Chakrabarty, J., *Theory of Plasticity*, McGraw–Hill, New York, 1987.

[14]Courtney, T. H., *Mechanical Behavior of Materials*, McGraw–Hill, New York, 1990.

[15]Miller, A. K. (ed.), *Unified Constitutive Equations for Creep and Plasticity*, Elsevier, New York, 1987.

[16]Allen, D. H., "Thermomechanical Coupling in Inelastic Solids," *Applied Mechanics Reviews*, Vol. 44, No. 8, 1991, pp. 361–373.

Problems

3.1. For a general continuum,

(a) Derive the nonlinear strain-displacement equations

$$\varepsilon_{ij} = \frac{1}{2}\left(\frac{\partial u_i}{\partial x_j} + \frac{\partial u_j}{\partial x_i} + \frac{\partial u_k}{\partial x_i}\frac{\partial u_k}{\partial x_j}\right) \qquad i, j = 1, 2, 3$$

(b) Write the linear strain-displacement equations in engineering notation: $u = u_1$, $v = u_2$, $w = u_3$ and $x = x_1$, $y = y_1$, $z = z_1$. Give the physical interpretation of normal and shearing strains using equations, figures, and words.

(c) Explain the difference between tensor shearing strains and engineering shearing strains.

3.2. Show that the stress tensor σ_{ij} is symmetric, that is, $\sigma_{ij} = \sigma_{ji}$.

3.3. Using Taylor series expansions and a small, rectangular parallelpiped, derive the three equations of motion,

$$\frac{\partial \sigma_{ij}}{\partial x_j} + B_i = \rho \frac{\partial^2 u_i}{\partial t^2} \qquad i = 1, 2, 3$$

3.4. Using a small tetrahedron, show that the surface tractions T_i are related to the stress tensor σ_{ij} by

$$T_i = \sigma_{ij} n_j$$

where n_j are the components of a unit vector normal to the surface of a continuum.

3.5. Using Taylor series expansions and a small rectangular parallelpiped, derive the conservation of energy equation for an undeformable body,

$$Q - \frac{\partial q_i}{\partial x_i} = \rho c(T) \frac{\partial T}{\partial t}$$

where $c(T)$ denotes that the specific heat c is a function of temperature.

3.6. Read D. H. Allen's paper.[16]

(a) Explain what is meant by thermomechanical coupling, and give an illustrative example.

(b) Write the energy equation for an isotropic, viscoplastic incompressible material, and define all symbols in the equation. Explain the physical significance of each term in the equation.

(c) If the elastic deformation of a bar under a uniaxial dynamic load was computed including thermomechanical coupling, would the predicted temperature increase or decrease? What would happen for inelastic deformation?

(d) Does Allen explain when thermomechanical coupling can be neglected? Can you give some examples?

3.7. Using the Selected Bibliography on Hyperbolic Heat Conduction, which follows, write a short paper describing constitutive relations that account for finite speed heat propagation. Your paper should:

(a) Include a short historical description, with primary references for the constitutive relations.

(b) Describe the physical meaning of each term in the constitutive relation.

(c) Discuss physical problems for which the constitutive relations apply.

(d) Discuss how material properties in the constitutive relations are determined.

Selected Bibliography: Hyperbolic Heat Conduction

For convenience of reference, a number of publications are listed that describe aspects of hyperbolic heat conduction. The publications are listed in chronological order.

Peshkov, V., "'Second Sound' in Helium II," *Journal of Physics*, Vol. 8, No. 6, 1944, p. 381.

Cattaneo, C., "A Form of Heat Conduction Equation Which Eliminates the Paradox of Instantaneous Propagation," *Comptes Rendus*, Vol. 247, 1958, p. 431.

Ackerman, C. C., Bertman, B., Fairbank, H. A., and Guyer, R. A., "Second Sound in Solid Helium," *Physical Review Letters*, Vol. 16, No. 18, 1966, pp. 789–791.

Weymann, H. D., "Finite Speed of Propagation in Heat Conduction, Diffusion, and Viscous Shear Motion," *American Journal of Physics*, Vol. 35, No. 6, 1967, pp. 488–766.

Gurtin, M. E., and Pipkin, A. C., "A General Theory of Heat Conduction with Finite Wave Speed," *Archives of Rational Mechanics and Analysis*, Vol. 31, No. 2, 1968, pp. 113–126.

Baumeister, K. J., and Hamill, T. D., "Hyperbolic Heat-Conduction Equation—A Solution for the Semi-Infinite Body Problem," *Journal of Heat Transfer*, Vol. 91, Nov. 1969, pp. 543–548.

Chan, S. H., Low, M. J. D., and Mueller, W. K., "Hyperbolic Heat Conduction in Catalytic Supported Crystallites," *AIChE Journal*, Vol. 17, No. 6, 1971, pp. 1499–1501.

Taitel, Y., "On the Parabolic, Hyperbolic and Discrete Formulation of the Heat Conduction Equation," *International Journal of Heat and Mass Transfer*, Vol. 15, No. 2, 1972, pp. 369–371.

Maurer, M. J., and Thompson, H. A., "Non-Fourier Effects at High Heat Flux," *Journal of Heat Transfer*, Vol. 95, May 1973, pp. 284–286.

Wilhelm, H. E., and Choi, S. H., "Nonlinear Hyperbolic Theory of Thermal Waves in Metals," *Journal of Chemical Physics*, Vol. 63, No. 5, 1975, pp. 2119–2123.

Berkovsky, B. M., and Bashtovoi, V. G., "The Finite Velocity of Heat Propagation from the Viewpoint of the Kinetic Theory," *International Journal of Heat and Mass Transfer*, Vol. 20, No. 6, 1977, pp. 621–626.

Wiggert, D. C., "Analysis of Early-Time Transient Heat Conduction by Method of Characteristics," *Journal of Heat Transfer*, Vol. 99, No. 1, 1977, pp. 35–40.

Kao, T., "Non-Fourier Heat Conduction in Thin Surface Layers," *Journal of Heat Transfer*, Vol. 99, No. 2, 1977, pp. 343–355.

Sadd, M. H., and Didlake, J. E., "Non-Fourier Melting of a Semi-Infinite Solid," *Journal of Heat Transfer*, Vol. 99, No. 1, 1977, pp. 25–28.

Cummings, J. C., Schmidt, D. W., and Wagner, W. J., "Experiments on Second-Sound Shock Waves in Superfluid Helium," *Physics of Fluids*, Vol. 21, No. 5, 1978, pp. 713–717.

Nowinski, J. L., *Theory of Thermoelasticity with Applications*, Sijhoff and Noordhoff, alphen aan der Rijn, The Netherlands, 1978, pp. 110–117.

Carey, G. F., and Tsai, M., "Hyperbolic Heat Transfer with Reflection," *Numerical Heat Transfer*, Vol. 5, No. 3, 1982, pp. 309–327.

DeSocio, L. M., and Gualtieri, G., "A Hyperbolic Stefan Problem," *Quarterly of Applied Mathematics*, Vol. 41, No. 1, 1983, pp. 253–259.

Turner, T. N., "Using Second-Sound Shock Waves to Probe the Intrinsic Critical Velocity of Liquid Helium II," *Physics of Fluids*, Vol. 26, No. 11, 1983, pp. 3227–3341.

Özisik, M. N., and Vick, B., "Propagation and Reflection of Thermal Waves in a Finite Medium," *International Journal of Heat and Mass Transfer*, Vol. 27, No. 10, 1984, pp. 1845–1854.

Torczynski, J. R., Gerthsen, D., and Roesgen, T., "Schlieren Photography of Second-Sound Shock Waves in Superfluid Helium," *Physics of Fluids*, Vol. 27, No. 10, 1984, pp. 2418–2423.

Hager, N. E. III, and Constable, J. H., "Heat Pulse Transmission from Sodium Fluoride into Helium I," *Journal of Low Temperature Physics*, Vol. 61, Nos. 5/6, 1985, pp. 455–470.

Glass, D. E., Özisik, M. N., and Vick, B., "Hyperbolic Heat Conduction with Surface Radiation," *International Journal of Heat and Mass Transfer*, Vol. 28, No. 10, 1985, pp.1823–1830.

Glass, D. E., Özisik, M. N., McRae, D. S., and Vick, B., "On the Numerical Solution of Hyperbolic Heat Conduction," *Numerical Heat Transfer*, Vol. 8, No. 4, 1985, pp. 497–504.

Glass, D. E., Özisik, M. N., McRae, D. S., and Vick, B., "Hyperbolic Heat Conduction with Temperature Dependent Thermal Conductivity," *Journal of Applied Physics*, Vol. 59, 1986, pp. 1861–1865.

Glass, D. E., Özisik, M. N., and Vick, B., "Non-Fourier Effects on Transient Temperature Resulting from Periodic On-Off Heat Flux," *International Journal of Heat Mass Transfer*, Vol. 30, No. 8, 1987, pp. 1623–1631.

Glass D. E., Özisik, M. N., and McRae, D. S., "Hyperbolic Heat Conduction with Radiation in an Absorbing and Emitting Medium," *Numerical Heat Transfer*, Vol. 12, No. 3, 1987, pp. 321–333.

Brorson, S. D., Fujimoto, J. G., and Ippen, E. P., "Femtosecond Electronic Heat-Transport Dynamics in Thin Gold Films," *Physical Review Letters*, Vol. 59, No. 17, 1987, pp. 1962–1965.

Joseph, D. D., and Preziosi, L., "Heat Waves," *Review of Modern Physics*, Vol. 61, No. 1, 1989, pp. 41–73.

Li, D., "The Well-Posedness of a Hyperbolic Stefan Problem," *Quarterly of Applied Mathematics*, Vol. 47, No. 2, 1989, pp. 221–231.

Tamma, K. K., and Railkar, S. B., "Specially Tailored Transfinite-Element Formulations for Hyperbolic Heat Conduction Involving Non-Fourier Effects," *Numerical Heat Transfer, Part B*, Vol. 15, No. 2, 1989, pp. 211–226.

Glass, D. E., Özisik, M. N., and Kim, W. S., "Hyperbolic Stefan Problem with Applied Surface Heat Flux and Temperature Dependent Thermal Conductivity," *Numerical Heat Transfer, Part A*, Vol. 18, No. 4, 1990, pp. 503–516.

Glass, D. E., and McRae, D. S., "Variable Specific Heat and Thermal Relaxation Parameter in Hyperbolic Heat Conduction," *Journal of Thermophysics and Heat Transfer*, Vol. 4, No. 2, 1990, pp. 252–255 (AIAA Paper 89-0317).

Kaminski, W., "Hyperbolic Heat Conduction Equation for Materials with a Nonhomogeneous Inner Structure," *Journal of Heat Transfer*, Vol. 112, No. 3, 1990, pp. 555–560.

Tamma, K. K., and Railkar, S. B., "Evaluation of Thermally Induced Non-Fourier Stress Wave Disturbances via Tailored Hybrid Transfinite Element Formulations," *Computers and Structures*, Vol. 34, No. 1, 1990, pp. 5–16.

Glass, D. E., Özisik, M. N., McRae, D. S., and Kim, W. S., "Formulation and Solution of Hyperbolic Stefan Problem," *Journal of Applied Physics*, Vol. 70, No. 3, 1991, pp. 1190–1197.

Tamma, K. K., and D'Costa, J. F., "Transient Modeling/Analysis of Hyperbolic Heat Conduction Problems Employing Mixed Implicit–Explicit α Method," *Numerical Heat Transfer, Part B*, Vol. 19, No. 1, 1991, pp. 49–68.

Glass, D. E., Tamma, K. K., and Railkar, S. B., "Hyperbolic Heat Conduction with Convection Boundary Conditions and Pulse Heating Effects," *Journal of Thermophysics and Heat Transfer*, Vol. 5, No. 1, 1991, pp. 110–116.

Tamma, K. K., and Namburu, R. R., "Hyperbolic Heat-Conduction Problems: Numerical Simulations via Explicit Lax–Wendroff–Based Finite Element Formulations," *Journal of Thermophysics and Heat Transfer*, Vol. 5, No. 2, 1991, pp. 232–239.

Chen, H.-T., and Lin, J.-Y., "Numerical Analysis for Hyperbolic Heat Conduction," *International Journal of Heat and Mass Transfer*, Vol. 36, No. 11, 1993, pp. 2891–2898.

4
Heat Transfer in Structures

4.1 Introduction

Here we describe basic concepts, equations, and solutions for heat-transfer in aerospace structures. We begin by presenting a general formulation for the heat-transfer problem in a continuum. The balance of the chapter describes the determination of temperatures in structural heat-transfer problems for high-speed aircraft and spacecraft. Solutions are given first for basic problems where temperature gradients due to conduction heat transfer are neglected. However, the majority of structural heat-transfer problems are conduction-dominated, with significant temperature gradients. Thus, later sections of the chapter emphasize conduction heat transfer with a variety of boundary conditions. Some sections emphasize analytical solutions that have practical value. Other sections deal with computational methods and their application to realistic, nonlinear problems. The chapter concludes with a description of computational models for convectively cooled structures. Readers interested in further details of the broad field of heat transfer should consult comprehensive texts on heat transfer, for example, Refs. 1–15.

4.2 Problem Formulation

For conduction heat transfer in aerospace structures, the classical heat-conduction equation is used. From Eq. (3.20), the conservation of energy equation is

$$-\frac{\partial}{\partial x_i}\left[k_{ij}\frac{\partial T}{\partial x_j}\right] + \rho c\frac{\partial T}{\partial t} = Q \tag{4.1}$$

which is a parabolic partial differential equation. This means that thermal disturbances propagate at infinite speeds through the structure. To address this anomaly, some authors have used a modified form of Fourier's law to derive a hyperbolic energy equation. In the hyperbolic energy equation, thermal disturbances propagate at a very high finite wave speed, which is called "second sound." The application of the hyperbolic energy equation for solids is controversial because the phenomenon of finite wave speeds of thermal disturbances has never been demonstrated for structural materials although it has been for gases. (A bibliography on hyperbolic heat conduction can be found at the end of Chapter 3). The parabolic energy equation is customarily used for structural heat transfer.

The heat-conduction equation is solved subject to an initial condition and boundary conditions on all portions of the surface. The initial condition specifies the temperature distribution at time zero. The boundary conditions may consist of specified surface temperature, specified heat flow, convective heat exchange, and

Heat Transfer Problems

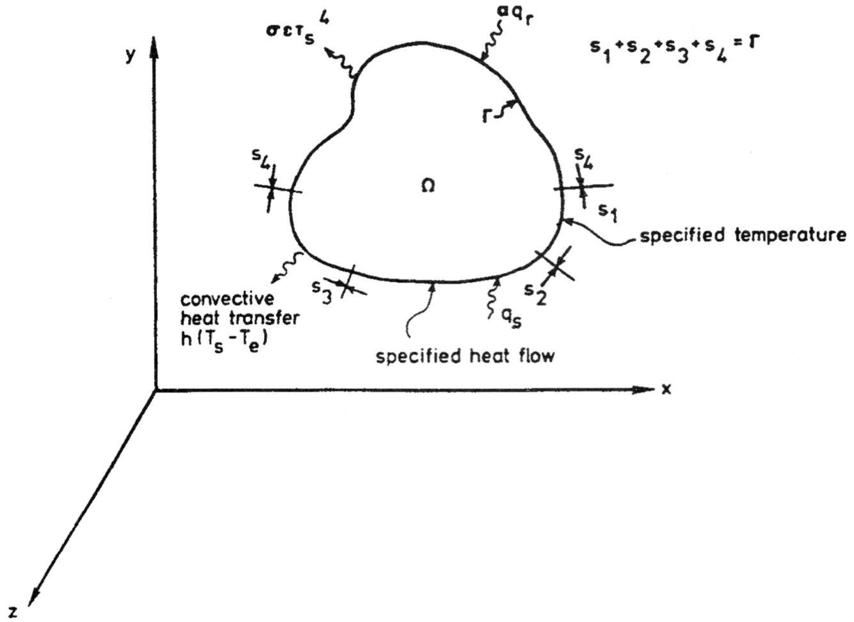

Fig. 4.1 Solution domain for a conduction problem.

radiation heat exchange. These may be written as

$$
\begin{aligned}
T_s &= T_1(x_1, x_2, x_3, t) && \text{on} \quad S_1 \\
q_n &= -q_s && \text{on} \quad S_2 \\
q_n &= h(T_s - T_e) && \text{on} \quad S_3 \\
q_n &= \sigma\varepsilon T_s^4 - \alpha_s q_r && \text{on} \quad S_4
\end{aligned}
\tag{4.2}
$$

where q_n denotes the heat flux normal to the surface and S_i $(i = 1, 4)$ portions of the surface (Fig. 4.1). The specified surface temperature is T_1, and specified surface heat flux is q_s (positive into the surface). In the convective boundary condition, the convective exchange temperature T_e is T_{aw} (see Sec. 1.3.2) for aerodynamic heating. In the radiation boundary condition, α_s is the surface absorptivity, and q_r is the incident radiation heat flux. For structural heat transfer with surfaces at significantly different, elevated temperatures, radiation exchanges between surfaces can occur. The determination of radiation exchanges between surfaces is complicated because 1) radiation emitted by a typical surface depends on its surface temperature, which is unknown, and 2) the geometric relationship between each surface must be considered. Often, the radiation problem is handled by discretizing the radiation boundary into N discrete surfaces, which are assumed isothermal. If a radiation heat flux on the ith surface is called H_i, a set of N simultaneous equations may be developed to determine H_i in terms of the temperatures T_i. In matrix form, these equations (see Sec. 8.2.3 and Eq. 8.30) may be written as

$$
\big[[I] - [F][(1-\varepsilon)]\big]\{H\} = [F]\{\varepsilon\sigma T^4\} \tag{4.3}
$$

where the components of the matrix $[F]$ are the view factors F_{ij}, and $[I]$ denotes the identity matrix. A view factor F_{ij} is the fraction of the radiation energy leaving surface i that arrives at surface j. The determination of view factors for complex three-dimensional structures is a formidable computational task.

4.3 Thermal Response with Infinite Thermal Conductivity

From time to time in applications, we encounter problems in which heat transfer occurs without significant temperature gradients. Small temperature gradients occur, for instance, in thin-walled members with high thermal conductivity. If the temperature gradients can be neglected, then the temperature is a function only of time, and the thermal analysis is simplified because we need to solve only an ordinary differential equation to obtain the thermal response. We present here some basic examples to illustrate the solution approach and to demonstrate fundamental features of the thermal response.

In formulating the pertinent ordinary differential equations, we begin with the basic conservation of energy equation (3.19) rather than the heat-conduction equation given in Eq. (4.1). Thus, we apply to each problem the basic conservation equation

$$\dot{E}_{\text{in}} + \dot{E}_{\text{gen}} = \dot{E}_{\text{sto}} + \dot{E}_{\text{out}} \tag{4.4}$$

which, in fact, we used to derive Eq. (4.1). In Eq. (4.4), recall that we are writing the energy balance for a control volume in rate form, where each $\dot{E}$ term has units in the SI system of $J/s = W$. The terms represent energy entering, generated, stored, and leaving the control volume, respectively.

4.3.1 Convectively Cooled Billet

A small billet of homogeneous material (Figure 4.2) initially has a uniform temperature T_0. At time $t = 0$, the billet is immersed in a fluid with constant temperature T_f. The fluid temperature T_f is lower than the initial temperature T_0 of the billet, so that the billet is being convectively cooled. We wish to determine the transient thermal response of the billet, i.e, $T(t)$.

Writing an energy balance with Eq. (4.4), we obtain

$$0 = \rho c V \frac{\mathrm{d}T}{\mathrm{d}t} + hA(T - T_f)$$

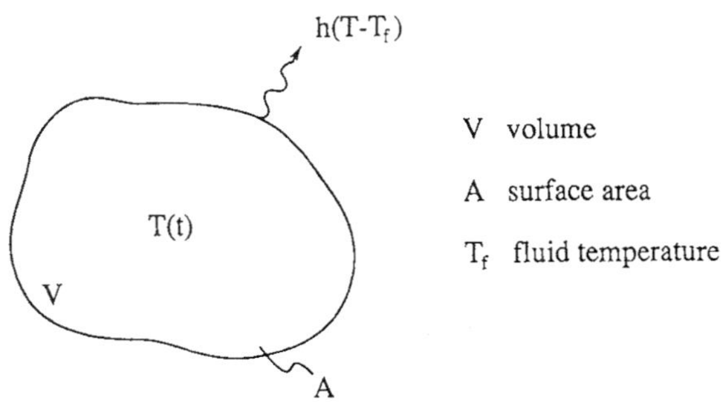

Fig. 4.2 Convectively cooled billet.

or

$$\frac{\mathrm{d}T}{\mathrm{d}t} + \frac{hA}{\rho c V}(T - T_f) = 0 \tag{4.5}$$

where V is the volume of the billet, and A is the surface area. The ratio of the volume of the billet to the surface area, V/A, is a characteristic length of the solid. For convenience, we define the coefficient of the temperature term as

$$B = hA/\rho c V$$

and let $\theta = T - T_f$, so that the energy equation becomes

$$\frac{\mathrm{d}\theta}{\mathrm{d}t} + B\theta = 0 \tag{4.6}$$

Equation (4.6) is a first-order, linear homogeneous equation that we may solve either by separation of variables or by the use of an integrating factor. The solution, after evaluation of the constant of integration with the initial condition, may be written as

$$\frac{T - T_f}{T_0 - T_f} = e^{-Bt} \tag{4.7}$$

As a result of the convective cooling, the billet temperature decreases smoothly as energy is lost to the surrounding fluid. After a long time, the billet temperature approaches the fluid temperature. The parameter B is a measure of the thermal response time of the system. Large values of B represent a rapid decrease in the billet temperature, and small values of B represent a slow decrease in temperature.

Note that, in the solution of Eq. (4.5), we introduced the value of $\theta = T - T_f$ as an intermediate step that led to Eq. (4.6). This substitution had two advantages: 1) It made Eq. (4.6) a homogeneous ordinary differential equation, and 2) it led to the nondimensional form of the temperature in the final solution given in Eq. (4.7). When convective boundary conditions are involved, this type of substitution is often used to derive temperature solutions.

4.3.2 Aerodynamically Heated Thin Plate

Consider an aerodynamically heated thin plate (Fig. 4.3) of thickness H and an arbitrary surface area A. We assume that the plate is thin and made of a high-conductivity material such as copper so that the temperature is uniform and is a function only of time. The plate experiences aerodynamic heating $q = h(T_{\mathrm{aw}} - T)$ on the upper surface where h is a constant convection coefficient, and T_{aw} is a constant adiabatic wall temperature. These values are determined for high-speed flight from the Mach number, altitude, and other aerodynamic parameters. Section 1.3.2 presents an approximate technique called the *reference temperature method* for prediction of h and T_{aw}. The other surfaces of the plate are perfectly insulated, and so there are no energy losses. At time zero, the plate has an initial temperature T_0. Then the aerodynamic heating commences, and we wish to determine the thermal response $T(t)$. All thermal properties are assumed constant.

Applying Eq. (4.4), with reference to Fig. 4.3a, we may write the energy balance as

$$h(T_{\mathrm{aw}} - T)A + 0 = \left(\rho c \frac{\mathrm{d}T}{\mathrm{d}t}\right) AH + 0$$

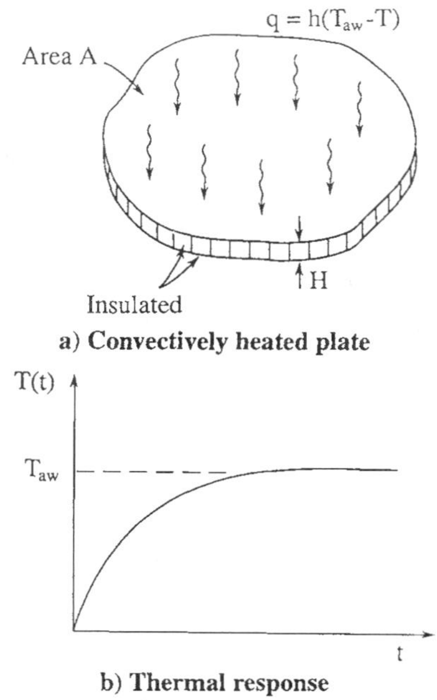

Fig. 4.3 Thermal response of an aerodynamically heated thin plate.

and we obtain the conservation equation

$$\frac{\mathrm{d}T}{\mathrm{d}t} + \beta T = \beta T_{\mathrm{aw}} \tag{4.8}$$

where $\beta = h/\rho c H$. Equation (4.8) is a first-order, linear nonhomogeneous ordinary differential equation that we may easily solve to obtain the thermal response,

$$T(t) = T_0 e^{-\beta t} + T_{\mathrm{aw}}(1 - e^{-\beta t}) \tag{4.9}$$

As a result of the aerodynamic heating, the temperature rises smoothly and, as time becomes large, the temperature asymptotically approaches the adiabatic wall temperature T_{aw} (Fig. 4.3b). The parameter β determines the rise time of the response.

4.3.3 Radiantly Heated Thin Plate

Consider a thin plate (Fig. 4.4) of thickness H with an arbitrary surface area A. As in the preceding problem, the plate is made of a high–thermal-conductivity material, so that temperature gradients may be neglected. The top surface of the plate is heated radiantly by a heat lamp and simultaneously emits radiant energy to space. The plate edges and bottom surface are perfectly insulated. At time equal to zero, the plate has a uniform initial temperature $T(0) = T_0$. At time zero, the heat lamp is turned on; thereafter, the plate is exposed to constant radiant heat flux q_r. We wish to determine the transient thermal response $T(t)$. For convenience in obtaining the analytical solution, all thermal properties are assumed constant.

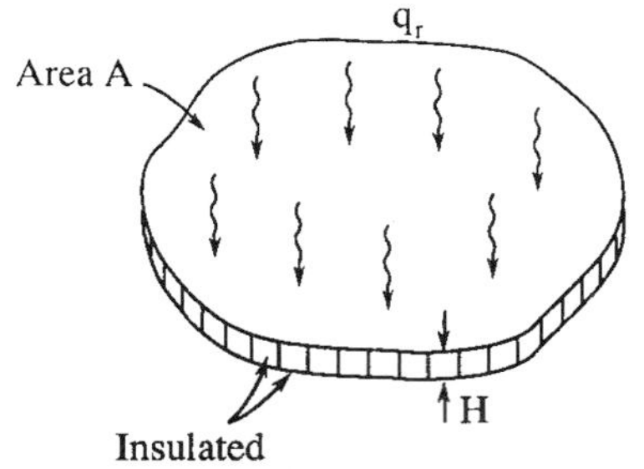

a) Radiantly heated thin plate

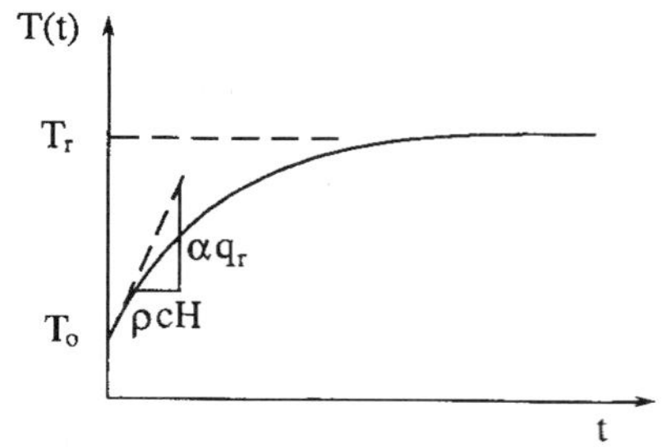

b) Thermal response

Fig. 4.4 Thermal response of a radiantly heated plate.

Applying Eq. (4.4) with reference to Fig. 4.4a, we may write the energy balance as

$$(\alpha q_r)A + 0 = \left(\rho c \frac{dT}{dt}\right) AH + (\sigma \varepsilon T^4)A$$

and we obtain the conservation equation

$$\rho c H \frac{dT}{dt} + \sigma \varepsilon T^4 = \alpha q_r \tag{4.10}$$

Equation (4.10) is a first-order, nonlinear, and non homogeneous ordinary differential equation. Before solving Eq. (4.10), let us first consider the initial and final behavior of the thermal response.

Initial response. At early times in the response, the radiation heat loss is small and can be neglected. Then we can integrate Eq. (4.10) directly to obtain

$$T(t) = T_0 + \frac{\alpha q_r}{\rho c H} t \tag{4.11}$$

Equation (4.11) shows that, initially, the temperature increases linearly with time at a rate directly proportional to the absorbed heat flux but inversely proportional to the material's capability to store energy, as expressed by the thermal capacitance $\rho c H$. Although, initially, the temperature linearly increases with time, the rate of increase begins to diminish as the radiation heat loss begins to mount.

Final response. With time, less and less energy is stored in the plate, and eventually, all the absorbed incident energy is radiated back to space. As time becomes large, the plate approaches a constant temperature—the radiation equilibrium temperature T_r. The value of the radiation equilibrium temperature T_r is obtained from Eq. (4.10) by setting $\mathrm{d}T/\mathrm{d}t$ equal to zero and solving for T_r,

$$T_r = \left[\frac{\alpha q_r}{\sigma\varepsilon}\right]^{\frac{1}{4}} \tag{4.12}$$

The radiation equilibrium temperature T_r is the horizontal asymptote for the thermal response sketched in Fig. 4.3b.

Although Eq. (4.10) is a nonlinear ordinary differential equation, we may obtain an analytical solution by separation of variables followed by direct integration. By manipulation, we may write

$$\frac{\sigma\varepsilon}{\rho c H}\,\mathrm{d}t = \frac{\mathrm{d}T}{T_r^4 - T^4}$$

where the definition of the radiation equilibrium temperature has been introduced in the denominator on the right-hand side. We can evaluate this integral directly with the help of integral tables. After integration, we use the initial condition $T(0) = T_0$ to evaluate the constant of integration. The final solution may be written

$$\frac{\sigma\varepsilon}{\rho c H}t + \frac{1}{4T_r^3}\left[\ln\frac{(T_r - T)(T_r + T_0)}{(T_r + T)(T_r - T_0)} + 2\tan^{-1}\frac{T_0}{T_r} - 2\tan^{-1}\frac{T}{T_r}\right] = 0 \tag{4.13}$$

Equation (4.13) does not give the temperature explicitly as a function of time; rather, it gives the inverse relation, time as a function of temperature. For numerical computations, we have two choices; we can either 1) specify values of T over the range from T_0 to T_r and solve Eq. (4.13) for the corresponding times or 2) specify values of time and use a numerical root-solving algorithm to solve Eq. (4.13) iteratively for the corresponding values of temperature. Either approach works well in practice.

We use Eq. (4.13) in several applications, such as for isothermal members in orbiting space structures. For applications, we need to keep in mind that it was derived for constant-radiation heat flux q_r. However, the equation can be used for applications in which q_r varies by time-marching with a relatively small time step Δt and assuming that q_r is constant over each time step.

4.4 Transient One-Dimensional Conduction Problems

We consider now transient conduction problems in one dimension. Thermal properties are assumed constant, and only linear boundary conditions are considered. Thus, the problems posed are classical linear boundary/initial-value problems that appear widely in engineering mathematics and heat-transfer texts. We give a brief review of basic solution techniques, but the emphasis is not intended to be on problem solving. Rather, the emphasis is on presentation of a few basic solutions for one-dimensional domains that are of value in solving thermal-structural problems. Along the way, we give physical interpretations of the predicted thermal response and identify parameters characteristic of the thermal behavior.

We return now to the conservation of energy equation (4.1) and specialize it for one dimension and constant properties. Neglecting internal heat generation, we may write the energy equation as

$$\frac{\partial^2 T}{\partial x^2} = \frac{1}{\kappa}\frac{\partial T}{\partial t} \tag{4.14}$$

where $\kappa = k/\rho c$ is the thermal diffusivity. We solve Eq. (4.14) subject to two boundary conditions and one initial condition. A large number of analytical solutions are available, and most heat-transfer texts[1–8] present a few analytical solutions. A classical reference containing a large number of solutions is the text by Carslaw and Jaeger.[16]

We demonstrate only two solution techniques. For one-dimensional domains of finite extent, we use separation of variables. For one-dimensional domains of semi-infinite extent, we use the Laplace transform technique. These techniques have been widely used for a variety of the linear problems posed by Eq. (4.1) and are covered in more detail in engineering mathematics courses.

4.4.1 Temperatures in a Thick Plate

We consider now a series of boundary/initial-value problems for transient temperature distributions $T(x, t)$ in a thick plate. In the first example, we use separation of variables to derive the solution for $T(x, t)$, but for the following examples, we present and discuss only the final solution.

Example 4.1: Specified Wall Temperatures

We consider a plate of thickness L that extends to infinity in the y and z directions (Fig. 4.5a). The plate has a uniform initial temperature T_0. At time equal to zero, the temperatures of the boundary surfaces at $x = 0$ and $x = L$ are instantaneously increased to new values, T_1 and T_2, respectively. Thus, the boundary and initial conditions are

$$T(0, t) = T_1 \tag{4.15a}$$

$$T(L, t) = T_2 \tag{4.15b}$$

$$T(x, 0) = T_0 \tag{4.15c}$$

In solving transient conduction problems, we should first determine, from examining the problem statement, if the problem possesses a steady-state solution. If it does, then we know that the thermal response must approach this solution as time becomes large. Then we can represent the solution for $T(x, t)$ as the sum of a steady-state solution and a transient solution. An advantage of this superposition is that the steady-state solution can be used to satisfy any nonhomogeneous boundary conditions; hence, the transient portion of the solution will have homogeneous boundary conditions.

For the present problem, we realize that, after a long time, the temperature will approach a steady-state solution, so that we represent the response as

$$T(x, t) = T_S(x) + T_T(x, t) \tag{4.16}$$

where $T_S(x)$ represents the final steady-state temperature distribution, and $T_T(x, t)$ denotes the transient portion of the solution. The transient solution diminishes to zero as time becomes large with $T(x, t)$ approaching $T_S(x)$.

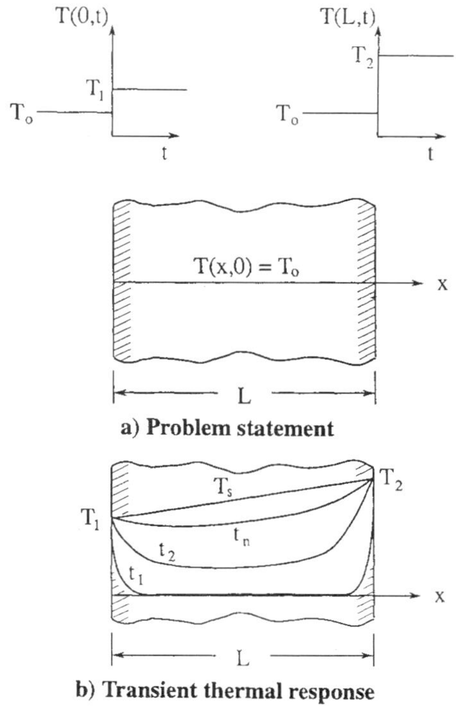

Fig. 4.5 Thick plate with specified wall temperatures.

Steady-state solution. We solve for the steady-state temperature distribution by setting the time derivative of $T_S(x)$ in Eq. (4.14) to zero, producing the ordinary differential equation,

$$\frac{\mathrm{d}^2 T_S}{\mathrm{d}x^2} = 0$$

We easily solve for $T_S(x)$ by integrating the equation twice and using the boundary conditions, Eqs. (4.15a) and (4.15b), to determine the constants of integration. Thus,

$$T_S(x) = T_1 + (T_2 - T_1)(x/L) \tag{4.17}$$

which shows that the steady temperature varies linearly between the specified wall temperatures.

Transient solution. By substituting Eq. (4.16) into the energy equation (4.14), we see that

$$\frac{\partial^2 T_T}{\partial x^2} = \frac{1}{\kappa}\frac{\partial T_T}{\partial t}$$

given that $\mathrm{d}^2 T_S/\mathrm{d}x^2 = 0$ and $\partial T_S/\partial t = 0$. The boundary conditions for T_T are determined from Eqs. (4.16), (4.15a), and (4.15b),

$$T_T(0,t) = T(0,t) - T_S(0) = T_1 - T_1 = 0$$

$$T_T(L,t) = T(L,t) - T_S(L) = T_2 - T_2 = 0$$

Notice, as we mentioned, that dividing the solution into two parts provides homogeneous boundary conditions for $T_T(x, t)$. The initial condition is not homogeneous but, instead, is

$$T_T(x, 0) = T(x, 0) - T_S(x)$$

and, using Eqs. (4.15c) and (4.17), we obtain

$$T_T(x, 0) = T_0 - (T_2 - T_1)(x/L)$$

We solve the new boundary/initial-value problem for $T_T(x, t)$ by separation of variables. The approach is to represent $T_T(x, t)$ as the product of two unknown functions,

$$T_T(x, t) = X(x)\theta(t)$$

and substitute into the partial differential equation for T_T. Separation of variables then produces two ordinary differential equations,

$$\frac{d\theta}{dt} - \lambda^2\kappa\theta = 0$$

$$\frac{d^2X}{dx^2} + \lambda^2 X = 0$$

In these equations, λ is a separation constant that is to be determined; the constant, in fact, will be identified with the eigenvalues for the problem. Solving these equations and combining the product produces the general solution,

$$T_T(x, t) = (c_1 \sin\lambda x + c_2 \cos\lambda x)e^{-\lambda^2\kappa t}$$

where c_1 and c_2 are constants of integration to be determined. Imposing the boundary conditions yields

$$T_T(0, t) = e^{-\lambda^2\kappa t}(0 + c_2) = 0$$

$$T_T(L, t) = e^{-\lambda^2\kappa t}(c_1 \sin\lambda L + c_2 \cos\lambda L) = 0$$

Thus, $c_2 = 0$ and $\sin\lambda L = 0$. The latter equation defines the eigenvalues because $\sin\lambda L$ will be zero only if

$$\lambda L = n\pi \qquad n = 1, 2, \ldots$$

Hence, there is an infinite set of eigenvalues computed from

$$\lambda_n = n\pi/L \qquad n = 1, 2, \ldots$$

Because the energy equation for T_T is linear, the most general solution is the linear superposition of all the possible solutions. Therefore, we may write

$$T_T(x, t) = \sum_{n=1}^{\infty} a_n \sin(n\pi x/L)\exp(-n^2\pi^2\kappa t/L^2)$$

where we have redefined the unknown constant c_1 as a_n.

We determine a_n from using the initial condition, Eq. (4.15c) . Thus,

$$T_T(x, 0) = \sum_{n=1}^{\infty} a_n \sin(n\pi x/L)$$

Next, we solve for a_n by multiplying by $\sin(m\pi x/L)$ and integrating over the domain from $x = 0$ to $x = L$,

$$\int_0^L T_T(x, 0) \sin(m\pi x/L)\,\mathrm{d}x = \sum_{n=1}^{\infty} a_n \int_0^L \sin(n\pi x/L) \sin(m\pi x/L)\,\mathrm{d}x$$

The integral on the right-hand side simplifies because the sin functions are *orthogonal*,

$$\int_0^L \sin(n\pi x/L) \sin(m\pi x/L)\,\mathrm{d}x = \begin{cases} 0 & m \neq n \\ L/2 & m = n \end{cases}$$

Therefore, in the infinite series, there is only one nonzero term, and that occurs when $m = n$. Then we can solve for a_n as

$$a_n = \frac{2}{L} \int_0^L T_T(x, 0) \sin(n\pi x/L)\,\mathrm{d}x$$

To obtain an explicit expression for a_n, we substitute for $T_T(x, 0)$ and evaluate the integral to obtain

$$a_n = \frac{2}{n\pi}\left[(-1)^n (T_2 - T_0) - (T_1 - T_0)\right]$$

Then the solution for $T_T(x, t)$comes from substituting this result into the series for $T_T(x, t)$ which yields

$$T_T(x, t) = \frac{2}{\pi} \sum_{n=1}^{\infty} \frac{1}{n}\left[(-1)^n (T_2 - T_0) - (T_1 - T_0)\right] \times \sin(n\pi x/L) \exp(-n^2\pi^2\kappa t/L^2) \tag{4.18}$$

We obtain the final solution to the original problem by superimposIng the steady-state solution, Eq. (4.17), and the transient solution, Eq. (4.18):

$$T(x, t) = T_1 + (T_2 - T_1)(x/L) + \frac{2}{\pi} \sum_{n=1}^{\infty} \frac{1}{n}\left[(-1)^n (T_2 - T_0) - (T_1 - T_0)\right] \sin(n\pi x/L) \exp(-n^2\pi^2\kappa t/L^2) \tag{4.19}$$

This response is sketched in Fig. 4.5b for successive instants of time $t_1, t_2, \ldots, t_n$. At very early times, we may anticipate convergence difficulties with the infinite series in Eq. (4.19) because it must represent the steep temperature gradients that propagate inward from the walls. Later, the series will converge more rapidly as the temperature distribution becomes easier to represent. As time increases, of course, the contribution of the series to the response diminishes as the temperature approaches the linear steady-state distribution.

Example 4.2: Specified Heat Flux and Insulated Surface

We consider the thick plate of Example 4.1 with uniform initial temperature but with different boundary conditions as shown in Fig. 4.6a. At time equal to zero, the boundary surface at $x = 0$ experiences a specified heat flux $q = q_s$. The boundary surface at $x = L$ is insulated so that it has $q = 0$. Thus, the boundary and initial conditions are

$$q(0, t) = -k\frac{\partial T}{\partial x}(0, t) = q_s \tag{4.20a}$$

$$q(L, t) = -k\frac{\partial T}{\partial x}(L, t) = 0 \tag{4.20b}$$

$$T(x, 0) = T_0 \tag{4.20c}$$

Unlike Example 4.1, we observe that this problem does not have a steady-state solution. Because the left-hand wall continuously receives thermal energy and the right-hand wall is insulated, the temperature of the plate must always increase.

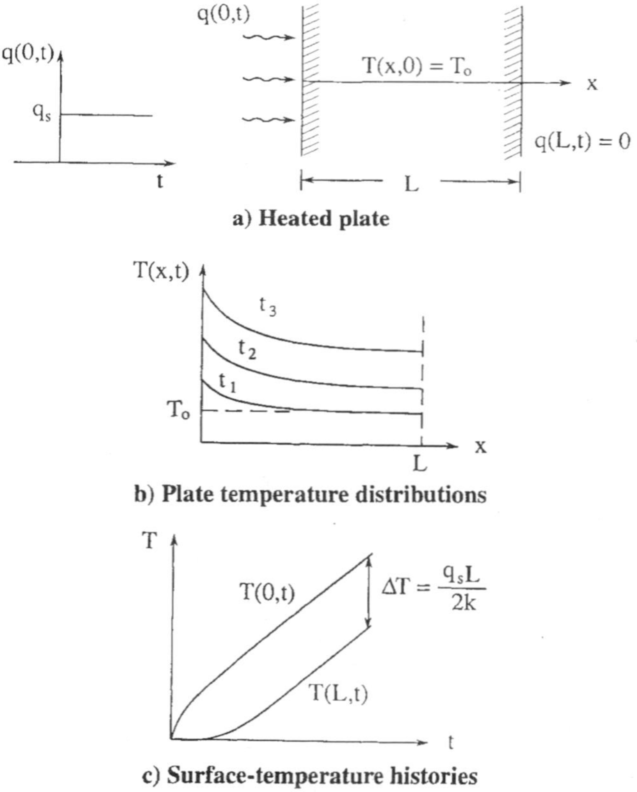

Fig. 4.6 Thermal response of a heated plate.

The solution to the boundary/initial-value problem is

$$T(x,t) = T_0 + \frac{q_s t}{\rho c L} + \frac{q_s L}{k}\left[\frac{1}{3} - \frac{x}{L} + \frac{x^2}{2L^2} - \frac{2}{\pi^2}\sum_{n=1}^{\infty}\frac{1}{n^2}\cos\frac{n\pi x}{L}\exp(-n^2\pi^2\kappa t/L^2)\right] \tag{4.21}$$

In the preceding solution, the first two terms on the right-hand side represent the temperature we would compute for an isothermal plate subject to a specified heat flux; see Eq. (4.11). Next, considering the terms within the square brackets, we notice that, as time becomes large, the infinite series approaches zero. Thus, for large time,

$$T(x,t) \to T_0 + \frac{q_s t}{\rho c L} + \frac{q_s L}{k}\left(\frac{1}{3} - \frac{x}{L} + \frac{x^2}{2L^2}\right) \tag{4.22}$$

From this solution, we see that, while the temperature continuously increases with time, the temperature variation across the plate maintains a uniform distribution. Plate temperature distributions are sketched in Fig. 4.6b. Moreover, if we evaluate the surface temperatures,

$$T(0,t) = T_0 + \frac{q_s t}{\rho c L} + \frac{q_s L}{3k} \qquad T(L,t) = T_0 + \frac{q_s t}{\rho c L} - \frac{q_s L}{6k}$$

we see that the temperature gradient ΔT across the plate becomes a constant,

$$\Delta T = \frac{q_s L}{2k} \tag{4.23}$$

The histories of the surface temperatures are graphed in Fig. 4.6c.

Example 4.3: Specified Heat Flux and Specified Surface Temperature

We consider the thick plate of Example 4.2 with uniform initial temperature T_0 and specified surface heat flux at $x = 0$. Now, however, we assume that the right-hand surface has a specified constant temperature equal to the initial temperature. Thus, the boundary and initial conditions are

$$q(0,t) = -k\frac{\partial T}{\partial x}(0,t) = q_s \tag{4.24a}$$

$$T(L,t) = T_0 \tag{4.24b}$$

$$T(x,0) = T_0 \tag{4.24c}$$

One basic difference between this problem and Example 4.2 is that, in this case, the problem has a steady-state solution. The steady-state solution occurs here because the incident thermal energy on the left-hand surface can be conducted through, and leave the plate through, the right-hand boundary. Thus, here the solution can be separated into two parts, the steady-state and transient solutions, as in Example 4.1. The solution is

$$T(x,t) = T_0 + q_s(L-x)/k - \frac{8q_sL}{\pi^2 k}\sum_{n=0}^{\infty}\frac{(-1)^n}{(2n+1)^2} \times \sin\left[\frac{(2n+1)\pi(L-x)}{2L}\right]\exp\left[-(2n+1)^2\frac{\pi^2\kappa t}{4L^2}\right] \tag{4.25}$$

where the first two terms on the right-hand side represent the steady-state solution, and the infinite series represents the transient solution. For large time, of course, the infinite series diminishes to zero, and the temperature approaches the steady-state solution.

Example 4.4: Convectively Heated Surface and Insulated Surface

Another plate problem of practical importance occurs when one surface is convectively heated and the other is insulated. The boundary and initial conditions are

$$q(0,t) = h[T_\infty - T(0,t)] = -k\frac{\partial T}{\partial x}(0,t) \tag{4.26a}$$

$$q(L,t) = 0 \tag{4.26b}$$

$$T(x,0) = T_0 \tag{4.26c}$$

where h is a constant convection coefficient, and T_∞ is a constant fluid temperature. The problem is similar to the aerodynamically heated thin plate of Sec. 4.3.2 except that here we consider conduction through the plate so that the temperature varies with x as well as t. As with the thin plate, we can expect a steady-state solution because, after a long time, the plate temperature becomes uniform, with a constant temperature of T_∞. The complete solution is

$$\frac{T-T_\infty}{T_0-T_\infty} = 2\sum_{n=1}^{\infty}\frac{\sin\beta_n\cos(1-x/L)}{\beta_n+\sin\beta_n\cos\beta_n}\exp(-\beta_n^2\kappa t/L^2) \tag{4.27}$$

where the characteristic values β_n are the roots of the transcendental equation

$$\beta_n\tan\beta_n = hL/k \tag{4.28}$$

Thus, this solution differs from the previous examples in that the eigenvalues depend on the problem parameters, so that Eq. (4.28) must be solved anew for each problem. However, solutions for Eq. (4.28) are tabulated in the literature for various values of hL/k; see Appendix IV in Carslaw and Jaeger.[16]

The parameter hL/k occurs frequently in conduction heat transfer with convection boundary conditions, and it is called the Biot number,

$$Bi = hL/k \tag{4.29}$$

The Biot number characterizes the importance of the surface heat transfer compared to internal-conduction heat transfer. High values of the Biot number correspond to specifying the surface temperature, whereas low values correspond to insulating the surface.

Another nondimensional number in heat transfer is the Fourier number Fo defined as

$$Fo = \kappa t / L^2 \tag{4.30}$$

We recognize the Fourier number from its appearance in the exponential term in our transient solutions, for example, in Eqs. (4.19), (4.21), (4.25), and (4.27). The Fourier number is dimensionless time, and an interpretation is that it characterizes the rate of heat transfer by conduction to the rate of energy stored by capacitance.

With the Biot and Fourier numbers, we may express the solution to this example, Eq. (4.27), in the nondimensional form

$$\frac{T - T_\infty}{T_0 - T_\infty} = f(Bi, Fo, x/L) \tag{4.31}$$

Equation (4.31) is used by some authors[4,7] to present general graphs of the thermal response.

Closure. The solutions presented in this section for temperature distributions in plates represent just a few of the many possible boundary conditions. And yet these examples include some important boundary conditions encountered in structural heat transfer, and we make use of these solutions later in the book. In Sec. 4.5, for example, we use these solutions to discuss temperature distributions in thin-walled structures.

4.4.2 Temperatures in Semi-Infinite Solids

In some applications, only the early phase of conduction heat transfer in a solid is of interest. Such applications are often called *thermal shock problems*. The actual solid may be a thick plate heated suddenly on one surface. At early times, temperatures on the unheated surface may be unchanged so that, for practical purposes, the plate may be assumed to be semi-infinite in extent. In theory, the series solutions presented for suddenly heated plates in the preceding section are applicable to many thick-plate problems. In practice, however, the infinite series typically have convergence difficulties for steep temperature gradients that may occur for early times. Thus, the alternative solutions that we present here for temperature in semi-infinite solids often have practical value.

The solution approach for solution of the heat-conduction equation (4.14) for semi-infinite domains is the Laplace transform method. There is an extremely rich mathematical literature on the Laplace transform method, and we will barely scratch the surface of the mathematics. We will use the definition of the Laplace transform to transform the partial differential equation and then use standard tables to perform the inverse operation. In the process, we think of the Laplace transform parameter s as a constant. The more elegant mathematical approach[17] is to regard s as a complex variable and perform the inversion process by integration in the complex plane.

The Laplace transform of a function $f(t)$ is denoted as

$$\mathcal{L}\{f(t)\} = \int_0^\infty e^{-st} f(t)\, \mathrm{d}t = \bar{f}(s)$$

where $\mathcal{L}\{\ \}$ denotes the transform operation, s is the transform parameter, and the bar over f denotes the transformed variable.

We wish to transform each partial derivative in Eq. (4.14). Starting with the space derivative,

$$\mathcal{L}\left\{\frac{\partial^2 T}{\partial x^2}\right\} = \int_0^\infty e^{-st}\frac{\partial^2 T}{\partial x^2}(x,t)\,\mathrm{d}t = \frac{\partial^2}{\partial x^2}\int_0^\infty e^{-st}T(x,t)\,\mathrm{d}t$$

$$= \frac{\partial^2}{\partial x^2}\bar{T}(x,s)$$

Hence,

$$\mathcal{L}\left\{\frac{\partial^2 T}{\partial x^2}\right\} = \frac{\mathrm{d}^2\bar{T}}{\mathrm{d}x^2}$$

where we can use an ordinary derivative because, after the Laplace transform with respect to time, the transformed temperature $\bar{T}(x,s)$ is only a function of x given that s is regarded as a parameter.

The transform of the time derivative in Eq. (4.14) is

$$\mathcal{L}\left\{\frac{\partial T}{\partial t}\right\} = \int_0^\infty e^{-st}\frac{\partial T}{\partial t}(x,t)\,\mathrm{d}t$$

which we evaluate by performing an integration by parts,

$$\mathcal{L}\left\{\frac{\partial T}{\partial t}\right\} = \left|e^{-st}T(x,t)\right|_0^\infty + s\int_0^\infty e^{-st}T(x,t)\,\mathrm{d}t$$

Thus, after the limits of integration are evaluated,

$$\mathcal{L}\left\{\frac{\partial T}{\partial t}\right\} = -T(x,0) + s\bar{T}(x,s)$$

The preceding equation shows that, in the process of evaluating the Laplace transform of the time derivative, we introduce the initial condition $T(x,0)$.

Finally, from the definition of the Laplace transform, the transform of a constant, say c_1, is simply

$$\mathcal{L}\{c_1\} = c_1\int_0^\infty e^{-st}\,\mathrm{d}t \qquad \text{or} \qquad \mathcal{L}\{c_1\} = \frac{c_1}{s}$$

With these basic equations and Laplace transform tables, we can solve unsteady heat-conduction problems.

Example 4.5: Sudden Change in Wall Temperature

We consider a semi-infinite solid that initially has a uniform temperature T_0. At time $t = 0$, we suddenly increase the temperature of thesurface to the constant

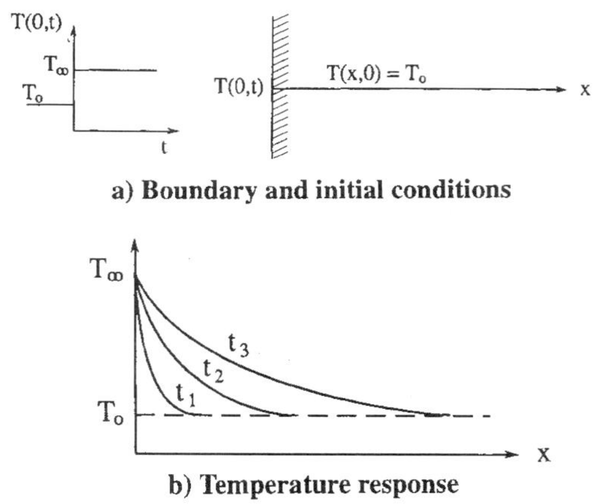

Fig. 4.7 Thermal response of a semi-infinite solid.

value T_∞ (Fig. 4.7a). The boundary and initial conditions are

$$T(0,t) = T_\infty \tag{4.32a}$$

$$\lim_{x\to\infty} T(x,t) = T_0 \tag{4.32b}$$

$$T(x,0) = T_0 \tag{4.32c}$$

Using Laplace transform results from the preceding paragraph, we take the Laplace transform of the heat-conduction equation (4.14) to obtain

$$\frac{d^2\bar{T}}{dx^2} = \frac{1}{\kappa}[-T(x,0) + s\bar{T}]$$

and, after using the initial condition, we obtain the ordinary differential equation

$$\frac{d^2\bar{T}}{dx^2} - \frac{s}{\kappa}\bar{T} = -\frac{T_0}{\kappa}$$

The Laplace transforms of the boundary conditions (4.32a) and (4.32b) are

$$\bar{T}(0,s) = \mathcal{L}\{T(0,T)\} = \mathcal{L}\{T_\infty\} = \frac{T_\infty}{s}$$

$$\lim_{x\to\infty} \bar{T}(x,s) = \mathcal{L}\left\{\lim_{x\to\infty} T(x,t)\right\} = \mathcal{L}\{T_0\} = \frac{T_0}{s}$$

Using these transformed boundary conditions, we next solve the differential equation for $\bar{T}$ to obtain

$$\bar{T}(x,s) = \frac{T_0}{s} + \frac{T_\infty - T_0}{s}\exp\left(-\frac{x}{\sqrt{\kappa}}\sqrt{s}\right)$$

We obtain the final solution for $T(x,t)$ by obtaining the inverse Laplace transform of this last result. Using a transform table from a standard mathematical handbook, we find that final solution is

$$T(x,t) = T_0 + (T_\infty - T_0)\left[1 - \operatorname{erf}\left(\frac{x}{2\sqrt{\kappa t}}\right)\right] \tag{4.33}$$

where erf( ) is the error function defined by the integral

$$\operatorname{erf}(z) = \frac{2}{\sqrt{\pi}} \int_0^z e^{-u^2} du \tag{4.34}$$

The error function is tabulated as a function of the argument z in standard mathematical tables. In addition, most computer systems have scientific subroutines for calculating the error function, so that temperature computations using Eq. (4.33) are relatively easy. We may write Eq. (4.33) alternatively as

$$T(x,t) = T_0 + (T_\infty - T_0)\operatorname{erfc}\left(\frac{x}{2\sqrt{\kappa t}}\right) \tag{4.35}$$

where $\operatorname{erfc}(N)$ is known as the complementary error function defined by

$$\operatorname{erfc}(z) = 1 - \operatorname{erf}(z) \tag{4.36}$$

The temperature response predicted by Eq. (4.35) is sketched in Fig. 4.7b.

Example 4.6: Sudden Change in Surface Heat Flux

As in the preceding example, we consider a semi-infinite solid with an initial uniform temperature. At time $t = 0$, we suddenly apply a heat flux q_s to the surface at $x = 0$. The boundary and initial conditions are

$$q(0,t) = q_s = -k\frac{\partial T}{\partial x}(0,t) \tag{4.37a}$$

$$\lim_{x\to\infty} T(x,t) = T_0 \tag{4.37b}$$

$$T(x,0) = T_0 \tag{4.37c}$$

The solution is given by Eckert and Drake[2] as

$$T(x,t) = T_0 + \frac{q_s}{k}\sqrt{\frac{4\kappa t}{\pi}}\left[\exp(-x^2/4\kappa t) - x\sqrt{\frac{\pi}{4\kappa t}}\operatorname{erfc}(x/\sqrt{4\kappa t})\right] \tag{4.38}$$

In Eq. (4.38), erfc( ) refers to the complementary error function defined in Eq. (4.36) in the preceding example.

Example 4.7: Sudden Change in Surface Convective Heating

To conclude the solutions for the thermal response of semi-infinite solids, we consider the sudden application of convective heating. The boundary and initial conditions are

$$q(0,t) = h[T_\infty - T(0,t)] = -k\frac{\partial T}{\partial x}(0,t) \tag{4.39a}$$

$$\lim_{x\to\infty} T(x,t) = T_0 \tag{4.39b}$$

$$T(x,0) = T_0 \tag{4.39c}$$

The solution is given by Kreith[4] as

$$\frac{T - T_0}{T_\infty - T_0} = \operatorname{erfc}\left(\frac{x}{2\sqrt{\kappa t}}\right) - \exp\left(\frac{hx}{k} + \frac{h^2 \kappa t}{k^2}\right)\operatorname{erfc}\left(\frac{x}{2\sqrt{\kappa t}} + \frac{h}{k}\sqrt{\kappa t}\right) \quad (4.40)$$

Kreith also presents a graphical representation of Eq. (4.40) in terms of a local Biot number and a nondimensional time parameter.

Closure. In this section, we present three solutions for the temperature response in semi-infinite solids for three suddenly applied surface boundary conditions: 1) temperature, 2) heat flux, and 3) convective heating. The three solutions, given in terms of the error function, represent "early time" temperature responses for the Examples 4.1–4.3 presented in Sec. 4.4.1. The solutions presented here often permit accurate computation of temperatures for early times in the response when series solutions encounter numerical difficulties.

4.4.3 Temperatures in Cylinders

All the previous one-dimensional examples have been in Cartesian coordinates. For finite solution domains, the eigenfunctions were simple trigonometric functions such as $\sin n\pi x/L$ in Example 4.1. For one-dimensional conduction problems in either cylindrical or spherical coordinates, the eigenfunctions are more advanced mathematical functions, including Bessel functions or Legendre polynomials. Bessel functions are encountered typically for problems with cylindrical boundaries. Solutions for both types of conduction problems are given by Carslaw and Jaeger.[16] We illustrate here two examples of one-dimensional conduction problems in cylindrical coordinates.

In cylindrical coordinates r, θ, z, the conservation of energy equation (4.1) for an isotropic material takes the form

$$\frac{1}{r}\frac{\partial}{\partial r}\left(r\frac{\partial T}{\partial r}\right) + \frac{1}{r^2}\frac{\partial^2 T}{\partial \theta^2} + \frac{\partial^2 T}{\partial z^2} = \frac{1}{\kappa}\frac{\partial T}{\partial t}$$

or

$$\frac{\partial^2 T}{\partial r^2} + \frac{1}{r}\frac{\partial T}{\partial r} + \frac{1}{r^2}\frac{\partial^2 T}{\partial \theta^2} + \frac{\partial^2 T}{\partial z^2} = \frac{1}{\kappa}\frac{\partial T}{\partial t} \quad (4.41)$$

We consider heat conduction in a circular cylinder whose axis coincides with the z axis. If the cylinder has axisymmetric boundary and initial conditions, independent of the coordinates θ and z, then the temperature becomes a function of only r and t, and the energy equation reduces to

$$\frac{\partial^2 T}{\partial r^2} + \frac{1}{r}\frac{\partial T}{\partial r} = \frac{1}{\kappa}\frac{\partial T}{\partial t} \quad (4.42)$$

In this case, the heat flow occurs in planes perpendicular to the axis of the cylinder along radial lines. Equation (4.42) is solved subject to two boundary and one initial condition.

We consider two boundary/initial-value problems for transient temperature distributions $T(r, t)$ in a solid, circular cylinder. In the first example, we use separation of variables to derive the solution for $T(r, t)$ but, for the second example, we present and discuss only the final solution.

Example 4.8: Specified Wall Temperature

We consider a cylinder of radius a that extends to infinity in the z direction; the cylinder has a constant initial temperature T_0 (Fig. 4.8a). At time equal to zero, the temperature of the boundary surface at $r = a$ is increased instantaneously to a new value T_a. Thereafter, a thermal transient ensues, and we know that, after a long time, the temperature within the cylinder for $0 \leq r \leq a$ approaches T_a. In formulating the problem, we can make the mathematics a little more convenient if we introduce a new dependent variable $\theta(x, t)$ defined by

$$\theta(r, t) = \frac{T(r, t) - T_a}{T_0 - T_a} \tag{4.43}$$

We choose this variable on the basis of our previous experience [see Eq. (4.27)] and because it will make the boundary condition at $r = a$ homogeneous. There is no actual boundary condition at the center of the cylinder since the temperature at $r = 0$ is unknown. The condition we impose at $r = 0$ is that the temperature is finite. In terms of $\theta(r, t)$, the problem formulation is

$$\frac{\partial^2 \theta}{\partial r^2} + \frac{1}{r}\frac{\partial \theta}{\partial r} = \frac{1}{\kappa}\frac{\partial \theta}{\partial t} \tag{4.44a}$$

$$\theta(0, t) \text{ is finite} \tag{4.44b}$$

$$\theta(a, t) = 0 \tag{4.44c}$$

$$\theta(r, 0) = 1 \tag{4.44d}$$

As in any one-dimensional separation of variables problem, we assume the solution in the form of the product of two unknown functions. Based on our previous derivations, we anticipate the form that the time function will take and

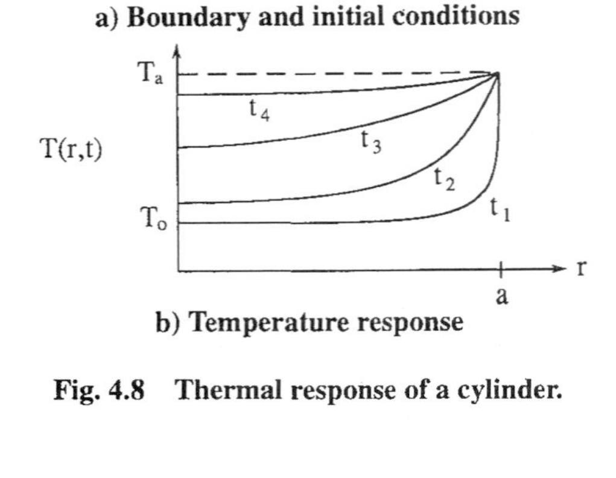

a) Boundary and initial conditions

b) Temperature response

Fig. 4.8 Thermal response of a cylinder.

seek the unknown spatial distribution. Thus, we begin by assuming a solution,

$$\theta(r,t) = \phi(r)e^{-\lambda^2 \kappa t}$$

where λ represents unknown eigenvalues to be determined. Substituting the preceding equation into the energy equation (4.44a) yields the ordinary differential equation for $\phi(r)$,

$$\frac{\mathrm{d}^2\phi}{\mathrm{d}r^2} + \frac{1}{r}\frac{\mathrm{d}\phi}{\mathrm{d}r} + \lambda^2\phi = 0$$

This last equation, a homogeneous, second-order ordinary differential equation with variable coefficients, is a special form of Bessel's equation. The solution given in books on applied mathematics and heat conduction (see Ref. 2) is

$$\theta(r,t) = AJ_0(\lambda r) + BY_0(\lambda r)$$

where $J_0(\lambda r)$ denotes the Bessel function of the first kind of order zero, and $Y_0(\lambda r)$ is the Bessel function of the second kind of order zero. Then we can write the general solution as

$$\theta(r,t) = [AJ_0(\lambda r) + BY_0(\lambda r)]e^{-\lambda^2 \kappa t}$$

where A and B are constants of integration to be determined. Imposing the first boundary condition that θ be finite at $r = 0$ requires that we take $B = 0$ since the Bessel function $Y_0(\lambda r)$ becomes infinite at $r = 0$. Then, imposing the second boundary condition, Eq. (4.44c), yields

$$\theta(a,t) = AJ_0(\lambda a)e^{-\lambda^2 \kappa t} = 0$$

For a nontrivial solution, we must require that

$$J_0(\lambda a) = 0$$

The preceding equation defines the eigenvalues since J_0 will be zero for an infinite set of values

$$\lambda_n a = \alpha_n$$

where α_n are the roots of the characteristic equation

$$J_0(\alpha_n) = 0 \qquad n = 1, 2, \ldots$$

These roots are tabulated in the literature, see for example Ref. 16. The final form of the solution is the superposition of the solutions. Therefore, for all values of n, we write

$$\theta(r,t) = \sum_{n=1}^{\infty} A_n J_0(\lambda_n r)e^{-\lambda_n^2 \kappa t}$$

where the set of unknown coefficients A_n remains to be determined.

As in previous examples, we evaluate the coefficients using the initial condition. Thus,

$$\theta(r, 0) = \sum_{n=1}^{\infty} A_n J_0(\lambda_n r)$$

and we solve for A_n by using the orthogonality condition. Mathematics books show that the Bessel functions $J_0(\lambda_n r)$ are orthogonal with respect to a weighting function r; hence,

$$\int_0^a r J_0(\lambda_m r) J_0(\lambda_n r)\,\mathrm{d}r = 0 \qquad m \neq n$$

Multiplying the equation for $\theta(r, 0)$ by $r J_0(\lambda_m r)$ and integrating over the domain from $r = 0$ to $r = a$ give

$$A_n = \frac{\int_0^a r\theta(r, 0) J_0(\lambda_n r)\,\mathrm{d}r}{\int_0^a r J_0^2(\lambda_n r)\,\mathrm{d}r}$$

since, by the orthogonality condition, all terms in the infinite series vanish except $m = n$. Next, substituting the initial condition given in Eq. (4.44d), we need to evaluate

$$A_n = \frac{\int_0^a r(1) J_0(\lambda_n r)\,\mathrm{d}r}{\int_0^a r J_0^2(\lambda_n r)\,\mathrm{d}r}$$

With some help from applied mathematics references, we find that

$$A_n = \frac{2}{a} \frac{1}{\lambda_n J_1(\lambda_n a)}$$

where $J_1(\lambda_n a)$ is the Bessel function of the first kind, of order one. Finally, we write the solution for $\theta(r, t)$ as

$$\theta(r, t) = \frac{2}{a} \sum_{n=1}^{\infty} \frac{J_0(\lambda_n r) e^{-\lambda_n^2 \kappa t}}{\lambda_n J_1(\lambda_n a)}$$

and referring to the definition for $\theta(r, t)$ in Eq. (4.43), we obtain the final solution as

$$\frac{T(r, t) - T_a}{T_0 - T_a} = \frac{2}{a} \sum_{n=1}^{\infty} \frac{J_0(\lambda_n r) e^{-\lambda_n^2 \kappa t}}{\lambda_n J_1(\lambda_n a)} \tag{4.45}$$

where λ_n, the eigenvalues, are the roots of the characteristic equation, $J_0(\lambda_n a) = 0$.

The response to be expected is sketched in Fig. 4.8b for successive times $t_1, t_2, \ldots, t_n$. To perform the computations on a computer, we need scientific subroutines for the Bessel functions, and these are widely available. As in Example 4.1, we may anticipate convergence problems with the infinite series at early times due to steep temperature gradients near the cylinder surface at $r = a$. Later as time becomes large, the series contribution approaches zero as the temperature approaches the constant value T_a. As in the previous examples, we see from Eq. (4.45) that the thermal diffusivity κ is the key material parameter in determining the duration of the thermal transient.

Example 4.9: Convectively Heated Surface

A related cylinder problem of practical value occurs when the surface $r = a$ has a convective heat exchange. The boundary and initial conditions are

$$T(0, t) \quad \text{is finite} \tag{4.46a}$$

$$-k\frac{\partial T}{\partial t}(a, t) = h[T(a, t) - T_\infty] \tag{4.46b}$$

$$T(r, 0) = T_0 \tag{4.46c}$$

where h is a constant convection coefficient, and T_∞ is a constant fluid temperature. The problem is similar to the preceding example except that the convective boundary condition leads to a different characteristic equation for the eigenvalues. The final solution is

$$\frac{T(r, t) - T_\infty}{T_0 - T_\infty} = 2\sum_{n=1}^{\infty} \frac{Bi}{\left[Bi^2 + (\lambda_n a)^2\right]} \frac{J_0(\lambda_n r)}{J_1(\lambda_n a)} e^{-\lambda_n^2 \kappa t} \tag{4.47}$$

where the eigenvalues λ_n are roots of the characteristic equation,

$$(\lambda_n a) J_1(\lambda_n a) = Bi\, J_0(\lambda_n a)$$

and Bi denotes the Biot number, defined as $Bi = ha/k$. In the last two equations, J_0 and J_1 denote the Bessel functions of the first kind, of order zero and one, respectively. The roots to the characteristic equation are a function of the Biot number and are tabulated in the literature; see Carslaw and Jaeger.[16] As in Example 4.4, a convectively heated plate, this solution can be represented in the nondimensional form:

$$\frac{T - T_\infty}{T_0 - T_\infty} = g(Bi, Fo, r/a)$$

where Fo is the Fourier number, $Fo = \kappa t/a^2$, a nondimensional time. Eckert and Drake[2] present nondimensional plots of the solution for a cylinder qiven in Eq. (4.47) and similar results for transient heat conduction in a sphere.

Closure. The results presented in this brief section on temperatures in cylinders illustrate two fundamental solutions for solid cylinders. The solutions have the same form as one-dimensional transient solutions for thick plates except for the eigenfunctions. For the circular conduction regions encountered, the eigenfunctions are Bessel functions. Except for esoteric mathematical details, Bessel functions present no difficulties in practical computations since scientific subroutines for their evaluation are widely available. A wide variety of other analytical solutions are available, including results for the practically important cases for hollow cylinders. See Carslaw and Jaeger[16] for a number of these solutions.

4.5 Temperatures in Thin-Walled Structures

Complex, thin-walled aerospace structures under service conditions rarely experience uniform temperatures. Often, the external heating is nonuniform. For example, leading edges of wings in high-speed flight experience high local heating at stagnation points. Moreover, as the external flow traverses the airfoil, aerodynamic heating changes with surface curvature and increases significantly at the transition from laminar to turbulent flow. In addition, shock interactions cause intense local heating.

Even if the external heating is nearly uniform, spatial variations in skin temperatures may occur at internal reinforcements such as ribs, stiffeners, and frames. At points or lines of attachment of local skin reinforcements, temperatures are lower than nearby skin temperatures because the reinforcement offers a greater material volume to absorb energy and a conduction path for heat transfer to the interior of the structure. The first analyses of temperature variations in thin-walled structures were made for supersonic aircraft in the 1950s.

In orbiting space structures, temperatures are unsteady because of the variation of incident heat fluxes with orbital position. Structural temperature may experience extremes from +250°F to −250°F for an orbit in which the spacecraft passes through the Earth's shadow. Temperatures vary spatially in orbiting structures because of nonuniform incident heat fluxes, structural conduction, and radiation heat losses from surfaces. In the vacuum of space, radiation heat transfer is particularly significant. For example, thin cylindrical shells may have major circumferential temperature variations since, typically, one-half of the cylindrical surface experiences heating from the solar flux while the remaining portion of the surface is shaded. All surfaces of the cylinder, heated and unheated, experience thermal-energy losses to space by radiation.

Finally, thermal properties of materials have a major effect on temperature distributions. Metallic and composite materials may have significant differences in thermal properties that yield substantially different temperature distributions.

4.5.1 Thickness Temperature Gradients

To gain insight into how properties of several common aerospace materials affect the temperature of a thin-walled structure, we examine a simple heat-conduction problem. Consider a thin plate of thickness H with the x, y, z coordinate system located at the middle surface. The initial temperature T_0 of the plate is zero. The surface $z = -H/2$ is perfectly insulated and, for time $t > 0$, the surface $z = H/2$ is uniformly heated by a constant heat flux q_s. To understand how material properties affect the temperature gradient through the plate thickness, consider the situation in which there is no temperature variation with x and y, i.e., $T(z, t)$. Thermal properties are assumed constant. Taking into account the different coordinate systems, the solution to the initial/boundary-value problem is given in Eq. (4.21) as

$$T(z,t) = \frac{q_s H}{k}\left[\frac{\kappa t}{H^2} + \frac{1}{2}\left(\frac{z}{H}+\frac{1}{2}\right)^2 - \frac{1}{6} - \frac{2}{\pi^2}\sum_{n=1}^{\infty}\frac{(-1)^n}{n^2}\cos n\pi\left(\frac{z}{H}+\frac{1}{2}\right)\exp(-n^2\pi^2\kappa t/H^2)\right] \tag{4.48}$$

where the thermal diffusivity is $\kappa = k/\rho c$. As we saw in Sec. 4.4.1, this solution predicts that, after an initial transient, temperatures of the top and bottom surfaces of the plate increase linearly with time according to the first term in the square brackets. Thus, the infinite series contributes to the temperature distribution only

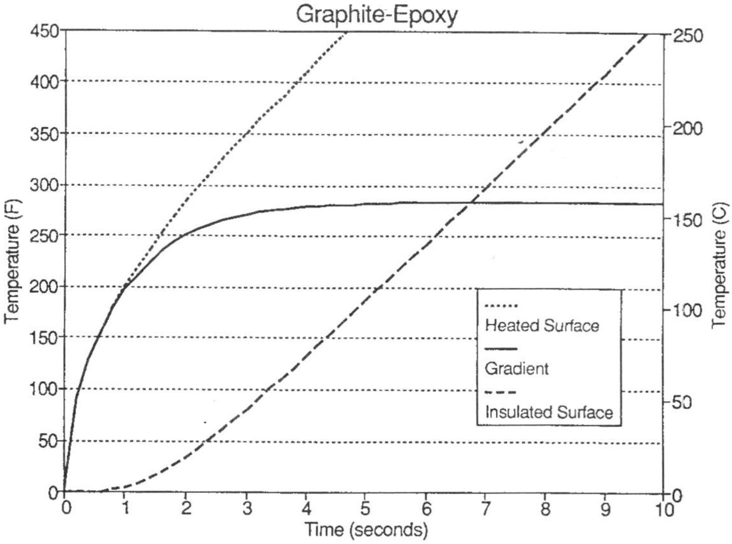

Fig. 4.9 Thermal response of a suddenly heated, thin graphite-epoxy plate, q_s = 0.1 Btu/in.^{2}-s (16 W/cm^2) and h = 0.125 in. (0.318 cm).

during the initial transient and, after early time, the series contribution becomes negligible. We showed, in Sec. 4.4.1, that neglecting the series permits the thickness temperature difference or gradient $\Delta T = T(H/2, t) - T(-H/2, t)$ to be expressed by the simple equation (4.23):

$$\Delta T = \frac{q_s H}{2k}$$

Equation (4.23) shows that the thickness temperature gradient varies linearly with the applied heat flux and plate thickness but inversely with the thermal conductivity. Table 4.1 gives thermal properties and thickness temperature gradients predicted for 10 metallic and composite plates. Figures 4.9 and 4.10 show transient temperatures for graphite–epoxy and aluminum plates, respectively. Graphite–epoxy and aluminum have the smallest and largest values of thermal conductivity, respectively, and produce extreme values for the temperature gradient. Graphite–epoxy has a very large thickness temperature gradient, and aluminum has a very small temperature gradient. Table 4.1 shows, in fact, that the graphite–epoxy composite is exceptional; its thermal conductivity is so low that an extraordinarily large thickness temperature gradient is predicted. Since the practical temperature range of this material is limited to less than 250°F (120°C), it is doubtful that such a large ΔT could ever occur in practice. The other metallics and metallic-based composites show relatively small thickness temperature gradients, with titanium showing the largest and aluminum the smallest gradients, respectively. The thickness temperature gradients shown may be regarded as upper bounds because thermal conductivity often increases with increasing temperature, and the insulated boundary condition on the bottom surface will not be realized because energy losses will inevitably occur as a result of surface convection or radiation heat transfer. In addition, any heat conduction in the x–y plane due to plate supports or nonuniform heating will lessen the thickness temperature gradient. For

Table 4.1 Thickness temperature gradients for various materials: q = 0.1 Btu/in.^{2}-s (16.4 W/cm^2); h = 0.125 in. (0.318 cm)

Material	Thermal conductivity, k		Thermal diffusivity, κ		Temperature gradient, ΔT	
	Btu/in.-s-°F	W/cm-°C	in.^{2}/s	cm^2/°F	°F	°C
1. Graphite–epoxy	2.2×10^{-5}	1.6×10^{-2}	1.5×10^{-4}	9.9×10^{-4}	280	160
2. Titanium (MIL-T-9047)	1.1×10^{-4}	8.3×10^{-2}	5.3×10^{-3}	3.4×10^{-2}	56	30
3. Carbon–carbon	2.0×10^{-4}	1.5×10^{-1}	1.1×10^{-2}	7.2×10^{-2}	31	17
4. Stainless (17-8PH)	2.2×10^{-4}	1.7×10^{-1}	7.4×10^{-3}	4.8×10^{-2}	28	15
5. Hastelloy-X	2.6×10^{-4}	1.9×10^{-1}	6.7×10^{-3}	4.3×10^{-2}	24	13
6. Copper (G3-heat treat)	5.6×10^{-4}	4.2×10^{-1}	2.1×10^{-2}	1.3×10^{-1}	11	6
7. Graphite–magnesium	5.9×10^{-4}	4.4×10^{-1}	2.8×10^{-2}	1.8×10^{-1}	11	6
8. Magnesium (AZ80A-F)	1.0×10^{-3}	7.6×10^{-1}	6.3×10^{-2}	4.1×10^{-1}	6.1	3.4
9. Graphite–aluminum	1.1×10^{-3}	8.3×10^{-1}	4.9×10^{-2}	3.1×10^{-1}	5.6	3.1
10. Aluminum (6061-T6)	2.2 × 10−3	1.7	9.9×10^{-2}	6.4×10^{-1}	2.8	1.6

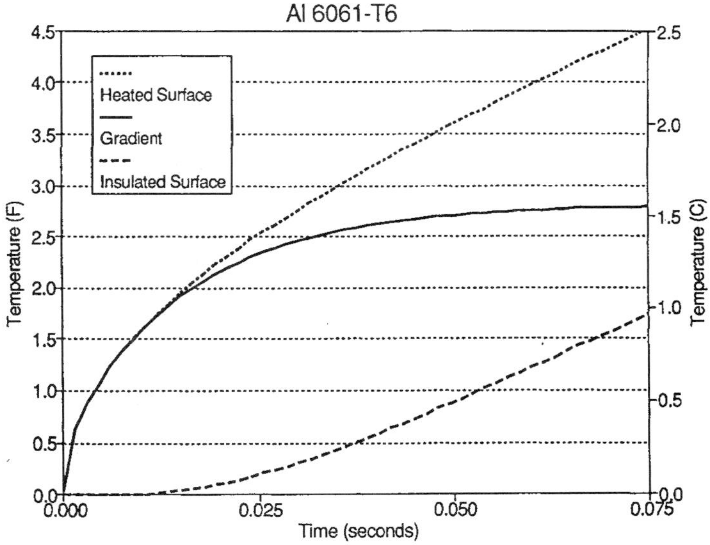

Fig. 4.10 Thermal response of a suddenly heated, thin aluminum alloy plate, q_s = 0.1 Btu/in.^{2}-s (16 W/cm^2) and h = 0.125 in. (0.318 cm).

thin-walled metallic structures, customary practice is to neglect thickness temperature gradients.

4.5.2 Spatial Temperature Distributions

Insight into the effect of thermal properties on the spatial variation of the thin-plate temperature, $T(x, y)$ is provided by solution of the linear, unsteady initial/boundary-value problem of a plate locally heated (Fig. 4.11). The edges of the plate at $y = \pm b$ are maintained at constant temperature $T_0 = 0$ by specified coolant flows. The bottom and top surfaces of the plate are perfectly insulated except along the centerline of the upper plate surface. For time $t > 0$, the plate experiences uniform heating q_s on a narrow strip along the x axis. Temperature is assumed uniform through the plate thickness. The solution to the symmetric, one-dimensional conduction problem is obtained by using Eq. (4.25) with appropriate modifications,

$$T(y,t) = \frac{q_0(b-y)}{k} - \frac{8q_0 b}{\pi^2 k}\sum_{n=0}^{\infty}\frac{(-1)^n}{(2n+1)^2}\sin\frac{\pi(2n+1)(b-y)}{2b} \times \exp[-\pi^2(2n+1)^2\kappa t/4b^2] \tag{4.49}$$

where $q_0 = q_s w/2H$, with w as the width of the heated strip, H is the plate thickness, and as before, κ is the thermal diffusivity. For large time, plate temperatures approach steady state, with the peak temperature at the centerline and a linear decrease to the specified zero values on the plate edges. As with the previous solution, Eq. (4.49) predicts plate transient temperature distributions that depend inversely on the material's thermal conductivity.

Using the thermal properties shown in Table 4.1, transient thermal responses were computed for graphite–epoxy, Hastelloy-X, and aluminum (Figs. 4.12–4.14). These results indicate the responses of low-, medium-, and high-thermal-

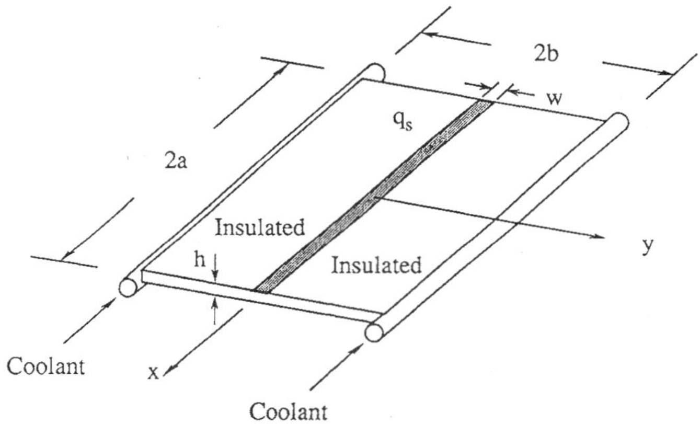

Fig. 4.11 Centrally heated, edge-cooled plate studied by Heldenfels and Roberts (1952).

conductivity material plates to local heating. The results show, for low-thermal-conductivity graphite–epoxy (Fig. 4.12), that temperatures rise rapidly with very steep temperature gradients and a highly localized heated region. On the other hand, for aluminum with high thermal conductivity (Fig. 4.14), temperatures rise more slowly with lower gradients as thermal energy is conducted into the cooler boundaries. The response of the Hastelloy-X, a high-temperature nickel-based alloy, falls somewhere between these two extremes, but the material demonstrates high local temperatures, steep gradients, and a relatively small heated region. Thin-

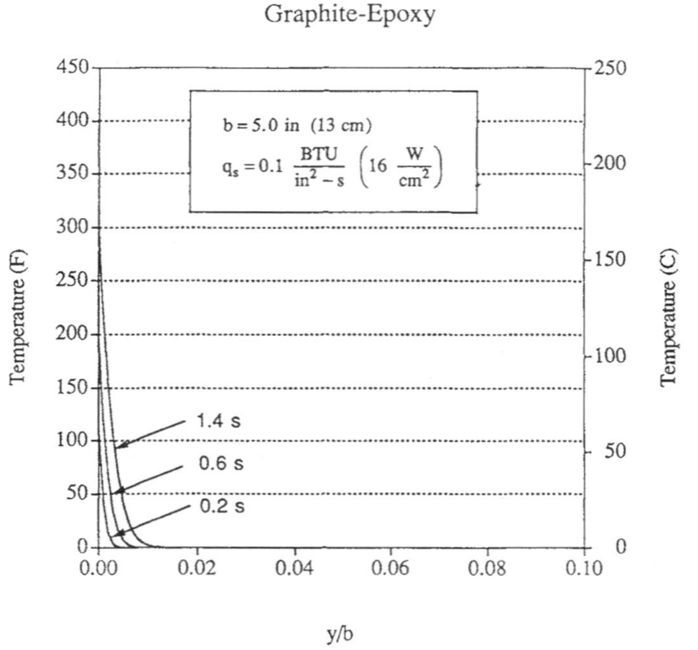

Fig. 4.12 Temperature distribution for a locally heated graphite–epoxy plate, $q_s = 0.1$ Btu/in.^{2}-s (16 W/cm^2) and $h = 0.125$ in (0.318 cm).

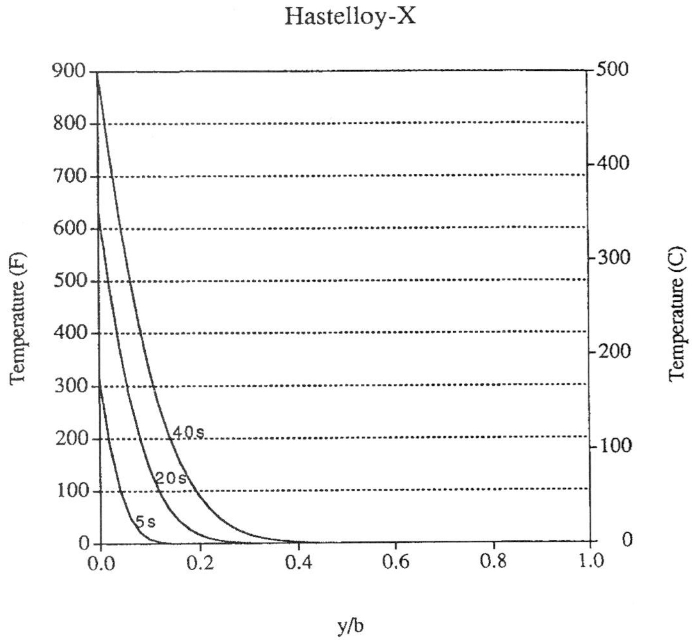

Fig. 4.13 Temperature distribution for a locally heated Hastelloy-X plate, q_s = 0.1 Btu/in.^{2}-s (16 W/cm^2) and h = 0.125 in (0.318 cm).

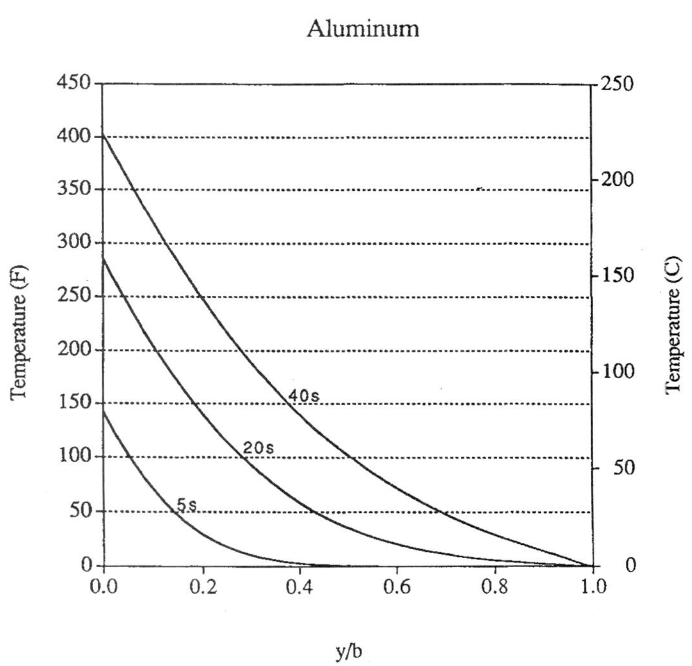

Fig. 4.14 Temperature distribution for a locally heated aluminum plate, q_s = 0.1 Btu/in.^{2}-s (16 W/cm^2) and h = 0.125 in (0.318 cm).

walled structures of this alloy subjected to localized heating are likely to experience inelastic buckling behavior as a result of the high local temperatures.

These examples demonstrate two types of thermal-gradient responses characteristic of thin-walled structures. For low-thermal-conductivity materials such as graphite–epoxy, high thickness temperature gradients will occur. However, for metals and metal matrix composites, thickness temperature gradients are negligible, and spatial temperature gradients are most likely to occur in applications.

4.6 Computational Methods

Over the last four decades, significant advances have been made in computational methods for structural heat transfer. The use of computers and computer graphics has made analyses of complex thermal structures a routine step in the design process. The proceedings of a conference on computational aspects of heat transfer in structures held at NASA Langley in 1981[18] describe the capability of computing temperatures of flight vehicles. Computer hardware and software have made, and continue to make, significant advances. Software is based on the finite difference or the finite volume method or the finite element method. Programs based on the finite difference and finite volume method include TRASYS, MITAS, and SINDA. Since 1972, TRASYS has been used extensively for U.S. spacecraft radiation heat-transfer analysis. There are several widely available finite element programs with heat-transfer capability, including ABAQUS, ANSYS, EAL, MARC, MSC/NASTRAN, and PATRAN. Current computer hardware trends include improving computational speeds by vector or parallel processing.

In the balance of this section, we highlight basic concepts of the computational methods used in temperature computations. Only basic concepts are presented to contrast the commonly used approaches, and readers interested in further details may consult the references.

The basic sequence followed in engineering modeling and solution methods for heat-transfer problems is shown in Fig. 4.15. A common step in all analyses is to replace the real system by a mathematical idealization. The real system may be a complete aircraft or spacecraft, or it may be a subsystem such as a wing section or a spacecraft appendage. The mathematical idealization is an engineering model including heat loads, boundary conditions, the geometrical configuration, heat-exchange modes, and material properties. After this step, we have two basic choices for developing a model to compute the desired temperatures—either a continuum model or a discrete model.

In the continuum model, we use basic principles, such as conservation of energy and Fourier's law, to develop governing ordinary or partial differential equations. If the model's geometry is simple and if the heat exchanges are linear, the heat loads not too complex, and the properties constant, we may follow a classical solution approach and obtain a mathematical solution. This is the approach we followed in Secs. 4.3 and 4.4, where we worked with simple models that led to linear ordinary and partial differential equations. We obtained classical closed-form solutions to the initial/boundary-value problems using standard mathematical techniques. In this classical solution approach, we obtain temperature as a continuous mathematical function of the spatial functions and time.

Often, however, we wish to solve a more complex problem to simulate a system's behavior more realistically. Complexities may arise from geometric irregularities, nonlinear heat transfer such as radiation, as well as temperature-dependent thermal properties and nonniform distributions of heat loads. In these circumstances, we cannot solve a differential equation formulation exactly, and we turn to methods

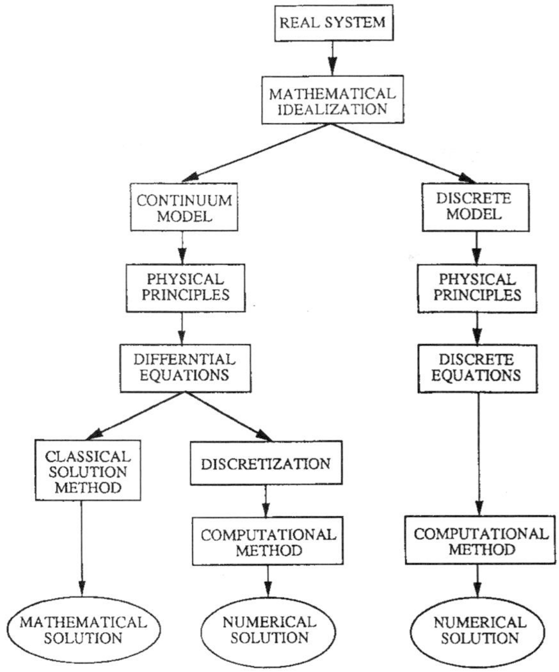

Fig. 4.15 Basic engineering modeling and solution methods.

for discretizing the problem. In these approaches, we approximate numerically the differential equation, the boundary and initial conditions, and seek temperatures at only a finite number of spatial points and at discrete values of time in the response. The two most popular discretization approaches for computational solutions to differential equations are the finite difference and finite element methods.

Alternatively, instead of developing a continuum model, we may elect to form a discrete model of the physical problem directly from the mathematical idealization. In this approach, the step of developing governing differential equations is bypassed. Physical principles are applied directly to yield equations that determine temperatures at discrete spatial points and times. The method may be called a *control volume* or *finite volume* approach.

In the following paragraphs, we discuss and contrast finite difference, finite element, and finite volume methods.

4.6.1 Finite Difference Method

The finite difference method is among the oldest methods for obtaining approximate solutions to partial differential equations. The method is described in numerous engineering and numerical methods textbooks. The textbook by Anderson et al.[19] develops and applies finite difference methods to the equations of heat transfer and fluid mechanics.

Application of finite difference methods to problems in heat transfer begins with replacing exact partial derivatives in the conservation of energy equation with approximate finite difference approximations. Suppose we subdivide the x axis in a Cartesian system into a discrete number of points $x_1, x_2, \ldots, x_n$, where the spacing between the points h_x is uniform. In a similar way, we divide the y

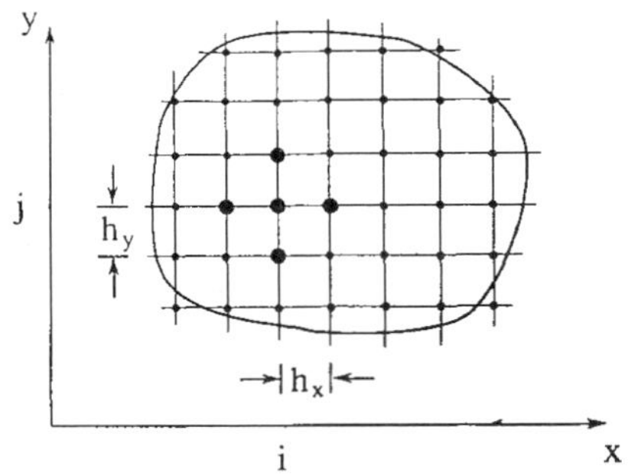

a) Typical finite difference grid

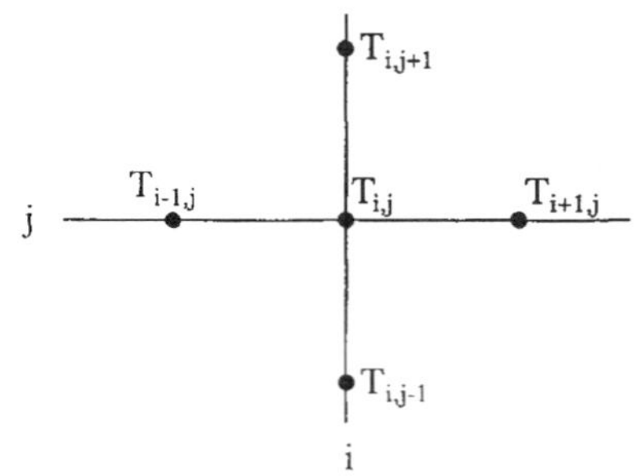

b) Typical interior node

Fig. 4.16 Finite difference discretization.

axis into a discrete number of points $y_1, y_2, \ldots, y_n$ where the spacing between the points h_y is also uniform. In this way, the typical finite difference grid shown in Fig. 4.16a is generated. Figure 4.16b shows that the temperature at a typical interior i, j is denoted by $T_{i,j}$ and the temperatures at surrounding points by similar subscripts. Points in the finite difference grid are often called *nodes, nodal points,* or *grid points.* The numerical solution consists of determining temperatures at the nodes.

Using Taylor series expansions, we may show that a derivative $\partial T/\partial x$ may be approximated in three ways.

Forward difference:

$$\left.\frac{\partial T}{\partial x}\right|_{i,j} = \frac{T_{i+1,j} - T_{i,j}}{h_x} + \mathcal{O}(h_x) \tag{4.50a}$$

Backward difference:

$$\left.\frac{\partial T}{\partial x}\right|_{i,j} = \frac{T_{i,j} - T_{i-1,j}}{h_x} + \mathcal{O}(h_x) \tag{4.50b}$$

Central difference:

$$\left.\frac{\partial T}{\partial x}\right|_{i,j} = \frac{T_{i+1,j} - T_{i-1,j}}{h_x} + \mathcal{O}\left(h_x^2\right) \tag{4.50c}$$

The terms indicated by $\mathcal{O}(\,)$ represent the truncation errors in the finite difference approximations. The notation $\mathcal{O}(h_x)$ indicates that, as h_x approaches zero, the error in the forward and backward difference approximations decreases linearly with h_x. These approximations are said to be first-order accurate. On the other hand, the central difference approximation has a truncation error $\mathcal{O}(h_x^2)$, and it is second-order accurate. With a second-order-accurate approximation, if the mesh spacing is decreased by 2, the truncation error decreases by 4. And so, for rapid solution convergence with mesh refinement, a higher-order difference approximation is desirable.

Finite difference approximations have also been developed for derivatives of higher order. One of the most popular approximations for second derivatives is the central difference approximation

$$\left.\frac{\partial^2 T}{\partial x^2}\right|_{i,j} = \frac{T_{i+1,j} - 2T_{i,j} + T_{i-1,j}}{h_x^2} + \mathcal{O}\left(h_x^2\right) \tag{4.51}$$

Ref. 18 tabulates a variety of difference approximations for first, second, and mixed (e.g., $\partial^2 T/\partial x \partial y$) partial derivatives.

We may use the finite difference approximations given in the preceding equations to obtain difference representations of the energy equation. As a first illustration, consider unsteady, one-dimensional conduction where, from Sec. 4.4, the energy equation (4.14) is

$$\frac{\partial^2 T}{\partial x^2} = \frac{1}{\kappa}\frac{\partial T}{\partial t}$$

and, as before, $\kappa = k/\rho c$ is the thermal diffusivity. One well-known finite difference approach is to use the central difference approximation given in Eq. (4.51) for the x derivative and the forward difference approximation given in (Eq. 4. 50a) for the t derivative. Thus, we write, for a typical node i at time n, the approximation,

$$\frac{T_{i+1}^n - 2T_i^n + T_{i-1}^n}{\Delta x^2} = \frac{1}{\kappa}\frac{T_i^{n+1} - T_i^n}{\Delta t} \tag{4.52}$$

This difference approximation to the partial differential equation represents a computational algorithm, for, at time n, we may use Eq. (4.52) to write a set of algebraic equations that will enable us to advance the solution to time $n+1$. Thus, starting with an initial distribution of nodal temperatures, we may use Eq. (4.52) to advance the solution recursively in time as long as we wish.

An advantage of the finite difference method is that we may generate the discretized equations rather easily using equations like (4.52). We should note that there are a number of issues such as solution convergence, stability, and conservation that we need to understand before applying the method. Ref. 18 and other texts discuss these in detail.

As a second illustration, if we wish to solve the steady, two-dimensional heat condition problem given by Laplace's equation,

$$\frac{\partial^2 T}{\partial x^2} + \frac{\partial^2 T}{\partial y^2} = 0 \tag{4.53}$$

we use the central difference approximation given in Eq. (4.51), first for the derivative $\partial^2 T/\partial x^2$ and then for $\partial^2 T/\partial y^2$. Then we obtain

$$\frac{T_{i+1,j} - 2T_{i,j} + T_{i-1,j}}{\Delta x^2} + \frac{T_{i,j-1} - 2T_{i,j} + T_{i,j+1}}{\Delta y^2} = 0 \tag{4.54}$$

where the mesh spacing in the x direction is Δx, and the mesh spacing in the y direction is Δy. If we assume $\Delta x = \Delta y$, the finite difference scheme has the form

$$4T_{i,j} - (T_{i-1,j} + T_{i+1,j} + T_{i,j-1} + T_{i,j+1}) = 0 \tag{4.55}$$

From Fig. 4.16b, we see that Eq. (4.55) relates the temperature at node i, j to temperatures at the four surrounding nodes. To solve for the temperatures in a two-dimensional domain, we apply Eq. (4.55) at each node to generate a set of simultaneous algebraic equations. At points on the boundary, Eq. (4.55) does not apply directly, but it may be used if we employ a set of fictitious nodes external to the boundary.

A disadvantage of finite difference methods is encountered when we wish to compute temperatures for regions with irregular geometries. For rectangular regions with straight boundaries, we may easily generate the difference approximations to the partial differential equations using equations like (4.55). But, for curved boundaries, for instance, we encounter problems. Finite difference books show two ways for dealing with these problems. One way is to use special difference approximations to represent derivatives with special irregular meshes. A second way is to use coordinate transformations to map the solution domain with curved boundaries into rectangular domains. These mappings are normally included in mesh-generation programs for finite difference solutions to heat-transfer and fluid-mechanics problems.

4.6.2 Finite Element Method

Another way to obtain approximate solutions to partial differential equations is the finite element method. The method was first developed for aircraft structural applications in the 1960s, but it has since been extended to heat-transfer and fluid-mechanics applications. There are numerous finite element textbooks, including Ref. 20, which emphasizes multidisciplinary applications. In this section, we introduce the finite element method and contrast it with the finite difference method. In Chapter 8, we will present the details of the finite element method for heat-transfer and thermal-stress computations.

Application of the finite element method to problems in heat transfer begins by subdividing the solution domain into a finite number of subdomains—finite elements—that are connected at the nodes. The elements may take a variety of shapes, but for two-dimensional regions, triangular or quadrilateral elements are popular. Figure 4.17a shows a solution domain discretized by a mesh of triangular and quadrilateral elements. Alternately, we may elect a mesh with only triangular or quadrilateral elements. An advantage of the finite element method is that nodal locations and element shapes may be selected to model irregular geometries conveniently.

The preceding section showed that, with the finite difference method, we write the discretized equations on a nodal basis. In other words, we write finite difference equations for each node for generating a set of simultaneous algebraic or ordinary differential equations for the entire mesh. In contrast, with the finite element method, we write the discretized equations on an element basis. In other

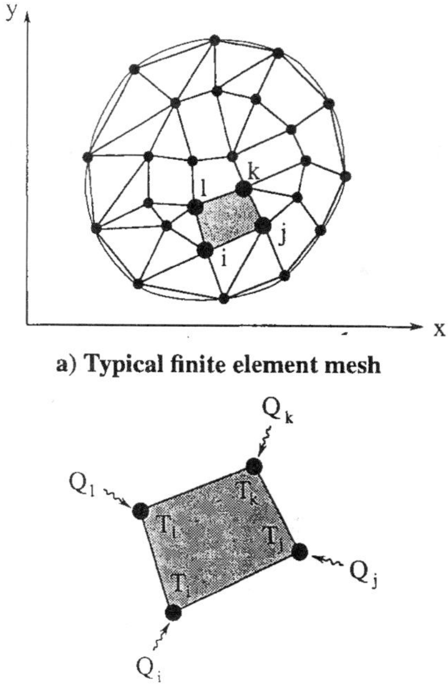

a) Typical finite element mesh

b) Typical finite element

Fig. 4.17 Finite element discretization.

words, we write equations for each element first. As a later step, we assemble element contributions to generate a set of simultaneous algebraic or ordinary differential equations for the entire mesh.

The mathematics for developing a finite element's equations is more involved than for finite difference approximations, and we defer consideration of these details until Chapter 8. We can, however, give some basic concepts of the method. Consider the typical quadrilateral element shown in Fig. 4.4b. The element connects four nodes, denoted by node numbers i, j, k, l. At each node, an unknown temperature and a nodal heat flow are denoted, respectively, by T_i, T_j, T_k, T_ℓ and Q_i, Q_j, Q_k, Q_ℓ.

A concept fundamental to the finite element method is the use of interpolation functions. For example, the temperature $T(x, y)$ at an arbitrary point within an element with four nodes is interpolated from the nodal temperatures by the equation,

$$T(x, y) = N_i(x, y)T_i + N_j(x, y)T_j + N_k(x, y)T_k + N_i(x, y)T_\ell \tag{4.56}$$

where $N_i(x, y), \ldots, N_\ell(x, y)$ denote the element's interpolation functions. These functions establish how the temperature varies within an element, and they are tabulated for popular elements such as triangles and quadrilaterals in finite element texts.

Element equations relating nodal temperatures and heat flows are derived in two ways: from a variational approach or from a weighted residual approach. In the variational approach, the partial differential equation expressing conservation of energy is replaced by an equivalent variational principle. Using the element temperature interpolation equation (4.56) and minimization techniques from the calculus of variations, the discretized conservation of energy equation for a typical element is derived. The second approach using the method of weighted residuals begins with the partial differential equation. The temperature interpolation equation is substituted into the partial differential equation to produce a residual function. The weighted residual is integrated over the element domain to produce the discretized equations. (The weighted residual approach is described with details in Chapter 8.) For steady heat transfer, a quadrilateral element's discretized equations are a set of four algebraic equations. Typically, the equations are written in the matrix form,

$$\begin{bmatrix} K_{ii} & K_{ij} & K_{ik} & K_{i\ell} \\ K_{ji} & K_{jj} & K_{jk} & K_{j\ell} \\ K_{ki} & K_{kj} & K_{kk} & K_{k\ell} \\ K_{\ell i} & K_{\ell j} & K_{\ell k} & K_{\ell\ell} \end{bmatrix} \begin{Bmatrix} T_i \\ T_j \\ T_k \\ T_\ell \end{Bmatrix} = \begin{Bmatrix} Q_i \\ Q_j \\ Q_k \\ Q_\ell \end{Bmatrix} \tag{4.57a}$$

or, more compactly,

$$[K]\{T\} = \{Q\} \tag{4.57b}$$

where the square coefficient matrix $[K]$ represents the element's conductance, $\{T\}$ is a vector of element nodal temperatures, and $\{Q\}$ is a vector of nodal heat flows. In computations, element conductance matrices are computed for each element in the mesh. These matrices are then added ("assembled" in finite element parlance) to yield the conductance matrix for the system. Assembly proceeds element by element in such a way that the elements are properly connected to appropriate neighboring elements to create the complete mesh.

The finite element method provides great generality in modeling complex shapes and also models nonhomogeneous and anisotropic materials conveniently. The price for the increased generality, compared with the finite difference method, is greater mathematical complexity. Finite element computer programs are more complicated than finite difference programs because more steps are required to generate the discretized equations. Computational costs for finite element analyses tend to be higher than those for finite difference analyses for the same reason. But with modern hardware and software, the difference in computational costs is often insignificant.

4.6.3 Finite Volume Method

In both the finite difference and the finite element methods, we form the discretized equations from a differential equation describing the conservation of energy. This discretization approach is shown in the left-hand path of Fig. 4.15. The finite volume approach follows the right-hand path in that figure. It is the approach introduced in many general heat-transfer texts, such as Ref. 4. Patankar[21] describes the finite volume approach in the general context of numerical methods for heat transfer and fluid flow.

Many engineers prefer the finite volume approach because it stems directly from the conservation of energy, fits well with an engineer's intuitive understanding of the various modes of heat transfer, and lends itself to direct physical interpretation. The physical domain is divided into a number of volumes, with a finite volume surrounding each node point. As with finite elements, the shape of each finite volume can be chosen at our convenience so that arbitrary irregular geometries are represented easily. A finite volume discretization for a two-dimensional domain is shown in Fig. 4.18a, and a typical finite volume is shown in Fig. 4.18b. For the interior finite volume shown, we may write conservation of energy using only conduction heat fluxes. We start with the basic conservation of energy equation given earlier in Eq. (4.4),

$$\dot{E}_{\text{in}} + \dot{E}_{\text{gen}} = \dot{E}_{\text{sto}} + \dot{E}_{\text{out}}$$

For our illustration, we neglect energy generated within the volume and assume all surface heat fluxes to be positive when entering the control volume. Then, referring to Fig. 4.18b and using a discrete approximation to Fourier's law, we may write the energy balance equation as

$$\frac{kA_{ij}}{L_{ij}}(T_j - T_i) + \frac{kA_{ik}}{L_{ik}}(T_k - T_i) + \frac{kA_{i\ell}}{L_{i\ell}}(T_\ell - T_i) + \frac{kA_{im}}{L_{ij}}(T_m - T_i) = \rho c V \frac{\mathrm{d}T_i}{\mathrm{d}t} \tag{4.58}$$

where a typical area such as A_{ij} is perpendicular to the line joining the nodes i and j, L_{ij} is the distance between nodes i and j, and V is the volume. By writing a similar equation for each control volume, we generate a set of simultaneous differential equations that describe the thermal response of the system. If a face of a finite volume on the surface of the domain has a convective or radiation heat exchange, we include it directly when we write the volume's energy balance.

In many traditional heat-transfer texts, the finite volume approach is explained in terms of an electrical circuit analogy in which heat-conduction paths between nodes are compared to resistors connecting nodes in an electrical circuit, and the

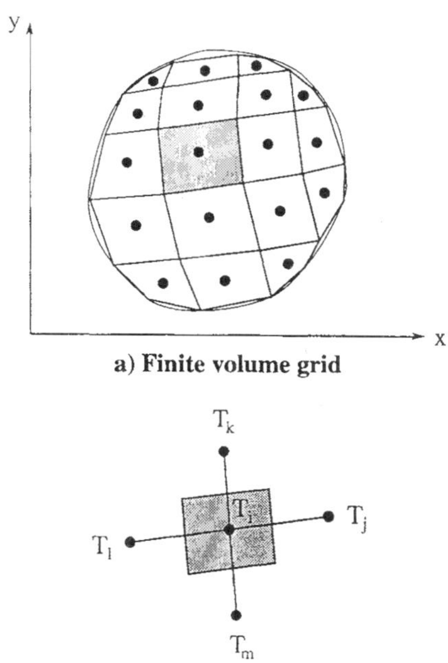

a) Finite volume grid

b) Typical finite volume

Fig. 4.18 Finite volume discretization.

thermal energy storage is compared to electrical capacitance. This analogy leads to diagrams with resistor symbols showing conduction paths, nodes at resistor junctions, and so forth. Sometimes, such models are called *lumped parameter models.*

Nowadays, within industry and government research laboratories, there is a trend to develop computational models directly from geometric computer models developed on a CAD system. This has led to finite volume heat-transfer models (as well as finite difference and finite element models) being developed directly from the geometric models. One approach is to generate finite element meshes where the nodes are at vertices of the control volumes and then generate the discretized equations using the finite volume approach. The result is similar to an element with constant temperature, but the discretized equations are easier and cheaper to generate.

This section has presented an overview of the three computational methods that are commonly used to compute temperatures in aerospace structures. The intent was to explain basic concepts of each method and to contrast differences between the methods without recourse to the details of mathematical and computer implementation. All three methods have been implemented in computer software and may be used to compute structural temperatures effectively. Often, a choice between methods is determined by the software available, traditional practices within the engineering office, or personal preference rather than any clear superiority of one method over the other.

4.7 Temperatures in Aircraft Structures

The determination of temperatures in aircraft structures began in the United States in the late 1940s with the advent of supersonic flight. At about the same time, temperature computations were conducted in support of test programs with V-2 rockets. Beginning about 1950 and for the next decade, extensive research was devoted to the prediction of temperatures and thermal stresses. Near the end of that decade and in the early 1960s, digital computers became available, and engineers began to analyze more complex structures using numerical methods, principally the finite difference method. Later in the 1960s, the development of the Concorde focused attention once again on the important role of airframe temperatures in supersonic flight. And, in the 1970s and 1980s, attention to airframe temperatures in the United States was motivated by the development of supersonic vehicles, such as the SR-71, and hypersonic vehicles, including the Space Shuttle and the National Aerospace Plane (X-30). Gatewood's 1957 text on thermal stresses[22] describes some of the early analytical approaches to computing structural temperatures. The 1960 book by Boley and Weiner[23] on the theory of thermal stresses also briefly describes analytical approaches. The 1965 book by Schuh,[24] which describes both analytical and numerical methods for determining temperatures in structures in high-speed flight, remains as one of the few, perhaps the only text devoted exclusively to heat transfer in structures. At the end of this chapter is a brief bibliography of papers on temperatures in aircraft structures. Only papers widely available in the open literature are cited. Other government, company, and university reports are cited in Refs. 22–24.

John S. Przemieniecki's summary of the major structural effect of aerodynamic heating[25] can be paraphrased as follows. The external surface of the structure of an aircraft or missile accelerated to a high Mach number is heated by forced convection until it attains its radiation equilibrium temperature, whereas the interior structure increases its temperature, mainly by conduction, at a much slower rate, depending on the thermal properties of the structural material. Thus, temperature gradients are set up in the structure, resulting in nonuniform thermal expansions that cannot proceed freely, and thermal stresses are induced.

Before digital computers were available, engineers faced with the task of estimating temperatures for the design of aircraft structures in high-speed flight had the alternative of analytical models or extremely laborious numerical methods. And yet computing temperatures in wing and fuselage structures with analytical models requires a number of simplifying assumptions about geometry as well as neglect of nonlinear effects such as temperature-dependent material properties and radiation heat transfer. These assumptions lead to inaccuracies in computed temperatures and a limited range of validity for the results. The approach requires analysts to have deep physical insight into the heat-transfer behavior, and there are benefits to be derived from this insight today in the era of large-scale computational models. Thus, analytical solutions are not only of historical interest, but also have a place as modern analytical tools for preliminary design as well as an independent means of validating results from computational models. Analytical solutions also have the advantage, as we have seen in previous sections, of identifying key parameters that determine a structure's thermal response. Thus, in contrast to computational models, analytical solutions are particularly convenient for parameter studies.

In this section, we describe briefly analytical and computational solutions for basic problems in aircraft structures. Illustrative examples are given to demonstrate how engineers have modeled aircraft structural heat transfer for high-speed flight. Some typical temperature distributions will be presented to provide insight into the thermal response of aircraft structures to aerodynamic heating.

Example 4.10: Temperatures in a Double Wedge-Shaped Wing

In a 1950 paper, Joseph Kaye[25] determined the transient temperature distribution in a solid, double-wedge (diamond-shaped) steel wing flying at supersonic speeds.

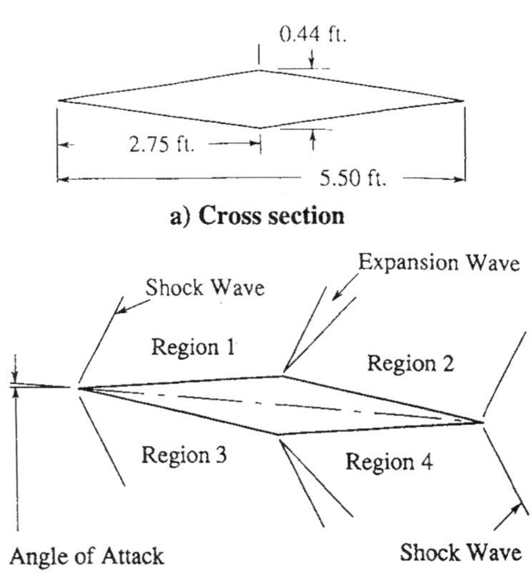

Fig. 4.19 Supersonic flow over a diamond-shaped wing.

The problem consists of a solid steel wing (Fig. 4.19a) flying at a Mach number $M = 1.4$ at 50,000-ft altitude. At time equal to zero, the wing begins a constant acceleration and, over a period of 138 s, accelerates to $M = 6.0$. The wing cross section is uniform, and conditions are constant in the spanwise direction. The objective is to determine the transient temperature distribution in the wing cross section.

Aerodynamic heating. The approach is first to compute the parameters required to represent aerodynamic heating. The heating is represented as a convective boundary condition given by Eq. (1.2),

$$q_w = h(T_{\text{aw}} - T_w)$$

where q_w is the local heating rate, h the convection coefficient, T_{aw} the adiabatic wall temperature, and T_w the wall temperature. The first step for the evaluation of h and T_{aw} is to compute the inviscid flow conditions above and below the wing (Fig. 4.19b). These conditions are evaluated using classical gasdynamics equations for the compression corners at the leading edges to compute flow parameters behind the shocks in regions 1 and 3. Then the expansions at midchord are analyzed to compute the flow conditions in regions 2 and 4.

With the inviscid flow conditions known in regions 1–4, heat-transfer theory for flow over a flat plate (Sec. 1.3.2) is used to compute the adiabatic wall temperatures and convection coefficients. The adiabatic wall temperatures for the four regions are computed from Eq. (1.6) as follows:

$$T_{\text{aw}} = T_\infty \left[1 + r\frac{\gamma - 1}{2} M^2 \right]$$

where T_∞ and M_∞ are the local temperature and Mach number, respectively, from the inviscid flow analysis. The recovery factor is computed from $r = Pr^{1/2}$ for laminar flow and $r = Pr^{1/3}$ for turbulent flow, and an average value $r = 0.9$ was used. The ratio of specific heats is denoted by γ as before. The differences between the inviscid flow temperatures for the four regions were small, and the adiabatic wall temperatures were so close together that the single curve shown in Fig. 4.20

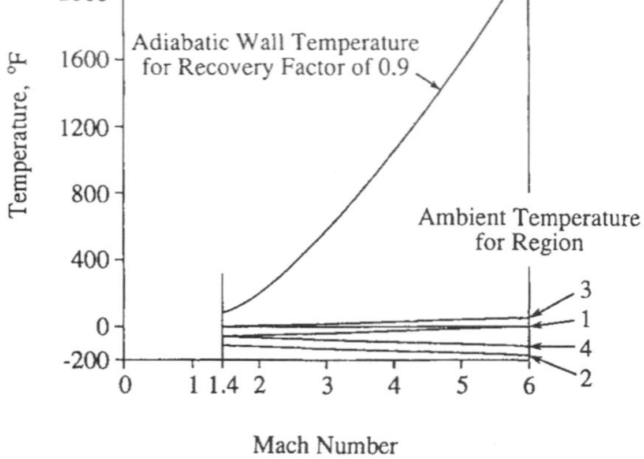

Fig. 4.20 Fluid temperatures vs Mach number (Ref. 26).

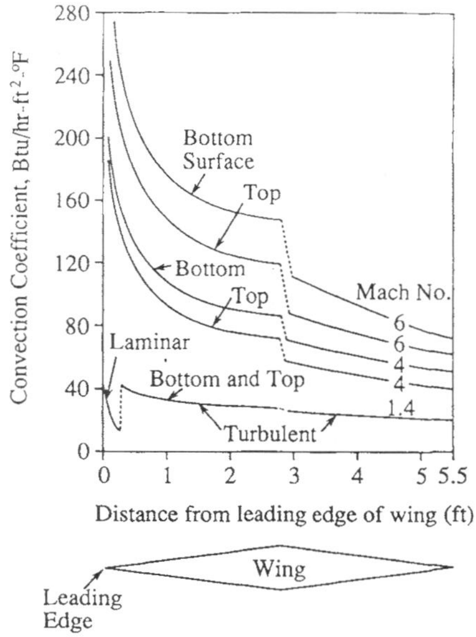

Fig. 4.21 Distributions of convection coefficients for three Mach numbers.

is used to represent the adiabatic temperature as a function of Mach number for all four regions.

The local convection coefficients were evaluated for a flat plate using a theory similar to the Eckert reference temperature method. Convection coefficients were calculated assuming a laminar and then turbulent boundary layer. Flow transition was assumed to occur for Reynolds number of 500,000. This led to a laminar boundary layer only for a few inches from a leading edge for each region and to a turbulent boundary layer thereafter. The distributions of the convective coefficients over the top and bottom surfaces of the, wing are shown for three times in the flight in Fig. 4.21. For $M = 1.4$ at time $t = 0$, the convection coefficients for the top and bottom surfaces of the wing are nearly the same and are represented by a single curve. At later times for higher Mach numbers, the coefficient is higher on the bottom, windward surface, as expected.

Temperatures. Temperatures in the cross section were computed by solving the conservation of energy equation assuming constant thermal properties. Surface-radiation heat transfer was neglected. The initial temperature of the cross section is assumed to be the adiabatic wall temperature for $M = 1.4$. Thus, the heat-conduction problem consisted of solving the conservation of energy equation subject to convective boundary conditions on the top and bottom surfaces of the wing. Since the convection coefficient varies along the wall and with time, an analytical solution was impractical. Instead, a finite volume numerical method of solution was used to compute the temperatures of the cross section. Two models were used: 1) a two-dimensional orthogonal network of conductors and 2) a series of one-dimensional models through the thickness of the wing. The two-dimensional model used 113 control volumes, and the solution was advanced in time over 32 time steps. In the precomputer days of 1950, Kaye characterizes the numerical approach as "long and laborious." For the one-dimensional models, he states that the time of calculation ran to as much as 3 man-days per temperature distribution!

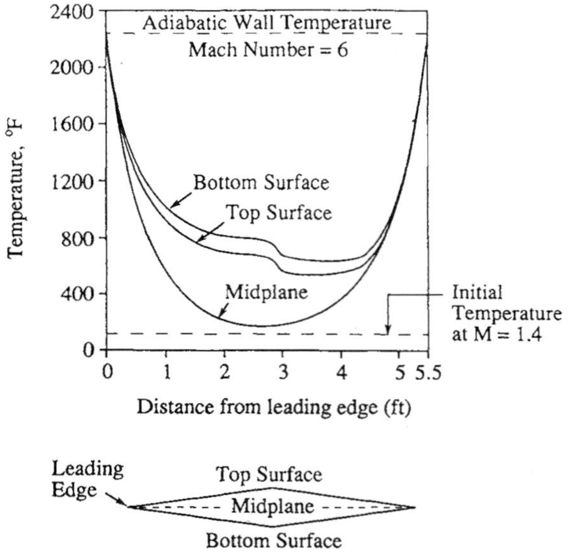

Fig. 4.22 Chordwise temperature distribution of a wing at M = 6 (Ref. 26).

The chordwise temperature distributions computed at $M = 6.0$ are shown in Fig. 4.22; the corresponding thickness temperature distributions at mid chord are shown for several Mach numbers in Fig. 4.23. Figure 4.22 shows that the maximum temperature occurs at the leading and trailing edges of the wing and is equal to the adiabatic wall temperature. As expected, the temperatures on the bottom, windward surface are hotter than the top, leeward surface. Both surface temperatures are considerably higher than midplane temperatures, with the maximum temperature difference occurring at midchord. Figure 4.23 shows the thickness temperature

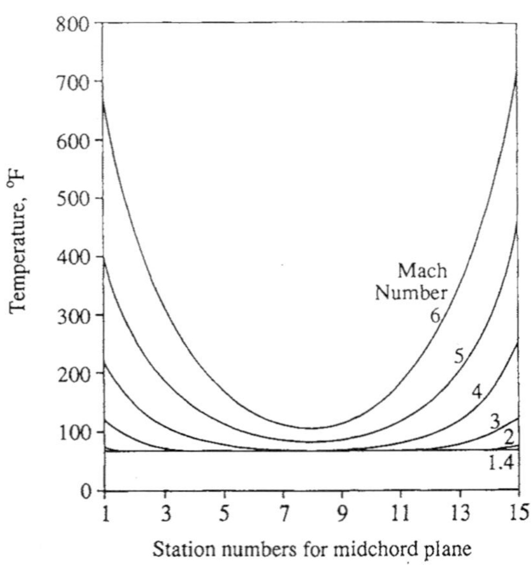

Fig. 4.23 Thickness temperature distribution in a wing at midchord.

gradients, which are considerably larger than the chordwise temperature gradients because of the relatively thin wing. The large difference in temperature gradients justifies the use of local one-dimensional thermal models through the thickness. The relatively low conductivity of the steel also contributed to the steep thickness gradients.

In a later paper with Yeh, Kaye[26] used one-dimensional analytical models to develop design charts to permit estimates of a double wedge-shaped wing's transient temperatures.

Example 4.11: Temperatures in a Multicell Wing

Another thermal-structural problem considered in the 1950s is the determination of temperatures and thermal stresses in a multicell wing. The wing structure considered is shown in Fig. 4.24a. Symmetry about the horizontal axis is assumed, and we restrict the analysis to one of the inner cells remote from the leading or trailing edge. The simplified model of a symmetrical section is shown in Fig. 4.24b. The aerodynamic skin and web are fabricated from metal, and their thicknesses are relatively small compared to the overall cross-sectional dimensions of a typical wing cell. Thus, a basic assumption is that temperature gradients through the skin and web thicknesses are small and may be neglected. Flight occurs at constant Mach number and altitude, and the adiabatic wall temperature and convection coefficient are known constants. The wing cross section is initially at a uniform temperature T_0. At time equal to zero, convective heating of the aerodynamic skin commences, and we wish to determine the subsequent temperature distributions in the skin and web as a function of time.

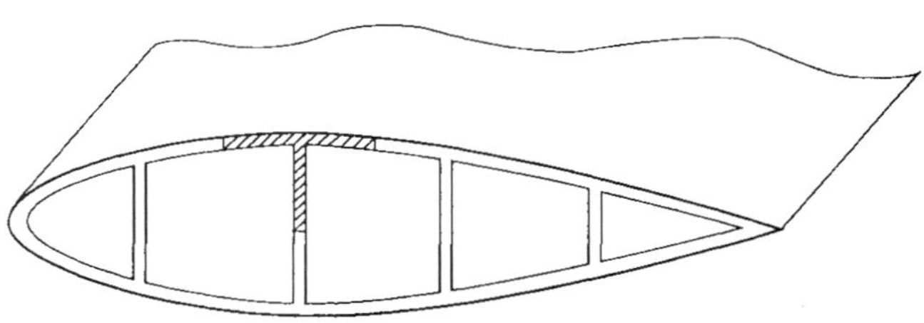

a) Wing section

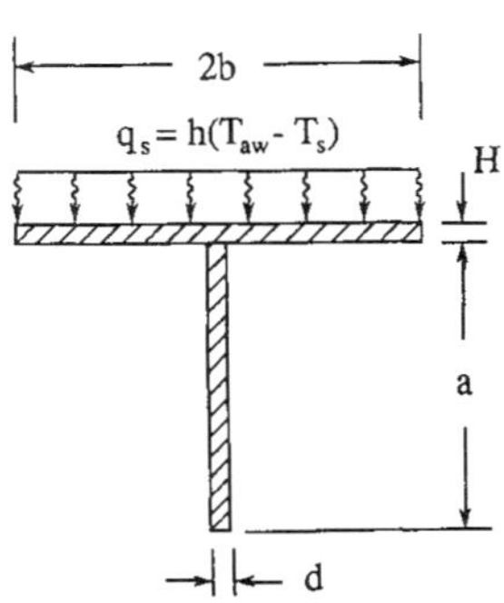

b) Aerodynamic skin and web geometry

Fig. 4.24 Idealized model of a multicell wing section.

As a result of the aerodynamic heating, the temperature of the aerodynamic skin will rise first. Then heat will be transferred by conduction through the skin/web junction to the web. As a result of the conduction, skin temperatures will be maximum midway between webs and have minimum values at the skin/web junctions. As energy is conducted into the webs from the skin, a temperature gradient will develop in the webs early in the response.

On the external aerodynamic surfaces, energy losses to the atmosphere by radiation may occur. Because of the temperature gradients on the internal surfaces, internal radiation exchanges and natural convection heat transfer are possible. In addition, depending on construction details, there may be joint thermal resistances at skin/web interfaces that affect the temperature distributions. With radiation and convection heat transfer, after the initial transient, temperatures will approach steady-state values, but with a nonuniform distribution over the cross section.

In the simplified 1950s analyses, these complexities were neglected. Both external and internal radiation heat transfer, as well as natural convection, were neglected. The joints were assumed to be perfect conductors. With these assumptions, after the initial transient, temperatures become uniform and approach the adiabatic wall temperature as time becomes large.

Two simplified thermal models were developed in the 1950s. The first model, developed by Hoff[27] in 1950, assumes the aerodynamic skin to have a uniform temperature. The second model, developed by Pohle and Oliver[28] 1954, allows the aerodynamic skin to have a nonuniform temperature distribution. We will describe these heat-transfer models, the mathematical formulations, and present typical results.

Uniform skin temperature model. The uniform skin model, Fig. 4.25a, assumes that the heat transfer from the skin to the web is negligible. In this way, the skin temperature $T_s(t)$ is independent of any space coordinate and is a function only of time. From an energy balance on the skin, we may write the ordinary differential equation

$$\frac{dT_s}{dt} + \beta T_s = \beta T_{aw}$$

where $\beta = h/\rho c H$. With the initial condition $T_s(0) = T_0$, the solution to the last equation is

$$T_s(t) = T_0 e^{-\beta t} + T_{aw}(1 - e^{-\beta t})$$

As time becomes large, the skin temperature approaches the adiabatic wall temperature as we expect. To determine the web temperature, the usual one-dimensional conservation of energy equation is solved subject to the condition that the temperature of the end of the web is prescribed by $T_s(t)$. With reference to Fig. 4.25a, the formulation of the conduction problem is

$$\frac{\partial^2 T_w}{\partial x^2} = \frac{1}{\kappa}\frac{\partial T_w}{\partial t} \qquad q_w(0, t) = 0$$

$$T_w(a, t) = T_s(t) \qquad T_w(x, 0) = 0$$

This is a linear, boundary/initial-value problem with a time-dependent boundary condition. The problem can be solved by separation of variables, and the solution appears in Ref. 28.

Nonuniform skin temperature model. To permit a temperature variation in the skin, the problem involves coupling of two, one-dimensional conservation of energy equations through continuity conditions at the web/skin interface. The

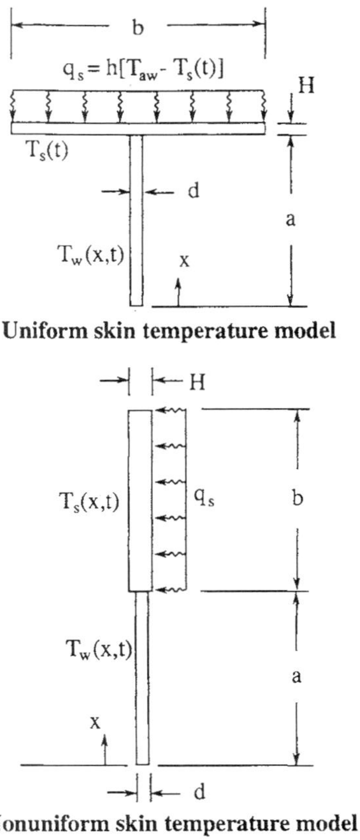

Fig. 4.25 Thermal models of a multicell wing section.

model is shown in Fig. 4.25b. The standard one-dimensional conservation of energy equation applies to the web, but a modified conservation of energy equation taking into account the aerodynamic heating applies to the skin. The problem formulation is

$$\frac{\partial^2 T_w}{\partial x^2} = \frac{1}{\kappa}\frac{\partial T_w}{\partial t} \qquad o \leq x \leq a$$

$$\frac{\partial^2 T_s}{\partial x^2} = \frac{1}{\kappa}\frac{\partial T_s}{\partial t} + \frac{h}{kH}(T_s - T_{\text{aw}}) \qquad a \leq x \leq b$$

$$\frac{\partial T_w}{\partial x}(0, t) = 0$$

$$T_w(a, t) = T_s(a, t)$$

$$d\frac{\partial T_w}{\partial x}(a, t) = H\frac{\partial T_s}{\partial x}(a, t)$$

$$\frac{\partial T_s}{\partial x}(b, t) = 0$$

$$T_w(x, 0) = T_s(x, 0) = T_0$$

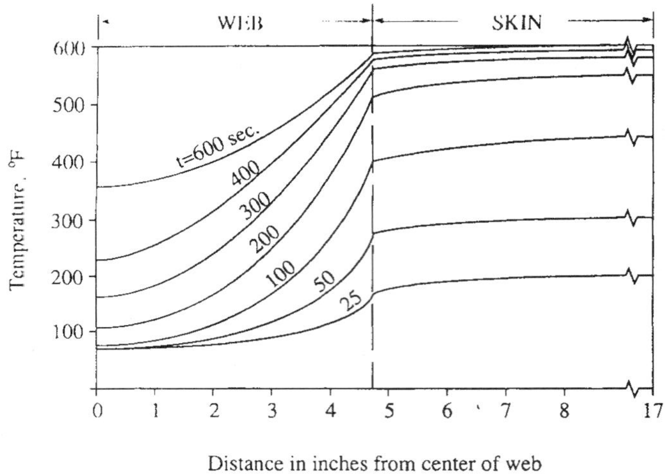

Fig. 4.26 Temperature distribution in web and skin (Ref. 29).

At $x = 0$ and $x = b$, the boundary conditions require that no heat flux occurs. At $x = a$, the continuity conditions require the temperature to be continuous and the energy flow to be continuous across the web/skin interface. The boundary/initial-value problem was solved by the Laplace transform method, and details are given in Ref. 28.

Pohle and Oliver[29] used the numerical data for a steel wing section from Hoff's paper[28] to compare the analytical models. The graphs in Fig. 4.26 show the variation in the temperature along the web and skin for various values of time. These temperature distributions from the second analytical model include the nonuniform skin temperatures.

Because of the relatively low thermal conductivity of the steel, the temperatures In the web rise slowly. There is only a small decrease in the temperature at the web/skin interface. Away from the web/skin interface, the temperatures approach the values predicted by the first analytical model, with uniform skin temperatures assumed. Pohle and Oliver attribute the small difference to the dimensions chosen for the cross section. For the results shown, the ratio of the area of the skin to the area of the web, $A_s/A_w = 10$, indicates that the skin area is much larger than the web area. Thus, less heat is conducted away by the web than if A_s/A_w were smaller. Pohle and Oliver argue that, if this ratio were smaller, the uniform skin temperature model would lead to serious errors for the web temperatures.

Example 4.12: Transient Temperatures in Fuselages

Closely related to the preceding wing example is the problem of determining temperatures in an aerodynamically heated fuselage. The problem was first analyzed by Przemieniecki in a 1956 paper.[24] The problem considered is shown in Fig. 4.27a. The fuselage is assumed to be a thin-walled, cylindrical shell, periodically reinforced by solid, circular, thin-walled bulkheads. Symmetry about the fuselage axis is assumed, and we consider a section of the shell with a frame at a representative interior location. Axisymmetric aerodynamic heating is assumed at constant Mach number and altitude so that the adiabatic wall temperature T_{aw}

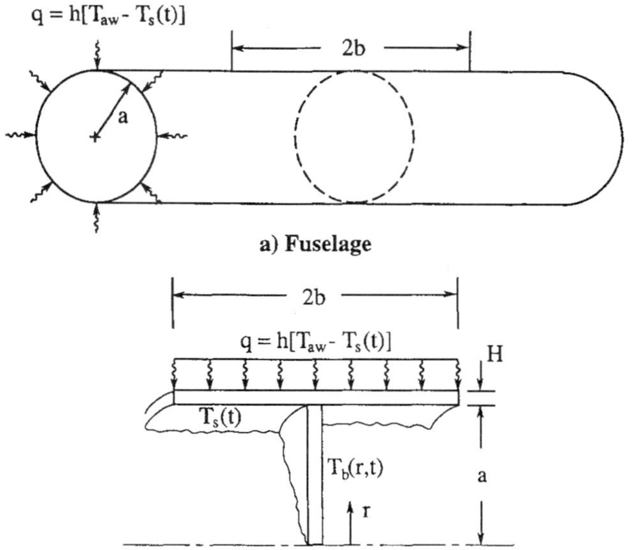

Fig. 4.27 Idealized model of fuselage.

and convection coefficient h are constant. Temperature gradients through the thin metallic shell and bulkhead are small and are neglected. Heat transfer by radiation and internal convection within the fuselage are neglected. At time equal to zero, convective heating of the aerodynamic skin begins, and we wish to determine the subsequent temperature distribution in the skin and bulkhead.

Aerodynamic skin. A uniform skin temperature thermal model is assumed as shown in Fig. 4.27b. As in the preceding example, this means that heat loss to the bulkhead is neglected and that the temperature of the skin is a function only of time, $T_s(t)$. By writing the conservation of energy for an axisymmetric, cylindrical section of the shell, we obtain exactly the same differential equation for the skin of the shell as we had for the wing in Example 4.11:

$$\frac{dT_s}{dt} + \beta T_s = \beta T_{aw}$$

where, as before, $\beta = h/\rho c H$. Assuming that β is constant, we obtain the same solution but, this time, we write the solution in the non-dimensional form

$$\frac{T_s - T_0}{T_{aw} - T_0} = (1 - e^{-\beta t})$$

where T_0 is the initial skin temperature. As time becomes large, the skin temperature approaches the adiabatic wall temperature.

Bulkhead. In the preceding wing example, we assumed perfect contact between the skin and web and imposed the skin temperature as a specified time-dependent boundary condition for the web. If there is a mechanical joint at the skin/bulkhead interface, thermal contact is not perfect, and a temperature jump occurs across the interface. Mathematically, this is equivalent to a discontinuity in the temperature distribution. The heat flux across the joint is assumed proportional to the difference in the skin and bulkhead temperatures at the interface. The constant of proportionality is the joint thermal contact conductance, noted here as h_c. Several sources in the bibliography at the end of this chapter deal with thermal conductance of aircraft structural joints.

With reference to Fig. 4.27b, the formulation of the conduction problem for the bulkhead temperature $T_b(r, t)$ is

$$\frac{\partial^2 T_b}{\partial r^2} + \frac{1}{r}\frac{\partial T_b}{\partial r} = \frac{1}{\kappa}\frac{\partial T_b}{\partial t}$$

$$T_b(0, t) \qquad \text{is finite}$$

$$k\frac{\partial T_b}{\partial r}(a, t) = h_c[T_s(t) - T_b(a, t)]$$

$$T(r, 0) = T_0$$

The conservation of energy equation was introduced previously in Eq. (4.42) . In fact, the boundary/initial-value problem is similar to the convectively heated cylinder problem, Example 4.9. The difference here is that we have a time-dependent boundary condition at $r = a$, where $T_s(t)$ is the skin temperature defined earlier. Przemieniecki[24] presents the solution, which we write in the nondimensional

$$\frac{T_b - T_0}{T_{aw} - T_0} = 1 - \frac{J_0(\mu r/a)e^{-\beta t}}{J_0(\mu) - (\theta\mu)J_1(\mu)} - 2\sum_{i=1}^{\infty}\frac{J_1(\lambda_i a)J_0(\lambda_i r)e^{-\lambda_i^2\kappa t}}{\lambda_i a[1 - (\lambda_i a/\mu)^2][J_0^2(\lambda_i a) + J_1^2(\lambda_i a)]}$$

where, as in Example 4.9, J_0 and J_1 denote Bessel functions of the first kind, of order zero and one, respectively. The nondimensional parameter $\mu^2 = \beta a^2/\kappa$, and $\theta = k/h_c a$. The eigenvalues λ_i are the roots of the characteristic equation

$$\theta(\lambda_i a)J_1(\lambda_i a) = J_0(\lambda_i a)$$

which was also encountered in Example 4.9. The roots of this equation are given by Carslaw and Jaeger.[16] Reference 25 shows that two extreme cases can be considered: 1) infinite thermal conductance, corresponding to a thermally continuous joint, and 2) zero thermal conductance, corresponding to perfect insulation at the joint. In the first case, $\theta \to 0$, and the temperature distribution is given by

$$\frac{T_b - T_0}{T_{aw} - T_0} = 1 - \frac{J_0(\mu r/a)e^{-\beta t}}{J_0(\mu)} - 2\sum_{i=1}^{\infty}\frac{J_0(\lambda_i r)e^{-\lambda_i^2\kappa t}}{\lambda_i a[1 - (\lambda_i a/\mu)^2][J_1(\lambda_i a)]}$$

where

$$J_0(\lambda_i a) = 0$$

defines the eigenvalues. In the second case, as $\theta \to \infty$ then $T_b \to T_0$, which means that the temperature of the bulkhead is not affected by the aerodynamic heating of the fuselage skin.

Numerical calculations for a fuselage skin/bulkhead give temperature distributions similar to the distributions predicted for the wing in the preceding example.

Table 4.2 Supersonic transport flight profile

Time, min	Altitude, ft	Mach number, M
—	—	—
25	60,000	2.4
265	70,000	2.4
290	—	—

By comparison, we may expect, because of the assumption of uniform skin temperatures, that these results will be accurate for cases in which the area of the skin is large compared with the area of the bulkhead. Numerical computations show that the aerodynamic skin temperatures are a maximum midway between bulkheads and decrease to a minimum at skin/bulkhead joints as thermal energy is conducted into the bulkhead.

Example 4.13: Supersonic Transport Wing Spar Thermal Cycle

In a 1967 paper, N'Guyen and Ripley[29] describe the design philosophy and fatigue testing of the Concorde. An essential difference between supersonic transports such as the Concorde and military aircraft is that supersonic transports are subjected to long periods of elevated temperature. In addition, the ascent into the cold environment of the upper atmosphere, the long exposure to aerodynamic heating, and the descent expose the airframe to a thermal cycle that has had a major impact on the design and fatigue test program of the Concorde.

N'Guyen and Ripley present the following scenario for a typical Concorde flight. Immediately after takeoff, the Concorde climbs subsonically and then accelerates to its cruising speed of Mach 2. In the initial climb at subsonic speeds, the external surfaces cool a little and then heat up in the acceleration phase, reaching a temperature of about 210°F, which is maintained during cruise. During deceleration and descent, the external structural temperature drops to about −4°F, rising again to ambient temperature on landing. Initially, the temperature of the internal structure lags behind the surface temperature on account of the time it takes for heat to be conducted internally. Thus, on the climb, the external structure will be hotter than the internal structure. During cruise, temperatures of the internal structure gradually approach the temperature of the external structure. The reverse takes place on the descent since skin temperatures drop during descent, and conduction now takes place from the hotter internal structure to the external surfaces. This reversal of temperature gradients and the associated compression/tension stress cycle subjects major portions of the airframe to a stress reversal and, suggests potential thermal-fatigue problems. N'Guyen and Ripley[29] describe some fatigue-testing programs conducted for the Concorde.

To illustrate this thermal cycle, we describe an analysis of a simplified wing spar for a hypothetical flight of a supersonic transport. The analysis uses the Eckert reference temperature method (see Sec. 1.3.2) to estimate aerodynamic heating and the uniform skin temperature thermal model described in Example 4.11.

The flight profile for the supersonic transport is given in Table 4.2. Between the given data points, the altitude and Mach number are assumed to vary linearly. To

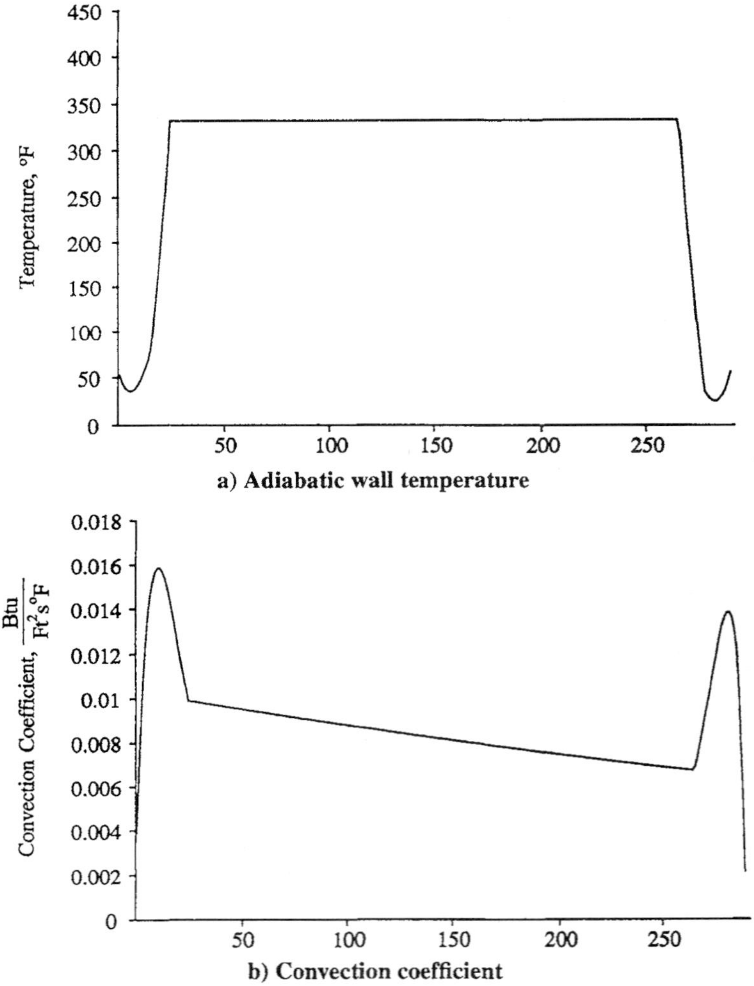

Fig. 4.28 Convective heating parameters for supersonic transport flight.

estimate the aerodynamic heating, the wing is assumed to be a flat plate at zero angle of attack. The flow is assumed turbulent, and the plate is assumed to be at the radiation equilibrium temperature throughout the flight. ,The atmosphere is assumed to be a perfect gas, standard atmospheric properties are assumed, and the Prandtl number Pr is assumed to be constant, $Pr = 0.71$. The variations of the adiabatic wall temperature and convection coefficient with time, as predicted by the flat-plate theory, are shown in Fig. 4.28.

To compute the temperature distributions, the uniform skin temperature model (Fig. 4.25a) is adopted. As in Example 4.11, the aerodynamic skin is convectively heated and is assumed to have uniform temperature. The web experiences one-dimensional conduction, and the top edge of the web has a prescribed temperature equal to the temperature of the aerodynamic skin. Radiation and internal convection are neglected. The conservation of energy equation for the skin is given by Eq.

(4.5) in Example 4.1

$$\frac{\mathrm{d}T_s}{\mathrm{d}t} + \frac{h(t)}{\rho c H} T_s = \frac{h(t)}{\rho c H} T_{\mathrm{aw}}(t)$$

but now the convection coefficient and the adiabatic wall temperatures are functions of time, as shown in Fig. 4.28.

The skin temperature was computed by time-marching the solution through the flight, assuming a constant h and linear variation of T_{aw} over each time step to compute the time history $T_s(t)$. An analytical solution based on constant h and linear T_{aw} was used recursively to advance the solution for T_s through the flight. To compute the evolution of the web temperature distribution, the skin temperature $T_s(t)$ was assumed to vary linearly over each time step, and an analytical solution for $T_w(t)$ was obtained. The web temperature problem formulation is,

$$\frac{\partial^2 T_w}{\partial x^2} = \frac{1}{\kappa}\frac{\partial T_w}{\partial t} \qquad q_w(0, t) = 0$$

$$T_w(a, t) = \frac{T_s(t_{n+1}) - T_s(t_n)}{\Delta t} t + T_s(t_n)$$

$$T_w(x, 0) = T_w(t_n)$$

where $n = 1, 2, \ldots,$ and Δt is the time step. The solution to the boundary/initial-value problem is used recursively for each time step where, for a given initial condition, the analytical solution is used to compute the web temperature at the end of the time step. This temperature distribution is used as the initial condition for the next time step, and the solution is advanced as far in time as desired. The solution approach is unconditionally stable numerically since no approximations have been used.

The approach was used to compute temperatures in a titanium wing spar with an area ratio of $A_s/A_w = 12$. The evolution of the spar temperature distributions is presented in Fig. 4.29. During the initial portion of the flight, Fig. 4.29a, the skin temperature first drops during ascent into the cold upper atmosphere but, then, as aerodynamic heating increases, the skin begins to warm. During the cruise portion of the flight, the skin is at a constant temperature of 330°F. During the early portion of the flight, the maximum temperature gradient from the skin to the center of the web (about 260°F) develops at about 50 min, and the gradient decreases thereafter as heat is conducted into the web. The gradient reverses during descent (Fig. 4.29b), as the aerodynamic heating diminishes rapidly and the skin cools. The maximum reversed gradient (about 230°F) occurs at the end of the flight when the aircraft has landed.

Example 4.14: Space Shuttle Wing

A modern example of heat transfer for a complex structure with aerodynamic heating is the Space Shuttle wing. William L. Ko and coauthors[31–34] have documented several numerical heat-transfer analyses and comparisons to flight-measured temperatures. The Space Shuttle airframe is constructed primarily of aluminum, which is protected from re-entry heating by a thermal protection system (TPS) from being overheated beyond the design temperature limit of 350°F (177°C).

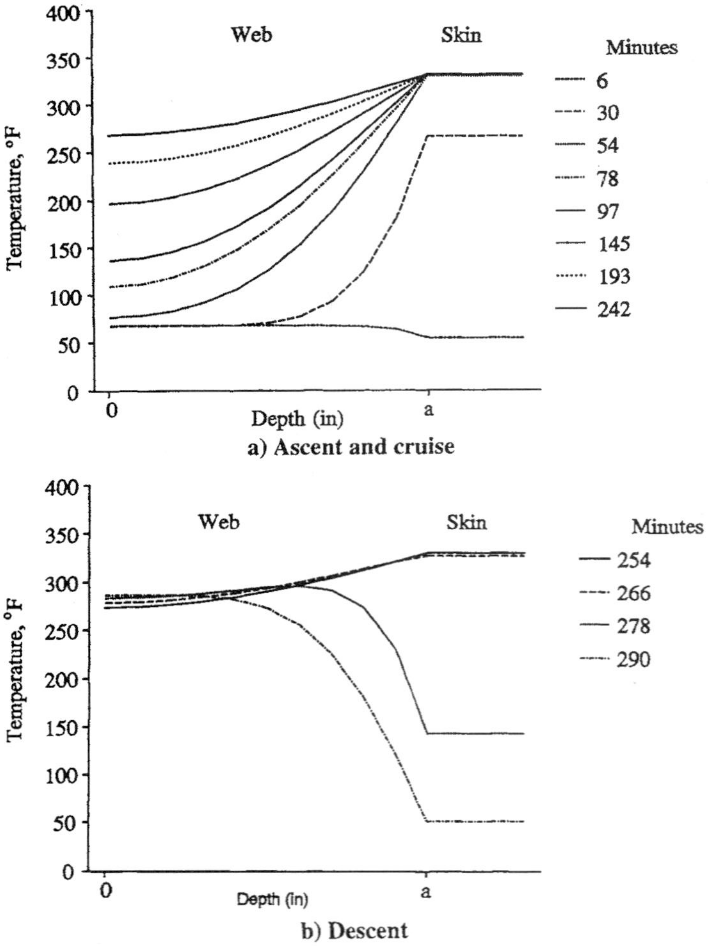

Fig. 4.29 Spar temperature distributions for 290-min flight.

The problem is to determine the evolution of the temperature distribution in the airframe from the beginning of re-entry until after the Space Shuttle has landed.

A plan and side view of the Space Shuttle showing two wing locations and a fuselage location where analyses were performed is shown in Fig. 4.30a. We describe calculations for wing station (WS) 240; the geometry of this section is shown in Fig. 4.30b. The upper and lower skins of bay 1 are made of aluminum honeycomb sandwich plates. The skins for bays 2, 3, and 4 are made of spanwise "hat" stringer–reinforced skins. The remaining spar webs are made of corrugated aluminum. The TPS for the lower wing skin consists of high-temperature reusable surface insulation (HSRI) with a strain isolation pad (SIP) lying between the HRSI and the skin. The purpose of the SIP is to isolate deformations of the aluminum wing structure from the relatively brittle HRSI. Most of the upper skin of bay 1 is protected by a low-temperature reusable surface insulation (LRSI) with a SIP layer. The remainder of the upper skin is covered with a felt reusable surface insulation (FRSI). A typical wing rib is a truss.

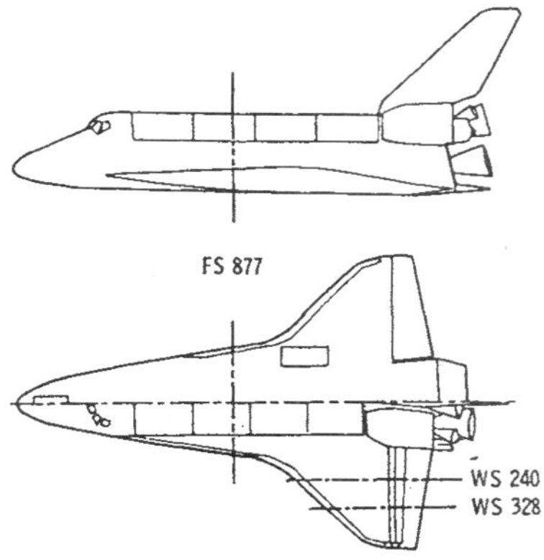

a) Wing location analyzed, WS 240

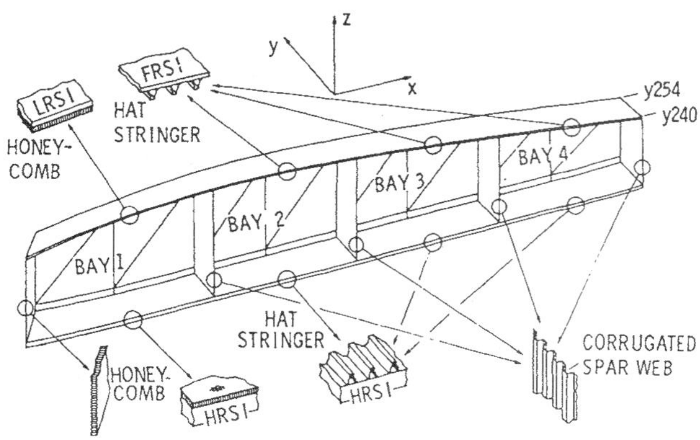

b) Geometry of the wing section

Fig. 4.30 Space Shuttle wing thermal analysis (Ref. 30).

Aerodynamic heating rates were computed using NASA software that takes into account the velocity–altitude–angle-of-attack history for a re-entry trajectory. Boundary-layer and other flow models are used to evaluate the complex re-entry heating. Three flow cases were assumed: 1) turbulent flow for the lower and upper aerodynamic surfaces, 2) laminar flow for both the lower and upper surfaces, and 3) laminar flow for the lower surface and the upper surface of bay 1, and separated flow for the remainder of the upper surface. Figure 4.31 shows the heat fluxes predicted for the three flow cases.

The thermal response was analyzed with both a finite difference model and a finite element model. The models represented conduction in the TPS, conduction in the aluminum airframe, external radiation losses on the aerodynamic surfaces, and internal radiation exchanges. The finite difference model used 410 nodes; the finite element model used 920 nodes. The finite element model is shown in Fig. 4.32. The finite element model uses three-dimensional heat-conduction elements for the TPS and one- and two-dimensional elements to represent conduction in the airframe. Because of gaps in the reusable surface insulation (RSI) tiles, heat conduction in the TPS was permitted only in the thickness direction. To account for the gaps, the thickness of the TPS was reduced by 80%. Surface elements were used in the finite element model to represent radiation exchanges. Internal convection was neglected.

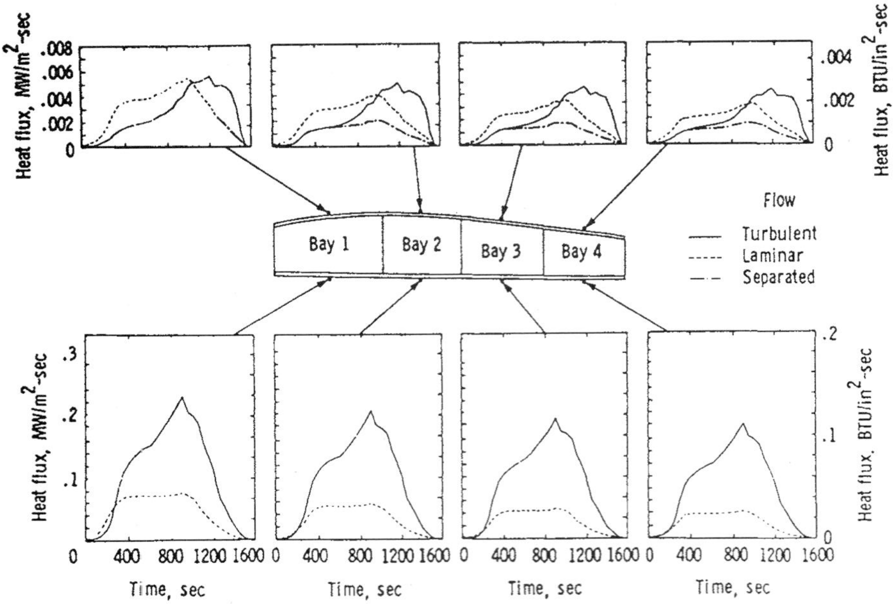

Fig. 4.31 Space Shuttle wing aerodynamic heating rates during re-entry (Ref. 30).

Predicted aluminum skin temperature histories from the finite element model are shown in Fig. 4.33 for the three heating cases mentioned earlier. Flight data are also shown for comparison. The significant variation of predicted skin temperatures with the three heating cases shows the importance of accurate prediction of aerodynamic heating. The flight data show the upper skin temperatures continuing to rise after touchdown, which suggests internal convective heat transfer from the hotter lower skin. Chordwise skin temperature distributions predicted by the finite difference and finite volume models for $t = 1600$ s are compared in Fig. 4.34. This distribution shows the characteristic maximum values midway between the spars, and minimum values at the skin/spar junctions, where heat is conducted into the spars. Agreement between predicted temperatures by the two models is generally good except for bay 1, which the authors attribute to differences in TPS modeling.

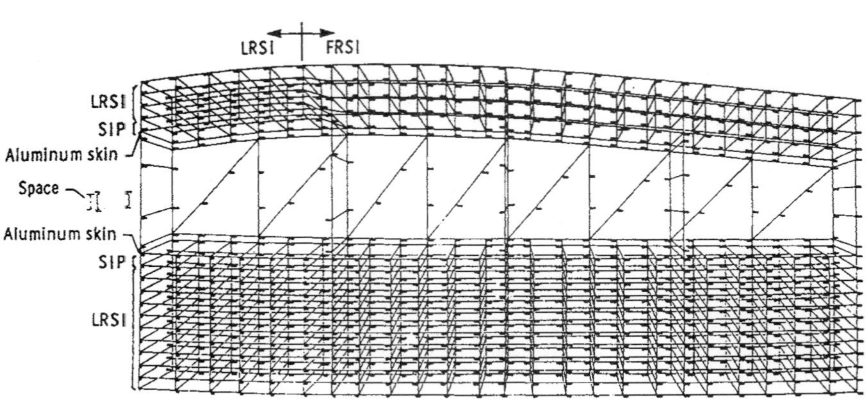

Fig. 4.32 Finite element thermal model of Space Shuttle wing (Ref. 30).

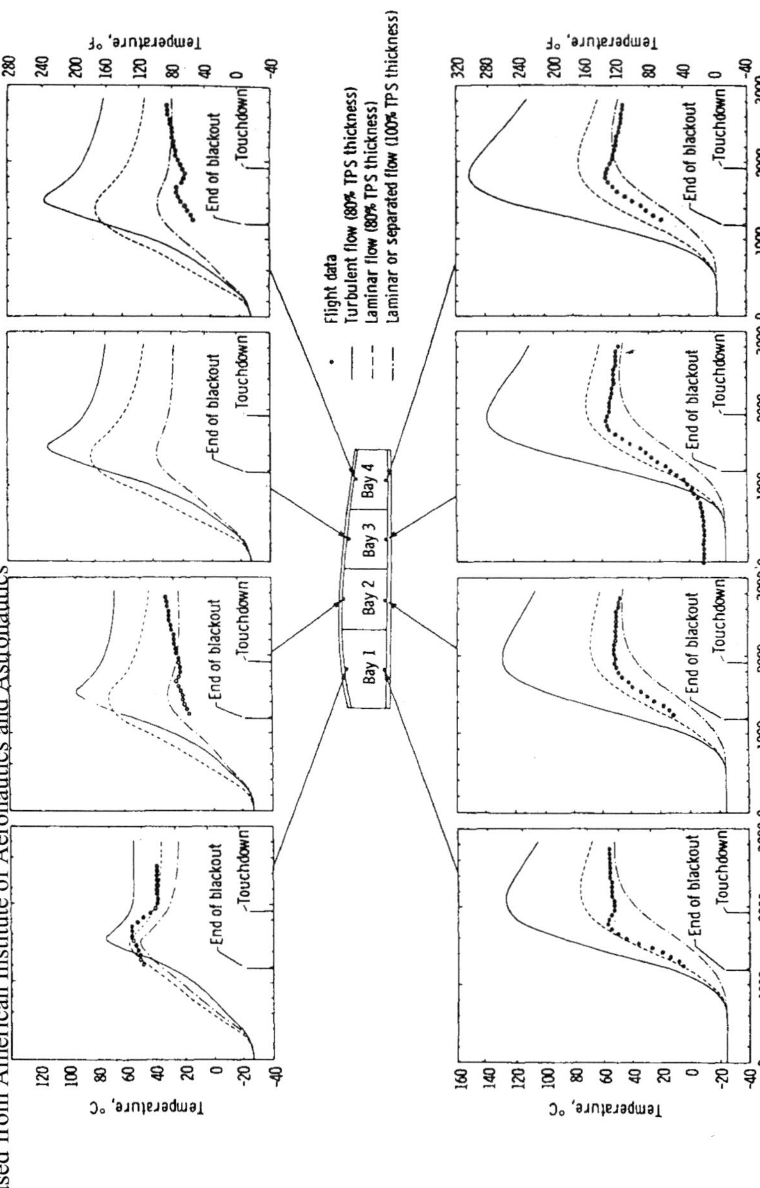

Fig. 4.33 Space Shuttle wing skin temperature histories (Ref. 30).

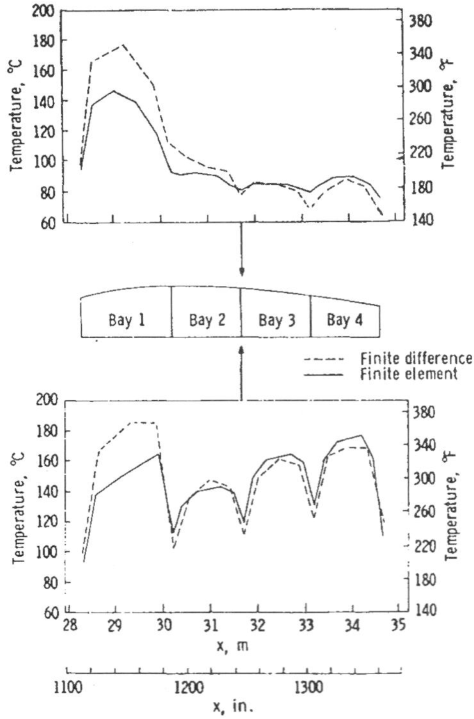

Fig. 4.34 Space Shuttle wing skin temperature distributions by finite element and finite difference models, t = 1600 s (Ref. 30).

Closure. The five examples presented in this section illustrate several aspects of temperature distributions in aircraft structures. They demonstrate the important interaction of aerodynamic heating with predicted temperatures and the role of the heating in establishing structural temperature gradients. The examples show how analytical models can predict some of the basic features of a structure's thermal response. The examples also show that computational models must be used when complex geometry and nonlinear effects have to be represented.

4.8 Thin-Walled Tube with Radiation

Thin-walled tubes are common structural members for orbiting space structures. Tubes are used as truss members as well as antenna booms. Tubes typically experience incident radiant heating and lose energy by radiation to deep space. In some applications (see Sec. 2.4), the axial variation of temperature is not significant, and the principal quantities of interest are the mean temperature of a tube cross section and the temperature gradient across the tube diameter. Although most conduction problems with radiation boundary conditions require numerical solutions, researchers[34,35] over the years have developed an analytical solution for this problem that is useful in applications.

4.8.1 Analytical Solution

We consider a thin-walled tube of radius R and wall thickness h (Fig. 4.35). The tube is subjected to a known incident heat flux S applied instantaneously at time

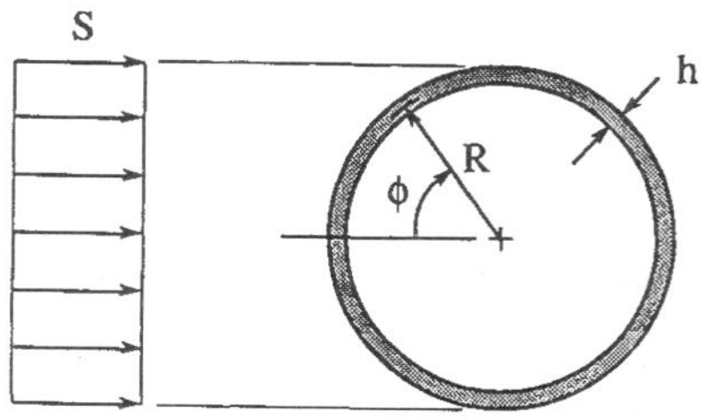

Fig. 4.35 Thin-walled tube with radiant heating.

equal to zero. We wish to obtain the subsequent transient temperature response of the tube, assuming that the tube has a known initial absolute temperature T_0.

To formulate the energy conservation equation, the thin-walled tube is idealized by the following assumptions:

1) Radiation heating is started at time zero, acquires its full value instantaneously, and is uniform along the tube length.

2) The tube wall thickness is so small that the temperature gradient across it can be negelected.

3) Heat conduction along the tube length is negligible.

4) Radiation heat transfer within the tube is negligible.

5) Convection heat transfer inside and outside the tube is negligible.

6) Thermal properties are assumed constant.

7) The incident heat flux and temperature distribution are symmetrical about a tube diameter.

With these assumptions, the absolute temperature at a point on the tube wall is $T = T(\phi, t)$ where ϕ is defined in Fig. 4.35, and t is time.

Conservation of energy for a small segment of the tube wall, including circumferential conduction and radiation from the tube's external surface, yields

$$\frac{\partial T}{\partial t} - \frac{k}{\rho c R^2}\frac{\partial^2 T}{\partial \phi^2} + \frac{\sigma \varepsilon}{\rho c h} T^4 = \frac{\alpha_s S}{\rho c h}\delta \cos\phi \tag{4.59}$$

where k is the thermal conductivity, ρ the density, c the specific heat, σ the Stefan–Boltzmann, constant, ε the surface emissivity, and α_s the surface absorptivity. The parameter δ is introduced to account for the fact that only the front of the tube is heated by S; δ is unity for the front, heated surface and zero for the back, unheated surface.

Although Eq. (4.59) is a nonlinear partial differential equation, an approximate solution may be obtained that is suitable for calculating the thermal response. The heat-flux distribution on the right-hand side of Eq. (4.59) is represented as a truncated Fourier series,

$$S\delta\cos\phi \approx S\left(\frac{1}{\pi} + \frac{1}{2}\cos\phi\right) \tag{4.60}$$

and the temperature distribution is separated into two parts: 1) a temperature $\bar{T}(t)$ independent of ϕ corresponding to the uniform heat flux S/π in Eq. (4.60), and 2) a nonuniform temperature distribution $T_m(t)\cos\phi$ corresponding to the circumferential variation of flux in Eq. (4.60) . Thus, the temperature is represented as

$$T(\phi, t) = \bar{T}(t) + T_m(t)\cos\phi \tag{4.61}$$

Experience shows that $T_m/\bar{T} < 1$ so that, when Eqs. (4.60) and (4.61) are substituted into Eq. (4.59), the nonlinear radiation term may be linearized. Thus, we approximate the radiation term by the truncated expansion

$$(\bar{T} + T_m \cos\phi)^4 \cong \bar{T}^4\left[1 + 4\frac{T_m}{\bar{T}}\cos\phi\right]$$

Substituting and matching coefficients

$$\frac{d\bar{T}}{dt} + \frac{\sigma\varepsilon}{\rho ch}\bar{T}^4 = \frac{1}{\pi}\frac{\alpha_s S}{\rho ch} \tag{4.62}$$

$$\frac{dT_m}{dt} + \left(\frac{k}{\rho c R^2} + \frac{4\sigma\varepsilon\bar{T}^3}{\rho ch}\right)T_m = \frac{1}{2}\frac{\alpha_s S}{\rho ch} \tag{4.63}$$

Equations (4.61) and (4.62) are two simultaneous ordinary differential equations. The initial conditions are obtained from Eq. (4.61) as

$$\bar{T}(x,0) = T_0 \qquad T_m(x,0) = 0 \tag{4.64}$$

where T_0 denotes the initial uniform tube temperature.

The differential equation (4.62) for $\bar{T}$ is nonlinear, but variables may be separated; the equation is then integrated directly to yield the solution (see Sec. 4.3.3)

$$\frac{1}{4\bar{T}_{ss}^3}\left[\ell n\frac{(\bar{T}_{ss} - \bar{T})(\bar{T}_{ss} + \bar{T}_0)}{(\bar{T}_{ss} + \bar{T})(\bar{T}_{ss} - \bar{T}_0)} + 2\tan^{-1}\frac{\bar{T}_0}{\bar{T}_{ss}} - 2\tan^{-1}\frac{\bar{T}}{\bar{T}_{ss}}\right] + \frac{\sigma\varepsilon}{\rho ch}t = 0 \tag{4.65}$$

where the steady-state value of the average tube temperature is

$$\bar{T}_{ss} = \left[\frac{1}{\pi}\frac{\alpha_s S}{\sigma\varepsilon}\right]^{1/4} \tag{4.66}$$

Note that the steady-state (radiation equilibrium) average tube temperature depends only on the incident heat flux and the ratio of the absorptivity to the emissivity; it is independent of the tube dimensions. Values for $\bar{T}(t)$ may be obtained by solving Eq. (4.65) numerically. The differential equation (4.63) for T_m depends nonlinearly on $\bar{T}(t)$. Experience has shown that, to solve (4.63) conveniently, a reasonable approximation is to assume $\bar{T}$ as a constant; we take $\bar{T}$ as the steady-state value $\bar{T}_{ss}$. After the initial condition is imposed, the solution may be written as

$$T_m(t) = \frac{1}{2}\frac{\alpha S}{\rho ch}\tau(1 - e^{-t/\tau}) \tag{4.67}$$

where τ is the thermal time constant defined by

$$\frac{1}{\tau} = \frac{k}{\rho c R^2} + \frac{4\sigma\varepsilon}{\rho ch}\left(\frac{\alpha_s S}{\pi\sigma\varepsilon}\right)^{3/4} \tag{4.68}$$

The steady-state value of T_m from Eq. (4.66) is

$$(T_m)_{ss} = \frac{1}{2}\frac{\alpha_s S}{\rho ch}\tau \tag{4.69}$$

Table 4.3 Tube properties for thermal response computations

	Tube			
Data	1 Graphite–epoxy	2 Aluminum 6061-T6	3 Stainless steel	4 Stainless steel
k, W/m-K	1.731	167	16.61	16.3
c, J/kg-K	1044	896	502	502.4
ρ, kg/m^3	1576	2700	7010	7930
ε	0.84	0.84	0.13	1
α	0.92	0.92	0.5	1
R, m	0.03812	0.03812	0.01092	0.00171
h, 10^{-4} m	7.14	7.14	2.35	0.8

The maximum temperature gradient ΔT across a tube diameter is then obtained as

$$\Delta T = 2T_m = \frac{\alpha S}{\rho c h}\tau \tag{4.70}$$

where τ is defined in Eq. (4.68).

4.8.2 Comparison with Finite Element Solution

The preceding analysis provides an approximate representation for the unsteady thermal response. The accuracy of the approximation rests heavily on the initial assumption that temperature varies with cosϕ around the tube circumference. The validity of this assumption depends on the properties of the material. To evaluate the accuracy of the analytical solutions, comparisons have been made with results from finite element numerical solutions to Eq. (4.59). The results suggest that the analytical solution is a good approximation for metals with relatively high thermal conductivity but a poor approximation for polymer matrix composites with low thermal conductivity.

Computations were performed for four tubes (Table 4.3) with $S = 7400$ W/m^2. The first tube is a graphite–epoxy composite, the second tube is aluminum, and third and fourth tubes are stainless steel but with different cross-sectional dimensions. Figures 4.36–4.39 present tube temperature distributions computed with Eq. (4.61) at selected times in the transient responses. More accurate temperatures computed with a detailed finite element model are shown for comparison. The results demonstrate that the approximate analytical temperature distribution, Eq. (4.61), does not accurately represent the composite tube's temperature-distribution, but it does give generally good to excellent results for the metallic tubes. Table 4.4 gives some additional results for the thermal response of the four tubes. The entry in the last row of Table 4.4, $k/\rho c R^2$, is a good parameter to characterize the accuracy of the approximate analytical solution. If this parameter is multiplied by time, the result is the Fourier number. The Fourier number is dimensionless and characterizes the ratio of the rate of heat transfer by conduction to the rate of energy stored. In this application, the accuracy of the approximate analytical solution improves with increasing Fourier number. The best agreement occurs for the fourth tube, where $k/\rho c R^2 = 0\ (1)$, and the worst agreement occurs for the composite tube where $k/\rho c R^2 = 0(10^{-3})$.

Table 4.4 Computed parameters for tube thermal response

	Tube			
	1 Graphite–epoxy	2 Aluminum 6061-T6	3 Stainless steel	4 Stainless steel
$\bar{T}_{ss}$, K	462	462	632	451
$1/\tau$, s^{-1}	0.0167	0.0584	0.0486	1.465
ΔT K	347	67.5	92.1	15.9
$k/\rho c R^2$, s^{-1}	0.000724	0.0475	0.0396	1.399

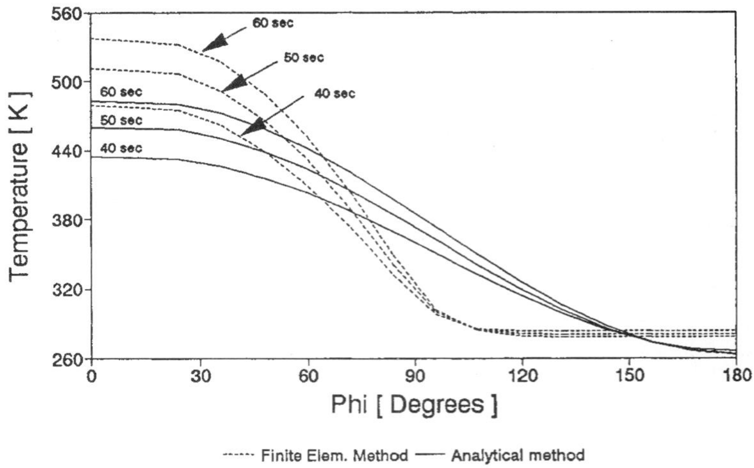

Fig. 4.36 Comparative thermal responses for graphite–epoxy, tube 1.

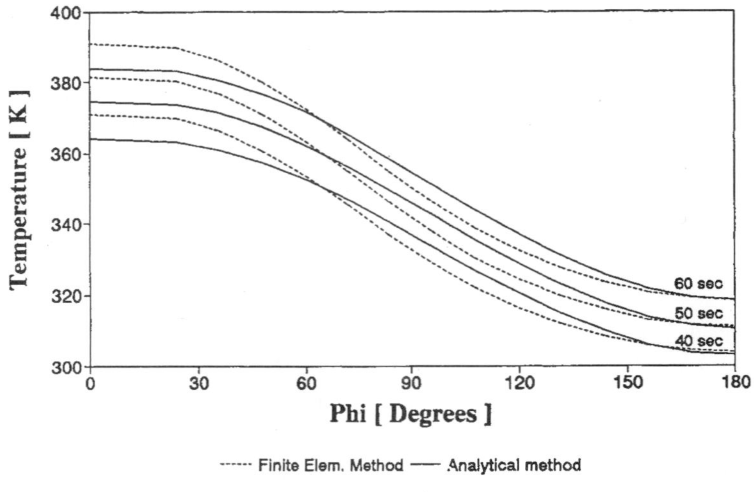

Fig. 4.37 Comparative thermal responses for aluminum, tube 2.

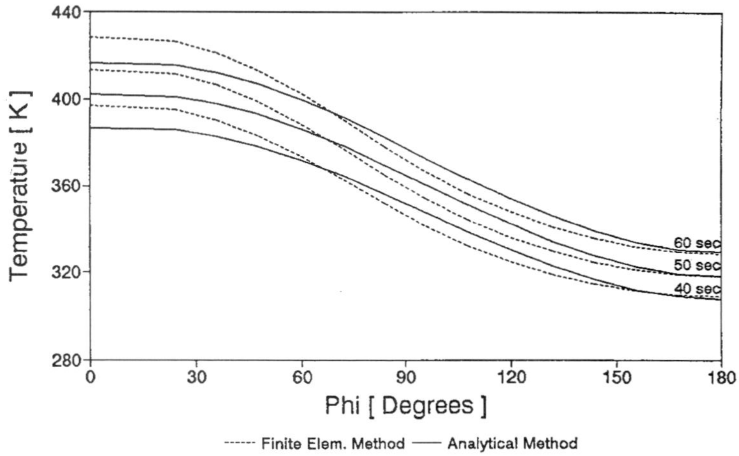

Fig. 4.38 Comparative thermal responses for stainless steel, tube 3.

4.9 Convectively Cooled Structures

Convectively cooled structures are strong candidates for use in hypersonic flight vehicles. For hypersonic flight, some leading edges and panels require active cooling systems to keep structural temperatures within acceptable ranges. The internal flow in the coolant passage has a predominant role in the thermal response of a hypersonic structure subject to external heating. A cross section of a typical convectively cooled structure is shown in Fig. 4.40. An aerodynamic skin and a coolant passage with internal heat exchanger protect the primary structure from the aerodynamic heating. The thin, typically metallic aerodynamic skin transfers the energy of the aerodynamic heating to a low-temperature coolant flow through the heat-exchanger fins that connect the aerodynamic skin to the primary structure. In typical engine structure, the coolant is cold hydrogen, which is later used as the propulsion system fuel.

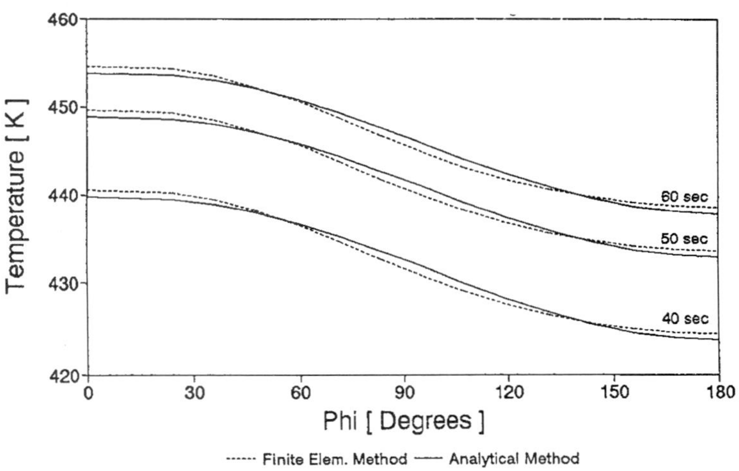

Fig. 4.39 Comparative thermal responses for stainless steel, tube 4.

Heat transfer in the aerodynamic skin consists of conduction combined with surface radiation. Heat transfer between the aerodynamic skin, the heat-exchanger surfaces, and the primary structure is by conduction at the solid/fluid interface. The dominant mode of heat transfer in the coolant flow is forced convection. The representation of the heat transfer in the coolant passage is the critical step in the heat-transfer analysis. Two basic representations can be employed. The first, denoted here as the engineering model, is based on a number of assumptions that greatly simplify the problem into a single energy equation with a specified mass flow rate. Detailed computation of the fluid velocity components and temperatures is not required. The second representation idealizes the coolant flow as a continuum model, and the partial differential equations describing conservation of mass, momentum, and energy are solved simultaneously to obtain fluid velocity and temperature distributions. This latter model is the most accurate, but it is also considerably more expensive than the engineering model. In this section, the engineering heat-transfer model is employed.

4.9.1 Engineering Model for Coolant Passages

The basic features of the engineering model are developed from the idealization shown in Fig. 4.41. A segment of the coolant passage of width w is shown; for simplicity, only the upper half of the coolant passage with the aerodynamic skin is shown. The engineering formulation[37–39] is based on the following assumptions:

1) The thermal-energy state of the fluid is characterized by the fluid bulk temperature T_F, which varies only in the flow direction, i.e., $T_F(x, t)$.

2) The flow is represented by the mass flow rate $\dot{m}$ in the coolant passage specified by $\dot{m} = \rho_F A_F V_F$, where ρ_F is the fluid density, A_F the cross-sectional area of the coolant passage, and V_F the coolant mean flow velocity.

3) A convection coefficient h is defined such that the heat flux $\dot{q}_F$ transferred between the structure and the coolant may be expressed as $\dot{q}_F = h(T_S - T_F)$, where $T_S(x, t)$ denotes the structural temperature at the fluid/solid interface.

4) The convection coefficient h may be expressed as a function of the fluid bulk temperature alone by using analytical/empirical equations for the Nusselt number, $Nu = hD/k_F$, where D is the hydraulic diameter of the coolant passage and k_F the thermal conductivity of the fluid coolant. The convection coefficient, as well as the solid and fluid thermal parameters, may, in general, be temperature-dependent.

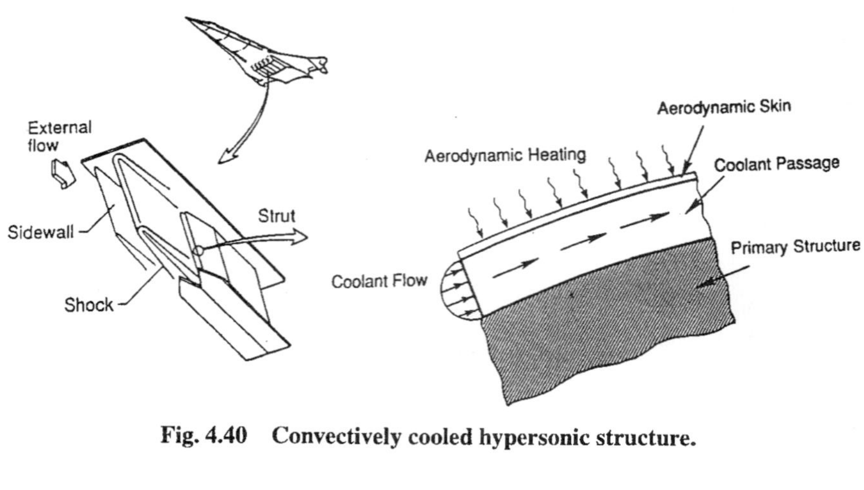

Fig. 4.40 Convectively cooled hypersonic structure.

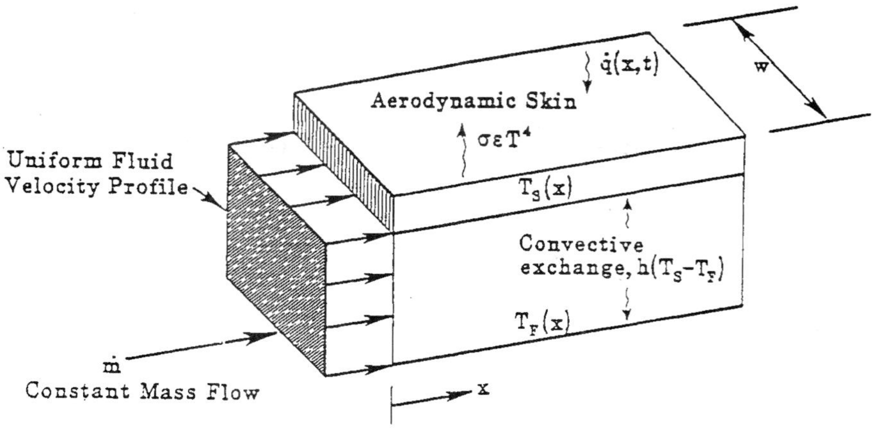

Fig. 4.41 Engineering model for coolant passage.

With these assumptions, energy balances on the aerodynamic skin and coolant give the governing conservation equations.

Fluid:

$$-\frac{\partial}{\partial x}\left(k_F A_F \frac{\partial T_F}{\partial x}\right) - wh(T_S - T_F) + \dot{m}c_F \frac{\partial T_F}{\partial x} + \rho_F c_F A_F \frac{\partial T_F}{\partial t} = 0 \quad (4.71a)$$

Solid:

$$-\frac{\partial}{\partial x}\left(k_S A_S \frac{\partial T_S}{\partial x}\right) + wh(T_S - T_F) + \sigma\varepsilon T_S^4 + \rho_S c_S A_S \frac{\partial T_S}{\partial t} = \dot{q} \quad (4.71b)$$

where the subscripts F and S denote the fluid and solid, respectively. In these equations, c denotes specific heat, σ is the Stefan–Boltzmann constant, and ε is the surface emissivity. Because of the temperature dependence of the thermal parameters and the radiation term, Eqs. (4.71a) and (4.71b) constitute a nonlinear set of partial differential equations.

The heat-exchanger fins are not included explicitly in this model. However, the heat transfer between the heat-exchanger fins and coolant can be taken into consideration through the use of an effective width w, which represents the area over which the convective heat exchange occurs.

4.9.2 Forced-Convection Problems

To illustrate basic features of heat transfer in convectively cooled structures, we present a few examples. The emphasis rests on analytical solutions for temperature distributions in the coolant flow because of the distinctive behavior that characterizes forced-convection heat transfer. We begin with steady-state solutions for temperatures in coolant passages with specified, constant wall temperatures $T_w = T_s$ to illustrate basic coolant temperature behavior and the parameters that characterize the heat transfer. These examples are followed by solutions to unsteady problems. These latter solutions have greater mathematical complexity, and we will make use of the Laplace transform technique. The analytical solutions are of limited value for practical applications, but they have an important role in the validation and evaluation of computational methods.

Example 4.15: Steady Heat Transfer in a Semi-Infinite Coolant Passage

We consider a coolant passage that extends to infinity in the positive x direction. We assume a steady-state solution for the temperature in the coolant fluid, i.e., $T_F = T_F(x)$. Then Eqs. (4.71a) and (4.71b) become uncoupled, and we may consider the fluid equation separately. Assuming constant thermal properties, we may write the fluid conservation of energy equation as

$$k_F A_F \frac{\mathrm{d}^2 T_F}{\mathrm{d}x^2} + \rho_F A_F c_F \frac{\mathrm{d}T_F}{\mathrm{d}x} + wh(T_F - T_w) = 0 \tag{4.72a}$$

or, alternatively,

$$-\frac{\mathrm{d}^2 T_F}{\mathrm{d}x^2} + \frac{Pe}{D}\frac{\mathrm{d}T_F}{\mathrm{d}x} + \frac{4Nu}{D^2}(T_F - T_w) = 0 \tag{4.72b}$$

where we have introduced the Peclet number Pe defined by

$$Pe = \frac{\rho_F c_F V_F D}{k_F} \tag{4.73}$$

where D is the hydraulic diameter of the coolant passage. By definition, $D = 4A_F/P_F$ is four times the ratio of the area of the flow cross section A_F to the perimeter P_F of the flow passage. The Peclet number characterizes the ratio of mass-transport convection heat transfer to axial conduction heat transfer in the fluid. High Peclet number flows correspond to convection-dominated flows; low Peclet numbers correspond to conduction-dominated flows.

Two boundary conditions are required to solve the second-order ordinary differential equation. We usually know the inlet temperature of the coolant flow but, normally, the outlet temperature is unknown. Instead, we specify that the conduction heat flux be zero at the outflow. By Fourier's law, zero heat flux is equivalent to specifying that the temperature gradient $\mathrm{d}T_F/\mathrm{d}x$ be zero at the outflow. Thus, for a semi-infinite coolant passage, we take the boundary conditions as

$$T_F(0) = T_0, \qquad \frac{\mathrm{d}T_F}{\mathrm{d}x}(\infty) = 0$$

where T_0 is the inlet fluid temperature. Using standard methods, the solution to Eq. (4.71b) subject to the preceding boundary conditions may be written

$$\frac{T_F(x) - T_w}{T_0 - T_w} = \exp\left[\left(Pe - \sqrt{Pe^2 + 16Nu}\right)x/2D\right] \tag{4.74}$$

Equation (4.74) shows that the coolant temperature increases smoothly from the inlet temperature T_0. This behavior is characteristic of many coolant temperature responses.

Example 4.16: Steady Heat Transfer in a Finite Length Coolant Passage

For a coolant passage of finite length L, we solve Eq. (4.73) subject to the boundary conditions

$$T_F(0) = T_0 \qquad \frac{dT_F}{dx}(L) = 0$$

The solution is

$$\frac{T_F(x) - T_w}{T_0 - T_w} = \cosh(m_1 x) - \frac{m_2 \sinh(m_2 L)}{m_1 \cosh(m_1 L)} \sinh(m_1 x) \tag{4.75a}$$

where

$$m_{1,2} = \frac{1}{2D}\left[Pe \pm \sqrt{Pe^2 + 16Nu}\right] \tag{4.75b}$$

In the two solutions given in Eqs. (4.74) and (4.75), if the wall temperature is greater than the fluid inflow temperature, i.e., $T_w > T_0$, the wall is convectively cooled with convection heat transfer from the wall to the fluid. Of course, these equations also apply when $T_0 > T_w$, and then the wall is convectively heated with convection heat transfer from the fluid to the wall.

Example 4.17: Unsteady Mass-Transport Convection

We once again consider a coolant passage that extends to infinity in the positive x direction. Convective heat transfer between the structure and coolant is neglected; i.e., we take the convection coefficient $h = 0$. As an additional assumption, we neglect axial conduction heat transfer in the coolant; i.e., we take the fluid thermal conductivity $k_F = 0$. With these assumptions, the conservation of energy equation for the fluid reduces to

$$V_F \frac{\partial T_F}{\partial x} + \frac{\partial T_F}{\partial t} = 0 \tag{4.76}$$

Equation (4.76) represents the transport of thermal energy due to the fluid mean flow velocity V_F. Alternatively, we may say that it is the convection of thermal energy due to mass-transport convection. Sometimes, Eq. (4.76) is called the *convection* or *advection equation*. In other words, it represents convection heat transfer due to the forced motion of the fluid. Mathematically, the equation is interesting because it is a first-order, *hyperbolic* partial differential equation. Because it is first-order, we need prescribe only one boundary condition. The hyperbolic character of the equation means that initial temperatures are propagated with the finite speed V_F of the fluid. As an illustration, consider the problem posed by Eq. (4.76) with the boundary condition

$$\frac{\partial T_F}{\partial x}(0, t) = 0$$

and the initial condition

$$T_F(x, 0) = T_0(x)$$

The solution is simply

$$T_F(x,t) = T_0(x - V_F t) \tag{4.77}$$

which, for $V_F > 0$, means that the initial temperature distribution is translated downstream as a result of the fluid motion. The shape and amplitude are unchanged because there is no heat-transfer mechanism such as a conduction to cause the temperature distribution to "spread" or diffuse through the fluid. The initial temperature "convects" with unaltered shape downstream.

Example 4.18: Unsteady Mass-Transport Convection with Fluid Conduction

If we continue to neglect the convective heat transfer between the structure and coolant but include the coolant axial conduction, Eq. (4.71a) reduces to

$$-k_F \frac{\partial^2 T_F}{\partial x^2} + \rho_F c_F V_F \frac{\partial T_F}{\partial x} + \rho_F c_F \frac{\partial T_F}{\partial t} = 0 \tag{4.78a}$$

or, alternatively,

$$-\frac{\partial^2 T_F}{\partial x^2} + \frac{Pe}{D}\frac{\partial T_F}{\partial x} + \frac{1}{\kappa}\frac{\partial T_F}{\partial t} = 0 \tag{4.78b}$$

where we use the Peclet number and hydraulic diameter as in Examples 4.15 and 4.16. As before, the thermal diffusivity is denoted by κ. Equation (4.78) is basically Eq. (4.76) with the additional second-order conduction heat-transfer term. Equation 4.78 is frequently called the *convective-diffusion equation* and, along with a two-dimensional counterpart, it is often used as a model equation for evaluating the performance of computational algorithms. In lighter moments, Eq. (4.78) has also been called the "defective-confusion" equation. Mathematically, for finite fluid thermal conductivity, Eq. (4.78) is a second-order, *parabolic* partial differential equation. Two boundary conditions and an initial condition are required.

Consider the boundary/initial-value problem posed by Eq. (4.78b), with the boundary conditions

$$T_F(0,t) = T_0 \qquad T_F(\infty, t) = 0$$

and initial conditions

$$T_F(x,0) = T_0 \qquad x \le 0$$

$$T_F(x,0) = 0 \qquad x > 0$$

The problem statement is given in Figs. 4.42a and 4.42b.

Equation (4.78b) is solved subject to the preceding boundary and initial conditions using the Laplace transform defined in Sec. 4.4.2. Performing the transform and then introducing the initial and boundary conditions yield

$$\bar{T}(x,s) = T_0 \exp\left(\frac{Pex}{D}\right) \exp\left[-\frac{1}{s}\frac{x}{\sqrt{\kappa}}\sqrt{s + \frac{\kappa Pe^2}{4D^2} + \frac{4Nu\kappa}{D^2}}\right]$$

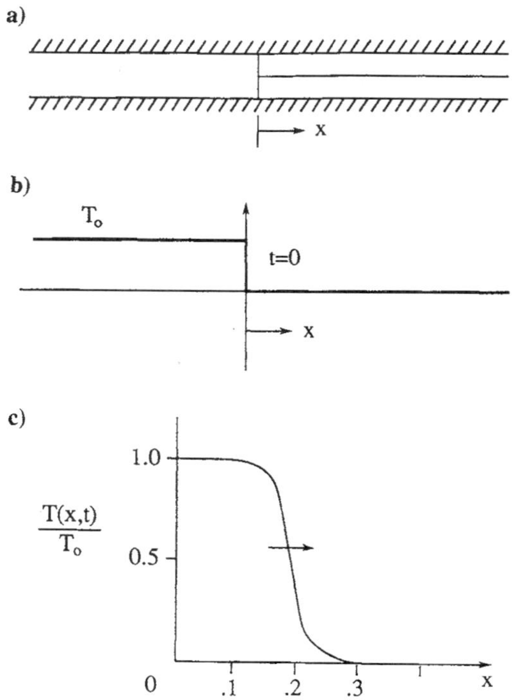

Fig. 4.42 Unsteady mass-transport convection with fluid convection, Pe = 10.

where $\bar{T}(x, s)$ denotes the Laplace transform of $T(x, t)$, and s denotes the transform parameter. The inversion is accomplished with the aid of a transform pair from Ref. 40, The final solution is

$$\frac{T_F(x,t)}{T_0} = \frac{1}{2}\left[\exp(Pe x/D)\operatorname{erfc}\left(\frac{x}{2\sqrt{\kappa t}} + \frac{Pe\sqrt{\kappa t}}{2D}\right)\right.$$
$$\left. + \operatorname{erfc}\left(\frac{x}{2\sqrt{\kappa t}} - \frac{Pe\sqrt{\kappa t}}{2D}\right)\right] \tag{4.79}$$

where erfc ( ) denotes the complementary error function defined in Eq. (4.36). This analytical solution represents a translating wave with diffusion at the leading and trailing edges (Fig. 4.42c).

Example 4.19: Unsteady Mass-Transport Convection, Conduction, and Surface Convection

We now consider the most general case described by Eq. (4.71b), the combination of mass-transport convection, conduction, and surface convection. In terms of the characteristic nondimensional parameter, the fluid-energy equation (4.70a) is

$$-\frac{\partial^2 T_F}{\partial x^2} + \frac{Pe}{D}\frac{\partial T_F}{\partial x} + \frac{4Nu}{D^2}(T_F - T_w) + \frac{1}{\kappa}\frac{\partial T_F}{\partial t} = 0 \tag{4.80}$$

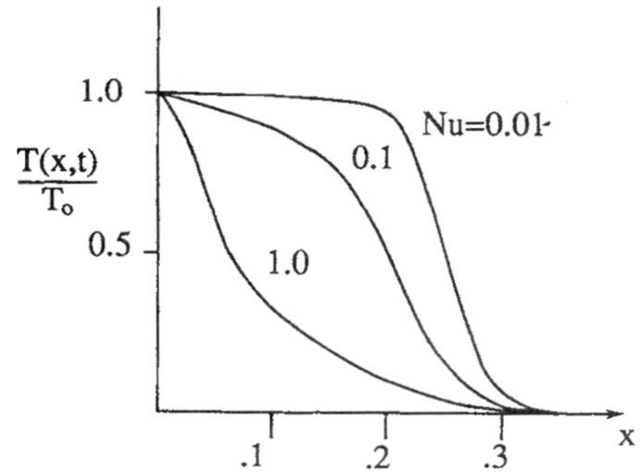

Fig. 4.43 Unsteady mass-transport convection, conduction, and surface convection, *Pe* = 10.

We solve Eq. (4.80) for the boundary and initial conditions of the preceding example. Using the Laplace transform technique as before, we obtain a similar but lengthier solution:

$$\begin{aligned}T(x,t) = {}& \frac{T_0 - T_w}{2}\exp(Pex/2D)\left[\exp(\lambda x/D)\operatorname{erfc}\left(\frac{x}{2\sqrt{\kappa t}} + \frac{\lambda\sqrt{\kappa t}}{D}\right)\right.\\ & \left.+ \exp(-\lambda x/D)\operatorname{erfc}\left(\frac{x}{2\sqrt{\kappa t}} - \frac{\lambda\sqrt{\kappa t}}{D}\right)\right]\\ & + \frac{T_w}{2}\exp(-4Nu\kappa t/D^2)\left[\exp(xPe/D)\operatorname{erfc}\left(\frac{x}{2\sqrt{\kappa t}} + \frac{Pe\sqrt{\kappa t}}{2D}\right)\right.\\ & \left.+ \operatorname{erfc}\left(\frac{x}{2\sqrt{\kappa t}} - \frac{Pe\sqrt{\kappa t}}{2D}\right)\right] + T_w[1 - \exp(-4Nu\kappa t/D^2)]\end{aligned} \tag{4.81}$$

where

$$\lambda = \sqrt{Pe^2/4 + 4Nu}$$

For $Nu = 0$ and $T_w = 0$, Eq. (4.81) reduces to the solution given in Example 4.18, Eq. (4.79). A typical thermal response predicted by Eq. (4.81) for $T_w = 0$ is sketched in Fig. 4.43. For small values of the Nusselt number, the thermal response is similar to the result shown in Fig. 4.42. As the Nusselt number increases, however, the character of the response changes from that of a translating wave with rounded edges to a response similar to a diffusion problem.

4.10 Closing Remarks

This chapter describes heat transfer in structures from the basic problem formulation to mathematical solutions for a variety of boundary/initial-value problems. Analytical solutions are given for several problems that find applications in aerospace structures. Computational methods are introduced, and a few numerical results are described, but the emphasis in this chapter rests on the analytical solution of conduction problems for a variety of regions and boundary conditions. Solutions are presented that are relevant to thin-walled aircraft and spacecraft components. Additional computational results will be presented in Chapter 8 where the finite element method is formulated and applied. Many of the analytical solutions developed in this chapter will be used in later chapters as we determine the response of continuum models and structural components to prescribed temperature distributions.

References

[1]Rohsenow, W. M., and Hartnett, J. P. (eds.), *Handbook of Heat Transfer*, McGraw–Hill, New York, 1973.

[2]Eckert, E. R. G., and Drake, R. M., Jr., *Analysis of Heat and Mass Transfer*, McGraw–Hill, New York, 1972.

[3]Chapman, A. J., *Heat Transfer*, 3rd ed., Macmillan, New York, 1974.

[4]Kreith, F., *Principles of Heat Transfer*, 3rd ed., Intext, New York, 1973.

[5]Pitts, D. R., and Sissom, L. E., *Theory and Problems of Heat Transfer*, Schaum's Outline Series, McGraw–Hill, New York, 1977.

[6]Özisik, M. N., *Boundary Value Problems of Heat Conduction*, International, Scranton, PA, 1968.

[7]Arpaci, V. S., *Conduction Heat Transfer*, Addison–Wesley, Reading, MA, 1966.

[8]Myers, C. E., *Analytical Methods in Conduction Heat Transfer*, McGraw–Hill, New York, 1971.

[9]Kays, W. N., *Convection Heat and Mass Transfer*, McGraw–Hill, New York, 1966.

[10]Bird. R. B., Stewart, W. E., and Lightfoot, E. N., *Transport Phenomena*, Wiley, New York, 1960.

[11]Ostrach, S., "Natural Convection in Enclosures," *Advances in Heat Transfer*, Vol. 8, edited by J. P. Hartnett and T. F. Irvine Jr., Academic, New York, 1972.

[12]Sparrow, E. N., and Cess, R. D., *Radiation Heat Transfer*, aug. ed., McGraw–Hill, New York, 1978.

[13]Love, T. J., *Radiative Heat Transfer*, Merrill, Columbus, OH, 1968.

[14]Siegel, R., and Howell, J. R., *Thermal Radiation Heat Transfer*, Vol. I-III, NASA SP-164, 1968; also available as *Thermal Radiation Heat Transfer*, McGraw–Hill, New York, 1980.

[15]Ozisik, M. N., *Radiative Transfer and Interactions with Conduction and Convection*, Wiley-Interscience, New York, 1973.

[16]Carslaw, H. S., and Jaeger, J. G., *Conduction of Heat in Solids*, 2nd ed., Oxford Univ., Press, Oxford, England, UK, 1959.

[17]Churchill, R. V., *Modern Operational Mathematics in Engineering*, 2nd ed., McGraw–Hill, New York, 1958.

[18]Adelman, H. M. (ed.), *Computational Aspects of Heat Transfer in Structures*, Proceedings of a symposium held at NASA Langley Research Center (Hampton, VA), Nov. 1981, NASA CP-2216.

[19]Anderson, D. A., Tannehill, J. C., and Pletcher, R. H., *Computational Fluid Mechanics and Heat Transfer*, Hemisphere, Washington, DC, 1984.

[20]Huebner, K. H., Thornton, E. A., and Byrom, T. G., *The Finite Element Method for Engineers*, 3rd ed., Wiley, New York, 1995.

[21]Patankar, S. V., *Numerical Heat Transfer and Fluid Flow*, Hemisphere, Washington, DC, 1980.

[22]Gatewood, B. E., *Thermal Stresses*, McGraw–Hill, New York, 1957.

[23]Boley, B., and Weiner, J. H., *Theory of Thermal Stresses*, Wiley, New York, 1960.

[24]Schuh, H., *Heat Transfer in Structures*, Pergamon, Oxford, England, UK, 1965.

[25]Przemieniecki, J. S, "Transient Temperature Distributions and Thermal Stresses in

Fuselage Shells with Bulkheads or Frames," *Journal of the Royal Aeronautical Society*, Vol. 60, Dec. 1956, pp. 799–804.

[26]Kaye, J., "The Transient Temperature Distribution in a Wing Flying at Supersonic Speeds," *Journal of the Aeronautical Sciences*, Vol. 17, No. 12, 1950, pp. 787–808.

[27]Kaye, J., and Yeh, V. C. M., "Design Charts for Transient Temperature Distribution Resulting from Aerodynamic Heating at Supersonic Speeds," *Journal of the Aeronautical Sciences*, Vol. 22, No. 11, pp. 765–762, 786, 1955.

[28]Hoff, N. J., "Structural Problems of Future Aircraft," *Proceedings of the Third Anglo-American Conference* (Brighton, England, UK), Royal Aeronautical Society, London, 1951, pp. 77–114.

[29]Pohle, F. V., and Oliver, H., "Temperature Distribution Thermal Stresses in a Model of a Supersonic Wing," *Journal of the Aeronautical Sciences*, Vol. 21, No. 1, 1954, pp. 8–16.

[30]N'Guyen, V. P., and Ripley, B. L., "Design Philosophy and Fatigue Testing of Concorde," *Proceedings of the Fifth ICAF Symposium*, Melbourne, Australia, May 1967. pp. 403–437.

[31]Ko, W. L., Quinn, R. D., Gong, L., Schuster, Lawrence S., and Gonzales, D., "Reentry Heat Transfer Analysis of the Space Shuttle Orbiter," *Computational Aspects of Heat Transfer in Structures*, edited by Howard M. Adelman, NASA CP-2216, Nov. 1981, pp. 295–325.

[32]Ko, W. L., Quinn, R. D., and Gong, L., "Finite Element Reentry Heat Transfer Analysis of Space Shuttle Orbiter," NASA, TP-2657, Dec. 1986.

[33]Ko, W. L., and Olona, T., "Effect of Element Size on the Solution Accuracies of Finite Element Heat Transfer and Thermal Stress Analyses of Space Shuttle Orbiter," NASA TM-88292, Aug. 1987.

[34]Ko, W. L., and Fields, R. A., "Thermal Stress Analysis of Space Shuttle Orbiter Subjected to Reentry Aerodynamic Heating," NASA TM-88286, Oct. 1987.

[35]Vigneron, F. R., "Thermal Curvature Time Constants of Thin-Walled Tubular Spacecraft Booms," *Journal of Spacecraft and Rockets*, Vol. 7, No. 10, 1970, pp. 1256–1259.

[36]Rimrott, R. P. J., and Abdel-Sayed, R., "Flexural Thermal Flutter under Laboratory Conditions," *Transactions of Canadian Society of Mechanical Engineers*, Vol. 4, No. 4, 1976–1977, pp. 189–196.

[37]Thornton, E. A., and Wieting, A. R., "Finite Element Methodology for Transient Conduction/Forced Convection Thermal Analysis," *Heat Transfer*, Thermal Control and Heat Pipes, edited by W. B. Olstad, Vol. 70, Progress in Astronautics and Aeronautics, AIAA, New York, 1980, pp. 77–103.

[38]Thornton, E. A., and Wieting, A. R., "Evaluation of Finite Element Formulations for Transient Conduction Forced Convection Analysis," *Numerical Heat Transfer*, Vol. 3, No. 3, 1980, pp. 281–295.

[39]Wieting, A. R., and Thornton, E. A., "Finite Element Thermal Analysis of Convectively-Cooled Aircraft Structures," *Numerical Methods in Heat Transfer*, edited by R. W. Lewis, K. Morgan, and O. C. Zienkiewicz, Wiley, New York, 1981, pp. 431–443.

[40]Erdélyl, A., Magnus, M., Oberhettinger, F., and Tricomi, F. G., *Tables of Integral Transforms*, McGraw–Hill, New York, 1954, p. 246, Tranform pair (10).

Problems

4.1. A segment of a thin-walled structure is heated aerodynamically on its external surface by a heat flux $q = h[T_{aw} - T(t)]$, where the convective coefficient h is constant, T_{aw} is the known adiabatic wall temperature, and $T(t)$ is the unknown wall temperature. The heated surface emits radiant energy to space and is cooled on its internal surface by forced convection.

(a) Assuming that the temperature of the thin-walled structure is uniform, i.e., $T(t)$, write an ordinary differential equation for determining the history for the unknown temperature.

(b) Neglecting radiation, obtain a solution for $T(t)$ assuming an initial condition, $T(0) = T_0$. What is the steady-state temperature of the structure?

(c) Investigate the possibility of obtaining an analytical solution for $T(t)$, including radiation.

4.2. A small billet of metal at a uniform initial temperature T_0 is placed in a container of a fluid with temperature T_f and convection coefficient h. Assume that the billet has uniform temperature that depends only on time, i.e., $T(t)$. Using an energy balance, derive an ordinary differential equation for $T(t)$. At time equal to zero, the temperature of the fluid T_f begins to increase linearly with time, say $T_f = at$ where a is a constant. Determine the solution for $T(t)$. Sketch the temperature histories for the fluid and the billet. Derive an equation for the temperature difference $T_f - T$ for large time.

4.3. Show that the one-dimensional conservation of energy equation

$$-k\frac{\partial^2 T}{\partial x^2} + \rho c\frac{\partial T}{\partial t} = Q(x, t)$$

is a parabolic partial differential equation.

(a) What is the physical significance of the equation's being parabolic?

(b) What is the mathematical significance?

4.4. An insulated rod of length L built-in between two walls is heated by constant internal heat generation. The boundary conditions are $T(0) = T_1$ and $T(L) = T_2$. Derive an equation for the steady temperature distribution $T(x)$.

4.5. The penetration depth for heat conduction in a semi-infinite solid is defined as $\delta(t)$. Reference 2, pp. 183–185, derives an equation for the penetration depth as $\delta = \sqrt{12\kappa t}$, where $\kappa = k/\rho c$ is the thermal diffusivity, and t is time.

(a) Write out the details of the derivation for $\delta(t)$.

(b) How do you explain the inconsistency between the infinite propagation speed of the parabolic heat-conduction equation and the penetration depth concept?

(c) Compute penetration depths for steel and aluminum for $t = 1, 10$, and 100 s. Express the penetration depths in meters and inches.

4.6. An insulated rod is cooled on its ends so that the boundary conditions are $T(0, t) = 0$ and $T(L, t) = 0$. For time less than or equal to zero, the rod is being heated so that, initially,

$$T(x, 0) = T_0 \sin(\pi x/L)$$

After $t = 0$, the heating is shut off. Derive an equation that gives $T(x, t)$ for $t > 0$.

4.7. A thin plate of thickness d is heated on the end $x = 0$ by a constant heat flux q and loses energy convectively along its length. Solve the steady heat-conduction problem

$$k\frac{\mathrm{d}^2T}{\mathrm{d}x^2} - \frac{h}{d}(T - T_\infty) = 0$$

with boundary conditions

$$-k\frac{\partial T}{\partial x}(0) = q \qquad k\frac{\partial T}{\partial x}(L) = 0$$

In the problem statement, k is the thermal conductivity, h a constant convection coefficient, and T_∞ the known convective exchange temperature. Sketch the expected temperature distribution.

4.8. Solve the unsteady heat-conduction equation

$$\frac{\partial^2 T}{\partial x^2} = \frac{1}{\kappa}\frac{\partial T}{\partial t}$$

for the boundary conditions

$$k\frac{\partial T}{\partial x}(0, t) = q \qquad T(L, t) = T_0$$

The thermal diffusivity is κ, and q is the heat flux. For time $t < 0$, $q = 0$, and q is constant for $t \geq 0$. The initial condition is $T(x, 0) = T_0$. Sketch the expected distribution for the temperature response for several time values.

4.9. Solve the preceding problem with the boundary condition of zero heat flux at $x = L$, i.e.,

$$k\frac{\partial T}{\partial x}(L, t) = 0$$

instead of specified temperature. Sketch the expected temperature response.

4.10. Solve the boundary/initial-value heat-conduction problem given by

$$k\frac{\partial^2 T}{\partial x^2} = \rho c\frac{\partial T}{\partial t}$$

where the boundary conditions are

$$q(0, t) = -k\frac{\partial T}{\partial x}(0, t) = 0$$

$$h[T_s(t) - T(L, t)] = k\frac{\partial T}{\partial x}(L, t)$$

$$T_s(t) = \frac{T_m}{t_1}t \qquad 0 < t < t_1$$

The initial condition is

$$T(x, 0) = T_0(x)$$

Use an approach in which the solution is separated into two parts.

4.11. Consider steady heat conduction into a hollow circular cylinder of inner radius a and outer radius b. If the inner radius $r = a$ is kept at temperature T_1 and the outer radius is kept at temperature T_2, show that the temperature distribution is given by

$$T(r) = \frac{T_1 \ln(b/r) + T_2 \ln(r/a)}{\ln(b/a)}$$

4.12. Consider steady heat conduction in a hollow circular cylinder where the inner radius $r = a$ is kept at temperature T_1, and the outer radius $r = b$ experiences a convective heat exchange. The boundary condition at $r = b$ is

$$-k\frac{dT}{dr}(b) = h[T(b) - T_\infty]$$

Derive an equation for $T(r)$.

4.13. A solid circular cylinder has an initial temperature T_0. For $t \geq 0$, the cylinder is heated by a constant heat flux q_s. Derive an equation for the transient temperature response $T(r, t)$. Use the separation of variables approach.

4.14. In Problem 1.4, aerodynamic heating rates were computed for a wing of a supersonic transport. For a 290-min mission, the convection coefficient h and the adiabatic wall temperature T_{aw} were computed for several times over the mission profile; both laminar and turbulent flows are assumed. The objective of this assignment is to compute the temperature history and temperature distributions in a wing spar using h and T_{aw} from the previous computations. The computational approach should:

(a) Assume turbulent flow. Assume the heating rate for $x = 60$ in.

(b) Use the Hoff model of the wing spar described in Sec. 4.7. In the Hoff model (Example 4.11), the aerodynamic skins are convectively heated and are assumed to have uniform temperature. The spar web experiences one-dimensional conduction and has a symmetrical temperature distribution; the top and bottom edges of the web have prescribed temperatures equal to the temperatures of the aerodynamic skins. Radiation and internal convection are neglected.

(c) For your computations, time-march the solution through the mission profile assuming constant h and a linear variation of T_{aw} over each time step, to compute the time history of the aerodynamic skin temperature $T_s(t)$. Over each time step, use an analytical solution to the differential equation based on constant h and linear T_{aw}. Use the solution recursively to advance the solution for T_s through the mission profile.

(d) To compute the web temperature distribution, assume that $T_s(t)$ varies linearly over each time step, and derive an analytical solution for the partial differential equation for $T_w(x, t)$. Use this solution recursively to time-march the solution through the mission computing temperature profiles in the web at key times.

(e) Discuss your results in a brief report that should include 1) a description of your solutions to the differential equations, 2) the temperature history for the aerodynamic skin, 3) web temperature distributions at key times in the mission profile, and 4) a brief discussion of the effectiveness of the computational approach and interpretation of the temperature distributions.

Data

Material:	Titanium	
Spar Dimensions:	Skin width	$= 2b = 120$ in.
	Skin thickness	$= H = 1.125$ in.
	Web depth	$= 2a = 30$ in.
	Web thickness	$= d = 0.375$ in.

Suggestion: Compute temperatures in SI units, but convert to degrees Fahrenheit for graphic displays.

4.15. Example 4.11 discusses analytical approaches for the determination of temperatures in a multicell wing. A formulation is given for a nonuniform, skin temperature model (Fig. 4.25b) is given in a series of equations. In the precomputer days of 1954, the problem was solved by Pohle and Oliver[28] using the Laplace transform technique, which led to a complicated analytical solution.

(a) Using a finite difference model, compute transient temperature distributions for the nonuniform skin temperature model using the data for a titanium wing given in the preceding problem.

(b) Plot transient distributions of the temperatures in the skin and web and compare your results with the Pohle and Oliver results for steel in Fig. 4.26. Discuss the validity of the uniform skin temperature model for the titanium wing.

4.16. Consider a thin-walled tube with radiation similar to the problem analyzed in Sec.4.8. The tube has a radius R and wall thickness h (Fig. 4.35). In a spin-stabilized spacecraft, the tube cross section is rotating counter clockwise at a constant angular velocity Ω. The tube experiences a constant solar incident heat flux denoted by S. We wish to determine the steady-state temperature distribution in the rotating tube.

(a) Write the appropriate conservation of energy for the tube.

(b) Following the procedure described in Sec. 4.8, assume $T(\phi, t) = \bar{T}(t) + \tilde{T}(\phi, t)$ and then derive governing ordinary differential equations for the tube temperature $\bar{T}(t)$ and the perturbation temperature $\tilde{T}(\phi, t)$.

(c) Show that the average temperature is independent of the angular velocity and is given by Eq. (4.66).

(d) Show that the perturbation temperature is given by

$$\tilde{T} = \tilde{T}_{\max} \cos(\phi + \Omega t - \phi_{\mathrm{d}})$$

where

$$\tilde{T}_{\max} = T_m / \sqrt{1 + \Omega^2 \tau^2}$$

$$\phi_{\mathrm{d}} = \tan^{-1}(\Omega t)$$

and where T_m is given by Eq.(4.69) and τ is defined by Eq. (4.66).

(e) Discuss the physical interpretation of the solution.

4.17. Steady heat transfer in a coolant passage is described by the boundary-value problem

$$-\frac{\mathrm{d}^2 T_F}{\mathrm{d}x^2} + \frac{Pe}{D}\frac{\mathrm{d}T_F}{\mathrm{d}x} = 0$$

with boundary conditions

$$T_F(0) = T_0 \qquad T_F(L) = 0$$

(a) Derive an analytical solution for the temperature $T_F(x)$.

(b) Select a hydraulic diameter D, and graph the temperature distributions T_F vs x for a range of Peclet numbers from $Pe = 0.1$ to $Pe = 25$. Discuss the character of the solution for very low and very high Peclet numbers.

4.18. The flow geometry and terminology for two steady merging flows of the same coolant are shown in Fig. P4.18. The flow passages are assumed to have the same cross-sectional areas and heated perimeters. The temperatures $T_i(x_i, t)$ in the ith flow satisfy

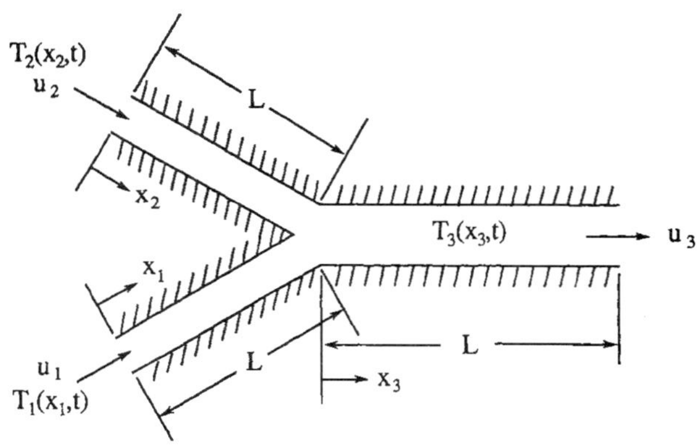

Fig. P4.18

$$-kA\frac{d^2T_i}{dx_i^2} + \rho c A u_i \frac{dT_i}{dx_i} + hP(T_i - T_w) = 0 \qquad i = 1, 2, 3$$

where P denotes the perimeter of a heated passage, and the wall temperature is the same for all three passages. At the inflow, the temperature boundary conditions are

$$T_1(0) = T - 1 \qquad T_2(0) = T_2$$

and, at the outflow,

$$\frac{\partial T_3}{\partial x_3}(L) = 0$$

(a) Write appropriate continuity conditions for the temperatures at the confluence of the flows at $x_1 = x_2 = L$ and $x_3 = 0$.

(b) Derive analytical solutions for the coolant passage temperature distributions; i.e., determine $T_i(x_i)$ for $i = 1, 2, 3$.

(c) Compute and graph the temperature distributions in the three coolant passages for

$$T_2 = 2T_1 \qquad u_1 = u_2 \qquad T_w = 0$$
$$Pe_1 = Pe_2 = \frac{\rho c u_1 D}{k} = 2.5$$

and $Nu = 0.1$, 1.0, and 10.0. Select a convenient hydraulic diameter.

Selected Bibliography: Temperatures in Aircraft Structures

For convenience of reference, a number of publications are listed that describe the determination of temperatures in aircraft structures. The publications are listed in chronological order.

Kaye, J., "The Transient Temperature Distribution in a Wing Flying at Supersonic Speeds," *Journal of the Aeronautical Sciences*, Vol. 17, No. 12, 1950, pp. 787–808.

Hoff, N. J., "Structural Problems of Future Aircraft," *Proceedings of the Third Anglo-American Aeronautical Conference* (Brighton, England, UK), Royal Aeronautical Society, London, 1951, pp. 77–114.

Eisenhardt, G. H., and Rohsenow, W. M., "Calculation of Thermal Stresses in a Wedge-Shaped Wing," *Journal of the Aeronautical Sciences*, Vol. 18, Feb. 1951, pp. 115–123.

Durham, F. P., "The Effect of Flight and Configuration Variables on Thermal Stresses in Diamond-Shaped Supersonic Wings," *Journal of the Aeronautical Sciences*, Vol. 18, Nov. 1951, pp. 755–766.

Barzelay, M. E., and Boison, J. C., "Investigation of Stresses Due to Thermal Gradients in Typical Aircraft Structures," NACA RM 51K06, Jan. 1952.

Parkes, E. W., "The Alleviation of Transient Thermal Stresses," *Aircraft Engineering*, Vol. 25, No. 28, 1953, pp. 51–53.

Parkes, E. W., "Transient Thermal Stresses in Wings," *Aircraft Engineering*, Vol. 25, No. 288, 1953, pp. 373–378.

Pohle, F. V., and Oliver, H., "Temperature Distribution and Thermal Stresses in a Model of a Supersonic Wing," *Journal of the Aeronautical Sciences*, Vol. 21, No. 1, 1954, pp. 8–16.

Barzelay, M., Tong, K. N., and Hollo, G., "Thermal Conductance of Contacts in Aircraft Joints," NACA Technical Note 3167, March 1954.

Schuh, H., "On the Calculation of Temperature Distribution and Thermal Stresses in Parts of Aircraft Structures of Supersonic Speeds," *Journal of the Aeronautical Sciences*, Vol. 21, No. 8, 1954, pp. 575–576.

Gerard, G., "Life Expectancy of Aircraft under Thermal Flight Conditions," *Journal of the Aeronautical Sciences*, Vol. 21, No. 10, 1954, pp. 67S–680.

Przemieniecki, J. S., "Transient Temperatures and Stresses in Plates Attained in High-Speed Flight," *Journal of the Aeronautical Sciences*, Vol. 22, No. 5, 1955, pp. 345–348.

Kotanchik, J. N., Johnson, A. E., Jr., and Ross, R. D., "Rapid Radiant-Heating Tests of Multiweb Beams," NACA T-3474, Sept. 1955.

Kaye, J., and Yeh, V. C. M., "Design Charts for Transient Temperature Distribution Resulting from Aerodynamic Heating at Supersonic Speeds," *Journal of the Aeronautical Sciences*, Vol. 22, No. 11, 1955, pp. 755–762, 786.

Schuh, H., "Transient Temperature Distributions and Thermal Stresses in a Skin-Shear Web Configuration at High-Speed Flight for a Wide Range of Parameters," *Journal of the Aeronautical Sciences*, Vol. 22, No. 12, 1955, pp. 829–836.

Griffith, G. E., and Miltonberger, G. H., "Some Effects of Joint Conductivity on the Temperatures and Thermal Stresses in Aerodynamically Heated Skin-Stiffener Combinations," NACA TN-3699, 1956.

Boley, B. A., "The Determination of Temperatures, Stresses and Deflections in Two-Dimensional Thermoelastic Problems," *Journal of the Aeronautical Sciences*, Vol. 23, No. 1, 1956, pp. 67–75.

Bisplinghoff, R. L., "Some Structural and Aeroelastic Considerations of High Speed Flight," (Nineteenth Wright Brothers Lecture), *Journal of the Aeronautical Sciences,* Vol. 23, No. 4, 1956, pp. 289–329.

Hoff, N. J., "Comparison of Radiant and Conductive Heat Transfer in a Supersonic Wing," *Journal of the Aeronautical Sciences*, Vol. 23, No. 7, 1956, pp. 694–696.

Goldberg, M. A., "Investigation of the Temperature Distribution and Thermal Stresses in a Hypersonic Wing Structure," *Journal of the Aeronautical Sciences*, Vol. 23, No. 11, 1956, pp. 981–990.

Przemieniecki, J. S., "Transient Temperature Distributions and Thermal Stresses in Fuselage Shells with Bulkheads or Frames," *Journal of the Royal Aeronautical Society*, Vol. 60, Dec. 1956, pp. 799–804.

Budiansky, B., and Mayers, J., "Influence of Aerodynamic Heating on the Effective Torsional Stiffness of Thin Wings," *Journal of the Aeronautical Sciences*, Vol. 23, No. 12, 1956, pp. 1081–1093.

Barber, A. D., Weiner, J. H., and Boley, B. A., "An Analysis of the Effect of Thermal Contact Resistance in a Sheet-Stringer Structure," *Journal of the Aeronautical Sciences*, Vol. 24, No. 3, 1957, pp. 232–234.

Gatewood, B. E., "Effect of Thermal Resistance of Joints upon Thermal Stresses," *Journal of the Aeronautical Sciences*, Vol. 24, No. 2, 1957, pp. 152–153.

Kosko, E., "The Numerical Determination of Transient Temperatures in Wings," *Canadian Aeronautical Journal*, Vol. 3, No. 3, 1957, pp. 87–95.

Singer, J., and Hoff, N. J., "Effect of the Change in Thermal Stresses Due to Large Deflections on the Torsional Rigidity of Wings," *Journal of the Aeronautical Sciences*, Vol. 24, No. 4, 1957, pp. 310–311.

Chen, S.-Y., "Comments on Investigation of the Temperature Distribution and Thermal Stresses in a Hypersonic Wing Structure," *Journal of the Aeronautical Sciences*, Vol. 24, No. 7, 1957, pp. 544–545.

Isakson, G., "A Simple Model Study of Transient Temperature and Thermal Stress Distribution Due to Aerodynamic Heating," *Journal of the Aeronautical Sciences*, Vol. 24, No. 8, 1957, pp. 611–619.

Kochanski, S. L., and Argyris, J. H., "Some Effects of Kinetic Heating on the Stiffness of Thin Wings," *Aircraft Engineering*, Vol. 29, No. 344, 1957, pp. 310–318.

Brooks, W. A., Jr., "Temperature and Thermal Stress Distributions in Some Structural Elements Heated at a Constant Rate," NACA TN-4306, 1958.

Legg, K. L. C., and Stevens, G., "Temperature Distributions in Aircraft Structures and the Influence of Mechanical and Physical Material Properties," *Journal of the Royal Aeronautical Society*, Vol. 62, No. 3, 1958, pp. 174–186.

Frank, I., "Transient Temperature Distribution in Aircraft Structures," *Journal of the Aeronautical Sciences*, Vol. 25, No. 4, 1958, pp. 265–267.

Chen, S.-Y., "Transient Temperature and Thermal Stresses in Skin of Hypersonic Vehicle with Variable Boundary Conditions," *Transactions of the ASME*, Vol. 80, No. 7, 1958, pp. 1389–1394.

Kitchenside, A. W., "The Effects of Kinetic Heating on Aircraft Structures," *Journal of the Aeronautical Sciences*, Vol. 62, Feb. 1958, pp. 105–117.

Turner, M. J., Dill, E. H., Martin, H. C., and Melosh, R. J., "Large Deflections of Structures Subjected to Heating and External Loads," *Journal of the Aeronautical Sciences*, Vol. 27, No. 2, 1960, pp. 97–106.

Barzeley, M. E., "Range of Interface Thermal Conductance for Aircraft Joints," NASA TN D 426, 1960.

N'Guyen, V. P., and Ripley, E. L., "Design Philosophy and Fatigue Testing of Concorde," *Proceedings of 5th ICAF Symposium* (Melbourne, Australia), May 1967, pp. 403–437.

Ko, W. L., Quinn, R. D., Gong, L., Schuster, L. S., and Gonzales, D., "Reentry Heat Transfer Analysis of the Space Shuttle Orbiter," *Computational Aspects of Heat Transfer in Structures*, edited by Howard M. Adelman, NASA CP 2216, Nov. 1981, pp. 295–325.

Ko, W. L., Quinn, R. D., and Gong, L., "Finite Element Reentry Heat Transfer Analysis of Space Shuttle Orbiter," NASA Technical Paper 2657, Dec. 1986.

Ko, W. L., and Olona, T., "Effect of Element Size on the Solution Accuracies of Finite Element Heat Transfer and Thermal Stress Analyses of Space Shuttle Orbiter," NASA TM 88292, Aug. 1987.

Ko, W. L., and Fields, R. A., "Thermal Stress Analysis of Space Shuttle Orbiter Subjected to Reentry Aerodynamic Heating," NASA TM 88286, Oct. 1987.

5
Thermoelasticity

5.1 Introduction

In this chapter, we begin our discussion of thermally induced deformations and stresses in aerospace structures by considering a classical elastic continuum. The determination of elastic deformations and stresses in a continuum under a prescribed temperature distribution is the subject of thermoelasticity. Thermoelasticity is an old subject, with origins dating to Duhamel in France in the 1830s. Timoshenko[1] describes Duhamel's contributions to thermoelasticity, which include formulation and solution of several basic problems. There is a broad, rich literature on thermoelasticity that is beyond our scope to address. Readers interested in going further than the brief treatment here will find several books on elasticity and thermoelasticity available for supplementary reading. The elasticity texts by Timoshenko and Goodier,[2] Wang,[3] and Sokolnikoff[4] describe thermoelastic formulations and present solutions to classical problems. The solid mechanics book by Fung[5] has a chapter and a bibliography on thermoelasticity. In addition, there are a number of books devoted exclusively to thermoelasticity or thermal stresses or both. These include works by Gatewood,[6] Boley and Weiner,[7] Nowacki,[8] Johns,[9] Parkus,[10] Burgreen,[11] and Nowinski.[12] Moreover, there are chapters on thermoelasticity in the more recent three-volume series on thermal stresses edited by Hetnarski.[13–15]

Our purpose here is to describe basic concepts, present fundamental equations, and give a few classical solutions to thermoelastic problems. The chapter bridges the transition between the basic concepts of continuum mechanics presented in Chapter 3 and basic engineering thermal stress analysis described next in Chapter 6. In this chapter, the two fundamental approaches for formulating classical thermoelastic problems are presented first. Then, selected solutions for dynamic thermoelasticity problems are described. The balance of the chapter presents solutions to quasistatic thermoelastic problems.

The chapter emphasizes the boundary/initial-value problem approach to thermoelasticity problems. From this viewpoint, we establish basic principles that guide us in developing engineering formulations in the next chapter. Moreover, the analytical solutions developed here provide a basic understanding of how temperature distributions affect structures. This knowledge is valuable when we strive to explain how complex structures respond in a real thermal environment. And, of course, analytical solutions are often valuable for evaluating computational models.

5.2 Thermoelasticity Formulations

The basic equations that describe thermoelastic problems are presented in the continuum mechanics concepts of Chapter 3. These equations include the equations of motion, the strain-displacement relations, and the constitutive equations. Additionally, we need appropriate boundary and initial conditions.

Our formulation assumes small displacement gradients, and we use the classical linear strain-displacement relations given in Eq. (3.10). Deformations are assumed small, and the equations of motion are referred to the undeformed configuration of the solid and are given by Eq. (3.18) The strain and stress tensors are symmetric. We assume one-way thermomechanical coupling, which means that the conservation of energy equation does not depend on deformations, and temperature may be determined independently from the displacements and stresses. We assume that temperatures have been determined by solving the uncoupled conservation of energy equation (3.20). We assume a linear elastic material, and we use as constitutive equations the generalized Hooke's law for a homogeneous isotropic material, Eq. (3.23). In solutions to classical thermoelastic problems, we assume that the material elastic properties—including the modulus of elasticity, Poisson's ratio, and the coefficient of thermal expansion—are constant, so that material properties are independent of temperature. The result of these assumptions is that we can formulate thermoelasticity problems as linear boundary/initial-value problems that have unique solutions.

The basic unknowns in a classical three-dimensional thermoelasticity problem may vary with position as denoted by the Cartesian coordinates x_1, x_2, x_3 and with time t. There are 15 basic unknowns:

Three displacement components:	u_1, u_2, u_3
Six strain components:	$\varepsilon_{11}, \varepsilon_{12}, \varepsilon_{13}, \varepsilon_{22}, \varepsilon_{23}, \varepsilon_{33}$
Six strain components:	$\sigma_{11}, \sigma_{12}, \sigma_{13}, \sigma_{22}, \sigma_{23}, \sigma_{33}$

The elastic solid may be subject to body forces $B_i, i = 1, 2, 3$, which are known functions of position and time. All points on the surface of the elastic solid have either specified displacement components or specified surface tractions $T_i, i = 1, 2, 3$, as known functions of position and time. The temperature T is a prescribed function of position and time. The initial values of all displacement and velocity components are known. With the problem formulated in terms of the basic 15 unknowns, we require 15 equations, which are

Six strain-displacement equations:	Eqs. (3.10)
Three equations of motion:	Eqs. (3.18)
Six constitutive equations:	Eqs. (3.23)

Thus, the problem is well formulated and, in principle, we may solve the 15 equations for 15 unknowns. Actually, the 15 equations are not solved simultaneously; instead, they are combined algebraically to formulate smaller sets of equations that we solve. Two of the traditional formulations are the displacement formulation and the stress formulation. We now write the equations for these formulations and, in the process, change notation from the index subscripts used in Chapter 3 to the traditional scalar notation that we employ in the balance of this book.

5.2.1 Displacement Formulation

In the displacement formulation, the 15 equations are combined algebraically into three equations in which the unknowns are the three displacement components u, v, w. The procedure consists of first substituting the strain-displacement equations into the constitutive equations so that the stresses are written in terms of derivatives of the displacement components and temperature. Then the stress components are substituted into the equations of motion to give three equations in terms of the displacement components:

$$
\begin{aligned}
(\lambda + G)\frac{\partial e}{\partial x} + G\nabla^2 u - (3\lambda + 2G)\alpha\frac{\partial}{\partial x}(T - T_{\text{ref}}) + B_x &= \rho\frac{\partial^2 u}{\partial t^2} \\
(\lambda + G)\frac{\partial e}{\partial y} + G\nabla^2 v - (3\lambda + 2G)\alpha\frac{\partial}{\partial y}(T - T_{\text{ref}}) + B_y &= \rho\frac{\partial^2 v}{\partial t^2} \\
(\lambda + G)\frac{\partial e}{\partial z} + G\nabla^2 w - (3\lambda + 2G)\alpha\frac{\partial}{\partial z}(T - T_{\text{ref}}) + B_z &= \rho\frac{\partial^2 w}{\partial t^2}
\end{aligned}
\tag{5.1a}
$$

where the dilatation is

$$
e = \frac{\partial u}{\partial x} + \frac{\partial v}{\partial y} + \frac{\partial w}{\partial z} \tag{5.1b}
$$

and the Lamé constants are related to the traditional engineering constants by

$$
\lambda = \frac{\nu E}{(1+\nu)(1-2\nu)} \tag{5.1c}
$$

$$
G = \frac{E}{2(1+\nu)} \tag{5.1d}
$$

where E is the modulus of elasticity, and ν is Poisson's ratio. The coefficient of thermal expansion is α, and T_{ref} is the reference temperature for zero thermal stress. Once the displacements have been determined by solving Eq. (5.1a), the strain components are evaluated from the strain-displacement equations:

$$
\begin{aligned}
&\varepsilon_x = \frac{\partial u}{\partial x} \qquad \varepsilon_{xy} = \frac{1}{2}\left(\frac{\partial u}{\partial y} + \frac{\partial v}{\partial x}\right) \qquad \varepsilon_{xz} = \frac{1}{2}\left(\frac{\partial u}{\partial z} + \frac{\partial w}{\partial x}\right) \\
&\varepsilon_y = \frac{\partial v}{\partial y} \qquad \varepsilon_{yz} = \frac{1}{2}\left(\frac{\partial v}{\partial z} + \frac{\partial w}{\partial y}\right) \\
&\varepsilon_z = \frac{\partial w}{\partial z}
\end{aligned}
\tag{5.2}
$$

The coefficients of one-half in the shearing strains indicate that we are using the tensor definition of shearing strain as opposed to the engineering definition, which does not include the one-half. After the strain components are evaluated using Eqs. (5.2), the stress components are calculated from the stress–strain equations:

$$
\begin{aligned}
\sigma_x &= \lambda e + 2G\varepsilon_x - (3\lambda + 2G)\alpha(T - T_{\text{ref}}) \\
\sigma_y &= \lambda e + 2G\varepsilon_y - (3\lambda + 2G)\alpha(T - T_{\text{ref}}) \\
\sigma_z &= \lambda e + 2G\varepsilon_z - (3\lambda + 2G)\alpha(T - T_{\text{ref}}) \\
\sigma_{xy} &= 2G\varepsilon_{xy} \qquad \sigma_{xz} = 2G\varepsilon_{xz} \qquad \sigma_{yz} = 2G\varepsilon_{yz}
\end{aligned}
\tag{5.3}
$$

The coefficients of two in the three equations for the shearing stress components compensate for the coefficients of one-half in the tensor definitions of the strain components in Eqs. (5.2). For later reference, note that

$$3\lambda + 2G = \frac{E}{1 - 2\nu} \tag{5.4}$$

The problem formulation will be complete if we specify the boundary and initial conditions. For most problems, the boundary conditions fall into two categories, we specify either the displacements or the surface tractions. Suppose that the surface consists of regions S_1 and S_2. Then we write prescribed displacement boundary conditions as

$$\left.\begin{aligned} u &= u_1(x, y, z, t) \\ v &= v_1(x, y, z, t) \\ w &= w_1(x, y, z, t) \end{aligned}\right\} \quad \text{on } S_1 \tag{5.5a}$$

where u_1, v_1, w_1 are known functions. The prescribed surface tractions T_x, T_y, T_z are written in terms of the stress components by using Eq. (3.12),

$$T_x = \sigma_x n_x + \sigma_{xy} n_y + \sigma_{xz} n_z$$

$$T_y = \sigma_{xy} n_x + \sigma_y n_y + \sigma_{yz} n_z$$

$$T_z = \sigma_{xz} n_x + \sigma_{yz} n_y + \sigma_z n_z$$

where n_x, n_y, n_z are components of a unit vector $\hat{n}$ normal to the surface. Hence, prescribed surface traction consists of specifying

$$\left.\begin{aligned} \sigma_x n_x + \sigma_{xy} n_y + \sigma_{xz} n_z &= T_{x1} \\ \sigma_{xy} n_x + \sigma_y n_y + \sigma_{yz} n_z &= T_{y1} \\ \sigma_{xz} n_x + \sigma_{yz} n_y + \sigma_z n_z &= T_{z1} \end{aligned}\right\} \quad \text{on } S_2 \tag{5.5b}$$

where T_{x1}, T_{y1}, T_{z1} are known functions. Other boundary conditions are also possible, for example, an elastic support where there is a relationship between the displacement and a traction component on the boundary. However, for a large number of problems, Eqs. (5.5a) and (5.5b) contain the basic possibilities.

The initial conditions consist of specifying initial values of the displacement components,

$$\left.\begin{aligned} u(x, y, z, 0) &= u_0(x, y, z) \\ v(x, y, z, 0) &= v_0(x, y, z) \\ w(x, y, z, 0) &= w_0(x, y, z) \end{aligned}\right. \tag{5.6a}$$

and the velocity components,

$$\begin{aligned}
\frac{\partial u}{\partial t}(x, y, z, 0) &= \dot{u}_0(x, y, z) \\
\frac{\partial v}{\partial t}(x, y, z, 0) &= \dot{v}_0(x, y, z) \\
\frac{\partial w}{\partial t}(x, y, z, 0) &= \dot{w}_0(x, y, z)
\end{aligned} \tag{5.6b}$$

where u_0, v_0, w_0 and $\dot{u}_0, \dot{v}_0, \dot{w}_0$ are known functions.

Equations (5.1–5.6) constitute the displacement formulation for three-dimensional thermoelasticity.

5.2.2 Stress Formulation

In the stress formulation, we wish to manipulate the equations of thermoelasticity into six equations where the basic unknowns are the six components of stress $\sigma_x, \sigma_y, \sigma_z, \sigma_{xy}, \sigma_{xz}, \sigma_{yz}$. Once these equations are solved for the stress components, we can use an inverse form of Hooke's law to solve for the strain components $\varepsilon_x, \varepsilon_y, \varepsilon_z, \varepsilon_{xy}, \varepsilon_{xz}, \varepsilon_{yz}$. At this point, we wish to integrate the six strain-displacement equations to obtain the three displacement components u, v, w. But, because there are six equations for three unknown displacements, we cannot always obtain a unique solution. Over the years, elasticians[2–4] have established that additional conditions must be imposed on the strain components to ensure that the six strain-displacement equations, when integrated, will yield unique solutions. These conditions are known as the *compatibility equations*. In terms of tensor shearing strains, Sokolnikoff[4] gives the compatibility equations as

$$\begin{aligned}
\frac{\partial^2 \varepsilon_x}{\partial y \partial z} &= \frac{\partial}{\partial x}\left(-\frac{\partial \varepsilon_{yz}}{\partial x} + \frac{\partial \varepsilon_{xz}}{\partial y} + \frac{\partial \varepsilon_{xy}}{\partial z}\right) \\
\frac{\partial^2 \varepsilon_y}{\partial z \partial x} &= \frac{\partial}{\partial y}\left(-\frac{\partial \varepsilon_{xz}}{\partial y} + \frac{\partial \varepsilon_{xy}}{\partial z} + \frac{\partial \varepsilon_{yz}}{\partial x}\right) \\
\frac{\partial^2 \varepsilon_z}{\partial x \partial y} &= \frac{\partial}{\partial z}\left(-\frac{\partial \varepsilon_{xy}}{\partial z} + \frac{\partial \varepsilon_{yz}}{\partial x} + \frac{\partial \varepsilon_{zx}}{\partial y}\right) \\
2\frac{\partial^2 \varepsilon_{xy}}{\partial x \partial y} &= \frac{\partial^2 \varepsilon_x}{\partial y^2} + \frac{\partial^2 \varepsilon_y}{\partial x^2} \\
2\frac{\partial^2 \varepsilon_{yz}}{\partial y \partial z} &= \frac{\partial^2 \varepsilon_y}{\partial z^2} + \frac{\partial^2 \varepsilon_z}{\partial y^2} \\
2\frac{\partial^2 \varepsilon_{xz}}{\partial z \partial x} &= \frac{\partial^2 \varepsilon_z}{\partial x^2} + \frac{\partial^2 \varepsilon_x}{\partial z^2}
\end{aligned} \tag{5.7}$$

The compatibility equations (5.7) apply to simply connected bodies, i.e., bodies without interior cavities or holes. For multiply connected bodies, supplementary relations are required; see Boley and Weiner[7] for a discussion of these conditions.

The basic six equations of the stress formulation originate from the compatibility relations in Eq. (5.7). The stress formulation typically assumes quasistatic behavior, zero body forces, and a constant reference temperature.

Quasistatic behavior means that we neglect the inertia forces in the equations of motion. The temperature may be a function of position and time so that the stresses, strains, and displacements may be a function of time as well. We will discuss the quasistatic assumption in later examples.

To express a typical compatibility equation in terms of the stress components, we use Hooke's law to introduce the stress components and temperature. Then the equations of equilibrium are introduced to simplify the resulting expressions. The basic six equations of the stress formulation are[7]

$$
\begin{aligned}
&(1+\nu)\nabla^2\sigma_x + \frac{\partial^2\Theta}{\partial x^2} + E\alpha\left(\frac{1+\nu}{1-\nu}\nabla^2 T + \frac{\partial^2 T}{\partial x^2}\right) = 0\\
&(1+\nu)\nabla^2\sigma_y + \frac{\partial^2\Theta}{\partial y^2} + E\alpha\left(\frac{1+\nu}{1-\nu}\nabla^2 T + \frac{\partial^2 T}{\partial y^2}\right) = 0\\
&(1+\nu)\nabla^2\sigma_z + \frac{\partial^2\Theta}{\partial z^2} + E\alpha\left(\frac{1+\nu}{1-\nu}\nabla^2 T + \frac{\partial^2 T}{\partial z^2}\right) = 0\\
&(1+\nu)\nabla^2\sigma_{xz} + \frac{\partial^2\Theta}{\partial x\partial z} + E\alpha\frac{\partial^2 T}{\partial x\partial z} = 0\\
&(1+\nu)\nabla^2\sigma_{yx} + \frac{\partial^2\Theta}{\partial x\partial y} + E\alpha\frac{\partial^2 T}{\partial x\partial y} = 0\\
&(1+\nu)\nabla^2\sigma_{yz} + \frac{\partial^2\Theta}{\partial y\partial z} + E\alpha\frac{\partial^2 T}{\partial y\partial z} = 0
\end{aligned}
\tag{5.8a}
$$

where

$$\Theta = \sigma_x + \sigma_y + \sigma_z \tag{5.8b}$$

Once the six stresses have been determined by solving Eqs. (5.8), the strains are evaluated from Hooke's law. The inverse form of Eqs. (5.3) is often called *generalized Hooke's law*:

$$
\begin{aligned}
\varepsilon_x &= \frac{1}{E}[\sigma_x - \nu(\sigma_y + \sigma_z)] + \alpha(T - T_{\text{ref}})\\
\varepsilon_y &= \frac{1}{E}[\sigma_y - \nu(\sigma_x + \sigma_z)] + \alpha(T - T_{\text{ref}})\\
\varepsilon_z &= \frac{1}{E}[\sigma_z - \nu(\sigma_x + \sigma_y)] + \alpha(T - T_{\text{ref}})\\
\varepsilon_{xy} = \sigma_{xy}/2G \qquad &\varepsilon_{xz} = \sigma_{xz}/2G \qquad \varepsilon_{yz} = \sigma_{yz}/2G
\end{aligned}
\tag{5.9}
$$

Finally, with the strain components as known functions of x, y, z, t, we integrate the strain-displacement equations to obtain the displacement components. From

Eqs. (5.2),

$$
\begin{aligned}
\frac{\partial u}{\partial x} &= \varepsilon_x(x, y, z, t) \\
\frac{\partial v}{\partial y} &= \varepsilon_y(x, y, z, t) \\
\frac{\partial w}{\partial z} &= \varepsilon_z(x, y, z, t) \\
\frac{\partial u}{\partial y} + \frac{\partial v}{\partial x} &= 2\varepsilon_{xy}(x, y, z, t) \\
\frac{\partial u}{\partial z} + \frac{\partial w}{\partial x} &= 2\varepsilon_{xz}(x, y, z, t) \\
\frac{\partial v}{\partial z} + \frac{\partial w}{\partial y} &= 2\varepsilon_{yz}(x, y, z, t)
\end{aligned}
\tag{5.10}
$$

Along the way, we need to impose specified displacement boundary conditions as indicated in Eqs. (5.5a) or specified surface-traction boundary conditions as indicated in Eq. (5.5b), or both. Initial conditions are not required because the inertia terms are neglected, and we do not integrate with respect to time.

5.2.3 Plane Stress and Plane Strain

The displacement and stress formulations of the preceding section state the mathematical formulations for thermoelasticity of a three-dimensional elastic solid. Often, however, the solution of these general equations is too difficult to carry out, even with computational methods and advanced computers. For many problems of practical interest, engineers have found that some simplifying assumptions can be made that render the problems more tractable. Two historically popular idealizations are *plane stress and plane strain.*

Plane stress formulation. The plane stress idealization applies to thin plates that are subject to temperature distributions that do not vary through the thickness of the plate. Section 4.5 shows that, for many heated thin-walled structures, temperature gradients through the thickness are small and may be neglected. Thus, for thin plates lying in the x–y plane, we may assume that the temperature is a function only of x, y, and time; hence, $T = T(x, y, t)$. Furthermore, if the only applied mechanical loads lie in the x–y plane, the lateral surfaces of the thick plate are traction-free. On these two surfaces, we know that $\sigma_z = \sigma_{xz} = \sigma_{yz} = 0$. Since the plate is thin, these stresses are also *assumed* to be zero through the thickness. Moreover, all displacements, stresses, and strains are assumed independent of z. The basic unknowns reduce to

Two displacement components:	$u(x, y, t)$, $v(x, y, t)$
Three strain components:	$\varepsilon_x(x, y, t)$, $\varepsilon_y(x, y, t)$, $\varepsilon_{xy}(x, y, t)$
Three stress components:	$\sigma_x(x, y, t)$, $\sigma_y(x, y, t)$, $\sigma_{xy}(x, y, t)$

The basic equations reduce to three strain-displacement equations, two equations of motion, and four stress–strain relations. From the three-dimensional relations, we may write the following simplified equations.

Strain-displacement equations:

$$\varepsilon_x = \frac{\partial u}{\partial x} \qquad \varepsilon_y = \frac{\partial v}{\partial y} \qquad \varepsilon_{xy} = \frac{1}{2}\left(\frac{\partial u}{\partial y} + \frac{\partial v}{\partial x}\right) \tag{5.11}$$

Equations of motion:

$$\begin{aligned} \frac{\partial \sigma_x}{\partial x} + \frac{\partial \sigma_{xy}}{\partial y} + B_x &= \rho \frac{\partial^2 u}{\partial t^2} \\ \frac{\partial \sigma_{xy}}{\partial x} + \frac{\partial \sigma_y}{\partial y} + B_y &= \rho \frac{\partial^2 v}{\partial t^2} \end{aligned} \tag{5.12}$$

In terms of the Lamé constants, the stress–strain equations are

$$\sigma_x = \lambda e + 2G\varepsilon_x - (3\lambda + 2G)\alpha(T - T_{\text{ref}})$$

$$\sigma_y = \lambda e + 2G\varepsilon_y - (3\lambda + 2G)\alpha(T - T_{\text{ref}})$$

$$\sigma_z = \lambda e + 2G\varepsilon_z - (3\lambda + 2G)\alpha(T - T_{\text{ref}}) = 0$$

$$\sigma_{xy} = 2G\varepsilon_{xy}$$

where the dilatation $e = \varepsilon_x + \varepsilon_y + \varepsilon_z$. Note that the strain ε_z is not zero but, rather, it is related to the strains ε_x, ε_y and to temperature through the third stress–strain equation that requires $\sigma_z = 0$. This equation is used to elimiate ε_z from the equations for σ_x and σ_y. After we perform this elimination, we write the results in terms of the engineering constants E and ν.

Stress–strain equations for plane stress:

$$\sigma_x = \frac{E}{1-\nu^2}(\varepsilon_x + \nu\varepsilon_y) - \frac{E\alpha}{1-\nu}(T - T_{\text{ref}})$$

$$\sigma_y = \frac{E}{1-\nu^2}(\varepsilon_y + \nu\varepsilon_x) - \frac{E\alpha}{1-\nu}(T - T_{\text{ref}}) \tag{5.13}$$

$$\sigma_{xy} = \frac{E}{(1+\nu)}\varepsilon_{xy}$$

Equations (5.11–5.13) represent eight equations that we can use to solve for the eight unknowns in the plane stress formulation.

In the displacement formulation, the eight equations are combined algebraically into two equations where the unknowns are the displacement components u and v. The procedure consists of substituting the strain-displacement relations into the stress–strain equations so that the stresses are written in terms of derivatives of the displacements and temperature. The stresses are then substituted into the equation of motion to yield

$$(1+\nu)\frac{\partial}{\partial x}\left(\frac{\partial u}{\partial x}+\frac{\partial v}{\partial y}\right)+(1-\nu)\nabla^2 u-2(1+\nu)\alpha\frac{\partial}{\partial x}(T-T_{\text{ref}})+B_x=\rho\frac{\partial^2 u}{\partial t^2}$$

$$(1+\nu)\frac{\partial}{\partial y}\left(\frac{\partial u}{\partial x}+\frac{\partial v}{\partial y}\right)+(1-\nu)\nabla^2 v-2(1+\nu)\alpha\frac{\partial}{\partial x}(T-T_{\text{ref}})+B_y=\rho\frac{\partial^2 v}{\partial t^2} \tag{5.14}$$

Equations (5.14) are solved subject to specifying either displacement or surface-traction boundary conditions at every point on the edges of the plate. These have the form

$$\left.\begin{aligned} u &= u_1(x,y,t) \\ v &= v_1(x,y,t) \end{aligned}\right\} \quad \text{on } S_1$$

$$\left.\begin{aligned} \sigma_x n_x + \sigma_{xy} n_y &= T_{x1} \\ \sigma_{xy} n_x + \sigma_y n_y &= T_{x2} \end{aligned}\right\} \quad \text{on } S_2 \tag{5.15}$$

The initial conditions consist of specifying initial values of displacement and velocity:

$$\begin{aligned} u(x,y,0) &= u_0(x,y) \\ v(x,y,0) &= v_0(x,y) \\ \frac{\partial u}{\partial t}(x,y,0) &= \dot{u}_0(x,y) \\ \frac{\partial v}{\partial t}(x,y,0) &= \dot{v}_0(x,y) \end{aligned} \tag{5.16}$$

where u_0, v_0 and $\dot{u}_0$, $\dot{v}_0$ are known functions. Once we solve Eqs. (5.14) for the displacement components u and v, the strain components can be evaluated using Eqs. (5.11), and the stress components are evaluated from Eqs. (5.13).

Plane strain formulation. The plane strain idealization applies to long, prismatic cylinders under mechanical loads uniformly distributed along the cylinder's lateral surface. The x–y plane lies in a typical cross-sectional plane, and z is along the cylinder's axis. The temperature is assumed independent of z, so that $T = T(x, y, t)$. The deformations of the cylinder at some distance from the ends are independent of z, and the displacements u and v are functions only of x, y, and t. If the ends of the cylinder are fully restrained, then w is zero there. Consequently, we also make the assumption that w is zero at every cross section of the cylinder. Thus, the plane strain idealization represents a "slice" out of a long body for which ε_z is zero. With these assumptions, the basic unknowns reduce to exactly the same eight unknowns that we had in the plane stress formulation. In addition, we have many of the same equations. The nonzero strain-displacement relations are defined by Eqs. (5.11), and the equations of motion are given by Eqs. (5.12). The principal difference lies in the stress–strain equations. In the plane stress case, we have $\sigma_z = 0$; in the plane strain case, we have $\varepsilon_z = 0$ because w is zero. Thus, the stress–strain equations reduce to

$$\sigma_x = \lambda(\varepsilon_x + \varepsilon_y) + 2G\varepsilon_x - (3\lambda + 2G)\alpha(T - T_{\text{ref}})$$

$$\sigma_y = \lambda(\varepsilon_x + \varepsilon_y) + 2G\varepsilon_y - (3\lambda + 2G)\alpha(T - T_{\text{ref}})$$

$$\sigma_z = \lambda(\varepsilon_x + \varepsilon_y) - (3\lambda + 2G)\alpha(T - T_{\text{ref}})$$

$$\sigma_{xy} = 2G\varepsilon_{xy}$$

The equations for σ_x and σ_y may be expressed in terms of the engineering constants E and ν using Eqs. (5.1c, 5.1d, and 5.4). The equation for σ_z can be expressed in terms of σ_x and σ_y by the elimination of $\varepsilon_x + \varepsilon_y$ between the three equations.

Stress–strain equations for plane strain:

$$\begin{aligned}
\sigma_x &= \frac{E}{(1+\nu)(1-2\nu)}[(1-\nu)\varepsilon_x + \nu\varepsilon_y] - \frac{E\alpha}{1-2\nu}(T - T_{\text{ref}}) \\
\sigma_y &= \frac{E}{(1+\nu)(1-2\nu)}[\nu\varepsilon_x + (1-\nu)\varepsilon_y] - \frac{E\alpha}{1-2\nu}(T - T_{\text{ref}}) \\
\sigma_{xy} &= \frac{E}{1+\nu}\varepsilon_{xy} \qquad \sigma_z = \nu(\sigma_x + \sigma_y) - E\alpha(T - T_{\text{ref}})
\end{aligned} \tag{5.17}$$

Comparing the stress–strain equations for plane stress (5.13) with the stress–strain relations for plane strain (5.17), we see that the equations for the normal stresses σ_x and σ_y have the same form but differ in the coefficients for ε_x, ε_y, and $(T - T_{\text{ref}})$. The equations for σ_{xy} are identical. In the plane strain formulation, there is an additional equation for σ_z that we can use if we wish, but it is not required to determine u and v. Thus, the mathematical formulations for plane stress and plane strain are identical except for coefficients in the constitutive relations.

Proceeding as we did for plane stress, we may write the equations of motion in terms of the displacements u and v. The boundary and initial conditions are the same as for plane stress. The derivation of the equations for the displacement formulation for plane strain appears in a problem at the end of the chapter.

5.2.4 Stress-Function Formulation for Plane Stress and Plane Strain

For the stress formulation in three-dimensional thermoelasticity, the six equations for strain compatibility (5.7) are required. These equations ensure that the three displacements derived by integrating the six strain-displacement equations will be unique, single-valued, and continuous. For a state of plane stress or plane strain, the six compatibility equations reduce to the single equation

$$\frac{\partial^2 \varepsilon_x}{\partial y^2} + \frac{\partial^2 \varepsilon_y}{\partial x^2} = 2\frac{\partial^2 \varepsilon_{xy}}{\partial x \partial y} \tag{5.18}$$

For plane stress and plane strain, a *stress-function* approach may be used as an alternative to the displacement formulation. The approach will be given for equilibrium problems with zero body forces. Then the equations of motion reduce to the two equilibrium equations

$$\frac{\partial \sigma_x}{\partial x} + \frac{\partial \sigma_{xy}}{\partial y} = 0 \qquad \frac{\partial \sigma_{xy}}{\partial x} + \frac{\partial \sigma_y}{\partial y} = 0 \tag{5.19}$$

These latter equations are satisfied identically if we introduce a stress function $F(x, y)$ so that the stresses are defined by

$$\sigma_x = \frac{\partial^2 F}{\partial y^2} \qquad \sigma_{xy} = -\frac{\partial^2 F}{\partial x \partial y} \qquad \sigma_y = \frac{\partial^2 F}{\partial x^2} \tag{5.20}$$

The function F is called the Airy stress function. If we introduce Hooke's law to express the strains in terms of the stresses, we may use the compatibity equation to derive the partial differential equation that $F(x, y)$ must satisfy. Because the stress–strain laws for plane stress and plane strain are different, we develop a separate stress-function equation for each case.

Plane stress. Returning now to the stress–strain equations for plane stress (5.13), we solve these for the strains. Then we may write

$$\begin{aligned} \varepsilon_x &= \frac{1}{E}(\sigma_x - \nu\sigma_y) + \alpha(T - T_{\text{ref}}) \\ \varepsilon_y &= \frac{1}{E}(\sigma_y - \nu\sigma_x) + \alpha(T - T_{\text{ref}}) \\ \varepsilon_{xy} &= \frac{(1+\nu)}{E}\sigma_{xy} \end{aligned} \tag{5.21}$$

Next, we substitute the strains in the preceding equation into the compatibility equation (5.18) to obtain

$$\frac{\partial^2}{\partial y^2}(\sigma_x - \nu\sigma_y) + \frac{\partial^2}{\partial x^2}(\sigma_y - \nu\sigma_x) + E\alpha\nabla^2 T = 2(1+\nu)\frac{\partial^2 \sigma_{xy}}{\partial x \partial y}$$

Finally, we introduce the stress function from Eqs. (5.20), and the last equation reduces to

$$\nabla^4 F = -E\alpha\nabla^2 T \tag{5.22a}$$

where

$$\nabla^4 F = \nabla^2(\nabla^2 F) = \frac{\partial^4 F}{\partial x^4} + 2\frac{\partial^4 F}{\partial x^2 \partial y^2} + \frac{\partial^4 F}{\partial y^4} \tag{5.22b}$$

Equation (5.22a) is a linear, nonhomogeneous partial differential equation that defines the stress function. Because Eq. (5.22a) is a fourth-order equation, two boundary conditions are required at each point on the boundary. Boley and Weiner[7] discuss appropriate boundary conditions and point out that the equation applies only for simply connected bodies. For plates with traction-free edges, they show that the stress function satisfies

$$F = \frac{\partial F}{\partial n} = 0 \tag{5.23}$$

at each point on the edge, where n denotes the direction normal to the edge.

Plane strain. In a similar manner, we may derive an equation that applies for plane strain. From Eqs. (5.17), the strains may be expressed as

$$\varepsilon_x = \frac{1+\nu}{E}[(1-\nu)\sigma_x - \nu\sigma_y] + (1+\nu)\alpha T$$

$$\varepsilon_y = \frac{1+\nu}{E}[(1-\nu)\sigma_y - \nu\sigma_x] + (1+\nu)\alpha T \tag{5.24}$$

$$\varepsilon_{xy} = \frac{1+\nu}{E}\sigma_y$$

Proceeding as we did for plane stress, we find that $F(x, y)$ must satisfy

$$\nabla^4 F = -\frac{E\alpha}{1-\nu}\nabla^2 T \tag{5.25}$$

Equation (5.25) is the same as the equation for plane stress (5.22a) except for the coefficient of the temperature term.

Thus, in two-dimensional problems, the stress-function approach leads to a single fourth-order partial differential equation. Once this equation is solved, the stresses are evaluated by differentiation using the definitions given in Eqs. (5.20). Knowing the stresses, we next evaluate the strains using the appropriate strain–stress equation; either Eqs. (5.21) for plane stress, or Eqs. (5.24) for plane strain. Finally, the displacements are determined by integrating the strain-displacement equations given by Eqs. (5.11).

5.2.5 Plane Problems in Cylindrical Coordinates

For thermoelastic problems with circular boundaries, we should formulate the problem in cylindrical coordinates. Many of the elasticity and thermoelasticity texts cited in the references give the 15 basic equations in cylindrical coordinates. Here, we will give without derivation, only the basic equations that apply for hollow cylinders (Fig. 5.1) with axisymmetric temperature distributions. Thus, we assume a temperature distribution that is a function only of the radial coordinate r; that is, we consider $T = T(r)$. Because of the radial symmetry, the problem can be formulated in terms of the radial displacement u, which is a function only of r. There are only two unknown planar stress components: σ_r, the radial stress, and σ_θ, the circumferential stress. The shearing stress $\sigma_{r\theta}$ is zero. The equilibrium equations reduce to the single radial equation,

$$\frac{d\sigma_r}{dr} + \frac{\sigma_r - \sigma_\theta}{r} = 0 \tag{5.26}$$

where body forces are neglected. The corresponding radial and circumferential strains are ε_r and ε_θ, respectively. The strain-displacement relations are

$$\varepsilon_r = \frac{du}{dr} \tag{5.27a}$$

$$\varepsilon_\theta = \frac{u}{r} \tag{5.27b}$$

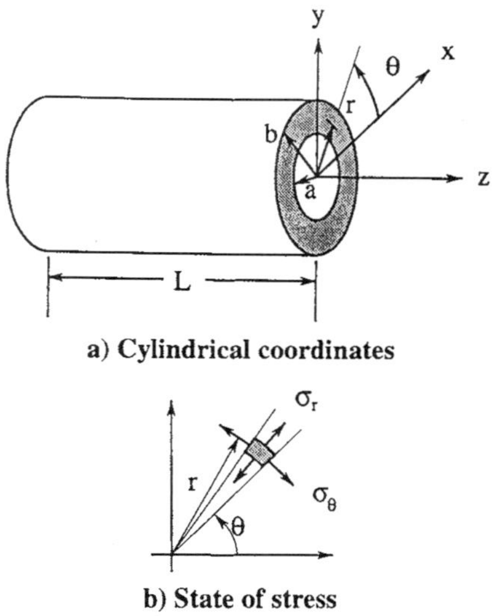

Fig. 5.1 Circular disk or cylinder.

With reference to Fig. 5.1, we see that we can formulate either a plane stress or a plane strain problem, depending on the length of the cylinder. If $L \ll b$, the cylinder reduces to a thin washer with the sides free from surface tractions, and we have a state of plane stress. On the other hand, if $L \gg b$ and we assume that the axial displacement w is zero, we have a state of plane strain.

Plane stress. In the r, θ coordinate system, the stress–strain equations for plane stress given in Eqs. (5.13) may be used by simply changing the subscripts. Thus,

$$\begin{aligned} \sigma_r &= \frac{E}{1-\nu^2}(\varepsilon_r + \nu\varepsilon_\theta) - \frac{E\alpha}{1-\nu}(T - T_{\text{ref}}) \\ \sigma_\theta &= \frac{E}{1-\nu^2}(\varepsilon_\theta + \nu\varepsilon_r) - \frac{E\alpha}{1-\nu}(T - T_{\text{ref}}) \end{aligned} \tag{5.28}$$

We may combine the three preceding equations to give a single ordinary differential equation for u,

$$\frac{\mathrm{d}^2 u}{\mathrm{d}r^2} + \frac{1}{r}\frac{\mathrm{d}u}{\mathrm{d}r} - \frac{u}{r^2} = (1+\nu)\alpha\frac{\mathrm{d}T}{\mathrm{d}r} \tag{5.29a}$$

or

$$\frac{\mathrm{d}}{\mathrm{d}r}\left[\frac{1}{r}\frac{\mathrm{d}(ru)}{\mathrm{d}r}\right] = (1+\nu)\alpha\frac{\mathrm{d}T}{\mathrm{d}r} \tag{5.29b}$$

assuming that the reference temperature is constant.

Plane strain. In the r, θ coordinates, Eqs. (5.17) are

$$\begin{aligned}\sigma_r &= \frac{E}{(1+\nu)(1-2\nu)}[(1-\nu)\varepsilon_r + \nu\varepsilon_\theta] - \frac{E\alpha}{1-2\nu}(T - T_{\text{ref}}) \\ \sigma_\theta &= \frac{E}{(1+\nu)(1-2\nu)}[\nu\varepsilon_r + (1-\nu)\varepsilon_\theta] - \frac{E\alpha}{1-2\nu}(T - T_{\text{ref}}) \\ \sigma_z &= \nu(\sigma_r + \sigma_\theta) - E\alpha(T - T_{\text{ref}})\end{aligned} \tag{5.30}$$

The displacement formulation for plane strain can then be derived as

$$\frac{\mathrm{d}}{\mathrm{d}r}\left[\frac{1}{r}\frac{\mathrm{d}(ru)}{\mathrm{d}r}\right] = \left(\frac{1+\nu}{1-\nu}\right)\alpha\frac{\mathrm{d}T}{\mathrm{d}r} \tag{5.31}$$

As we have seen on previous occasions, the only difference between plane stress and plane strain is the coefficients of the temperature term.

Given a temperature distribution, the solution procedure is to solve the appropriate ordinary differential equation, either Eq. (5.29) or Eq. (5.31) for the displacement $u = u(r)$. We will need two boundary conditions: either specified displacement or specified stress, or both. Once $u(r)$ is determined, the strain components are evaluated from Eqs. (5.27), and then the stresses are evaluated from either Eq. (5.28) or Eq. (5.30).

In this section, we have described thermoelastic formulations for important two-dimensional idealizations. We will use the equations to obtain analytical solutions to classical problems and for computational solutions. The formulations are important for determining stresses and displacements that arise naturally for problems with only "in-plane" stresses. The formulations are also important in other more complicated problems. For instance, the plane stress formulation is important for determining the *membrane* forces that occur in the bending of thin plates and shells.

In the next two sections, we solve a few problems, beginning with these formulations. Unfortunately, the mathematics of these formulations is too complex to permit a general solution procedure for arbitrary geometric shapes. Indeed, only a few problems with relatively simple shapes and boundary conditions permit analytical solutions. We will present a few such solutions that have practical value.

5.3 Dynamic Response in Semi-infinite Solids

A class of dynamic problems in thermoelasticity that permits analytical solutions is the response of an elastic half-space to suddenly applied thermal boundary conditions. The problems were first addressed by V. I. Danilovskaya[16,17] in the early 1950s, and we refer to them as the Danilovskaya problems. Boley and Weiner[7] describe the solution to the second Danilovskaya problem and present additional references. Other references include a very early finite element solution by Nickell and Sackman[18] and recent numerical solutions by Tamma.[19]

The problem consists of an elastic solid occupying the half-space $x > 0$ with the plane $x = 0$ free of surface tractions. In the first Danilovskaya problem, the temperature of the surface $x = 0$ is suddenly changed at time equal to zero to a new specified value. In the second Danilovskaya problem, the surface at $x = 0$ is convectively heated by suddenly changing the ambient temperature to a new value.

The dynamic thermoelasticity problem is made tractable by assuming the displacement components as

$$u = u(x, t) \qquad v = 0 \qquad w = 0 \tag{5.32}$$

For these displacements, Eqs. (5.2) show that the only nonzero strain component is $\varepsilon_x = \partial u/\partial x$ and the dilatation $e = \partial u/\partial x$. Hooke's law, Eqs. (5.3), shows that the three normal stresses are given by

$$\sigma_x = (\lambda + 2G)\frac{\partial u}{\partial x} - \beta(T - T_{\text{ref}}) \tag{5.33a}$$

$$\sigma_y = \sigma_z = \lambda\frac{\partial u}{\partial x} - \beta(T - T_{\text{ref}}) \tag{5.33b}$$

and all shearing stresses are zero. In Eqs. (5.33), we have introduced the parameter β defined as

$$\beta = (3\lambda + 2\mu)\alpha \tag{5.33c}$$

In terms of the engineering elastic constants,

$$\beta = E\alpha/(1 - 2\nu) \tag{5.33d}$$

The three equations of motion (3.18) reduce to the single equation

$$\frac{\partial \sigma_x}{\partial x} = \rho\frac{\partial^2 u}{\partial t^2}$$

At this point, we could express the equation of motion in terms of the displacement u as the dependent variable as in Eq. (5.1a). However, Danilovskaya chose to eliminate the displacement u in the preceding equation to obtain an equation of motion with the stress σ_x as the dependent variable. By differentiating the last equation, we may write

$$\frac{\partial^2 \sigma_x}{\partial x^2} = \rho\frac{\partial^2}{\partial t^2}\left(\frac{\partial u}{\partial x}\right)$$

and then, using Eq. (5.33a), we substitute for $\partial u/\partial x$ to obtain

$$a^2\frac{\partial^2 \sigma_x}{\partial x^2} = \frac{\partial^2 \sigma_x}{\partial t^2} + \beta\frac{\partial^2 T}{\partial t^2} \tag{5.34a}$$

where

$$a = \sqrt{(\lambda + 2G)/\rho} \tag{5.34b}$$

The parameter a is the propagation speed of a dilatation wave in an elastic medium. Timoshenko and Goodier[2] show that an elastic medium has two types of elastic waves. The first is a wave that produces only a volume change, i.e., a dilatation wave, with a propagation speed given by Eq. (5.34b); the second is a wave of shearing distortion that produces no dilatation. The speed of propagation of a distortion wave is $\sqrt{(G/\rho)}$. The Danilovskaya displacement assumptions permit only dilatation waves and preclude distortion waves.

We recognize Eq. (5.34a) as a nonhomogeneous, hyperbolic wave equation. Thus, we may anticipate that suddenly applied thermal boundary conditions will induce a stress wave that propagates from the boundary into the solid with a propagation speed computed from Eq. (5.34b). For Eq. (5.34a), we need two boundary conditions. The first comes from requiring that the surface at $x = 0$ be traction-free, and the second comes from requiring that the stress be zero for large values of x. Thus, we may write

$$\sigma_x(o, t) = 0 \tag{5.35a}$$

$$\lim_{x \to \infty} \sigma_x(x, t) = 0 \tag{5.35b}$$

For Eq. (5.34a), we need two initial conditions, which we take as

$$\sigma_x(x, o) = 0 \tag{5.36a}$$

$$\frac{\partial \sigma_x}{\partial x}(x, o) = 0 \tag{5.36b}$$

After finding the temperature $T(x, t)$ by solving the uncoupled energy equation, the Danilovskaya problem consists of solving Eqs. (5.34) subject to the boundary conditions given by Eqs. (5.35) and initial conditions given by Eqs. (5.36). Analytical solutions to the boundary/initial-value problem are obtained by using the Laplace transform approach.

5.3.1 First Danilovskaya Problem

The first Danilovskaya problem[16] consists of an elastic half-space subjected to a prescribed surface-temperature change. We solve the uncoupled energy equation for this problem in Sec. 4.4.2 to obtain $T(x, t)$. From Eq. (4.35),

$$T(x, t) = T_0 + (T_\infty - T_0)\,\mathrm{erfc}\left[x/(2\sqrt{\kappa t})\right]$$

where T_0 is the initial solid temperature, and T_∞ is the suddenly applied temperature of the wall (see Fig. 4.7). In the equation above erfc(z) denotes the complementary error function of argument z, and erfc(z) $= 1 -$ erf(z), where erf(z) is the error function. The argument of the complementary error function depends on the thermal diffusivity κ, a property of the material. The thermal diffusivity is defined by $\kappa = k/\rho c$, here k is the material thermal conductivity, ρ the density and c the specific heat.* The approach for solving Eq. (5.34a) is first to take the Laplace transform to obtain

$$a^2 \frac{\mathrm{d}^2 \bar{\sigma}_x}{\mathrm{d}x^2} = s^2 \bar{\sigma}_x + \beta\left(s^2 \bar{T} - s T_0\right) \tag{5.37}$$

---

*For convenience of reference, the following derivatives are sometimes useful:

$$\frac{\partial}{\partial t}\left[\mathrm{erfc}\left(\frac{x}{2\sqrt{\kappa t}}\right)\right] = \frac{x}{2\sqrt{\pi}(\kappa t)^{3/2}} \exp(-x^2/4\kappa t)$$

$$\frac{\partial^2}{\partial x^2}\left[\mathrm{erfc}\left(\frac{x}{2\sqrt{\kappa t}}\right)\right] = \frac{x}{2\sqrt{\pi}(\kappa t)^{3/2}} \exp(-x^2/4\kappa t)$$

where s is the transform parameter, $\bar{\sigma}_x(x, s)$ denotes the Laplace transform of the stress $\sigma_x(x, t)$, and $\bar{T}(x, s)$ denotes the Laplace transform of the temperature $T(x, t)$. In performing the Laplace transform, we have employed the initial conditions given in Eqs. (5.36) and the initial condition $\partial T/\partial t(x, 0) = 0$. The Laplace transform of Eq. (4.35), $\bar{T}$, may be evaluated with tables, or we may use the transform given in Example 4.5. After this substitution, we obtain the transformed equation as

$$\frac{d^2\bar{\sigma}_x}{dx^2} - \frac{s^2}{a^2}\bar{\sigma}_x = \frac{\beta(T_\infty - T_0)s}{a^2}\exp\left(-x\sqrt{\frac{s}{\kappa}}\right) \tag{5.38}$$

The preceding equation is a nonhomogeneous, second-order ordinary equation, which we may solve easily in terms of two constants of integration. These constants are obtained by transforming the boundary conditions in Eqs. (5.35) to obtain two boundary conditions for $\bar{\sigma}_x$. After these steps, we obtain

$$\bar{\sigma}_x(x, s) = \beta(T_\infty - T_0)\left[\frac{1}{s - (a^2/\kappa)}\exp(-xs/a) - \frac{1}{s - (a^2/\kappa)}\exp\left(-x\sqrt{\frac{s}{\kappa}}\right)\right] \tag{5.39}$$

The final solution for $\sigma_x(x, t)$ is found by obtaining the inverse Laplace transform of Eq. (5.39). With the help of the integral transform tables given in Ref. 20, the final solution is

$$\frac{\sigma_x(x, t)}{\beta(T_\infty - T_0)} = \exp\left[\frac{a}{\kappa}(at - x)\right] - \frac{1}{2}D_1(x, t) \qquad x < at \tag{5.40a}$$

$$\frac{\sigma_x(x, t)}{\beta(T_\infty - T_0)} = -\frac{1}{2}D_1(x, t) \qquad x > at \tag{5.40b}$$

where we introduce the first Danilovskaya function $D_1(x, t)$, defined by

$$D_1(x, t) = \exp\left[\frac{a(at + x)}{\kappa}\right]\text{erfc}\left(\frac{2at + x}{2\sqrt{\kappa t}}\right) + \exp\left[\frac{a(at - x)}{\kappa}\right]\text{erfc}\left(\frac{-2at + x}{2\sqrt{\kappa t}}\right) \tag{5.40c}$$

Equations (5.40) represent the propagation of the stress σ_x into the elastic medium for the sudden change in the surface temperature. The equations represent a propagating stress wave with a discontinuity at the wave front. To see this, we evaluate Eq. (5.40a) at $x = at^-$ and Eq. (5.40b) at $x = at^+$. From Eq. (5.40a),

$$\frac{\sigma_x(x = at^-, t)}{\beta(T_\infty - T_0)} = 1 - \frac{1}{2}D_1(at^-, t) \qquad x < at$$

and, from Eq. (5.40b),

$$\frac{\sigma_x(x = at^+, t)}{\beta(T_\infty - T_0)} = -\frac{1}{2}D_1(at^+, t) \qquad x > at$$

Thus, there is a jump discontinuity $\Delta\sigma_x$ in the stress at the wave front given by

$$\frac{\Delta\sigma_x}{\beta(T_\infty - T_0)} = \sigma_x(at^-, t) - \sigma_x(at^+, t) = 1 \tag{5.41}$$

In the response, this discontinuity is propagated along the characteristic $x = at$.

5.3.2 Second Danilovskaya Problem

The second Danilovskaya problem[17] consists of the elastic half-space subjected to convective heating for a sudden increase in the ambient temperature. The temperature solution for this problem is given in Sec. 4.4.2, Example 4.7, Eq. (4.40),

$$\frac{T(x,t)-T_C}{T_\infty - T_0} = \operatorname{erfc}\left(\frac{x}{2\sqrt{\kappa t}}\right) - \exp\left(\frac{hx}{k}+\frac{h^2\kappa t}{k^2}\right)\operatorname{erfc}\left(\frac{x}{2\sqrt{\kappa t}}+\frac{h}{k}\sqrt{\kappa t}\right)$$

where T_0 is the initial solid temperature, and T_∞ is the ambient temperature of the fluid to the left of the half-space. Other than the new prescribed temperature distribution, the problem statement is the same as that for the first Danilovskaya problem. We seek a solution to the equation of motion (5.34a), subject to boundary conditions of Eqs. (5.35) and initial conditions of Eqs. (5.36).

To proceed, we first take the Laplace transform of Eq. (4.40) to obtain

$$\bar{T}(x,s) = \frac{T_0}{s} + \frac{T_\infty - T_0}{s}\left[\exp\left(-\frac{x}{k}\sqrt{s}\right)\right] \Big/ \left(1+\frac{h}{k}\sqrt{\frac{s}{\kappa}}\right) \tag{5.42}$$

and then substitute the result into Eq. (5.37) to obtain the ordinary differential equation for $\bar{\sigma}_x(x,s)$. After solving this equation and imposing the boundary conditions,

$$\bar{\sigma}_x(x,s) = \beta(T_\infty - T_0)\left\{\left[\exp\left(-\frac{x}{a}\sqrt{s}\right)\right]\Big/\left[\left(s-\frac{a^2}{\kappa}\right)\left(1+\frac{k}{h}\sqrt{\frac{s}{\kappa}}\right)\right]\right.$$
$$\left.-\left[\exp\left(-x\sqrt{\frac{s}{\kappa}}\right)\right]\Big/\left[\left(s-\frac{a^2}{\kappa}\right)\left(1+\frac{k}{h}\sqrt{\frac{s}{\kappa}}\right)\right]\right\}$$

The solution for $\sigma_x(x,t)$ is obtained by inverting this last equation with inverse transform pairs found in Ref. 21. The solution is

$$\frac{\sigma_x(x,t)}{\beta(T_\infty - T_0)} = \left\{\exp\left[\frac{a(at-x)}{\kappa}\right]\right\}\Big/\left[1-\left(\frac{k}{h}\frac{a}{\kappa}\right)^2\right]$$
$$\times\left[1-\frac{k}{h}\frac{a}{\kappa}\operatorname{erf}\left(\sqrt{\frac{a(at-x)}{\kappa}}\right)\right] - \left\{\exp\left[\frac{h^2}{k^2}\frac{\kappa}{a}(at-x)\right]\right\}\Big/\left[1-\left(\frac{k}{h}\frac{a}{\kappa}\right)^2\right]$$
$$\times\operatorname{erfc}\left(\sqrt{\frac{h^2}{k^2}\frac{\kappa}{a}(at-x)}\right) - \frac{1}{2}D_2(x,t) \qquad x < at \tag{5.43a}$$

$$\frac{\sigma_x(x,T)}{\beta(T_\infty - T_0)} = -\frac{1}{2}D_2(x,t) \qquad x > at \tag{5.43b}$$

where the second Danilovskaya function $D_2(x, t)$ is defined by

$$D_2(x,t) = \left\{\exp\left[\frac{a(at-x)}{\kappa}\right] \Big/ \left(1+\frac{k}{h}\frac{a}{\kappa}\right)\right\}\operatorname{erfc}\left(\frac{x-2at}{2\sqrt{\kappa t}}\right)$$
$$+\left\{\exp\left[\frac{a(at+x)}{\kappa}\right] \Big/ \left(1-\frac{k}{h}\frac{a}{\kappa}\right)\right\}\operatorname{erfc}\left(\frac{x+2at}{2\sqrt{\kappa t}}\right)$$
$$-\left\{\exp\left(\frac{hx}{k}+\frac{h^2}{k^2}\kappa t\right) \Big/ \left[1-\left(\frac{k}{h}\frac{a}{\kappa}\right)^2\right]\right\}\operatorname{erfc}\left(\frac{x}{2\sqrt{\kappa t}}+\frac{h\sqrt{\kappa t}}{k}\right) \tag{5.43c}$$

Equations (5.43) represent the propagation of the stress σ_x into the elastic medium as a result of the sudden application of convective heating. In this case, there is no jump discontinuity at $x = at$. This can be seen by evaluating Eq. (5.43a) at $x = at^-$ and Eq. (5.43b) at $x = at^+$. This analysis shows that equations give the same result, namely, that

$$\sigma_x(at, t) = -\tfrac{1}{2} D_2(at, t)$$

Thus, the amplitude of the stress wave that propagates into the interior along the characteristic $x = at$ is continuous. Note also that the stress amplitude is not constant but, rather, varies with time t according to the D_2 function defined in Eq. (5.43c).

As a final comment on these results, we should mention that the solution we derive in Eqs. (5.43) is identical to the equations that appear in Danilovskaya's original paper,[17] Boley and Weiner[7] solve the problem in nondimensional variables and present their solution in a different form. We have been unable to replicate their nondimensional equations although we believe their numerical results to be correct.

5.3.3 Numerical Results

To understand the dynamic response further, we present numerical results for the dynamic and quasistatic responses for the first Danilovskaya problem. Table 5.1 gives material properties for several metallic materials. Table 5.2 lists values for several parameters used in computing the dynamic response of thermoelastic solids. A key parameter in stress waves is the propagation speed for a dilatation wave denoted by a in Eqs. (5.34b). Table 5.2 shows, for metals, that the propagation speed is very high; it is of the order of 2.5×10^5 in./s (6×10^3 m/s). Thus, we may expect thermal stress wave phenomena to occur in very short times.

The relevant equations for the first Danilovskaya problem include the temperature $T(x, t)$ given in Eq. (4.35). The dynamic response, including inertia forces, is characterized by a triaxial state of stress, where

$$\sigma_x(x, t) = \text{Eq. (5.40)}$$

$$\sigma_y(x, t) = \sigma_z(x, t) = \frac{\nu}{1-\nu}\sigma_x - \frac{E\alpha}{1-\nu}(T - T_{\text{ref}})$$

where the second equation comes from combining Eqs. (5.33a) and (5.33b). We obtain the quasistatic response by neglecting the inertia force in the equation of

Table 5.1 Material properties

Material	Young's modulus E		Poisson's ratio ν	Coefficient of thermal expansion α		Density ρ		Specific heat c		Thermal conductivity k	
	10^6 psi	10^9 N/m^2		10^{-6}/°F	10^{-6}K	10^{-4} lb s^2/in.^{4}	g/cm^3	Btu ft/lb s^2 °F	kJ/kg K	Btu/h ft °F	W/mK
Stainless 17-7PH	29	200	0.32	11	19	7.2	8.0	0.1	0.5	9.7	17
Aluminum 2024-T4	11	72.4	0.33	13	23	2.6	2.8	0.2	0.9	80	140
Aluminum 5086-H32	10	70.3	0.33	13	24	2.6	2.8	0.2	1	72	130
Aluminum 6061-T6	9.9	68.3	0.33	13	23	2.6	2.8	0.2	1	96	170
Magnesium AZ80A-F	6.5	44.8	0.35	14	25	1.6	1.7	0.3	1.1	92.4	76
Titanium MIL-T-9047	16	107	0.32	5.2	9.4	4.4	4.5	0.1	0.5	4.8	8.3
Hastelloy-X	29	200	0.32	8.2	15	7.7	8.0	0.1	0.5	11.2	19
Copper G-3	16	110	0.32	9	16	7.0	8.0	0.1	0.4	24.2	42

Table 5.2 Thermoelastic quantities

Material	Specific weight ρg		Elastic expansion coefficient β		Thermal diffusivity κ		Dilatational wave speed a		Lamé constants λ		Lamé constants μ	
	lb/in.^{3}	10^3 N/m^3	psi/°F	10^6 N/m^2 K	10^{-3} in.^{2}/s	10^{-6} m^2/s	10^3 in./s	10^3 m/s	10^6 psi	10^9 N/m^2	10^6 psi	10^9 N/m^2
Stainless 17-7PH	0.28	76.4	886	11	7.4	4.8	241	6.12	20	134	11	76
Aluminum 2024-T4	0.10	27.4	420	4.9	84	54	245	6.22	7.7	53.1	3.9	27
Aluminum 5086-H32	0.10	27.4	382	4.9	76	49	247	6.26	7.4	51.0	3.8	26
Aluminum 6061-T6	0.10	27.4	378	4.7	99	64	240	6.11	7.2	49.6	3.7	26
Magnesium AZ80A-F	0.063	17.1	303	3.8	63	41	249	6.33	5.6	38.6	2.4	17
Titanium MIL-T-9047	0.16	44.2	231	2.8	5.3	3.4	230	5.84	10	71.7	5.9	41
Hastelloy-X	0.30	81	660	8.2	6.7	4.3	232	5.90	20	164	11	76
Copper G-3	0.27	74.0	400	5.0	21	13	180	4.58	11	74.5	6.1	42

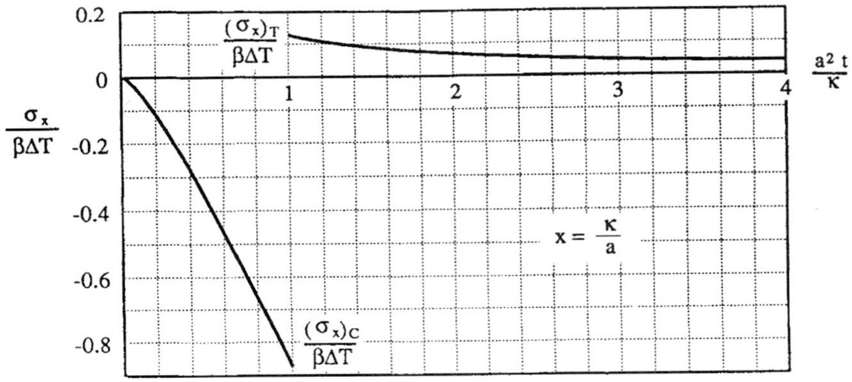

Fig. 5.2 Sketch of dynamic response for first Danilovskaya problem.

motion. This shows that $\partial \sigma_x / \partial x = 0$ and, hence, $\sigma_x = \sigma_x(t)$. However, the free surface of the half-space requires that $\sigma_x = 0$ at $x = 0$; hence, $\sigma_x \equiv 0$. Then, from combining Eq. (5.33a) and (5.33b), we obtain the solution for other stress components. Thus, the quasistatic response is given by

$$\sigma_x(x, t) \equiv 0 \qquad \sigma_y(x, t) = \sigma_z(x, t) = -\frac{E\alpha}{1 - \nu}(T - T_{\text{ref}})$$

Comparing the two solutions, we see that the role of the inertia forces is to introduce the stress wave defined by Eqs. (5.40).

To characterize the stress wave, Danilovskaya,[16] as well as Boley and Weiner,[7] plots the stress history at a typical point on the interior of the half-space. Figure 5.2 shows a typical stress history in nondimensional variables. The figure shows that, as the wave reaches the interior point, the stress rises to a maximum in compression. As the stress wave passes the point, there is a discontinuous change from compression to a smaller tensile stress. This change is the jump discontinuity we identify in Eq. (5.41). Thereafter, the figure shows that the tensile stress subsides to zero.

To gain an understanding of the stress magnitudes and the time scale for the dynamic response, see Fig. 5.3, which presents numerical results for the dynamic response of a stainless steel half-space subjected to a sudden surface-temperature change of $\Delta T = T - T_{\text{ref}} = 10°\text{F}$. Figure 5.3a gives the problem statement and Fig. 5.3b shows the stress history for a point P at $x = 0.125$ in. The computations show that because of the high propagation speed, the stress wave reaches point P in a very short time, of the order of 10^{-7} s. The computations also show that, very early in the response, the compressive stress rises to a maximum of 3190 psi and the tensile stress diminishes so that, by the time the wave reaches P, the compressive stress is at a maximum but the tensile stress is negligibly small. Note that the duration of the stress wave is extremely small, of the order of 10^{-13} s. At point P, the stresses $\sigma_y = \sigma_z$ also occur and have values of about 40% of σ_x. At this very early time in the response, the temperature change at point P is still virtually zero, and so the second term in the $\sigma_y = \sigma_z$ equation that is proportional to $E\alpha\Delta T$ does not contribute to these stresses. The preceding equation shows that the second term, in fact, is the quasistatic response.

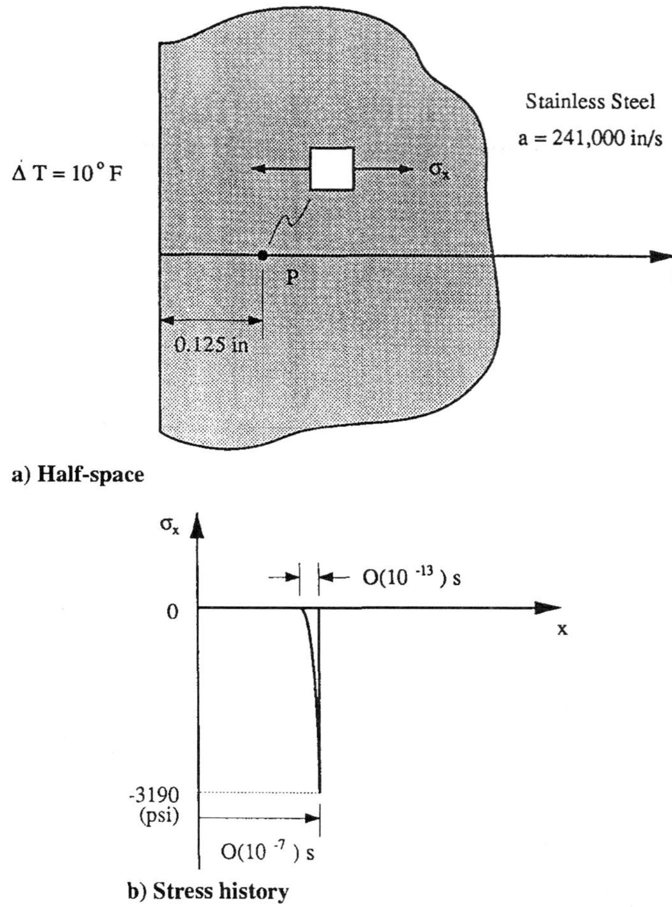

Fig. 5.3 Dynamic response of σ_x at x = 0.125 in. for first Danilovskaya problem.

Figure 5.4 presents the quasistatic response at point P. The figure shows that the interior temperature change rises slowly in contrast to the time scale for wave propagation. Perceptible temperature changes and quasistatic stresses are sensed for times of the order of seconds. After a relatively long time, the interior temperature rises to ΔT, and the maximum quasistatic stresses rise to

$$\sigma_y = \sigma_z|_{\max} = -\frac{E\alpha}{1-\nu}\Delta T = -4560\,\text{psi}$$

The first Danilovskaya solution for a typical metallic material shows that the thermal-stress response due to a sudden change in surface temperature occurs on two distinct time scales. The dynamic stresses induced by the inertia forces consist of an extremely sharp stress wave that propagates at a very high wave speed. With customary instrumentation in the laboratory, we would have great difficulty detecting the stress wave. The second part of the response, as predicted by the quasistatic model, occurs at a much slower rate than is consistent with our intuition. We may easily measure the quasistatic response with customary

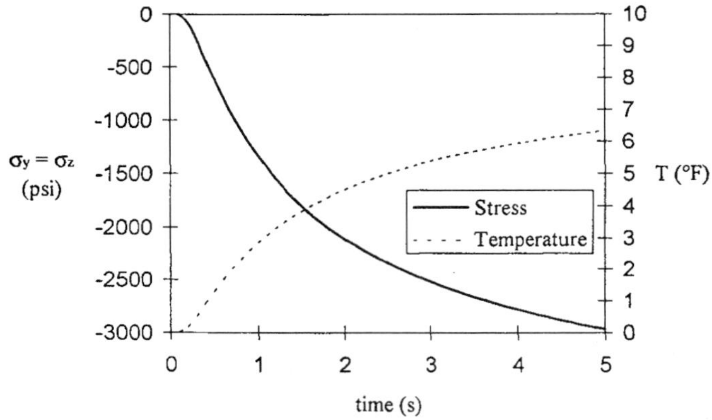

Fig. 5.4 Quasistatic response at $x = 0.125$ in. for first Danilovskaya problem.

laboratory thermocouple and strain-gauge instrumentation. The maximum values of the dynamic stresses are significant but are smaller than the maximum quasistatic stresses.

5.4 Fundamental Quasistatic Problems

We consider now two important fundamental problems in thermoelasticity. The first problem is that of a fully restrained solid with a uniform temperature change, and the second problem is an unrestrained solid with a linear temperature distribution. In each problem, we seek the thermally induced stresses.

5.4.1 Fully Restrained Solids

Consider now an elastic solid subject to a uniform temperature change $\Delta T(t) = T(t) - T_{\text{ref}}$, where we assume a constant reference temperature. The body forces are assumed to be zero, and inertia forces are neglected so that the thermally induced elastic response is quasistatic. Displacements at every point on the surface of the body are restrained to be zero so that expansion (or contraction) of the body cannot occur. Because deformations of points on the boundary are restrained, we show that displacements of points on the interior are also zero. Let us now consider the thermal stresses for three-dimensional and then two- and one-dimensional solids.

Three-dimensional solid. For the present assumptions of $\Delta T(t)$, zero body forces, and quasistatic behavior, the displacement formulation for a general three-dimensional solid given by Eqs. (5.1) reduces to the homogeneous set of equations,

$$
\begin{aligned}
(\lambda + G)\frac{\partial e}{\partial x} + G\nabla^2 u &= 0 \\
(\lambda + G)\frac{\partial e}{\partial y} + G\nabla^2 v &= 0 \\
(\lambda + G)\frac{\partial e}{\partial z} + G\nabla^2 w &= 0
\end{aligned}
\tag{5.44}
$$

The three equilibrium equations given in Eqs. (5.44) do not depend on temperature (we assume that λ and G are constant), nor do the equations contain any mechanical loads. Because the three displacements are specified as zero at all points on the boundary, the only possible solution is that the solid remain undeformed:

$$u(x, y, z, t) = v(x, y, z, t) = w(x, y, z, t) \equiv 0$$

Given that the displacements are identically zero, then, by the strain-displacement relations (5.2), all strains are zero,

$$\varepsilon_x = \varepsilon_y = \varepsilon_z = \varepsilon_{xy} = \varepsilon_{xz} = \varepsilon_{yz} = 0$$

at interior points. Referring now to the stress–strain relations (5.3), we see that

$$\sigma_x = \sigma_y = \sigma_z = -(3\lambda + 2G)\alpha\Delta T$$

By using Eq. (5.4), we may write this last equation in terms of E and ν as

$$\sigma_x = \sigma_y = \sigma_z = -\frac{E\alpha}{1 - 2\nu}\Delta T \tag{5.45}$$

and all shearing stresses are identically zero. Thus, a fully restrained solid with a uniform temperature change will experience a hydrostatic state of stress, with the magnitude of the stress components given by Eq. (5.45). Note that the material parameter that determines the magnitude of the thermal stress is the *product* $E\alpha$ since Poisson's ratio ν is about the same for most materials. We see that an increase in temperature above the reference temperature, $\Delta T = T - T_{\text{ref}} > 0$, causes compressive thermal stress, whereas a decrease in temperature below the reference temperature causes tensile stress.

Thin plate. Suppose that, instead of a three-dimensional solid, we have a thin plate that is fully restrained on its boundary. The lateral surfaces of the plate are traction-free, and the plate is in a state of plane stress. Following the same arguments as for the three-dimensional solid, we conclude that the displacements are identically zero:

$$u(x, y, t) = v(x, y, t) \equiv 0$$

and, therefore, the strain components are zero:

$$\varepsilon_x = \varepsilon_y = \varepsilon_{xy} = 0$$

at interior points. Using the stress–strain law for plane stress, Eqs. (5.13), we find that

$$\sigma_x = \sigma_y = -\frac{E\alpha}{1 - \nu}\Delta T \tag{5.46}$$

and the shearing stress $\sigma_{xy} = 0$. Thus, a fully restrained thin plate in a state of plane stress will experience a hydrostatic state of stress with the magnitude of stress given by Eq. (5.46). The only differences from the stress state in three dimensions, Eq. (5.45), are the absence of σ_z because the plate is thin and the coefficient of Poisson's ratio in the denominator.

One-dimensional rod. In the simplest case, we consider a rod for which the cross-sectional dimensions are small, so that we have only one displacement component u, one strain component ε_x, and one stress component σ_x. If the ends of the rod are fully restrained, the interior displacements and strains will be zero at every point, and the stress will be given by

$$\sigma_x = -E\alpha\Delta T \tag{5.47}$$

Comparing Eqs. (5.45–5.47), we see that a fully restrained elastic solid has thermal stresses given by the equation

$$\sigma = -kE\alpha\Delta T \tag{5.48}$$

where k is a constant depending on the idealization. For $\nu = 0.3$, the equations give $k = 1.0$, 1.4, and 2.5 for one, two, and three dimensions, respectively.

Neglecting stress concentrations, Goodier[22] states that we can use Eq. (5.48) to estimate maximum thermal stresses in bodies with temperature variations if we use the maximum local ΔT. In general, he suggests that the greatest thermal stress is estimated by

$$\sigma_{\max} = -kE\alpha\Delta T_{\max} \tag{5.49}$$

where $0.5 \leq k \leq 2.5$. For example, a local "hot spot" in a thin plate will cause compressive thermal stresses σ_x and σ_y with an approximate magnitude of $1.4E\alpha\Delta T_{\max}$, where $\Delta T_{\max}$ is the maximum temperature difference between the hot spot and the cooler part of the plate.

5.4.2 Unrestrained Solid with Linear Temperature Distribution

Consider now an elastic three-dimensional solid that is completely *unrestrained*. Once again, the body forces are assumed to be zero, and the thermally induced elastic response is quasistatic. Because thermal deformations are completely unrestrained, the surface of the solid is free from surface tractions. Now we assume a special temperature distribution that is linear in the x, y, z Cartesian coordinates:

$$T(x, y, z, t) = a(t) + b(t)x + c(t)y + d(t)z \tag{5.50}$$

where the coefficients are known functions of time.

Let us refer now to the six equations for the stress formulations given in Eqs. (5.8). These equations assume zero body forces and constant reference temperature. Temperature enters the equations through second derivatives with respect to the spatial coordinates. Thus, if we substitute Eq. (5.50) into Eqs. (5.8), the temperature terms vanish. The result will be six homogeneous equations for the six unknown stresses. Since all surface tractions are zero on the boundary, all stresses must be zero there as well. Thus, the only possible solution to the set of homogeneous equations with homogeneous boundary conditions is that all stresses must be zero. And thus, for an *unrestrained* body with a linear temperature distribution of the form of Eq. (5.50),

$$\sigma_x = \sigma_y = \sigma_z = \sigma_{xy} = \sigma_{xz} = \sigma_{yz} \equiv 0 \tag{5.51}$$

Note that the temperature distribution must be linear in *Cartesian* coordinates. The stress would *not* be zero if the temperature were linear in another coordinate system such as cylindrical coordinates.

Thus, an unrestrained solid with a linear temperature distribution will have zero stresses as the solid deforms with the temperature change. And, although the stresses are zero, there will be strains and displacements. This conclusion is drawn from a three-dimensional elastic solid, but it applies to other structural idealizations as well. For example, we see this behavior illustrated for beams. A statically determinant beam with a linear temperature gradient through the beam depth experiences bending deformations, but the thermal stresses are zero.

In this section, we have seen that fully *restrained* elastic bodies experience states of homogeneous stress without strain and deformation. In contrast, *unrestrained* bodies with linear temperature distributions have zero stress but experience strain and deformation. Thus, we have examples of "stress without strain" and "strain without stress." If we confine our thinking to the *isothermal* Hooke's law, we reach an impasse in trying to explain these apparently contradictory statements. The resolution of the anomaly, of course, lies in the fact that Hooke's law for thermoelasticity problems contains $\alpha \Delta T$ terms.

We should keep in mind that the conclusions reached here assume an isotropic, homogeneous elastic material with constant material properties.

5.5 Two-Dimensional Thermoelastic Problems

Using the thermoelastic formulations developed in Sec. 5.2, we now solve some basic thermoelastic problems assuming quasistatic behavior. For convenience, we will analyze the problems for temperatures independent of time. The solutions will be exact within the assumptions of plane stress or plane strain, but they will be restricted in geometry and boundary conditions. The solutions are valuable for the insight they provide into basic thermoelastic behavior of structural components with bending. The solutions provide the foundations for engineering thermoelastic models that are applicable to more general geometry and boundary conditions.

5.5.1 Unrestrained Thin Rectangular Beam

Consider a long, thin rectangular beam[7] (Fig. 5.5) subjected to a prescribed temperature distribution $T(y)$. The beam is unrestrained, so that all edges are traction-free. The beam length is much larger than the depth, i.e., $L \gg 2h$. In addition, the beam thickness b is small compared to the length L, so that we have plane stress. We wish to develop equations for the displacements, stresses, and strains within the beam as a function of x, y for the prescribed temperature distribution $T(y)$.

With some previous knowledge of beam behavior, an effective approach is to use the stress-function formulation for plane stress. From Sec. 5.2.4, Eq. (5.22),

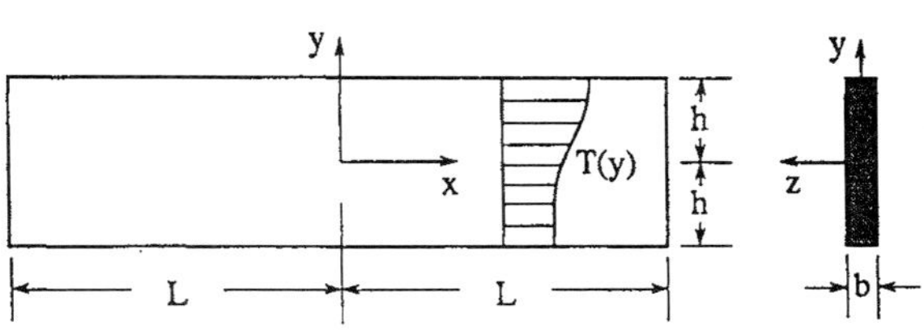

Fig. 5.5 Unrestrained, thin, rectangular beam.

the stress function $F(x, y)$ for plane stress must satisfy

$$\frac{\partial^4 F}{\partial x^4} + 2\frac{\partial^4 F}{\partial x^2 \partial y^2} + \frac{\partial^4 F}{\partial y^4} + E\alpha\nabla^2 T = 0$$

where

$$\sigma_x = \frac{\partial^2 F}{\partial y^2} \qquad \sigma_{xy} = -\frac{\partial^2 F}{\partial x \partial y} \qquad \sigma_y = \frac{\partial^2 F}{\partial x^2}$$

Rather than try to find a general solution for $F(x, y)$, we use engineering insight to obtain a solution specifically for this problem. Because the beam is unrestrained and the prescribed temperature is independent of x, the problem reminds us of pure bending of beams in strength of materials. In pure bending, there is only a normal stress σ_x, and the shearing stress σ_{xy} is zero. Moreover, the top and bottom edges of the beam are stress-free, so that there is a possibility that σ_y may also be zero on the beam interior. With these considerations, we assume the stress state

$$\sigma_x = \sigma_x(y) \qquad \sigma_{xy} = 0 \qquad \sigma_y = 0 \tag{5.52}$$

and see if we can satisfy the equations of thermoelasticity. If we can, then the assumed stresses are the exact solution.

Based on the assumed state of stress given in Eqs. (5.52), an examination of Eqs. (5.20) shows that the stress function F is independent of x; i.e., $F = F(y)$. Then the governing differential equation reduces to

$$\frac{\mathrm{d}^2}{\mathrm{d}y^2}\left(\frac{\mathrm{d}^2 F}{\mathrm{d}y^2} + E\alpha T\right) = 0$$

Using the definition of σ_x, we write this last equation as

$$\frac{\mathrm{d}^2}{\mathrm{d}y^2}(\sigma_x + E\alpha T) = 0$$

which shows that

$$\sigma_x + E\alpha T = c_1 + c_2 y$$

where c_1 and c_2 are constants. The constants are determined from the boundary conditions at the ends of the beam. Because the beam is unrestrained, the boundary conditions require that the ends be stress-free. For an arbitrary variation of temperature T with y, the preceding equation shows that we cannot find constants c_1 and c_2 to make the stress zero at every point through the depth. Hence, with the assumed solution, we cannot satisfy the traction-free boundary condition exactly. Instead, we resort to an approximation that will limit the validity of the solution to a very long beam. We find values for c_1 and c_2 that make the resultant force and resultant moment of the stress distribution zero. This approach ensures that the beam is in equilibrium.

For the resultant force and moment to be zero, we integrate the stress and the first moment of the stress over the beam's cross-sectional area A and set the results equal to zero. Thus, we require

$$\int_A \sigma_x \, \mathrm{d}A = 0 \qquad \int_A \sigma_x y \, \mathrm{d}A = 0$$

Substituting for σ_x and performing the integrations give two equations that we can solve to obtain c_1 and c_2. The solution is found to be

$$\sigma_x = \frac{E\alpha}{2bh} \int_A T(y) \, \mathrm{d}A + \left[\frac{3}{2} \frac{E\alpha}{bh^3} \int_A T(y) y \, \mathrm{d}A \right] y - E\alpha T(y)$$

which we write as

$$\sigma_x = \frac{N_T}{A} + \frac{M_T y}{I} - E\alpha T(y) \tag{5.53a}$$

with

$$N_T = E\alpha \int_A T(y) \, \mathrm{d}A \tag{5.53b}$$

$$M_T = E\alpha \int_A T(y) y \, \mathrm{d}A \tag{5.53c}$$

where $A = 2bh$ is the cross-sectional area, and $I = 2bh^3/3$ is the moment of inertia of the cross section. The quantity N_T has units of force, and we will call it the *thermal force*. Similarly, M_T has units of moment, and we will call it the *thermal moment*. Note that the stress σ_x in Eq. (5.53a) has the form of the flexure formula for an isothermal beam (i.e., $P/A + Mc/I$), but with the addition of the term $-E\alpha T(y)$. Thus, the thermal stress is the sum of a constant, a linear function of y, and a constant times the temperature variation through the depth. Unlike an isothermal beam, the thermal stress σ_x will not be a linear function of y because of the temperature term.

Using the expression given in Eq. (5.53) for σ_x, we can now derive equations for the strains ε_x, ε_y, and ε_{xy} and then integrate the strains to obtain the displacements u and v. From the strain–stress equations (5.21), we find that

$$\varepsilon_x = \frac{N_T}{EA} + \frac{M_T}{EI} y$$

$$\varepsilon_y = -\frac{\nu}{E} \left(\frac{N_T}{A} + \frac{M_T y}{I} \right) + (1 + \nu) \alpha T(y) \tag{5.54}$$

$$\varepsilon_{xy} = 0$$

Then we may integrate the strains to obtain the displacements. First,

$$\frac{\partial u}{\partial x} = \varepsilon_x = \frac{N_T}{EA} + \frac{M_T}{EI} y$$

so that

$$u(x, y) = \left(\frac{N_T}{EA} + \frac{M_T}{EI}y\right)x + f_1(y)$$

where $f_1(y)$ is a function to be determined. With the coordinate system located at the beam center, the y axis lies in a plane of symmetry so that we have the boundary condition $u(0, y) = 0$. Imposing this condition requires $f_1(y) = 0$. Thus, the u displacement is given by

$$u(x, y) = \left(\frac{N_T}{EA} + \frac{M_T y}{I}\right)x \tag{5.55}$$

Note that at a typical section, $x =$ const along the beam, and the u displacement varies linearly with y. This means that "plane sections remain plane." Next, we integrate ε_y to obtain the v displacement

$$\frac{\partial v}{\partial y} = \varepsilon_y = -\frac{\nu}{E}\left(\frac{N_T}{A} + \frac{M_T y}{I}\right) + (1+\nu)\alpha T(y)$$

so that

$$v(x, y) = -\frac{\nu}{E}\left(\frac{N_T y}{A} + \frac{M_T y^2}{2I}\right) + (1+\nu)\alpha \int T(y)\,\mathrm{d}y + f_2(x)$$

where $f_2(x)$ is a function to be determined. To find f_2, recall that the shearing strain ε_{xy}, Eqs. (5.11), is zero; thus, we have the condition

$$\frac{\partial u}{\partial y} + \frac{\partial v}{\partial x} = 0$$

Substituting u from Eq. (5.55) and the expression for v in the preceding equation gives the condition

$$\frac{M_T x}{I} + \frac{\mathrm{d}f_2}{\mathrm{d}x} = 0$$

Hence,

$$f_2 = -\frac{M_T x^2}{2I} + c_2$$

The constant c_2 is determined from the condition of symmetry at the origin that $v(0, 0) = 0$. The final solution for $v(x, y)$ may then be written as

$$v(x, y) = -\frac{\nu}{E}\left(\frac{N_T y}{A} + \frac{M_T y^2}{2I}\right) + G(y) - \frac{M_T x^2}{2EI} + G(0) \tag{5.56a}$$

where we define $G(y)$ as the indefinite integral

$$G(y) = (1+\nu)\alpha \int T(y)\,\mathrm{d}y \tag{5.56b}$$

Equations (5.53–5.56) define the thermoelasticity solution for the stresses, strains, and displacements for the unrestrained, thin rectangular beam. The solution satisfies all the equations of thermoelasticity, but it is not an "exact" solution because the traction-free boundary conditions at the ends of the beam are not satisfied exactly. Instead, the resultant force and moment were required to be zero so that global equilibrium is assured. By Saint Venant's principle,* the solution should be quite accurate at a distance away from the beam ends. Thus, for a long thin beam, the solution is nearly an exact solution.

Example 5.1

To illustrate the character of thermal stresses for an unrestrained thin beam of rectangular cross section, let us determine the stress distribution for a particular variation of temperature. We consider the family of symmetrical temperatures introduced by Rao and Johns,[23]

$$T(y) = (T_0 + T_1) - T_1(y/h)^n \qquad y > 0$$

$$T(-y) = T(y) \qquad y < 0$$

when $n = 1, 2, 3, \ldots$. The sketch of the temperatures in Fig. 5.6a shows that the preceding equation predicts a peak temperature of $T_0 + T_1$ at the beam center and a temperature T_0 at the top and bottom surfaces for $y = \pm h$. We wish to determine the thermal-stress distribution by evaluating Eqs. (5.53),

$$\sigma_x = \frac{N_T}{A} + \frac{M_T y}{I} - E\alpha T(y)$$

where N_T is the thermal force, and M_T the thermal moment. For the rectangular cross section shown in Fig. 5.1,

$$N_T = E\alpha b \int_{-h}^{h} T(y)\,\mathrm{d}y = 2E\alpha b \int_{0}^{h} T(y)\,\mathrm{d}y$$

where we integrate over only the upper one-half of the beam since T is symmetric. Substituting $T(y)$ and evaluating the integral yields

$$N_T = 2E\alpha bh\left[(T_1 - T_0) - \frac{T_1}{n+1}\right]$$

The thermal moment,

$$M_T = E\alpha b \int_{-h}^{h} T(y)y\,\mathrm{d}y = 0$$

*According to Timoshenko and Goodier,[2] Saint Venant's principle states that if the forces acting on a small portion of the surface of an elastic body are replaced by another statically equivalent system of forces acting on the same portion of the surface, this redistribution of loading produces substantial changes in the stresses locally but has a negligible effect on the stresses at distances which are large in comparison with the linear dimensions of the surface on which the forces are changed.

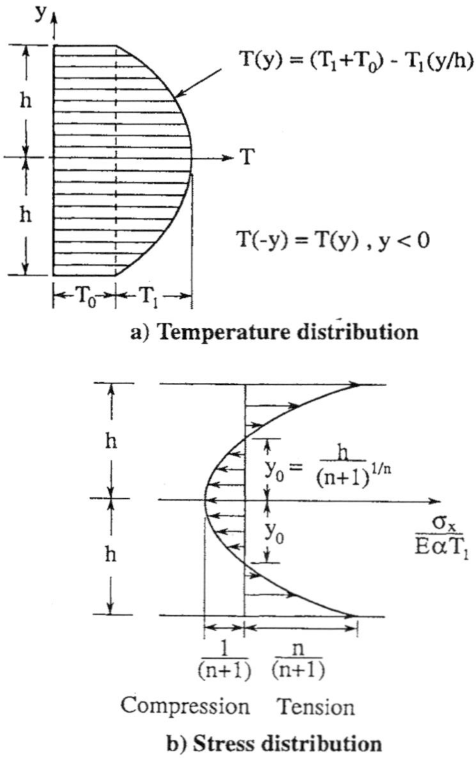

Fig. 5.6 Thermal stresses in an unrestrained, thin, rectangular beam.

where the integral is zero because the product of $T(y)$ times y is an antisymmetric function. In general, we can say that a symmetrical temperature distribution does not produce a thermal moment. Substituting these results in the equation for σ_x gives the thermal stress,

$$\sigma_x = -E\alpha T_1 \left[\frac{1}{n+1} - \frac{y^n}{h^n}\right] \qquad y > 0$$

$$\sigma_x(-y) = \sigma_x(y) \qquad y < 0$$

where the stress distribution is symmetrical because of the symmetrical temperatures. Figure 5.6b shows a sketch of a typical stress distribution. There are several important points to observe. Note first that the stress magnitude depends only on T_1; the uniform part of the temperature distribution T_0 does not induce thermal stress because the beam is free to expand. Second, we see that, in the central "hot" core of the beam between $\pm y_0$, the thermal stress is compressive. For the two regions outside the hot core for $y_0 < |y| < h$, the thermal stresses are tensile. Tensile stresses are required since the beam is unrestrained, and the stress distribution must be self-equilibrating. For $n = 1$, the maximum tensile and compressive stresses are equal but, for $n \geq 2$, the maximum stresses are always tensile.

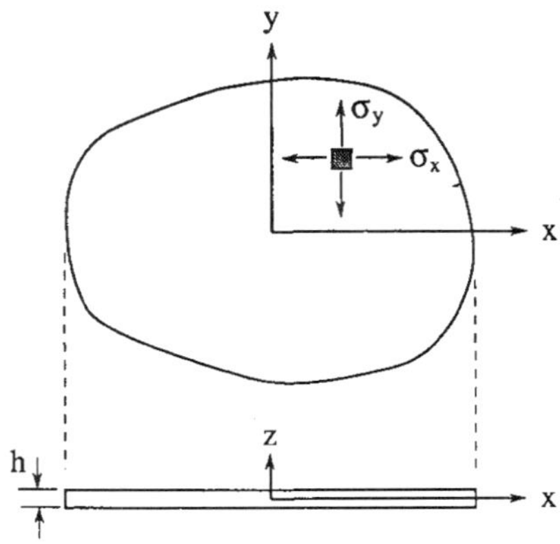

Fig. 5.7 Unrestrained thin plate with thickness temperature gradient.

The paper of Rao and Johns[23] calls the thermal stresses evaluated earlier "infinitely long beam stresses" since the stresses are incorrect at the end of a finite beam, where the stresses should be zero. Practical experience gained from comparing this analytical solution with finite element plane stress analyses indicates that the analytical solution predicts acceptable stresses near the beam center for even relatively short beams, say, $L \geq 3(2h)$.

5.5.2 Unrestrained Thin Plate

A problem similar to the preceding beam problem is an unrestrained thin plate[7] (Fig. 5.7). We assume that the thickness of the plate is small compared to its other dimensions and that the top and bottom surfaces of the plate, as well as the plate edges, are traction-free. The body forces are zero. The plate is subjected to a specified temperature distribution that varies through the thickness but is independent of the spatial coordinates x and y; thus, $T = T(z)$.

Although the plate is thin, the problem differs from the preceding beam problem because we do not have plane stress. Here, because of the temperature distribution $T(z)$, the stresses will vary through the thickness, contrary to the basic assumption of plane stress. We will need to analyze this problem with the three-dimensional equations of elasticity. However, as in the beam solution, we greatly simplify the problem by assuming a relatively simple state of stress.

From the behavior of a thin plate, we expect that on account of the temperature $T(z)$, the plate will bend, giving a displacement w in the z direction. We also expect the plate to expand (or contract) in the x–y plane, giving u and v displacements. When the plate bends, it behaves differently from a beam because it has a deformed shape with curvatures in both the x–z and y–z planes. Thus, we may expect a biaxial state of stress with components σ_x and σ_y and, because of bending, we expect these stresses to vary through the plate thickness. Since the plate is thin, we can neglect stresses with a z subscript; that is, we take $\sigma_z = \sigma_{xz} = \sigma_{yz} = 0$. We assume that the biaxial stresses σ_x and σ_y have the form

$$\sigma_x = \sigma_y = f(z) \tag{5.57a}$$

where $f(z)$ is a function to be determined, and we take all other stresses to be zero:

$$\sigma_{xy} = \sigma_z = \sigma_{yz} = \sigma_{xz} = 0 \tag{5.57b}$$

We consider now the stress function formulation of three-dimensional thermoelasticity as expressed in Eqs. (5.8). Substituting the assumed stresses from Eqs. (5.57) and the prescribed temperature distribution $T = T(z)$, we obtain, from the first three equations,

$$(1+\nu)\frac{d^2 f}{dz^2} + E\alpha\left(\frac{1+\nu}{1-\nu}\right)\frac{d^2 T}{dz^2} = 0$$

$$(1+\nu)\frac{d^2 f}{dz^2} + E\alpha\left(\frac{1+\nu}{1-\nu}\right)\frac{d^2 T}{dz^2} = 0$$

$$2\frac{d^2 f}{dz^2} + E\alpha\left(\frac{1+\nu}{1-\nu}\frac{d^2 T}{dz^2} + \frac{d^2 T}{dz^2}\right) = 0$$

and the last three equations are satisfied identically. The three preceding equations are actually identical, and we may write them in the form

$$\frac{d^2}{dz^2}\left(f + \frac{E\alpha}{1-\nu}T\right) = 0$$

By direct integration,

$$f + \frac{E\alpha}{1-\nu}T = c_1 + c_2 z$$

and, thus, we can write the stresses as

$$\sigma_x = \sigma_y = f(z) = c_1 + c_2 z - \frac{E\alpha}{1-\nu}T$$

where c_1 and c_2 are constants of integration. These constants are determined from the boundary conditions on the edge of the plate. As in the preceding beam problem, we cannot select c_1 and c_2 so that the plate edges are stress-free. Instead, we determine c_1 and c_2 so that the resultant force and moment on the edge are zero. These conditions require

$$\int_{-(h/2)}^{h/2} \sigma_x \, dz = \int_{-(h/2)}^{h/2} \sigma_x z \, dz = 0$$

and similarly for σ_y. Substituting for σ_x and performing the integrations give two equations for c_1 and c_2. Solving for these constants and substituting them into Eq. (5.57a) yields

$$\sigma_x = \sigma_y = \frac{1}{1-\nu}\left[\frac{N_T}{h} + \frac{12M_T}{h^3}z - E\alpha T\right] \tag{5.58a}$$

where

$$N_T = E\alpha \int_{-(h/2)}^{h/2} T(z)\, dz \tag{5.58b}$$

and

$$M_T = E\alpha \int_{-(h/2)}^{h/2} T(z)z \, dz \tag{5.58c}$$

The quantity N_T has units of force/length, and it is the thermal force. Similarly, M_T has units of moment/length, and it is the thermal moment. Equations (5.58) represent the plate thermal stresses for an arbitrary temperature distribution through the thickness.

The remaining steps in the solution determine the strain components ε_x, ε_y, and ε_z and then the displacement components u, y, and w. The results are

$$\varepsilon_x = \varepsilon_y = \frac{1}{E}\left[\frac{N_T}{h} + \frac{12M_T}{h^3}z\right] \tag{5.59a}$$

$$\varepsilon_z = -\frac{2\nu}{1-\nu}\frac{1}{E}\left[\frac{N_T}{h} + \frac{12M_T}{h^3}z\right] + \left(\frac{1+\nu}{1-\nu}\right)\alpha T \tag{5.59b}$$

and all shearing strains are zero. The displacements are found by integrating the strain components and are given by

$$u(x, y, z) = \frac{1}{E}\left[\frac{N_T}{h} + \frac{12M_T}{h^3}z\right]x \tag{5.60a}$$

$$v(x, y, z) = \frac{1}{E}\left[\frac{N_T}{h} + \frac{12M_T}{h^3}z\right]y \tag{5.60b}$$

$$w(x, y, z) = -\frac{2\nu}{1-\nu}\frac{1}{E}\left[\frac{N_T}{h}z + \frac{6M_T}{h^3}z^2\right] + \left(\frac{1+\nu}{1-\nu}\right)\alpha \int T(z)\, dz + g(x, y) \tag{5.60c}$$

where

$$g(x, y) = -\frac{6M_T}{h^3 E}(x^2 + y^2) \tag{5.60d}$$

In the derivation of the displacement components, the constants of integration were evaluated by imposing the conditions that the displacements be zero at the origin. The function $g(x, y)$ was determined by imposing the zero shearing strain conditions that require

$$\varepsilon_{xz} = \frac{\partial w}{\partial x} + \frac{\partial u}{\partial z} = 0 \qquad \varepsilon_{yz} = \frac{\partial w}{\partial y} + \frac{\partial v}{\partial z} = 0$$

Equations (5.58–5.60) constitute the thermoelasticity solution for the unrestrained thin plate with a thickness temperature distribution. The plate thermal response is characterized by a biaxial state of stress with $\sigma_x = \sigma_y$. The stress equations are the sum of a uniform component determined by the thermal force N_T, a component that varies linearly with z determined by the thermal moment M_T, and a component that varies directly with the temperature $T(z)$. In general, the thermal stress varies nonlinearly with z. If the temperature distribution is symmetric about the middle x–y plane, the thermal moment will be zero, and the stress distribution will also be symmetric. If the thermal moment is zero, $w = w(z)$, and the plate does not bend.

There are two equal in-plane strain components $\varepsilon_x = \varepsilon_y$ and a strain component through the thickness ε_z. The in-plane strains ε_x and ε_y measure extension or contraction of fibers parallel to the middle plane in the x and y directions, respectively. These strains reflect deformations induced by the thermal force and thermal moment. The strain ε_z measures the extension or contraction of fibers perpendicular to the middle plane of the plate. This strain measures deformation per unit length through the thickness of the plate.

The displacements u and v reflect in-plane extensions or contractions, as well as the effect of bending. For $z = 0$, the terms $N_T x/Eh$ and $N_T y/Eh$ are deformations of the middle plane. For an increase in temperature, N_T is positive, and the middle plane expands linearly in the x and y directions. The second terms in the u and v displacements are the contributions of bending of the plate. Because these displacements vary linearly with z, a straight line normal to the middle plane before deformation remains a straight line during deformation. This behavior occurs because the shearing strains ε_{xz} and ε_{yz} are zero. The w displacement consists of the sum of three terms. The first two terms are independent of x and y and are a function of z. These terms are a consequence of the plate thickness changing with temperature. The last term $g(x, y)$ is independent of z and characterizes bending displacement. Note that it is directly proportional to M_T. The displacement contribution varies with $x^2 + y^2$, and so the deformed shape of the plate is a paraboloid of revolution. The direction of bending, up or down, is determined by the sign of M_T; the sign of M_T depends on the temperature distribution. Intuitively, if the bottom surface of the plate is hotter than the top surface, we expect that $w > 0$ and that the deformed shape will be curved upward. If the top surface is hotter than the bottom surface, we expect that $w < 0$ and that the deformed shape will be curved downward.

The solution satisfies all the equations of thermoelasticity, but it is not an exact solution because the stresses σ_x and σ_y are not zero on the plate edges. The stresses satisfy global equilibrium and, at distances away from the edges, the stresses are accurate in accordance with Saint Venant's principle.

5.5.3 Hollow Cylinder in Plane Strain

Consider a long, hollow cylinder[2] with inside radius a and outside radius b (Fig. 5.1). The cylinder experiences an axisymmetric temperature variation $T(r)$. The ends of the cylinder are restrained against thermal expansion, and we assume a state of plane strain. From Sec. 5.2.5, we use the basic equations for a displacement formulation. The first step is to solve Eq. (5.31),

$$\frac{\mathrm{d}}{\mathrm{d}r}\left[\frac{1}{r}\frac{\mathrm{d}(ru)}{\mathrm{d}r}\right] = \left(\frac{1+\nu}{1-\nu}\right)\alpha\frac{\mathrm{d}T}{\mathrm{d}r}$$

for the displacement u. Because the left-hand side is written in an exact differential form, we may integrate Eq. (5.31) twice to obtain u. This gives

$$u(r) = \left(\frac{1+\nu}{1-\nu}\right)\alpha\frac{1}{r}\int Tr\,\mathrm{d}r + c_1 r + \frac{c_2}{r}$$

where c_1 and c_2 are constants of integration. Because we have a hollow cylinder, the boundary conditions are that the surface tractions vanish on the inner and outer surfaces of the cylinder. In this case, the boundary conditions require that the radial stress be zero on these surfaces; thus,

$$\sigma_r(a) = 0 \qquad \sigma_r(b) = 0$$

From Hooke's law, Eqs. (5.30), we may express the radial stress σ_r in terms of the strain components. Then using the strain components given by Eqs. (5.27), we may write the radial stress in terms of the displacement $u(r)$ and its derivative. Finally, substituting the equation above for $u(r)$ and evaluating the stress on the boundaries yields,

$$-\frac{E\alpha}{1-\nu}\frac{f(a)}{a^2} + \frac{E}{(1+\nu)(1-2\nu)}c_1 - \frac{E}{1+\nu}\frac{c_2}{a^2} = 0$$

$$-\frac{E\alpha}{1-\nu}\frac{f(b)}{b^2} + \frac{E}{(1+\nu)(1-2\nu)}c_1 - \frac{E}{1+\nu}\frac{c_2}{b^2} = 0$$

where

$$f(r) = \int T(r)r\,dr$$

Solving the algebraic equations gives

$$\frac{E}{(1+\nu)(1-2\nu)}c_1 = \frac{E\alpha}{1-\nu}\frac{f(b)-f(a)}{b^2-a^2}$$

$$\frac{E}{1+\nu}c_2 = \frac{E\alpha}{1-\nu}\left[\frac{a^2 f(b) - b^2 f(a)}{b^2-a^2}\right]$$

With these constants, we can write a general solution in terms of integrals of the temperature. After some algebra, we find

$$\sigma_r = \frac{E\alpha}{1-\nu}\left[\frac{1}{b^2-a^2}\int_a^b Tr\,dr - \frac{a^2}{b^2-a^2}\frac{1}{r^2}\int_a^b Tr\,dr - \frac{1}{r^2}\int_a^r Tr\,dr\right]$$

$$\sigma_\theta = \frac{E\alpha}{1-\nu}\left[\frac{1}{b^2-a^2}\int_a^b Tr\,dr + \frac{a^2}{b^2-a^2}\frac{1}{r^2}\int_a^b Tr\,dr + \frac{1}{r^2}\int_a^r Tr\,dr - T\right]$$

$$\sigma_z = \nu(\sigma_r + \sigma_\theta) - E\alpha T \tag{5.61}$$

In writing Eqs. (5.61), we have introduced the definite integral of the temperature through the relations

$$f(b) - f(a) = \int_a^b Tr\,dr$$

and

$$f(r) - f(a) = \int_a^r Tr\,dr$$

Note that, in the stress–strain equations (5.30), we have taken the reference temperature T_{ref} to be zero. This means that, when we apply these equations, the temperature $T(r)$ should be taken as the *change* from the reference temperature. The stress σ_z is the normal stress on the cylinder to keep the displacement $w(r)$ and the strain $\varepsilon_z(r)$ as zero. Timoshenko and Goodier[2] show how to add a constant stress to the stress σ_z so that the end stresses are self-equilibrating. The result is to alter σ_z, the radial displacement u, and the axial displacement w, but σ_x and σ_y are unchanged.

Example 5.2

The preceding solution for the stresses in a hollow cylinder apply for steady-state heat conduction where $T(r)$, or it can be applied for quasistatic applications where $T(r, t)$. In Chapter 4, the heat-conduction problem in cylindrical coordinates is formulated, and a few problems are solved. To illustrate typical thermal stresses in a hollow cylinder, consider the steady heat-flow case in which the the inner surface at $r = a$ has a specified constant temperature T_1 and the outer surface at $r = b$ has a specified constant temperature T_2. The temperature solution (see Problem 4.11) is

$$T(r) = \frac{T_1 \ln(b/r) + T_2 \ln(r/a)}{\ln(b/a)} \tag{5.62}$$

At $r = a$ and $r = b$, we can see that the specified boundary temperatures are met. In applying the equation to compute the terms stresses, we take T_1 and T_2 as the numerical values above (or below) the reference temperature for zero stress. Substituting $T(r)$ in Eqs. (5.61) and evaluating the integrals, we find the following expressions for the thermal stresses:

$$\begin{aligned}
\sigma_r &= \frac{E\alpha}{2(1-\nu)\ln(b/a)}\left\{\left(T_2b^2 - T_1a^2\right)\left(1 - \frac{a^2}{r^2}\right)\frac{\ln(b/a)}{b^2 - a^2}\right. \\
&\quad \left. + T_1\left[\ln(r/b) + \frac{a^2}{r^2}\ln(b/a)\right] + T_2\ln(a/r)\right\} \\
\sigma_\theta &= \frac{E\alpha}{2(1-\nu)\ln(b/a)}\left\{\left(T_2b^2 - T_1a^2\right)\left(1 + \frac{a^2}{r^2}\right)\frac{\ln(b/a)}{b^2 - a^2}\right. \\
&\quad \left. - \left[T_1\ln(b/r) + T_2\ln(r/a)\right] - \frac{T_1a^2}{r^2}\ln(b/a) + (T_1 - T_2)\right\} \\
\sigma_z &= \nu(\sigma_r + \sigma_\theta) - E\alpha T
\end{aligned} \tag{5.63}$$

The distributions of the thermal stresses over the wall of a cylinder with $a/b = 0.3$ and $\nu = 0.33$ for $T_1/T_2 = 5$ are shown in Fig. 5.8. The graph shows that the stresses σ_θ and σ_z are the largest stresses, and the radial stress σ_r is small in comparison. The radial stress is zero on the boundaries as required, is compressive at all points and, typically, is much smaller than the other stresses. The stress σ_z is compressive over the entire cross section; this is the distribution required to maintain $w = \varepsilon_z = 0$. Note that σ_θ is compressive on the hotter inner surface but that it changes signs and becomes tensile on the outer surface. The maximum compressive stress occurs at

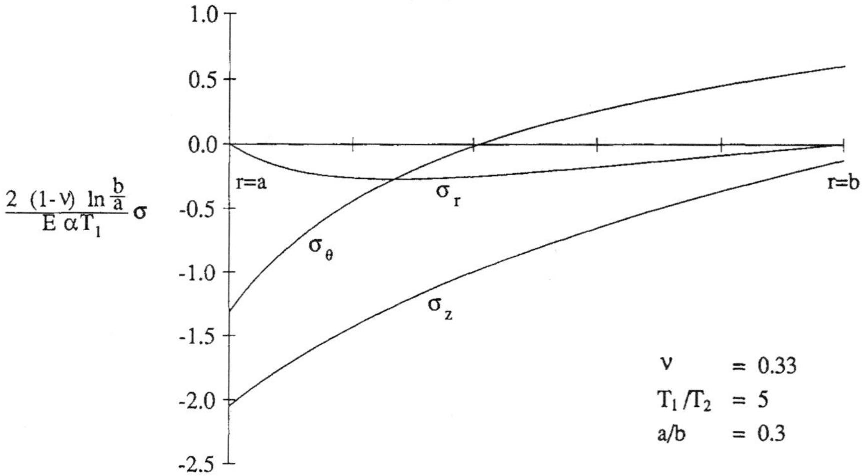

Fig. 5.8 Thermal stress distribution in hollow cylinder.

$r = a$ and is larger than the maximum tensile stress, which occurs at $r = b$. The tensile stress is significant, however. A tube of a brittle material that is weak in tension carrying a hot fluid could fail as a result of tensile cracks starting on the outer surface. The sense of the stresses reverses if T_2 is greater than T_1.

5.6 Torsion

Elasticity textbooks[2–4] develop the theory for torsion of long prismatic bars with noncircular cross sections. The formulation is an elegant illustration of classic elasticity. The problem of pure torsion is of limited interest in thermoelasticity because of the absence of thermoelastic effects. For a homogeneous isotropic material Hooke's law (Eqs. 5.3) shows that temperature changes do not affect the shearing stresses directly. Although large temperature changes alter the shearing modulus G, we assume that elastic properties are constant in classical thermoelasticity formulations. Thus, according to classical thermoelasticity, a temperature distribution does not alter the behavior of a bar under pure torsion.

Later in the next chapter, when we consider behavior of beams under thermal and mechanical loads, we have occasion to use some relations for torsion from elasticity. Thus, for completeness and later reference, we highlight the torsion problem here. Full details are presented in Refs. 3 and 4 as well as in numerous other texts.

The classical elasticity formulation considers a long, slender prismatic bar fixed at one end and subject to a torque M_x at the other end (Fig. 5.9a). Each cross section of the bar rotates through an angle of twist θ_x. We assume that the rotations are small and that the rotation at any section is proportional to the distance from the free end. Thus,

$$\theta_x = \alpha x \tag{5.64}$$

where α is the angle of twist per unit length. On a typical cross section shown in Fig. 5.9b, we locate an arbitrary point P by polar coordinates r, β. The in-plane

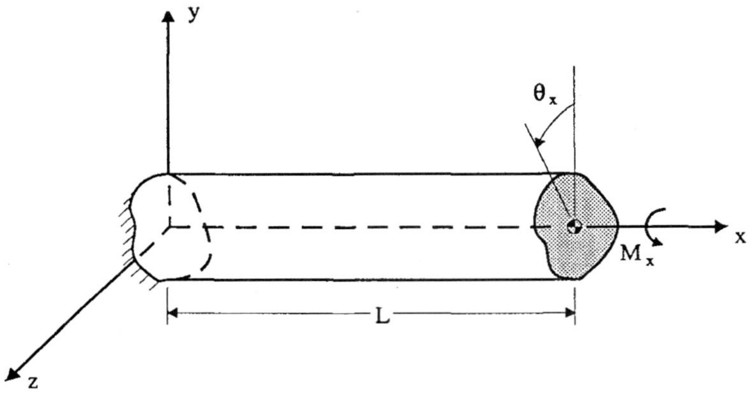

a) Bar with noncircular cross section

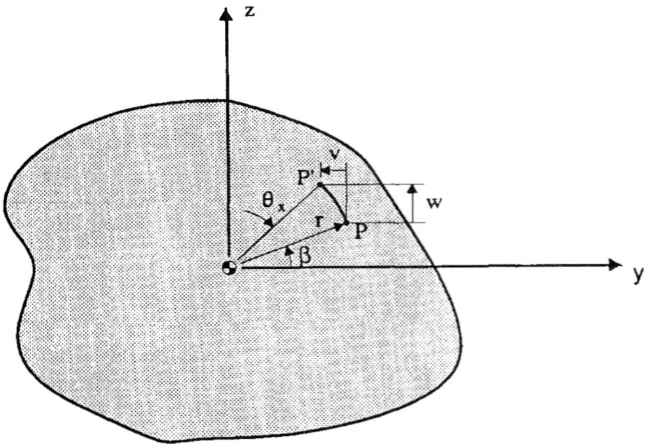

b) Bar rotation and displacements

Fig. 5.9 Torsion of bar with noncircular cross section.

displacements v and w are given by

$$v = r\cos(\beta + \theta_x) - r\cos\beta$$

$$w = r\sin(\beta + \theta_x) - r\sin\beta$$

Assuming that the angle of twist is small enough so that $\cos\theta_x \cong 1$ and $\sin\theta_x \cong \theta_x$, the preceding equations reduce to

$$v = -\theta_x r\sin\beta = -\theta_x z \qquad w = \theta_x r\cos\beta = -\theta_x y$$

Combining with Eq. (5.64), we obtain

$$v = -\alpha x z \qquad w = \alpha x y$$

Because of the twisting of the bar, the point P also experiences a displacement u along the x axis. For general shapes, the u displacement varies over the cross section. This deformation is called *warping*, and it occurs for noncircular cross sections. For the special case of a circular cross section, warping does not occur and u is zero. Thus, for a circular cross section, plane sections remain plane. In general, we assume that $u = \alpha\phi(x, y)$, where $\phi(x, y)$, is known as the *warping function*. The warping function becomes the basic unknown for the problem. To summarize, the displacements are

$$u = \alpha\phi(x, y) \qquad v = -\alpha xz \qquad w = \alpha xy \tag{5.65}$$

For the given displacements, the equations of elasticity show that the warping function must satisfy Laplace's equation,

$$\nabla^2\phi = \frac{\partial^2\phi}{\partial y^2} + \frac{\partial^2\phi}{\partial z^2} = 0 \tag{5.66}$$

Equation (5.66) is solved subject to the boundary condition that the lateral surface of the bar be free of surface tractions. This condition yields

$$\left(\frac{\partial\phi}{\partial y} - z\right)n_y + \left(\frac{\partial\phi}{\partial z} + y\right)n_z = 0 \tag{5.67}$$

where n_y and n_z are direction cosines of a normal vector to the bar surface. The only nonzero stresses are the shearing stresses σ_{xy} and σ_{xz}, which lie in the plane of a cross section. The stresses are given by

$$\sigma_{xy} = \alpha G\left(\frac{\partial\phi}{\partial y} - z\right) \tag{5.68a}$$

$$\sigma_{xz} = \alpha G\left(\frac{\partial\phi}{\partial z} + y\right) \tag{5.68b}$$

On the ends of the bar, elasticity texts show that the resultant forces obtained by integrating the stresses σ_{xy} and σ_{xz} over the cross-sectional area are zero. In addition, the resultant torque M_x due to the stresses is

$$M_x = \int_A (y\sigma_{xy} - z\sigma_{xz})\,\mathrm{d}A$$

Substituting Eq. (5.68), we may write the torque integral as

$$M_x = G\alpha\int_A \left(y^2 + z^2 + y\frac{\partial\phi}{\partial z} - z\frac{\partial\phi}{\partial y}\right)\mathrm{d}A$$

or

$$M_x = GJ\alpha \tag{5.69a}$$

with

$$J = \int_A \left(y^2 + z^2 + y\frac{\partial\phi}{\partial z} - z\frac{\partial\phi}{\partial y}\right)\mathrm{d}A \tag{5.69b}$$

where J is the *torsion constant*. Note that J depends on the warping function as well as the shape of the cross section. For circular cross sections, the warping function $\phi \equiv 0$, and J is the polar moment of inertia of the circular area. Otherwise, for general cross sections, J is *not* the polar moment of inertia but, instead, defined as part of the solution to the elasticity problem. The torsion constant J is tabulated widely for commonly encountered cross sections.

From the preceding solution, we see that the torque M_x is proportional to the angle of twist per unit length. The product GJ measures the stiffness or torsional rigidity of the bar. We may express the angle of twist per unit length α in terms of the actual rotation angle θ_x by $\alpha = \mathrm{d}\theta_x/\mathrm{d}x$. Thus, we may write the torque equation as

$$M_x = GJ\frac{\partial \theta_x}{\partial x} \tag{5.70}$$

where we use partial derivatives to permit M_x and θ_x to be a function of x and time t.

Equations (5.65–5.70) define the torsion problem of classical elasticity. The solution of the boundary-value problem posed in Eqs. (5.66) and (5.67) gives the warping function $\phi(y, z)$. From the warping function, the stresses are evaluated using Eqs. (5.68), and the torsion constant is evaluated using Eqs. (5.69). The basic torque–twist relationship for the bar is defined by Eq. (5.70). We will use this result later as we develop thermoelastic beam theory. The classical formulation for torsion described here is often called *Saint Venant torsion* for the French elastician who developed it in 1853.[1]

Looking back at Eqs. (5.65–5.70), we see that they are indeed independent of the bar temperature. Some one hundred years after Saint Venant, in the 1950s, engineers became concerned with torsion of thin wings subject to aerodynamic heating. During these investigations, Saint Venant's torsion theory was modified to account for a loss in wing torsional rigidity due to the temperatures encountered in high-speed flight. We discuss this behavior in the next chapter.

5.7 Closing Remarks

This chapter summarizes basic concepts, equations, and solutions for classical linear thermoelasticity problems. The general equations of continuum mechanics are specialized for elastic solids. Displacement and stress formulations are posed for three-dimensional solids, as well as the specialized cases of plane stress and plane strain. Classical dynamic thermoelasticity problems are solved analytically to establish the role of inertia forces. Basic quasistatic problems are solved for fully restrained and unrestrained bodies. The chapter provides a transition between the very general principles of continuum mechanics highlighted in Chapter 3 and the engineering mechanics thermal-stress analyses to be presented in Chapters 6 and 7.

References

[1]Timoshenko, S. P., *History of Strength of Materials*, McGraw–Hill, New York, 1953, pp. 242–243.

[2]Timoshenko, S., and Goodier, J. N., *Theory of Elasticity*, McGraw–Hill, New York, 1951.

[3]Wang, C.-T., *Applied Elasticity*, McGraw–Hill, New York, 1953.

[4]Sokolnikoff, I. S., *Mathematical Theory of Elasticity*, McGraw–Hill, New York, 1956.

[5]Fung, Y. C., *Foundations of Solid Mechanics*, Prentice–Hall, Englewood Cliffs, NJ, 1965.

[6]Gatewood, B. E., *Thermal Stresses*, McGraw–Hill, New York, 1957.

[7]Boley, B. A., and Weiner, J. H., *Theory of Thermal Stresses*, Wiley 1960; reprinted edition with corrections available from Krieger, Malabar, FL.

[8]Nowacki, W., *Thermoelasticity*, Addison–Wesley, Reading, MA, 1962.

[9]Johns, D. J., *Thermal Stress Analysis*, Pergamon, Oxford, England, UK, 1965.

[10]Parkus, H., *Thermoelasticity*, Blaisdell, Waltham, MA, 1968.

[11]Burgreen, D., *Elements of Thermal Stress Analysis*, C. P. Press, Jamaica, NY, 1971.

[12]Nowinski, J. L., *Theory of Thermoelasticity with Applications*, Sijhoff and Noordhoff, Alphen aan den Rijn, The Netherlands, 1978.

[13]Hetnarski, R. (ed.), *Thermal Stresses I*, North Holland, Amsterdam, 1986.

[14]Hetnarski, R. (ed.), *Thermal Stressess II*, North-Holland, Amsterdam, 1987.

[15]Hetnarski, R. (ed.), *Thermal Stresses III*, North-Holland, Amsterdam, 1989.

[16]Danilovskaya V. I., "Thermal Stresses in an Elastic Half-Space Arising After a Sudden Heating of Its Boundary," *Prikladnaya Matematika i Mekhanika*, Vol. 14, May–June 1950, pp. 316–318 (in Russia).

[17]Danilovskaya, V. I., "On a Dynamical Problem of Thermoelasticity," *Prikladnaya Matematika i Mekhanika*, Vol. 16, No. 3, 1952, pp. 341–344 (in Russian).

[18]Nickell, R. E., and Sackman, J. L., "Approximate Solutions in Linear, Coupled Thermoelasticity," *Journal of Applied Mechanics*, Vol. 35, June 1968, pp. 255–265.

[19]Tamma, K. K., and Railkar, S. B., "Evaluation of Thermally Induced Non-Fourier Stress Wave Disturbances via Tailored Hybrid Transfinite Element Formulations," *Computers and Structures*, Vol. 34, No. 1, 1990, pp. 5–16.

[20]Erdélyi, A., Magnus, W., Oberhettinger, F., and Tricomi, F. G., *Tables of Integral Transforms*, Vol. I, McGraw–Hill, New York, 1954, p. 246, Pair (10).

[21]Campbell, G. A., and Foster, R. M., *Fourier Integrals for Practical Applications*, Van Nostrand, New York, 1951, p. 56, pair 545.5, and p. 93, pair 812.

[22]Goodier, J. N., "Thermal Stresses and Deformation," *Journal of Applied Mechanics*, Vol. 24, No. 3, 1957, pp. 467–474.

[23]Rao, R., and Johns, D. J., "Some Thermal Stress Analyses for Rectangular Plates," *Journal of Aeronautical Society of India*, Vol. 13, No. 4, 1961, pp. 99–104.

Problems

5.1. Derive the three basic equations (5.1) for the displacement formulation of dynamic thermoelasticity problems.

5.2. Derive the first and fourth equations of the stress formulation for thermoelasticity given in Eqs. (5.8a). Then, use cyclic permutation of the subscripts to write the remaining equations.

5.3. For a *thick* rectangular plate, the temperature distribution is given as

$$T(x, y) = T_1 + \frac{x}{a}(T_2 - T_1)$$

where a is the length of the plate along the x axis, and b is the length of the plate along the y axis. The temperatures T_1 and T_2 are constant. Body forces are zero.

(a) Verify that the following displacements are a solution to the three-dimensional thermoelasticity problem:

$$u(x, y, z) = \alpha x T_1 + \frac{\alpha}{2a}(x^2 - y^2 - z^2)(T_2 - T_1)$$

$$v(x, y, z) = \alpha y T_1 + \frac{\alpha}{a} xy(T_2 - T_1)$$

$$w(x, y, z) = \alpha z T_1 + \frac{\alpha}{a} xz(T_2 - T_1)$$

(b) What boundary conditions are satisfied?
(c) What are the strains in the thick plate?
(d) What are the stresses?

5.4. Derive Eqs. (5.14) for the displacement formulation for plane stress. What assumptions are made?

5.5. For the case of plane strain, derive the basic equations that u and v must satisfy.

5.6. The displacement functions

$$u(x, y) = \alpha x T_1 + \frac{\alpha}{2a}(x^2 - y^2)(T_2 - T_1)$$

$$v(x, y) = \alpha y T_1 + \frac{\alpha}{a} xy(T_2 - T_1)$$

$$T(x, y) = T_1 + \frac{x}{a}(T_2 - T_1)$$

are proposed as the solution for a thin rectangular plate of dimensions a, b.
(a) Is this a solution?
(b) What boundary conditions are satisfied?
(c) What are the strains?
(d) What are the stresses?

5.7. In Sec. 4.4.2, a solution is given for the temperature distribution in a semi-infinite elastic solid due to a suddenly applied heat flux q_s on the surface $x = 0$. The solution given in Eq. (4.38) is

$$T(x, t) = T_0 + \frac{q_s}{k}\sqrt{\frac{4\kappa t}{\pi}}\left[\exp(-x^2/4\kappa t) - x\sqrt{\frac{\pi}{4\kappa t}}\,\mathrm{erfc}(x/\sqrt{4\kappa t})\right] \tag{4.38}$$

(a) Verify that Eq. (4.38) is the solution for the one-dimensional heat-conduction equation.
(b) Derive the solution for the stress $\sigma_x(x, t)$ in an elastic semi-infinite solid. Use the boundary and initial conditions given for the Danilovskaya problems in Sec. 5.3. Discuss the possibility of a jump discontinuity in the stress response, and compute the stress response for stainless steel.

5.8. Calculate the ΔT required to yield an isothermal, fully restrained body if the material is a) 6061-T6 aluminum, b) titanium, c) Hastelloy-X, d) Inconel 718, and

e) copper. Present your results in a table. Give the results in degrees Fahrenheit.

5.9. For a state of plane strain, derive equations for the stress components if the body is fully restrained.

5.10. For an unrestrained, thin rectangular beam, assume a linear temperature distribution $T(y, t) = a(t) + b(t)y$, and show that the stress $\sigma_x(y)$ is zero in accordance with the principle described in Sec. 5.4.2.

5.11. For an unrestrained thin plate, assume a linear temperature distribution $T(z, t) = a(t) + b(t)z$, and show that the stresses $\sigma_x(z, t) = \sigma_y(z, t)$ are zero in accordance with the general principle described in Sec. 5.4.2.

5.12. For the unrestrained thin plate described in Sec. 5.5.2,

(a) Derive the strain components ε_x, ε_y, and ε_z

(b) Derive the displacement components u, v, and w, and show that $g(x, y) = -6M_T(x^2 + y^2)/h^3E$.

5.13. An infinitely long, thin plate is subjected to *bilinear* temperature distribution of the type represented in Fig. 5.6. For an aluminum plate with the following conditions, determine the stress distribution, and compute the maximum tensile and compressive stresses. Use $2h = 24$ in., $b = 0.25$ in., $T_0 = 95°$ F, $T_1 = 245°$F, $E = 10^7$ psi, and $\alpha = 13 \times 10^{-6}$ 1/°F. Sketch the temperature and stress distributions.

5.14. In Sec. 4.5.1, the transient temperature distribution for a thin plate subjected to a suddenly applied heat flux is given in Eq. (4.48).

(a) For this transient temperature distribution, derive an equation for the transient response of the stresses $\sigma_x = \sigma_y$ in the unrestrained thin plate of Sec. 5.5.2.

(b) For a titanium plate of thickness $h = 0.125$ in. (0.318 cm), use the thermal property data given in Table 4.1 and the elastic property data given in Table 5.1 to compute the transient thermal-stress response for the plate subjected to a suddenly applied heat flux $q_s = 0.1$ Btu/in.^{2}-s (16.4 W/cm^2). Where do the maximum stresses occur?

(c) If the material is changed to Hastelloy-X, how do the maximum temperature and thermal stresses change?

5.15. For the Saint Venant torsion problem discussed in Sec. 5.6, use the displacements given in Eqs. (5.65) to derive Eqs. (5.66–5.68).

6
Thermal Structures Problems: Rods and Beams

6.1 Introduction

In the preceding chapter, we began our discussion of thermally induced deformations and stresses in aerospace structures by considering classical thermoelasticity problems. These formulations provide insight into basic features of a thermoelasticity problem, but the equations are much too complicated to apply to real structures. In this chapter, we begin the study of aerospace thermal structures with analysis of basic structural components. We consider the formulation of the engineering theory for thermal-stress analysis of rods and beams. We consider each of these in turn as an individual structural component and concentrate on boundary/initial-value problems. When the individual structural components are synthesized to construct a built-up structure, we will abandon the analytical approach and turn to finite element analysis. Some of the theory developed here for basic thermal-structural components will be used in Chapter 8, where we consider finite element analysis.

References available on thermal-structures problems include the books by Gatewood,[1] Boley and Weiner,[2] Johns,[3] and Burgreen.[4] Other referenceable material on thermal-structures problems appears in papers that we cite in this and succeeding chapters. We will, however, make use of structural mechanics texts, such as Refs. 5–8, which describe closely related isothermal problems.

In formulating engineering mechanics theories for the behavior of rods and beams, we build on the thermoelasticity foundation established in the preceding chapter. But we discard the general three-dimensional equations and replace them with specialized equations that apply to the structural component at hand. In doing this, we are guided by the rich mechanics technology that has evolved over 200 years. In every application, we are guided by three fundamental principles applicable to all solid mechanics problems: 1) equilibrium, 2) compatibility, and 3) constitution. We mean *equilibrium* in the sense of D'Alembert so that, by inclusion of body forces, we may convert equilibrium equtions easily into equations of motion. By *compatibility*, we mean that the structural component deforms so that all parts fit together continuously. (We do not mean the equations of strain compatibility (5.7) although the fundamental notions are the same.) By *constitution*, we mean that there is a relationship between stress and strain, such as Hooke's law characterizing material behavior. These principles, of course, apply to isothermal structures as well. There is a fourth principle for thermal structures—*conservation of energy*. We assume that it has been satisfied a priori in determination of the temperature.

6.2 Rods

By a rod, we mean a long, slender member that experiences a one-dimensional state of normal stress σ_x (Fig. 6.1). All other stresses are small and are neglected.

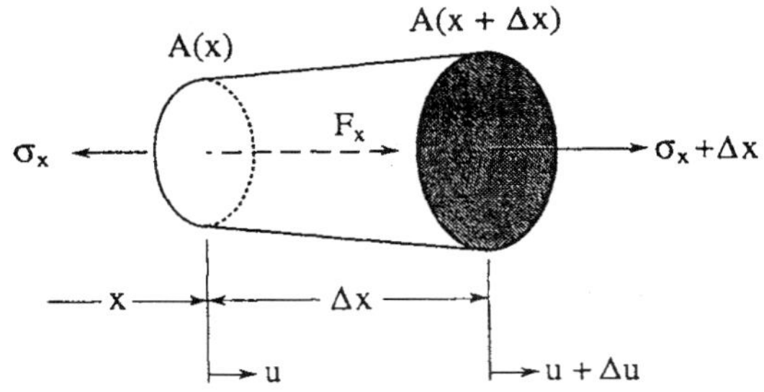

Fig. 6.1 Segment of a rod.

The stress σ_x is assumed uniform over the cross section, and it may be a function of position x and time t. The rod experiences a thermal environment with a known temperature $T(x, t)$. The cross-sectional area of the rod may vary with x, i.e., $A = A(x)$, although, in many problems, A is constant. The stress σ_x may be tensile or compressive, but we assume that transverse bending is precluded by appropriate constraints. The theory also applies to cables subjected to tensile forces.

In a rod, there is only a single axial displacement component $u(x, t)$. Although the displacement may vary with position x and time t, the displacement does not vary over the cross section. Thus, a cross section translates in the x direction. Although we may model more complex material behavior (see Chap. 11), we assume that Hooke's law applies.

6.2.1 Problem Formulation

Acting on a segment of a rod shown in Fig. 6.1 are the stress σ_x and a body force per unit volume F_x. We write the equation of motion

$$\sum F_x = ma_x$$

$$-\sigma_x A + (\sigma_x + \Delta\sigma_x)(A + \Delta A) + \left(F_x + \frac{1}{2}\Delta F_x\right)\left(A + \frac{1}{2}\Delta A\right)\Delta x = \rho A \Delta x \frac{\partial^2 u}{\partial t^2}$$

so that

$$\frac{\Delta(\sigma_x A)}{\Delta x} + F_x A + \frac{1}{2}\Delta(F_x A) + \frac{1}{4}\Delta F_x \Delta A = \rho A \frac{\partial^2 u}{\partial t^2}$$

In the limit as $\Delta x \to 0$, we obtain

$$\frac{\partial}{\partial x}(\sigma_x A) + F_x A = \rho A \frac{\partial^2 u}{\partial t^2} \tag{6.1}$$

In Eq. (6.1), ρ is the material density, which we assume to be constant. The only nonzero strain-displacement equation is

$$\varepsilon_x = \frac{\partial u}{\partial x} \tag{6.2}$$

and Hooke's law is

$$\sigma_x = E\varepsilon_x - E\alpha(T - T_{\text{ref}}) \tag{6.3}$$

where E is the elastic modulus, α the coefficient of thermal expansion, and T_{ref} the reference temperature. We assume that these parameters are constant. If we combine Eqs. (6.1–6.3), we obtain the equation of motion with the displacement u as the dependent variable:

$$\frac{\partial}{\partial x}\left[EA\frac{\partial u}{\partial x}\right] + P(x,t) - P_T(x,t) = \rho A(x)\frac{\partial^2 u}{\partial t^2} \tag{6.4a}$$

were we have introduced forces per unit length,

$$P(x,t) = F_x(x,t)A(x) \qquad P_T(x,t) = \frac{\partial}{\partial x}[EA(x)\alpha(T - T_{\text{ref}})] \tag{6.4b}$$

We regard P_T as the thermal force per unit length.

Since Eq. (6.4a) is a second-order equation in x, we must have one boundary condition at each end of the rod. The possibilities are to specify either the displacement or the stress. Thus, for a rod of length L, we specify

$$\begin{aligned} u(0,t) &= u_1(t) \quad \text{or} \quad \sigma_x(0,t) = \sigma_1(t) \\ u(L,t) &= u_2(t) \quad \text{or} \quad \sigma_x(L,t) = \sigma_2(t) \end{aligned} \tag{6.5a}$$

In addition, we must have two initial conditions,

$$u(x,0) = u_0(x) \qquad \frac{\partial u}{\partial t}(x,0) = \dot{u}_0(x) \tag{6.5b}$$

Equations (6.4) and (6.5) constitute a classical linear boundary/initial-value problem for determination of the axial displacement $u(x,t)$.

The equation of motion includes the inertia force which we neglect for quasistatic problems. In applications, we may start the solution with the more fundamental equation (6.1) in terms of the stress σ_x.

6.2.2 Quasistatic Problems

We consider now two special rod problems where, for quasistatic behavior, we can solve the boundary-value problem to obtain analytical solutions. In each example, we assume a rod of length L, zero body forces, and constant cross-sectional area. With these assumptions, the equation of motion (6.1) reduces to

$$\frac{\partial \sigma_x}{\partial x} = 0$$

so that, if we permit the temperature to vary with x and t, $T = T(x,t)$, then,

$$\sigma_x = f(t)$$

where the function $f(t)$ is to be determined. From Eqs. (6.2) and (6.3), we next write

$$\varepsilon_x = \frac{\partial u}{\partial x} = \frac{f(t)}{E} + \alpha(T - T_{\text{ref}})$$

or, after integration,

$$u(x, t) = \frac{f(t)x}{E} + \alpha \int_0^x (T - T_{\text{ref}})\,\mathrm{d}x + c_1$$

where c_1 is a constant. Let us consider two common sets of boundary conditions.

One end clamped and one end free. If the end at $x = 0$ is clamped, we have the boundary condition $u(0, t) = 0$, which shows that $c_1 = 0$. If the end at $x = L$ is free, then, using Eq. (6.3), we can write

$$\sigma_x(L, t) = f(t) + E\alpha(T - T_{\text{ref}}) - E\alpha(T - T_{\text{ref}}) \doteq 0$$

hence, $f(t) = 0$. Thus, the final solution becomes

$$\sigma_x(x, t) = 0 \qquad \varepsilon_x(x, t) = \alpha(T - T_{\text{ref}}) \qquad u(x, t) = \alpha \int_0^x (T - T_{\text{ref}})\,\mathrm{d}x$$

This solution is, of course, exactly what we expect. Since the end at $x = L$ is free, equilibrium requires that every cross section have zero stress. Moreover, the rod is free to expand so that the strain and displacement respond directly with the prescribed temperature.

Both ends clamped. With both ends clamped, the boundary conditions are

$$u(0, t) = 0 \qquad u(L, t) = 0$$

Using the equation for $u(x, t)$,

$$c_1 = 0 \qquad f(t) = -E\frac{\alpha}{L} \int_0^L (T - T_{\text{ref}})\,\mathrm{d}x$$

or

$$f(t) = -E\alpha T_{\text{avg}}$$

where T_{avg} is the average temperature,

$$T_{\text{avg}} = \frac{1}{L} \int_0^L (T - T_{\text{ref}})\,\mathrm{d}x$$

With c_1 and $f(t)$ known, the solution for clamped ends is

$$\sigma_x(t) = -E\alpha T_{\text{avg}} \qquad \varepsilon_x(x, t) = \alpha[-T_{\text{avg}} + (T - T_{\text{ref}})]$$

$$u(x, t) = \alpha\left[-T_{\text{avg}}x + \int_0^x (T - T_{\text{ref}})\,\mathrm{d}x\right]$$

Thus, for a temperature increase above the reference temperature, the rod experiences uniform compressive stress. The variations in the strain and axial displacement variations with x are determined by the temperature distribution. To illustrate how ε_x and u may vary, consider a prescribed steady-state linear temperature distribution,

$$T - T_{\text{ref}} = T_1(1 - x/L) + T_2 x/L$$

For this change in temperature, ΔT_{avg} is

$$\Delta T_{\text{avg}} = \frac{1}{L}\int_0^L \left[T_1\left(1 - \frac{x}{L}\right) + T_2\frac{x}{L}\right] \mathrm{d}x = \frac{T_1 + T_2}{2}$$

Substituting, we find that

$$\sigma_x = -E\alpha\frac{T_1 + T_2}{2} \qquad \varepsilon_x = \frac{\alpha(T_1 - T_2)}{2}\left(1 - \frac{2x}{L}\right)$$

$$u = \frac{\alpha(T_1 - T_2)}{2}x\left(1 - \frac{x}{L}\right)$$

From these results, we see that if we prescribe a uniform temperature change in the rod by setting $T_1 = T_2 = \Delta T$, then, as we might expect,

$$\sigma_x = -E\alpha\Delta T \qquad \epsilon_x = 0 \qquad u = 0$$

This is the case of the fully restrained elastic solid that we discussed in Sec. 5.4. For $T_1 \neq T_2$, the strain varies linearly with x, and the displacement varies quadratically with x. The sense of ϵ_x and u depends on the relative magnitude of T_1 and T_2. The two possibilities are sketched in Fig. 6.2. Consider the case when $T_1 > T_2$, where the left-hand portion of the rod is hotter than the right-hand portion. The strain variation shows that, for $0 < x < L/2$, strains are positive and fibers extend. But, since there is no overall change in the length of the rod, for $L/2 < x < L$, fibers must contract with negative strains. The displacement graph shows that all points for $0 < x < L$ move to the right as indicated by positive $u(x)$. The maximum displacement occurs at $x = L/2$. For $T_1 < T_2$, Fig. 6.2 shows that the strain-displacement behavior is reversed.

This simple but clarifying example shows how thermal-structural behavior may be counterintuitive. We might expect in this fully restrained problem that, because the stress is constant, the strain and displacements would be zero. However, because the temperature varies over the rod, with some parts hotter than others, fiber lengths change locally to accommodate the thermal expansions. The accumulation of these strains along the rod must satisfy the global restraints. The local variations in the strains and deformations illustrate the important role of temperature gradients.

6.2.3 Space Bridge Tether

In a 1979 novel,[9] Arthur C. Clarke introduced the concept of the orbital space bridge. One version of the concept involves a geostationary satellite with a tether from the satellite to Earth. Cargo travels up and down the tether, reducing the costs of transporting people and goods to Earth orbit. In addition, spacecraft may be launched from the tower at modest costs by extracting energy from the Earth's rotation. The design and construction of such a system is an enormous challenge because of the huge size and great expense. Some of the basic technical problems were addressed in a 1975 paper by Jerome Pearson.[10] One technical aspect of the

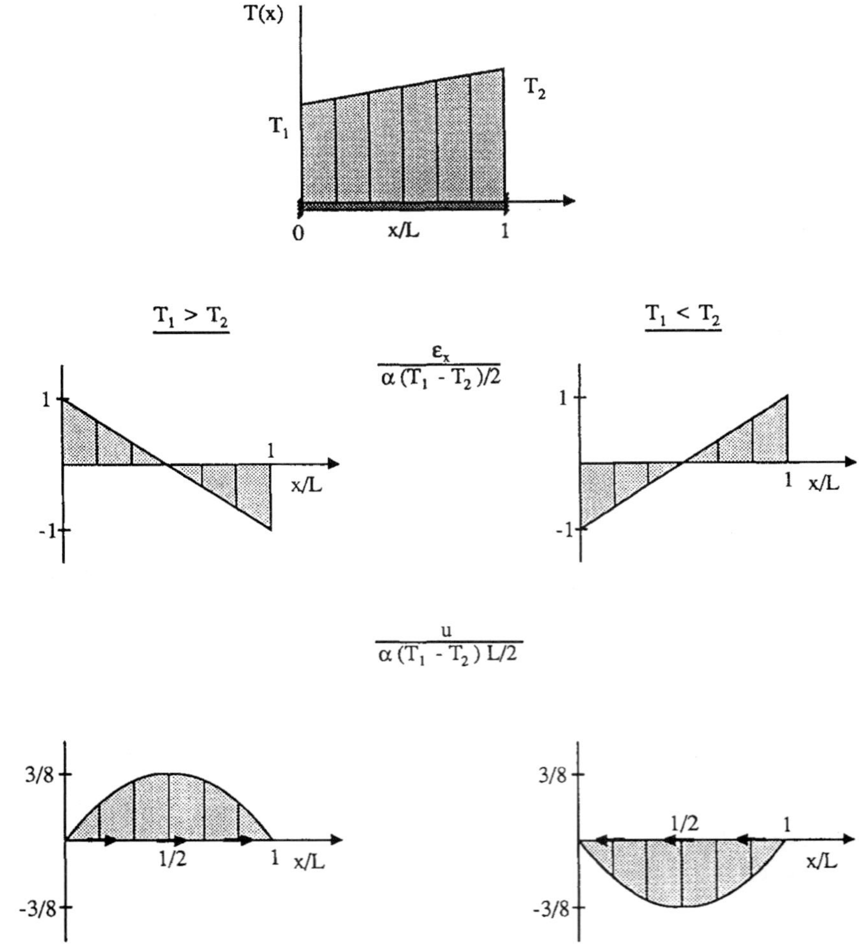

Fig. 6.2 Strains and displacement in a fully restrained rod with linear temperature distribution.

space bridge that has received little attention is the thermal-structural behavior of the portion of the tether that extends into the Earth's upper atmosphere. Studies of tether applications[11] have shown significant aerodynamic heating and elevated tether temperatures.

The geometry of a space bridge concept is shown in Fig. 6.3. The concept consists of a central platform in geostationary orbit. Tethers join the platform to lower and upper termini. All points on the space bridge are located by radial coordinates measured from the Earth's center. As the space bridge orbits the Earth, the tether experiences gravitational and centrifugal body forces. For increasing r, the gravitational force decreases with $1/r^2$, and the centrifugal force increases with r. Two important considerations guide the tether design. The first is that the tether should be in tension to avoid buckling. This can be achieved by selecting the positions r_1 and r_2 of the lower and upper termini so that the resultant of the gravitational and centrifugal forces for the entire assembly is zero. The balanced

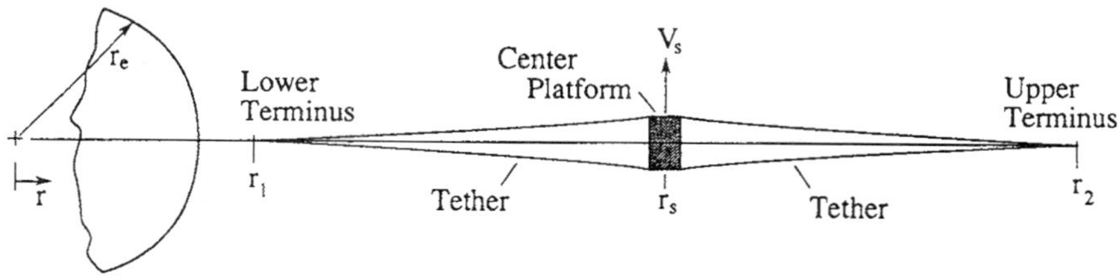

Fig. 6.3 Geometry of a space bridge.

force system has a maximum tensile force at the platform, and the tensile forces diminish to minimum values at the termini. The second consideration is to minimize the tether weight by tapering the tether cross section while maintaining constant stress.

In addition to the mechanical loads, the tethers experience aerodynamic and environmental heating. The lower tether extends into the upper atmosphere, and a portion of the tether for $r > r_1$ experiences aerodynamic heating. The entire tether will experience environmental heating throughout the orbit. As discussed in Sec. 2.3, we should consider heating from solar radiation, Earth radiation, and Earth-reflected radiation (albedo). We recall that heating will be greatly reduced during shadow transit since only Earth-emitted heating is present. Consequently, the first step is to determine the heat-flux distribution along the tether for each orbital position. Then the tether temperature distribution is evaluated for the orbit. The temperature varies along the tether length and with time. Depending on construction details, there may be a temperature gradient over the cross section. Here we assume that $T = T(r, t)$, where T is the average cross-sectional temperature. Our objective is to determine the variation in the cross-sectional area $A = A(r)$ for minimum tether weight and to evaluate tether elongations due to mechanical and thermal loads.

If quasistatic behavior is assumed, the equation of motion (6.1) reduces to

$$\frac{\mathrm{d}}{\mathrm{d}r}(\sigma_r A) + F_r A = 0$$

where the area varies with r, $A = A(r)$, and F_r is the body force per unit volume. The first step is to write an equation for the body force. Pearson[10] shows that

$$F_r = \rho g_e r_e^2 \left(-\frac{1}{r^2} + \frac{r}{r_s^3} \right)$$

where ρ is the density of the tether, g_e the acceleration of gravity on the Earth's surface, and r_e the earth's radius. The first term inside the parenthesis, $-1/r^2$, represents the gravitational attraction; the second term is the centrifugal force, where r_s is the radius of the geostationary orbit. To determine the distribution for A, we set the stress σ_r equal to an allowable stress σ_a and separate variables,

$$\frac{\mathrm{d}A}{A} + \frac{F_r}{\sigma_n}\mathrm{d}r = 0$$

Substituting for F_r and performing the integration, we obtain

$$\ln A = \frac{\rho g_e r_e^2}{\sigma_a}\left(-\frac{1}{r} - \frac{r^2}{2r_s^3}\right) + c_1$$

where c_1 is a constant. By setting $A = A_s$ at $r = r_s$, we may solve for c_1. Finally, the equation for the area distribution is

$$A(r) = A_s \exp\left(\frac{3r_e^2}{2hr_s}\right) \exp\left[-\frac{r_e^2}{h}\left(\frac{1}{r} + \frac{r^2}{2r_s^3}\right)\right]$$

where $h = \sigma_a/\rho g_e$. This result shows that the minimum weight tether has an exponential area taper from a maximum at the radius r_s to a minimum at a terminus radius. Note that the area distribution is determined completely by the mechanical forces and is not affected by the temperature distribution. The tether elongations are affected by temperature, however, and we return to Eqs. (6.2) and (6.3) to determine these quantities. We may write the axial strain as

$$\frac{\partial u}{\partial r} = \frac{\sigma_r}{E} + \alpha[T(r, t) - T_{\text{ref}}]$$

Now, because we have selected the area distribution so that $\sigma_r = \sigma_a$, a constant, we may integrate this last equation directly to obtain

$$u(r, t) = \frac{\sigma_a}{E}(r - r_1) + \alpha \int_{r_1}^{r} [T(r, t) - T_{\text{ref}}]\,\mathrm{d}r + g(t)$$

where $g(t)$ is to be determined. Using the condition that there be zero tether displacement at $r = r_s$

$$g(t) = \frac{\sigma_a(r_s - r_1)}{E} - \alpha \int_{r_s}^{r_a} [T(r, t) - T_{\text{ref}}]\,\mathrm{d}r$$

Finally,

$$u(r, t) = \frac{\sigma_a(r - r_s)}{E} + \alpha \int_{r_1}^{r} [T(r, t) - T_{\text{ref}}]\,\mathrm{d}r - \alpha \int_{r_1}^{r_s} [T(r, t) - T_{\text{ref}}]\,\mathrm{d}r$$

At the termini,

$$u(r_1, t) = -\frac{\sigma_a(r_s - r_1)}{E} - \alpha \int_{r_1}^{r_s} [T - T_{\text{ref}}]\,\mathrm{d}r$$

$$u(r_2, t) = \frac{\sigma_a(r_2 - r_s)}{E} + \alpha \int_{r_s}^{r_2} [T - T_{\text{ref}}]\,\mathrm{d}r$$

The minus signs in the equation for $u(r_1, t)$ mean that the lower terminus will move toward the Earth as the tether elongates in response to the tensile stress σ_a and a positive temperature change $T - T_{\text{ref}}$. At the same time, the equation for $u(r_2, t)$ shows that the upper terminus moves away from the Earth as a result of tether elongation.

6.3 Beams

We use the term *beam* to describe a long, slender, straight member that undergoes significant bending deformations. Consider the beam shown in Fig. 6.4, where we take the x axis along the beam length. The beam cross-sectional area is assumed constant, and the x axis passes through the centroids of the cross sections. The shape of the cross section is arbitrary. We permit the beam to experience simultaneously extension, bending, shearing, and torsion. Thus, the beam deformation is measured by the displacements u, v, w. In addition, because of torsion, the beam twists and, because of bending, cross sections rotate. We measure these rotations by the angles $\theta_x, \theta_y, \theta_z$, respectively. The positive sense of the rotations is defined by the right-hand rule. The beam experiences a prescribed temperature distribution $T(x, y, z, t)$. In addition, the beam supports distributed mechanical loads of two types: distributed forces per unit length f_x, f_y, f_z and distributed moments per unit length m_x, m_y, m_z. We must specify appropriate boundary conditions at each end of the beam and appropriate initial conditions for the dynamic response. The problem is to determine stresses and deformations throughout the beam as a function of time.

Beam theory rests on engineering assumptions based on the physical response of beams to loads, and there is a hierarchy of theories, starting with the linear formulations of strength of materials and proceeding upward to the nonlinear formulations of advanced solid mechanics. We cannot describe in detail all these theories, but we will develop theories that have practical value in thermal-structural applications. For a long and slender beam, we may assume that shearing deformations are small compared to bending deformations, and that they may be neglected. This assumption is reasonable for many aerospace applications. If we do not have a long, slender member, then we will probably not wish to pursue an analytical solution and turn to a finite element model. The type of beam cross section is important. We begin with the beam theory for a compact cross section where the classical assumption of "plane sections remain plane" is applicable. This formulation is known as Bernoulli–Euler beam theory.

The majority of applications of beam theory are for the range of deflections and rotations that may be assumed small enough to permit us to formulate linear boundary/initial-value problems. But, in some applications, such as buckling, we need to consider the geometric effects of larger motions, which may lead to nonlinear equations. We briefly develop a nonlinear formulation and contrast it

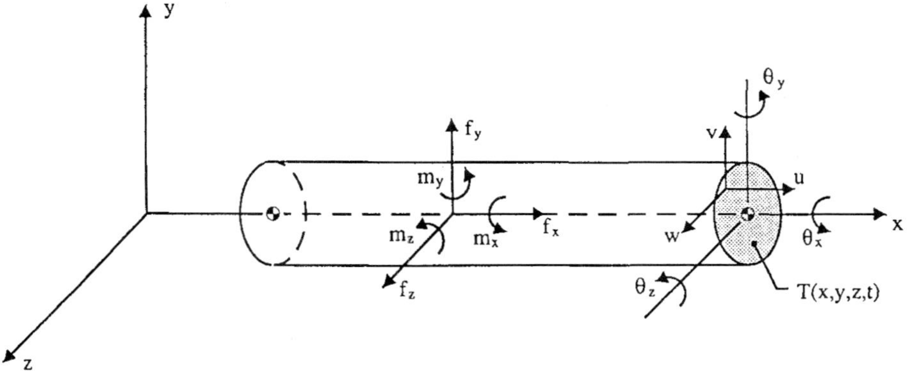

Fig. 6.4 Slender beam showing senses of deformations and distributed loads.

with the previous linear theory. Later, we present a formulation to account for the cross-sectional warping that characterizes the behavior of thin-walled members common to many aerospace applications. Throughout these theoretical developments, we assume quasistatic behavior. In a later chapter, we consider thermally induced motion, including inertia effects.

Throughout the theoretical developments, we assume that material properties are constant. The material is assumed to be linear and elastic, and we use Hooke's law for a homogeneous isotropic material. The books on aerospace structures by Rivello,[6] Allen and Haisler,[7] and Donaldson[8] develop beam theory for nonhomogeneous materials using an approach based on modulus-weighted section properties.

After completion of the beam formulations, we solve several problems, including linear, statically determinant, and indeterminant beams, a geometrically nonlinear problem, and a torsion–flexure buckling problem.

6.3.1 Formulation for Compact Cross Sections

The formulation begins with the definitions of the stress resultants. Then, the displacements are defined, and the beam normal stress is written in terms of displacement derivatives. Next, the normal stress is expressed in terms of the stress resultants. Then, the equilibrium equations for a general beam segment are written in terms of the applied forces and stress resultants. Consequently, the equilibrium equations are developed in terms of the displacements. Finally, the boundary conditions are written to complete the problem formulation.

Stress Resultants. Figure 6.5a shows the normal and shearing stresses acting on a small element of area $\mathrm{d}A$ at a typical beam cross section. The stresses consist of a normal stress σ_x and two shearing stresses σ_{xy} and σ_{xz}. Let us multiply these stresses by $\mathrm{d}A$ and integrate the forces and the moments of these forces over the cross sectional area A to produce the stress resultants:

$$N_x = \int_A \sigma_x \,\mathrm{d}A \tag{6.6a}$$

$$V_y = \int_A \sigma_{xy} \,\mathrm{d}A \tag{6.6b}$$

$$V_z = \int_A \sigma_{xz} \,\mathrm{d}A \tag{6.6c}$$

$$M_x = \int_A (\sigma_{xz} y - \sigma_{xy} z) \,\mathrm{d}A \tag{6.6d}$$

$$M_y = \int_A \sigma_A z \,\mathrm{d}A \tag{6.6e}$$

$$M_z = \int_A \sigma_x y \,\mathrm{d}A \tag{6.6f}$$

The stress resultants are shown in their positive senses in Fig. 6.5b. The resultant force N_x is the beam axial force; the resultants V_y and V_z are the beam shear forces. The resultant M_x is the beam torque; the resultants M_y and M_z are the beam bending moments. Note that the positive senses of the bending moments M_y and M_z are defined by the moments generated from the tensile stress σ_x located

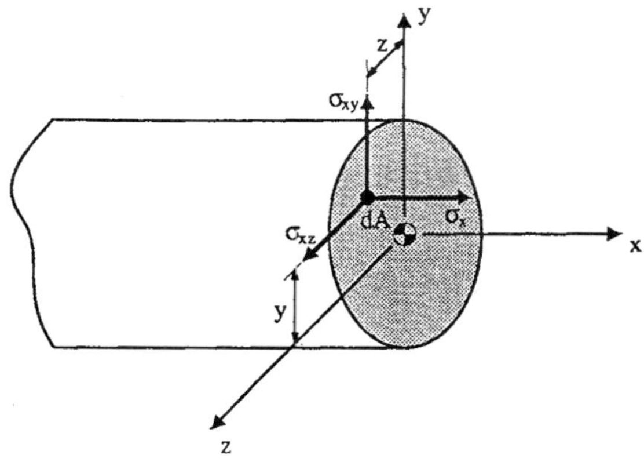

a) Stresses

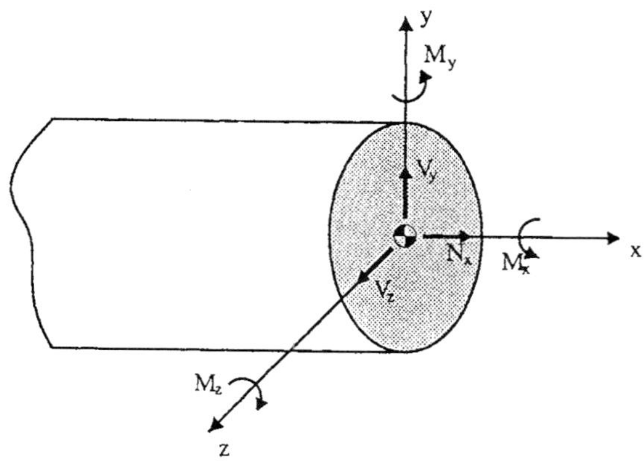

b) Stresses resultants

Fig. 6.5 Stresses and stress resultant at a beam cross section.

by positive coordinates y and z on the cross section. The torque M_x follows the right-hand rule, as does the bending moment M_y; but the positive sense of M_z does *not* follow the right-hand rule.

Stresses due to extension and bending. In Chapter 5 in our discussion of thermoelasticity, we develop a solution for the stresses and displacements for a thin beam of rectangular cross section subjected to a temperature variation through the depth of the beam. For a long beam, the solution satisfied the equations of thermoelasticity exactly. We can use the solution as a guide in making assumptions for an engineering beam theory.

We make three basic assumptions:

1) Plane cross sections of the beam before bending remain planes after bending.

2) The in-plane displacement u in Fig. 6.4 may be expressed as the sum of deformations due to extension (or contraction) and bending.

3) Lateral beam displacements v and w result from bending deformations only; shearing deformations and changes in the transverse beam dimensions are negligibly small.

With these assumptions, we write the beam's displacements as

$$u(x, y, z, t) = u_0(x, t) - y\frac{\partial v}{\partial x}(x, t) - z\frac{\partial w}{\partial x}(x, t) \tag{6.7a}$$

$$v(x, y, z, t) = v(x, t) \tag{6.7b}$$

$$w(x, y, z, t) = w(x, t) \tag{6.7c}$$

where $u_0(x, t)$ denotes the extension (or contraction) of the beam's fiber on the centroidal axis at $y = z = 0$. The derivative $\partial v/\partial x$ denotes the beam's slope due to bending in the x–y plane, and the derivative $\partial w/\partial x$ denotes the beam's slope due to bending in the x–z plane. Equation (6.7a) follows directly from displacement assumptions 1 and 2. Equation (6.7c) follow directly from displacement assumption 3. The first equation expresses the assumption that, after deformation, a cross section may translate and rotate but must remain plane. The second and third equations express the assumption that, for a given x, all points through the beam depth undergo the same displacements.

For the assumed displacements of Eqs. (6.7), the linear strain-displacement relations (5.2) reduce to

$$\begin{aligned}
\varepsilon_x &= \frac{\partial u}{\partial x} = \frac{\partial u_0}{\partial x} - y\frac{\partial^2 v}{\partial x^2} - z\frac{\partial^2 w}{\partial x^2} \\
\varepsilon_y &= \frac{\partial v}{\partial y} = 0 \\
\varepsilon_z &= \frac{\partial w}{\partial z} = 0 \\
\epsilon_{xy} &= \frac{1}{2}\left(\frac{\partial u}{\partial y} + \frac{\partial v}{\partial x}\right) = \frac{1}{2}\left(-\frac{\partial v}{\partial x} + \frac{\partial v}{\partial x}\right) = 0 \\
\varepsilon_{xz} &= \frac{1}{2}\left(\frac{\partial u}{\partial z} + \frac{\partial w}{\partial x}\right) = \frac{1}{2}\left(\frac{\partial w}{\partial x} + \frac{\partial w}{\partial x}\right) = 0 \\
\varepsilon_{yz} &= \frac{1}{2}\left(\frac{\partial v}{\partial z} + \frac{\partial w}{\partial y}\right) = 0
\end{aligned} \tag{6.8}$$

Thus, the only nonzero normal strain is ε_x, and all shearing strains are zero.

The thermoelasticity solution showed that, for a beam with only a prescribed temperature, the normal stresses σ_y and σ_z, as well as all shearing stresses are zero. For engineering beam theory, we likewise assume that σ_y and σ_z are small and may be neglected. For the shearing stresses, however, we have a contradiction. To support transverse mechanical loads, the beam must develop shearing stresses σ_{xy} and σ_{xz}, which lead to the beam shear forces V_y and V_z. The contradiction comes from the fact that the assumed displacements yielded ε_{xy} and ε_{xz} as zero in Eqs. (6.8). By Hooke's law, the associated stresses are zero. The contradiction is inherent in elementary beam theory as a consequence of the assumption that the plane sections remain plane. To resolve the contradiction, beam theory derives appropriate shearing stresses to satisfy equilibrium.

We express the only nonzero normal stress σ_x in terms of the strain ε_x and temperature T by the uniaxial Hooke's law,

$$\sigma_x(x, y, z, t) = E\varepsilon_x(x, y, z, t) - E\alpha[T(x, y, z, t) - T_{\text{ref}}] \tag{6.9a}$$

Substituting for ε_x using Eqs. (6.8) gives

$$\sigma_x = E\left(\frac{\partial u_0}{\partial x} - y\frac{\partial^2 v}{\partial x^2} - z\frac{\partial^2 w}{\partial x^2}\right) - E\alpha(T - T_{\text{ref}}) \tag{6.9b}$$

From this last equation, we can see how the normal stress varies with deformation. There is a uniform component from the strain $\partial u_0/\partial x$ along the x-axis, and there are two components that vary linearly through the beam depth with y and z. These latter components depend on the bending curvatures as measured by $\partial^2 v/\partial x^2$ and $\partial^2 w/\partial x^2$. Finally, there is the direct thermal stress $E\alpha\Delta T$, which varies over the beam cross-section depth with the temperature.

The final step in this phase of our analysis is to express the stress resultants in terms of the deformation derivatives. To do this, we substitute Eq. (6.9b) into Eqs. (6.6a), (6.6e), and (6.6f) and evaluate the integrals over the beam cross section,

$$N_x = \int_A \sigma_x \, \mathrm{d}A = EA\frac{\partial u_0}{\partial x} - N_T \tag{6.10a}$$

$$M_y = \int_A \sigma_x z \, \mathrm{d}A = -EI_{yz}\frac{\partial^2 v}{\partial x^2} - EI_y\frac{\partial^2 w}{\partial x^2} - M_{yT} \tag{6.10b}$$

$$M_z = \int_A \sigma_x y \, \mathrm{d}A = -EI_z\frac{\partial^2 v}{\partial x^2} - EI_{yz}\frac{\partial^2 w}{\partial x^2} - M_{zT} \tag{6.10c}$$

where we have introduced the following properties of the cross section:

$$I_y = \int_A z^2 \, \mathrm{d}A \qquad I_z = \int_A y^2 \, \mathrm{d}A \qquad I_{yz} = \int_A y^2 \, \mathrm{d}A \tag{6.11}$$

The first two properties, I_y and I_z, are the moments of inertia of the cross-sectional area about the y and z axes, respectively; I_{yz} is the product of inertia of the cross-sectional area. The product of inertia may be positive, negative, or zero. It will be zero if the y and z axes are principal axes. In the derivation of Eqs. (6.10), the integrals of the first moments of area,

$$\int_A y \, \mathrm{d}A = \int_A z \, \mathrm{d}A = 0$$

because the origin of the coordinate system is at the centroid of the cross section. In addition, we introduced

$$N_T = E\alpha \int_A (T - T_{\text{ref}})\mathrm{d}A \tag{6.12a}$$

$$M_{yT} = E\alpha \int_A (T - T_{\text{ref}})z \, \mathrm{d}A \tag{6.12b}$$

$$M_{zT} = E\alpha \int_A (T - T_{\text{ref}})y \, \mathrm{d}A \tag{6.12c}$$

where N_T is the thermal force, and M_{yT} and M_{zT} are thermal moments about the y and z axes, respectively.

From Eqs. (6.10), we may solve for the displacement derivatives to obtain

$$\frac{\partial u_0}{\partial x} = \frac{N_x + N_T}{EA} \tag{6.13a}$$

$$\frac{\partial^2 v}{\partial x^2} = \frac{I_{yz}(M_y + M_{yT}) - I_y(M_z + M_{zT})}{E(I_y I_z - I_{yz}^2)} \tag{6.13b}$$

$$\frac{\partial^2 w}{\partial x^2} = \frac{I_{yz}(M_z + M_{zT}) - I_z(M_y + M_{yT})}{E(I_y I_z - I_{yz}^2)} \tag{6.13c}$$

As mentioned earlier, we recognize $\partial u_0/\partial x$ as the beam strain along the centroidal axis. The second derivatives, $\partial^2 v/\partial x^2$ and $\partial^2 w/\partial x^2$, are the bending curvatures in the two coordinate planes. Finally, we substitute Eqs. (6.13) into Eq. (6.9b) to express the stress σ_x in terms of the stress resultants,

$$\sigma_x = \frac{N_x + N_T}{A} + \frac{[I_y(M_z + M_{zT}) - I_{yz}(M_y + M_{yT})]y}{I_y I_z - I_{yz}^2} + \frac{[I_z(M_y + M_{yT}) - I_{yz}(M_z + M_{zT})]z}{I_y I_z - I_{yz}^2} - E\alpha(T - T_{\text{ref}}) \tag{6.14}$$

Equations (6.13) and (6.14) express the beam deformation derivatives and the normal stress in terms of the stress resultants, thermal force, and thermal moments. If we have a statically determinant beam problem, these equations are sufficient to solve for the beam's deformations and normal stress. For, if a beam is statically determinant, we may solve for the internal forces at any beam section by three $\Sigma F = 0$ and three $\Sigma M = 0$ global equilibrium equations of statics. Knowing the internal stress resultants as a function of x and having a prescribed temperature distribution, we will know the right-hand sides of Eqs. (6.13) completely. Then we obtain the displacements u, v, w by integration with the proper boundary conditions.

If the beam problem is statically indeterminant, we will not be able to evaluate the stress resultants in Eqs. (6.13) in terms of only known applied loads because we will have more beam reactions than we have global equilibrium equations. For statically indeterminant beam problems, we must first solve for the beam displacements before we evaluate the stress resultants. Thus, we must develop partial differential equations for beam deformations that result from the equilibrium equations for a local beam element.

Equilibrium equations. To develop the beam equilibrium equations, let us now consider a differential length Δx of the beam as shown in Fig. 6.6. Figure 6.6a shows the stress resultants acting on two beam cross sections. In response to the applied body forces and moments, as well as the temperature, the stress resultants change from the left face to the right face of the beam segment. The torque M_x and bending moments M_x and M_y are represented by the double-headed arrows. Figure 6.6b shows the body forces and moments per unit length acting at the center of the element. For the very small deformations that we consider, the equilibrium

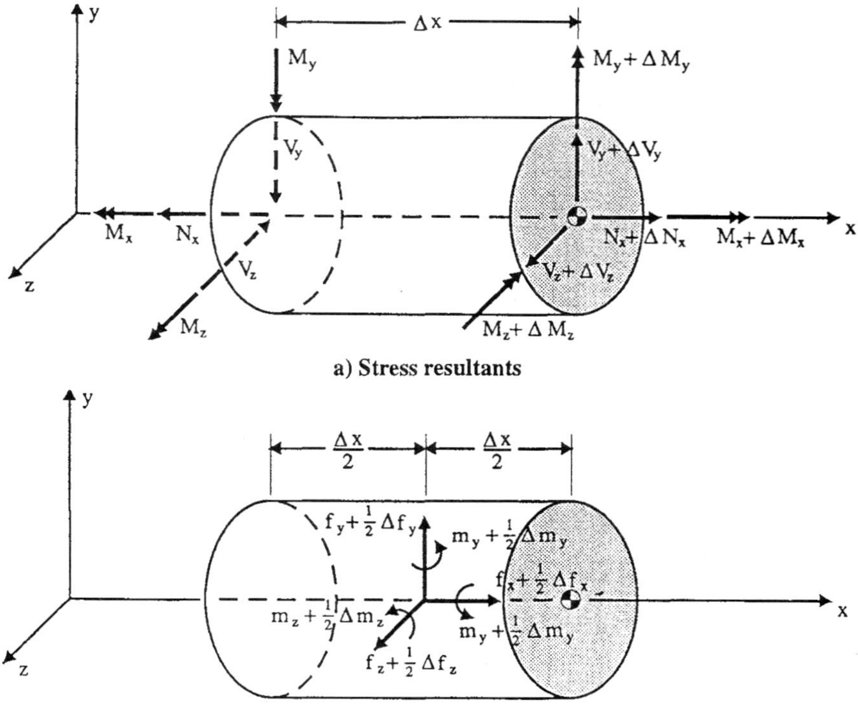

a) Stress resultants

b) Distributed loads

Fig. 6.6 Stress resultants and distributed loads acting on a differential beam length.

equations may be written for the undeformed configuration of the beam. This means that we neglect the relative rotations between the left-hand and right-hand faces of the beam segment. Using unit vectors $\hat{i}, \hat{j}, \hat{k}$ in the global Cartesian coordinate system, the force equilibrium is

$$
\begin{aligned}
\vec{F} = &\left[N_x + \Delta N_x - N_x + \left(f_x + \tfrac{1}{2}\Delta f_x\right)\Delta x\right]\hat{i} \\
&+\left[V_y + \Delta V_y - V_y + \left(f_y + \tfrac{1}{2}\Delta f_y\right)\Delta x\right]\hat{j} \\
&+\left[V_z + \Delta V_z - V_z + \left(f_z + \tfrac{1}{2}\Delta f_z\right)\Delta x\right]\hat{k} = 0
\end{aligned}
$$

which leads to the three scalar equations,

$$\frac{\Delta N_x}{\Delta x} + f_x + \frac{1}{2}\Delta f_x = 0$$

$$\frac{\Delta V_y}{\Delta x} + f_y + \frac{1}{2}\Delta f_y = 0$$

$$\frac{\Delta V_z}{\Delta x} + f_z + \frac{1}{2}\Delta f_z = 0$$

In the limit as $\Delta x \to 0$, the force equilibrium equations become

$$\frac{\partial N_x}{\partial x} + f_x = 0 \tag{6.15a}$$

$$\frac{\partial V_y}{\partial x} + f_y = 0 \tag{6.15b}$$

$$\frac{\partial V_z}{\partial x} + f_z = 0 \tag{6.15c}$$

Moment equilibrium requires that

$$\vec{M} = \Sigma(\vec{r} \times \vec{F}) = 0$$

Hence, by taking moments from the left-hand face, we may write

$$\begin{aligned}&\left[M_x + \Delta M_x - M_x + \left(m_x + \tfrac{1}{2}\Delta m_x\right)\Delta x\right]\hat{\imath} + [M_y + \Delta M_y - M_y \\ &+ \left(m_y + \tfrac{1}{2}\Delta m_y\right)\Delta x\right] j + \left[M_z - (M_z + \Delta M_z) + \left(m_z + \tfrac{1}{2}\Delta m_z\right)\Delta x\right]\hat{k} \\ &+ \tfrac{1}{2}\Delta x\,\hat{\imath}\left[\left(f_x + \tfrac{1}{2}\Delta f_x\right)\Delta x\,\hat{\imath} + \left(f_y + \tfrac{1}{2}\Delta f_y\right)\Delta x\,\hat{\jmath} + \left(f_z + \tfrac{1}{2}\Delta f_z\right)\Delta x\,\hat{k}\right] \\ &+ \Delta x\,\hat{\imath}\left[(m_x + \Delta N_x)\hat{\imath} + (V_y + \Delta V_y)\hat{\jmath} + (V_z + \Delta V_z)\hat{k}\right] = 0\end{aligned}$$

We work out the cross-product terms and write the three scalar equations as

$$\frac{\Delta M_x}{\Delta x} + m_x + \frac{1}{2}\Delta m_x = 0$$

$$\frac{\Delta M_y}{\Delta x} + m_y + \frac{1}{2}\Delta m_y - \frac{1}{2}\Delta x\left(f_z + \frac{1}{2}\Delta f_z\right) - (V_z + \Delta V_z) = 0$$

$$-\frac{\Delta M_z}{\Delta x} + m_z + \frac{1}{2}\Delta m_z - \frac{1}{2}\Delta x\left(f_y + \frac{1}{2}\Delta f_y\right) + (V_y + \Delta V_y) = 0$$

Then, in the limit as $\Delta x \to 0$, the moment equilibrium equations are

$$\frac{\partial M_x}{\partial x} + m_x = 0 \tag{6.16a}$$

$$\frac{\partial M_y}{\partial x} + m_y = V_z \tag{6.16b}$$

$$\frac{\partial M_z}{\partial x} - m_z = V_y \tag{6.16c}$$

Looking back, we can see that the moments of the body forces were of order Δx^2 so that, in the limit, these terms vanish, and the body forces do not contribute to moment equilibrium. Equations (6.15) and (6.16) are the equilibrium equations of linear beam theory. With these equations and previous results, we can derive the partial differential equations that govern thermoelastic behavior for both statically determinant and indeterminant beams.

Governing differential equations. We can now develop the differential equations that describe 1) extensional deformation of the centroidal axis, 2) bending in the x–y plane, and 3) bending in the x–z plane.

Extension. Equation (6.10a) relates the stress resultant N_x to the axial displacement u_0 of the centroidal axes. Substituting Eq. (6.10a) into Eq. (6.15a) yields

$$EA\frac{\partial^2 u_0}{\partial x^2} - \frac{\partial N_T}{\partial x} + f_x = 0 \tag{6.17}$$

Thus, the extension of the centroidal axis depends on the gradient of the thermal force and the distributed force f_x. Equation (6.17) has terms similar to Eq. (6.4a), which describes the behavior of a rod. The major difference is that a rod assumes a temperature that is uniform over the cross section where, for the beam, the temperature may vary over the cross section. This variation is taken into account in Eq. (6.12a), where the thermal force N_T is derived by integration of the temperature over the cross-sectional area. Note that the extension $u_0(x, t)$ is independent of the transverse bending.

For a prescribed temperature distribution, we first evaluate the thermal force N_T using Eq. (6.12a). Given the distributed force f_x and appropriate boundary conditions, we can solve Eq. (6.17) for $u_0(x, t)$. Finally, from Eq. (6.10a) we evaluate $N_x(x, t)$.

Bending in the x–y plane. First, for convenience of reference, we recall, from Eq. (6.10c), the definition of the bending moment M_z. Then we substitute the bending moment M_z into the equilibrium equation (6.16c) to produce an equation for the shear force V_y in terms of derivatives of the displacements. Finally, we substitute the equation for V_y in the equilibrium equation (6.15b) to produce the governing differential equation for deflections for the x–y plane. The results are as follows:

$$M_z = -EI_z\frac{\partial^2 v}{\partial x^2} - EI_{yz}\frac{\partial^2 w}{\partial x^2} - M_{zT} \tag{6.18a}$$

$$V_y = -EI_z\frac{\partial^3 v}{\partial x^3} - EI_{yz}\frac{\partial^3 w}{\partial x^3} - \frac{\partial M_{zT}}{\partial x} - m_z \tag{6.18b}$$

$$EI_z\frac{\partial^4 v}{\partial x^4} + EI_{yz}\frac{\partial^4 w}{\partial x^4} + \frac{\partial^2 M_{zT}}{\partial x^2} + \frac{\partial m_z}{\partial x} = f_y \tag{6.18c}$$

Bending in the x–z plane. In a similar way, we may write the three equations describing bending in the x–z plane:

$$M_y = -EI_{yz}\frac{\partial^2 v}{\partial x^2} - EI_y\frac{\partial^2 w}{\partial x^2} - M_{yT} \tag{6.19a}$$

$$V_z = -EI_{yz}\frac{\partial^3 v}{\partial x^3} - EI_y\frac{\partial^3 w}{\partial x^3} - \frac{\partial M_{yT}}{\partial x} + m_y \tag{6.19b}$$

$$EI_{yz}\frac{\partial^4 v}{\partial x^4} + EI_y\frac{\partial^4 w}{\partial x^4} + \frac{\partial^2 M_{yT}}{\partial x^2} - \frac{\partial m_y}{\partial x} = f_z \tag{6.19c}$$

To solve a statically indeterminant beam problem, we begin with the partial differential equations for the deflections v and w, Eqs. (6.18c) and (6.19c). For a prescribed temperature distribution, we evaluate the thermal moments M_{yT} and M_{zT} using Eqs. (6.12b) and (6.12c). Given the distributed loads and appropriate boundary conditions, we next solve the coupled differential equations for $v(x,t)$ and $w(x,t)$. The differential equations are coupled through the terms containing the product of inertia I_{yz}. For a general orientation of the coordinate axes y and z, the product of inertia is not zero, and the differential equations are coupled. However, if the orientation of the y–z axes represents principal axes, then $I_{yz} = 0$, and the differential equations become uncoupled. This occurs, for example, if the cross section has an axis of symmetry.

Once the displacements have been determined, the bending moments M_z and M_y, as well as the shearing forces V_y and V_z, be evaluated using Eqs. (6.18a) and (6.19a) and Eqs. (6.18b) and (6.19b), respectively. The normal stress σ_x is given by Eq. (6.14) in terms of the axial force N_x, the bending moments M_z and M_y, and the temperature distribution.

Torsion. From Sec. 5.6, the torque M_x is related to the angle of twist θ_x by Eq. (5.70),

$$M_x = GJ\frac{\partial \theta_x}{\partial x}$$

where G is the shearing modulus and J the torsional constant. Substituting into the equilibrium equation (6.16a),

$$GJ\frac{\partial^2 \theta_x}{\partial x^2} + m_x = 0 \tag{6.20}$$

From Eq. (6.20), we see that the torsion problem is independent of temperature as well as uncoupled from extension and bending.

Boundary conditions. We may separate the boundary conditions into the cases of extension, bending, and torsion. Since we consider quasistatic behavior, initial conditions are not required.

Extension. Since Eq. (6.17) is a second-order equation in x, we must have two boundary conditions, one at each end of the beam. The possibilities are to specify either the axial displacement or the axial force. Thus, for a beam of length L, we specify

$$\begin{aligned} u_0(0,t) &= u_1(t) \quad \text{or} \quad N_x(0,t) = F_1(t) \\ u_0(L,t) &= u_2(t) \quad \text{or} \quad N_x(L,t) = F_2(t) \end{aligned} \tag{6.21}$$

For a free end, the axial force N_x is specified as zero.

Bending. The two differential equations (6.18a) and (6.19a) are fourth-order equations in x and, altogether, we will need eight boundary conditions: four conditions must be prescribed at each end of the beam. Normally, these conditions consist of specifying either displacement or shear force, or specifying either slope or bending moment. Thus, at a typical end, say $x = a$, where $a = 0$ or $a = L$, we specify four boundary conditions from the following possibilities:

$$\begin{aligned} v(a,t) &= v_1(t) \quad \text{or} \quad V_y(a,t) = V_{y1}(t) \\ \frac{\partial v}{\partial x}(a,t) &= \theta_{z1}(t) \quad \text{or} \quad M_z(a,t) = M_{z1}(t) \\ w(a,t) &= w_1(t) \quad \text{or} \quad V_z(a,t) = V_{z1}(t) \\ \frac{\partial w}{\partial x}(a,t) &= \theta_{y1}(t) \quad \text{or} \quad M_y(a,t) = M_{y1}(t) \end{aligned} \tag{6.22}$$

These possibilities include the commonly encountered cases of clamped, simple support, and free-end conditions. There are other possibilities, of course, such as an elastic support.

Torsion. The boundary conditions for torsion are similar to those for extension. The possibilities are to specify either the angle of twist or the torque. Thus, we specify that

$$\begin{aligned} \theta_x(0,t) &= \theta_1(t) \quad \text{or} \quad M_x(0,t) = G_1(t) \\ \theta_x(L,t) &= \theta_2(t) \quad \text{or} \quad M_x(L,t) = G_2(t) \end{aligned} \tag{6.23}$$

For a free end, the torque M_x is specified as zero.

In this section, we present the classical formulation for quasistatic behavior of beams with compact cross sections. Actually, the beam cross sections do not have to be solid. The cross sections can be hollow but, because we have assumed that plane sections remain plane, the theory does not allow for warping of the cross sections. In addition, we have made geometric assumptions that limit applications to small deflections. The principal geometric assumptions are the use of the linear strain displacement relations and small rotations. In addition, the equilibrium equations were developed for the undeformed beam.

6.3.2 Formulation for Large Deflections

In the preceding section, we develop the classical linear formulation for long, slender beams. The theory predicts that bending is uncoupled from extension and torsion. For example, according to this theory, an axial force either in tension or compression applied to a beam causes only an extension or compression. Furthermore, if we apply a lateral load that induces bending, axial deformations are not affected. Similar uncoupled behavior occurs for loads that induce torsional deformations. We encounter similar behavior for thermal loads.

Such limitations in behavior contradict our physical intuition and preclude modeling important thermally induced behavior, including buckling. The limitations in the theory arise mainly because the equilibrium equations were written for the undeformed beam; for, in writing the equilibrium equations, beam deflections and rotations were ignored. In addition, classical beam theory is based on the linear strain-displacement equations, where products and powers of displacement derivatives are neglected. This means that the equations are limited to very small deformations. Thus, two geometric assumptions are implicit in classical beam theory, and the theory is limited to small rotations and small strains. Some authors say that classical beam theory is limited to infinitesimal deflections. More practically, we say that classical beam theory is applicable for small deflections. As a rough rule, small-deflection classical beam theory applies when transverse bending displacements v and w are less than the beam depth. A precise definition is difficult to provide because the accuracy of the theory depends on loading and boundary conditions.

The assumption of small deflections means that all mechanical and thermal load terms in the governing partial differential equations [see Eqs. (6.17), (6.18c), and (6.19c)] are independent of deformation. The assumption of small strains means that only linear displacement gradients appear in the equations. Consequently, a small-deflection beam is desirable mathematically because the partial differential equations are linear. There is extensive literature on classical beam theory, and it finds wide application in analysis and design.

We now consider a formulation for beam problems where the two basic geometric limitations are removed. The theory we develop permits consideration of problems that involve coupling between extension, bending, and torsion. We call it large-deflection beam theory, but we shall see that theory will still be limited to moderate rotations. The most general beam formulation for unlimited deformations we denote as finite deflection theory. Such large deformations are rarely encountered in aerospace applications and will not be considered here.

We begin by developing a new set of equilibrium equations that take into account beam deformations. Next, we formulate the equations applicable to problems involving coupling between extension, bending, and torsion.

Equilibrium equations. When we derive the equilibrium equations given in Eqs. (6.15) and (6.16), we assume that the differential beam segment Δx shown in Fig. 6.6 is undeformed. This means that relative rotations between the left-hand and right-hand faces of the beam segment are neglected. Let us now rederive the equilibrium equations, taking into account the relative rotations. We follow an approach presented by Rivello[6] and Donaldson.[8] To begin, consider the deformed beam segment shown in Fig. 6.7. (Body forces and moments are omitted for clarity.)

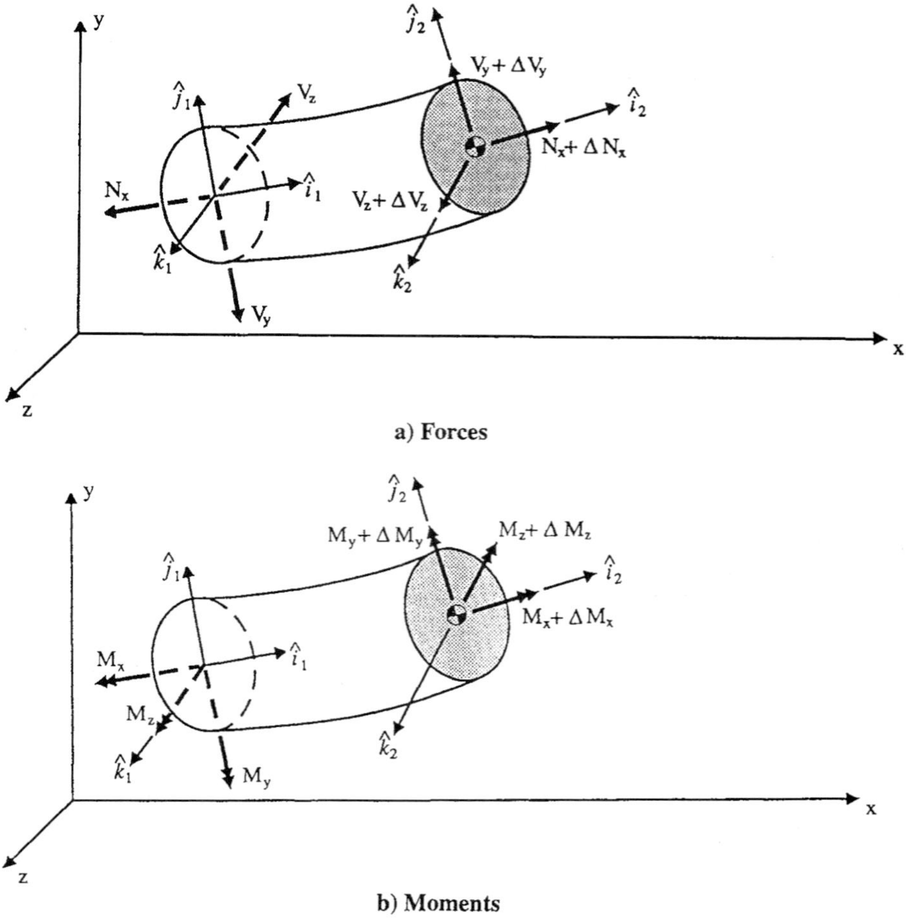

Fig. 6.7 Deformed beam segment.

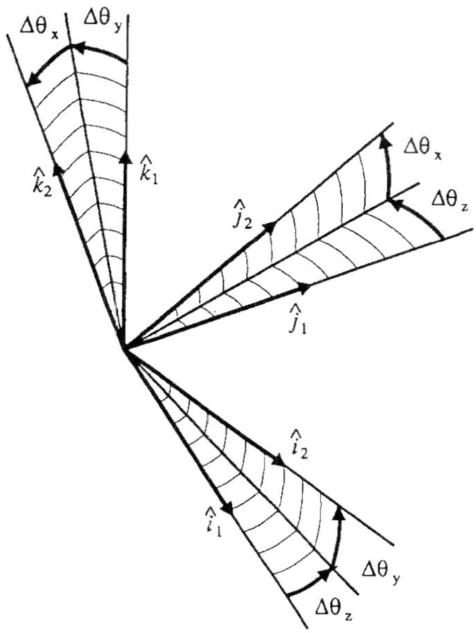

Fig. 6.8 Relative rotations of unit vectors (from Ref. 12).

In the deformed configuration, two sets of unit vectors are shown. On the left-hand face, we denote the unit vectors as $\hat{\imath}_1, \hat{\jmath}_1, \hat{k}_1$ and, on the right-hand face, we denote the unit vectors as $\hat{\imath}_2, \hat{\jmath}_2, \hat{k}_2$. The rotations of the left-hand face are $\theta_x, \theta_y, \theta_z$, and the rotations of the right-hand face are $\theta_x + \Delta\theta_x, \theta_y + \Delta\theta_y, \theta_z + \Delta\theta_z$. The incremental rotations $\Delta\theta_x, \Delta\theta_y, \Delta\theta_z$ represent relative rotations between faces. We wish to take these rotations into account when we write the equilibrium equations. Our approach will be to write force and moment equations for the incremental rotations and then pass to the limit as the incremental rotations approach zero.

A key step is to establish the geometric relationship between the two sets of unit vectors. Figure 6.8 shows the relative positions of the unit vectors after a particular sequence of angular increments. (For infinitesimal angular increments, the sequence is not important because infinitesimal rotations are commutative.[13]) From this figure, we may write the equations relating the unit vectors,

$$\begin{aligned} \hat{\imath}_2 &= \hat{\imath}_1 + \Delta\theta_z\hat{\jmath}_1 + \Delta\theta_y\hat{k}_1 \\ \hat{\jmath}_2 &= -\Delta\theta_z\hat{\imath}_1 + \hat{\jmath}_1 + \Delta\theta_x\hat{k}_1 \\ \hat{k}_2 &= -\Delta\theta_y\hat{\imath}_1 - \Delta\theta_x\hat{\jmath}_1 + \hat{k}_1 \end{aligned} \tag{6.24}$$

Next, we write the equilibrium equations for the deformed beam segment in terms of both sets of unit vectors. Then we introduce Eqs. (6.24) to combine the terms for a single set of unit vectors. Finally, we write the scalar equations and pass to the limit to produce the equilibrium equations. For simplicity, we omit higher-order terms that vanished in the limit during our previous derivation, such as the small increments of the body forces and moments. The force equilibrium equation is

$$\begin{aligned} \vec{F} = &(-N_x\hat{\imath}_i - V_y\hat{\jmath}_i - V_z\hat{k}_1) + (f_x\hat{\imath}_1 + f_y\hat{\jmath}_i + f_z\hat{k}_1)\Delta x \\ &+ (N_x + \Delta N_x)\hat{\imath}_2 + (V_y + \Delta V_y)\hat{\jmath}_2 + (V_z + \Delta V_z)\hat{k}_2 \end{aligned}$$

After substituting Eqs.(6.24) for $\hat{\imath}_2, \hat{\jmath}_2, \hat{k}_2$, we can then combine terms as coefficients of $\hat{\imath}_1, \hat{\jmath}_1, \hat{k}_1$. These steps lead to the three scalar equations,

$$\frac{\Delta N_x}{\Delta x} - (V_y + \Delta V_y)\frac{\Delta\theta_y}{\Delta x} - (V_z + \Delta V_z)\frac{\Delta\theta_z}{\Delta x} + f_x = 0$$

$$(N_x + \Delta N_x)\frac{\Delta\theta_z}{\Delta x} + \frac{\Delta V_y}{\Delta x} - (V_z + \Delta V_z)\frac{\Delta\theta_z}{\Delta x} + f_y = 0$$

$$(N_x + \Delta N_x)\frac{\Delta\theta_y}{\Delta x} + \frac{\Delta V_z}{\Delta x} - (V_y + \Delta V_y)\frac{\Delta\theta_x}{\Delta x} + f_z = 0$$

In the limit as $\Delta x \to 0$, the force equilibrium equations become

$$\frac{\partial N_x}{\partial x} - V_y\frac{\partial\theta_y}{\partial x} - V_z\frac{\partial\theta_z}{\partial x} + f_x = 0$$

$$N_x\frac{\partial\theta_z}{\partial x} + \frac{\partial V_y}{\partial x} - V_z\frac{\partial\theta_z}{\partial x} + f_y = 0$$

$$N_x\frac{\partial\theta_y}{\partial x} + \frac{\partial V_y}{\partial x} + V_y\frac{\partial\theta_x}{\partial x} + f_z = 0$$

For a long, slender beam, the shear forces are usually much smaller than the axial force. For this reason, the terms that involve products of shear forces and derivatives of rotations are much smaller than other terms and are neglected. Thus, the force equilibrium equations reduce to

$$\frac{\partial N_x}{\partial x} + f_x = 0$$

$$\frac{\partial V_y}{\partial x} + N_x\frac{\partial\theta_z}{\partial x} + f_y = 0$$

$$\frac{\partial V_z}{\partial x} + N_x\frac{\partial\theta_y}{\partial x} + f_z = 0$$

The moment equilibrium equations are derived in a similar way. Taking moments from the left-hand face, we may write

$$\begin{aligned}(-M_x\hat{\imath}_1 - M_y\hat{\jmath}_1 + M_z\hat{k}_1) + (M_x + \Delta M_x)\hat{\imath}_2 + (M_y + \Delta M_y)\hat{\jmath}_2 \\ - (M_z + \Delta M_z)\hat{k}_2 + (m_x\hat{\imath}_1 + m_y\hat{\jmath}_1 + m_2\hat{k}_1)\Delta x \\ + \Delta x\,\hat{\imath} \times [(N_x + \Delta N_x)\hat{\imath}_2 + (V_y + \Delta V_z)\hat{\jmath}_2 + (V_z + \Delta V_z)\hat{k}_2] = 0\end{aligned}$$

Writing out the cross products, introducing Eqs. (6.24), and then taking the limit lead to the three scalar equations,

$$\frac{\partial M_x}{\partial x} - M_y\frac{\partial\theta_z}{\partial x} + M_z\frac{\partial\theta_y}{\partial x} + m_x = 0$$

$$\frac{\partial M_y}{\partial x} - M_x\frac{\partial\theta_z}{\partial x} + M_z\frac{\partial\theta_x}{\partial x} + m_y = V_z$$

$$\frac{\partial M_z}{\partial x} - M_x\frac{\partial\theta_y}{\partial x} - M_y\frac{\partial\theta_x}{\partial x} - m_z = V_y$$

The last two sets of scalar equations are the six equilibrium equations for large deflections that take into account the change in rotations along the beam. By referring to the small-deflection equilibrium equations (6.15) and (6.16), we see that the effect of the rotations is to introduce new terms that involve derivatives of the three rotations. Of these, the derivative $\partial\theta_x/\partial x$ is associated with torsion and represents the angle of twist per unit length. The derivatives $\partial\theta_y/\partial x$ and $\partial\theta_z/\partial x$ are associated with bending curvature. From calculus, these derivatives are defined in terms of v and w as

$$\frac{\partial\theta_z}{\partial x} = \left(\frac{\partial^2 v}{\partial x^2}\right)\Big/\left[1+\left(\frac{\partial v}{\partial x}\right)^2\right]^{3/2} \qquad \frac{\partial\theta_y}{\partial x} = \left(\frac{\partial^2 w}{\partial x^2}\right)\Big/\left[1+\left(\frac{\partial w}{\partial x}\right)^2\right]^{3/2}$$

Thus, if we make the assumptions used in classical small-deflection beam theory that

$$\left(\frac{\partial v}{\partial x}\right)^2 \ll 1 \qquad \left(\frac{\partial w}{\partial x}\right)^2 \ll 1$$

then the slope derivatives represent the beam curvatures by the linear equations

$$\frac{\partial\theta_z}{\partial x} = \frac{\partial^2 v}{\partial x^2} \qquad \frac{\partial\theta_y}{\partial x} = \frac{\partial^2 w}{\partial x^2}$$

Introducing these definitions leads to the six large-deflection equilibrium equations,

$$\frac{\partial N_x}{\partial x} + f_x = 0 \tag{6.25a}$$

$$\frac{\partial V_y}{\partial x} + \underline{N_x\frac{\partial^2 v}{\partial x^2}} + f_y = 0 \tag{6.25b}$$

$$\frac{\partial V_z}{\partial x} + \underline{N_x\frac{\partial^2 w}{\partial x^2}} + f_z = 0 \tag{6.25c}$$

$$\frac{\partial M_x}{\partial x} \underline{- M_y\frac{\partial^2 v}{\partial x^2} + M_z\frac{\partial^2 w}{\partial x^2}} + m_x = 0 \tag{6.25d}$$

$$\frac{\partial M_y}{\partial x} + \underline{M_x\frac{\partial^2 v}{\partial x^2} + M_z\frac{\partial\theta_x}{\partial x}} + m_y = V_z \tag{6.25e}$$

$$\frac{\partial M_z}{\partial x} \underline{- M_x\frac{\partial^2 w}{\partial x^2} - M_y\frac{\partial\theta_x}{\partial x}} - m_z = V_y \tag{6.25f}$$

The underlined terms are the contributions introduced by considering the deformed beam segment. A typical new term consists of the product of a stress resultant and a displacement derivative. For example, Eq. (6.25b) contains a new term that is the product of the axial force resultant N_x and the bending curvature in the x–y plane, $\partial^2 w/\partial x^2$. Similarly, the torque equilibrium equation (6.25d) contains products of the bending moments and curvatures. If the stress resultants are not constant, the underlined contributions may introduce nonlinear terms in the final partial differential equations. We shall see how this arises as we next develop formulations that apply for problems with extension, bending, and torsion coupling.

Governing differential equations. We can now use the large-deflection equilibrium equations to develop a new set of differential equations that describe extension, bending, and torsion of a beam. To simplify the equations, we assume that the y–z axes are principal axes so that the product of inertia I_{yz} is zero.

Extension of centroidal axis. The stress resultant N_x is the integral of the normal stress σ_x over the cross section, as shown in Eq. (6.6a). The stress σ_x is related to the strain ε_x and temperature T by Hooke's law, as shown in Eq. (6.9a). Thus, we may write

$$N_x(x,t) = \int_A \sigma_x \, dA = E\int_A \varepsilon_x(x,y,z,t)\, dA - E\alpha \int_A [T(x,y,z,t) - T_{\text{ref}}]\, dA$$

or

$$N_x(x,t) = E\int_A \varepsilon_x(x,y,z,t)\, dA - N_T(x,t)$$

where the thermal force N_T is defined by Eq. (6.12a). For large deflections, we use the nonlinear strain-displacement equations from continuum mechanics. From Eq. (3.9), we express the strain ε_x by the general equation,

$$\varepsilon_x = \frac{\partial u}{\partial x} + \frac{1}{2}\left[\left(\frac{\partial u}{\partial x}\right)^2 + \left(\frac{\partial v}{\partial x}\right)^2 + \left(\frac{\partial w}{\partial x}\right)^2\right]$$

For a long, slender beam, the lateral displacements v and w are much larger than the axial displacement u. Thus, we neglect the displacement gradient $(\partial u/\partial x)^2$ in comparison to the displacement gradients $(\partial v/\partial x)^2$ and $(\partial w/\partial x)^2$. With this simplification, we substitute for ε_x to obtain

$$N_x = E\int_A \frac{\partial u}{\partial x}\, dA + \frac{1}{2}E\int_A \left[\left(\frac{\partial v}{\partial x}\right)^2 + \left(\frac{\partial w}{\partial x}\right)^2\right] dA - N_T$$

Finally, we introduce the u, v, w displacements for a beam from Eq. (6.7). The integral of the first term on the right-hand side simplifies because the origin of the coordinate system is located at the centroid. The nonlinear term in the second term may be brought outside the integral because the beam deflections are independent of y and z; i.e., $v = v(x,t)$ and $w = w(x,t)$. Consequently, we obtain

$$N_x = EA\frac{\partial u_0}{\partial x} + \frac{1}{2}EA\left[\left(\frac{\partial v}{\partial x}\right)^2 + \left(\frac{\partial w}{\partial x}\right)^2\right] - N_T \tag{6.26}$$

Recall that u_0 is the axial displacement along the centroidal axis. Thus, for large deflections, transverse bending contributes to the axial force through the nonlinear terms involving the beam slopes. These nonlinear effects are most important in applications where the in-plane motion of the beam is restrained so that the beam must stretch to deform laterally. We will illustrate this behavior in a later example. The partial differential equation for u_0 is obtained by substituting Eq. (6.26) into the equilibrium equation (6.25a). This step yields

$$EA\frac{\partial^2 u_0}{\partial x} + \frac{1}{2}EA\frac{\partial}{\partial x}\underline{\left[\left(\frac{\partial v}{\partial x}\right)^2 + \left(\frac{\partial w}{\partial x}\right)^2\right]} - \frac{\partial N_T}{\partial x} + f_x = 0 \tag{6.27}$$

Comparing Eq. (6.27) with the earlier equation for small deflections, Eq. (6.17), we see that the equations differ by the underlined term. As mentioned, this nonlinear term represents the stretching of the centroidal axis due to transverse bending. In comparison to Eq. (6.17), the new equation is coupled to the bending displacements and can no longer be solved separately. It is also a nonlinear equation since the beam slopes $\partial v/\partial x$ and $\partial w/\partial x$ appear squared.

Bending in the x–y plane. We follow the same steps as we did for small deflections. We recall from Eq. (6.10c) the definition of the bending moment M_z. Then we substitute the bending moment into the equilibrium (6.25f) to produce an equation for the shear force V_y. Finally, we substitute the shear force V_y into the equilibrium equation (6.25b). With principal axes assumed, the results are

$$M_z = -EI_z \frac{\partial^2 v}{\partial x^2} - M_{zT} \tag{6.28a}$$

$$V_y = -EI_z \frac{\partial^3 v}{\partial x^3} \underline{- M_x \frac{\partial^2 w}{\partial x^2} - M_y \frac{\partial \theta_x}{\partial x}} - \frac{\partial M_{zT}}{\partial x} - m_z \tag{6.28b}$$

$$EI_z \frac{\partial^4 v}{\partial x^4} \underline{- N_x \frac{\partial^2 v}{\partial x^2} + \frac{\partial}{\partial x}\left(M_x \frac{\partial^2 w}{\partial x^2}\right) \frac{\partial}{\partial x}\left(M_y \frac{\partial \theta_x}{\partial x}\right)} + \frac{\partial^2 M_{zT}}{\partial x^2} + \frac{\partial m_z}{\partial x} = f_y \tag{6.28c}$$

Comparing Eqs. (6.28) with the earlier equation for small deflections Eqs. (6.18), we see that the equations differ by the underlined terms. The underlined terms couple bending in the x–y plane with bending in the x–z plane and torsion about the x axis.

Bending in the x–z plane. In a similar way, we write the equations for the x–z plane:

$$M_y = -EI_y \frac{\partial^2 w}{\partial x^2} - M_{yT} \tag{6.29a}$$

$$V_z = -EI_z \frac{\partial^3 w}{\partial x^3} \underline{+ M_x \frac{\partial^2 v}{\partial x^2} + M_z \frac{\partial \theta_x}{\partial x}} - \frac{\partial M_{yT}}{\partial x} + m_y \tag{6.29b}$$

$$EI_y \frac{\partial^4 w}{\partial x^4} \underline{- N_x \frac{\partial^2 w}{\partial x^2} - \frac{\partial}{\partial x}\left(M_x \frac{\partial^2 v}{\partial x^2}\right) - \frac{\partial}{\partial x}\left(M_z \frac{\partial \theta_x}{\partial x}\right)} - \frac{\partial^2 M_{yT}}{\partial x^2} + \frac{\partial m_y}{\partial x} = f_z \tag{6.29c}$$

Once again, the underlined terms represent the new contributions introduced for large deflections. As before, the new terms represent coupling between bending in the orthogonal coordinate planes as well as coupling with torsion about the x axis.

Torsion. From Sec. 5.6, the torque M_x is related to the angle of twist θ_x by Eq. (5.70),

$$M_x = GJ \frac{\partial \theta_x}{\partial x}$$

Substituting into Eq. (6.25d),

$$GJ \frac{\partial^2 \theta_x}{\partial x^2} \underline{- M_y \frac{\partial^2 v}{\partial x^2}} + M_z \frac{\partial^2 w}{\partial x^2} + m_x = 0 \tag{6.30}$$

where the underlined terms represent the large-deflection contributions that couple bending and torsion.

Equations (6.27–6.30) constitute the set of partial differential equations that describe large-deflection, coupled extension, bending, and torsion. They represent four partial differential equations with four unknowns: 1) the displacement $u_0(x, t)$ of the centroidal axis, 2) the bending displacement $v(x, t)$, 3) the bending displacement $w(x, t)$, and 4) the angle of twist $\theta_x(x, t)$. The underlined terms represent coupling between the four equations. Note that the underlined terms are nonlinear because they include products or powers of derivatives of the dependent variables.

The equations are solved subject to the boundary conditions described in Sec. 6.3.1.

6.3.3 Formulation for Thin-Walled Open Cross Sections

We consider now a beam with a thin-walled open cross section. The beam experiences combined axial, bending, and torsional loads, as well as an arbitrary temperature distribution. Open cross sections, in comparison to closed cross sections, are relatively weak in torsion. When subjected to torsional loads, a beam with an open cross section demonstrates significant warping. The assumption that plane sections remain plane is no longer valid. If a cross section is free to warp, the torsion is said to be *uniform*. On the other hand, if a cross section is restrained from warping, the torsion is said to be *nonuniform*. When a cross section of a thin-walled beam with an open cross section is restrained from warping, significant warping stresses may develop. Moreover, the effects of warping stresses do not "die out" rapidly, so that their effect is felt some distance away from the boundary and, hence, Saint Venant's principle (Sec. 5.5.1) is no longer valid. Thus, warping complicates a beam's thermal-stress formulation and requires special consideration.

Beams with open cross sections are used frequently in aerospace applications because of their light weight and accessibility for maintenance. Because of its small torsional stiffness, a beam with an open cross section has low torsional natural frequencies and can buckle in torsion or combined flexure and torsion.

The development of the theory of thin-walled beams is generally attributed to Vlasov.[15] We follow presentations given by Oden,[5] Rivello,[6] and Donaldson.[8] Related discussions of thin-walled structures in a civil engineering context appear in the books by Galambos[16] and Murray.[17]

We develop here a bending and torsion formulation for thin-walled beams with open cross sections. We assume, as before, that the beam experiences elastic behavior. For simplicity, we assume that the y–z coordinate axes are principal axes for the cross section; the origin of the coordinate system is at the centroid. For convenience, we assume that f_x, the body force per unit length in the x direction, is zero and that m_y and m_z, the body moments per unit length about the y and z axes, are zero. We retain the body forces per unit length f_y and f_z, as well as the body torque m_x per unit length.

Stresses and torque. A typical small element of a thin-walled beam with open cross section experiences a combined state of normal and shearing stress (Fig. 6.9). We regard the stress components as having constituents from Bernoulli–Euler beam theory and Saint Venant torsion theory. In addition, there are constituents associated with restrained warping. In this paragraph, we provide an overview of the state of stress and describe how these constituents contribute to the torque developed at a cross section. In succeeding paragraphs, we develop the theory for the new contributions.

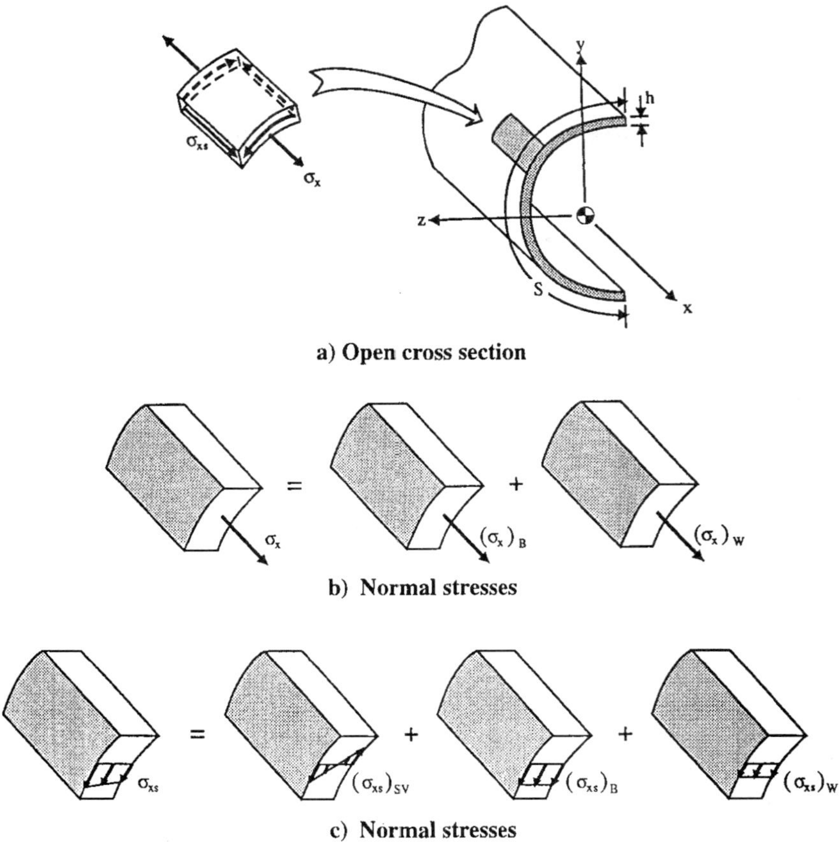

Fig. 6.9 Stresses in a thin-walled beam with open cross section.

Normal stress. We represent the normal stress σ_x by two constituents (Fig. 6.9b): the normal stress $(\sigma_x)_B$ from Bernoulli–Euler beam theory that we developed in Sec. 6.3.1, and the warping stress $(\sigma_x)_W$ that we develop later in this section. The bending stress $(\sigma_x)_B$ arises from extension and bending where plane sections remain plane. The warping stress $(\sigma_x)_W$ arises from restrained warping of the cross section induced by torsion. Thus, we may write

$$\sigma_x = (\sigma_x)_B + (\sigma_x)_w \tag{6.31}$$

where, from Eq. (6.14), with principal axes assumed,

$$(\sigma_x)_B = \frac{N_x + N_{xT}}{A} + \frac{M_z + M_{zT}}{I_z} y + \frac{M_y + M_{yT}}{I_y} z - E\alpha\Delta T \tag{6.32}$$

where $\Delta T = T - T_{\text{ref}}$. Recall that the axial force N_x is assumed positive for tension; the bending moments are M_y and M_z, and their positive senses are shown in Fig. 6.5b. The thermal force N_{xT} and the thermal moments are defined by Eqs. (6.12). An equation for $(\sigma_x)_w$ in terms of the torque M_x and cross-sectional properties remains to be developed.

Shearing stress. We represent the shearing stress σ_{xs} by three constituents (Fig. 6.9c): the shearing stress $(\sigma_{xs})_{SV}$ from Saint Venant uniform torsion theory, the shearing stress for Bernoulli–Euler beam bending theory $(\sigma_{xs})_B$, and the shearing stress $(\sigma_{xs})_W$ associated with warping. Thus, we write

$$\sigma_{xs} = (\sigma_{xs})_{\text{SV}} + (\sigma_{xs})_B + (\sigma_{xs})_W \tag{6.33}$$

Elasticity and mechanics of solids texts show that the shearing stress in a thin-wall section in uniform torsion varies linearly through the thickness, as shown in Fig. 6.9c. The maximum stress is given by

$$(\sigma_{xs})_{\text{SV}} = \frac{M_x h}{J} \tag{6.34}$$

where h is the wall thickness, and J the torsion constant. For the thin-walled cross section shown in Fig. 6.9a, $J = Sh^3/3$, where S is the length of the centerline of the "unrolled" thin cross section, and h is the wall thickness. Equations for $(\sigma_{xs})_B$ and $(\sigma_{xs})_W$ are to be developed.

Torque. Similarly, we regard the torque M_x developed on any cross section as the sum of three constituents: the torque $(M_x)_{\text{SV}}$ resulting from the Saint Venant shearing stresses, the torque $(M_x)_W$ resulting from the warping shearing stresses, and the torque $(M_x)_{\sigma_x}$ arising from the normal stress α_x. In addition, we will see that there is a torque induced by the variation of temperature with x that we denote as $(M_x)_T$. Thus,

$$M_x = (M_x)_{\text{SV}} + (M_x)_W + (M_x)_{\sigma_x} + (M_x)_T \tag{6.35}$$

Section 5.6 shows that the Saint Venant torque may be represented as

$$(M_x)_{\text{SV}} = GJ\frac{\partial\theta_x}{\partial x} \tag{6.36}$$

where G is the shear modulus, and J the torsion constant of the cross section. The derivative $\partial\theta_x/\partial x$ represents the angle of twist per unit length, and we assume that $\theta_x = \theta_x(x, t)$. Equations for the torques from warping, the normal stress, and variation in temperature are to be developed.

Shearing stress due to bending. For the long, slender beams with compact cross sections that we consider in Sec. 6.3.1, shearing stresses are small and may usually be neglected. For thin-walled beams, however, shearing stresses are much more important and must be considered. On account of the *thin* wall of thickness h, shearing stresses tend to be larger in magnitude. Furthermore, thin-walled members are more susceptible to local buckling. Because states of pure shearing stress cause compressive normal stresses on planes at a 45-deg angle from the planes of shearing stress, large shearing stresses may induce local buckling.

In isothermal beam bending, shear stresses $(\sigma_{xs})_B$ are caused by variations in the normal stress $(\sigma_x)_B$ along the beam, that is, by the normal stress varying with x. That is also true for beams with temperature distributions but, additionally, the variation in the temperature with x contributes to the shearing stress.

To begin, let us examine the equilibrium condition of a typical wall element such as the one shown in Fig. 6.10. For convenience, we work with the shear flow q,

$$q = h\sigma_{xs} \tag{6.37}$$

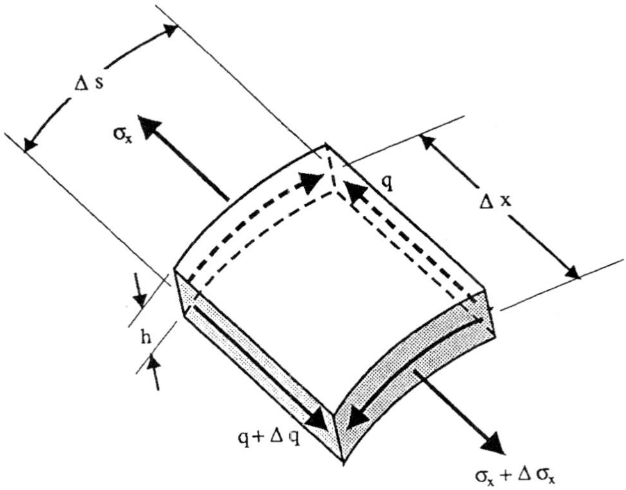

Fig. 6.10 Shear flow and stresses on a typical wall element of a thin-walled beam.

where the shear flow has units of force per length. Summing forces in the x direction and taking the limit as Δx and Δs approach zero, we obtain

$$\frac{\partial q}{\partial s} + h\frac{\partial \sigma_x}{\partial x} = 0 \tag{6.38}$$

Multiplying Eq. (6.38) by ds and integrating along the wall centerline gives

$$q(s) = q(0) - \int_0^s \frac{\partial \sigma_x}{\partial x} h \, \mathrm{d}s$$

At $s = 0$, we have a free edge, $q(0) = 0$. Introducing Eq. (6.32) for σ_x, we find that the shear flow due to bending, q_B, is

$$q_B(s) = -\frac{1}{A}\frac{\partial}{\partial x}(N_x + N_T)\int_0^s h \, \mathrm{d}s - \frac{1}{I_z}\left(\frac{\partial M_z}{\partial x} + \frac{\partial M_{zT}}{\partial x}\right)\int_0^s yh \, \mathrm{d}s$$
$$- \frac{1}{I_y}\left(\frac{\partial M_y}{\partial x} + \frac{\partial M_{yT}}{\partial x}\right)\int_0^s zh \, \mathrm{d}s + E\alpha \int_0^s \frac{\partial(\Delta T)}{\partial x} h \, \mathrm{d}s$$

From the equilibrium equations (6.15) and (6.16), we have

$$\frac{\partial N_x}{\partial x} = 0 \qquad \frac{\partial M_y}{\partial x} = V_z \qquad \frac{\partial M_z}{\partial x} = V_y$$

since we assume that the body force $f_x = 0$ and that the body moments $m_y = m_z = 0$. Now, let

$$f_{xT} = \frac{\partial N_T}{\partial x} \tag{6.39a}$$

$$V_{zT} = \frac{\partial M_{yT}}{\partial x} \tag{6.39b}$$

$$V_{yT} = \frac{\partial M_{zT}}{\partial x} \tag{6.39c}$$

and define

$$Q_z(s) = \int_0^s yh\,ds \tag{6.40a}$$

$$A(s) = \int_0^s h\,ds = hs \tag{6.40b}$$

$$Q_y(s) = \int_0^s zh\,ds \tag{6.40c}$$

$$q_T = E\alpha \int_0^s \frac{\partial(\Delta T)}{\partial x} h\,ds \tag{6.40d}$$

where $A(s)$ is the area of the beam wall of length s, and Q_z and Q_y are the familiar first moments of this area with respect to the y and z axes. With these substitutions, the shear flow is given by

$$q_B = -\frac{f_{xT}A(s)}{A} - \frac{(V_y + V_{yT})Q_z}{I_z} - \frac{(V_z + V_{zT})Q_y}{I_y} + q_T \tag{6.41}$$

In Eq. (6.41), the first and last terms on the right-hand side are shear flows introduced by the temperature gradient along the x axis. The "thermal shear forces" are defined by analogy with the actual shear forces and have no clear physical interpretation. They, too, depend on the temperature gradient along the x axis. If the temperature is independent of x, the quantities defined by Eqs. (6.39) are zero, and q reduces to the standard equation used in isothermal problems,

$$q_B = -\left(\frac{V_y Q_z}{I_z} + \frac{V_z Q_y}{I_y}\right) \tag{6.42}$$

Hence, if the temperature does not vary along the beam, the shear flow due to bending is independent of temperature, and we may apply Eq. (6.42). Thus, even though the temperature may vary with y and z, the shear flow is not affected; we use the same equation that applies for isothermal problems. If the temperature varies with x, we must use Eq. (6.41), and the effect of temperature will be greatest where the axial temperature gradient is a maximum.

Shear center. In our beam discussions, we have located the origin of our Cartesian coordinate system at the centroid of the cross section. The x axis is the locus of the centroids of all cross sections. Another axis of importance in beam deformations is the *elastic axis*. The elastic axis is the locus of the shear centers of all cross sections. The *shear center* is the point on a cross section through which the resultant of the transverse shear forces must pass to avoid twisting. To illustrate this concept, imagine a cantilever beam with a tip force F. For any cross-sectional shape, we can find a point, the shear center, at which to attach F so that the beam will undergo only bending without twisting; for, if F passes through the shear center, the beam may have v and w displacements, but the angle of twist θ_x will be zero. Another significance of the elastic axis is that it acts as the twist axis for torsion. Torques applied about the elastic axis cause twisting of cross sections but no bending of the elastic axis. (We prove this in a later paragraph.) For large deflections and rotations, these statements are not valid because even bending loads

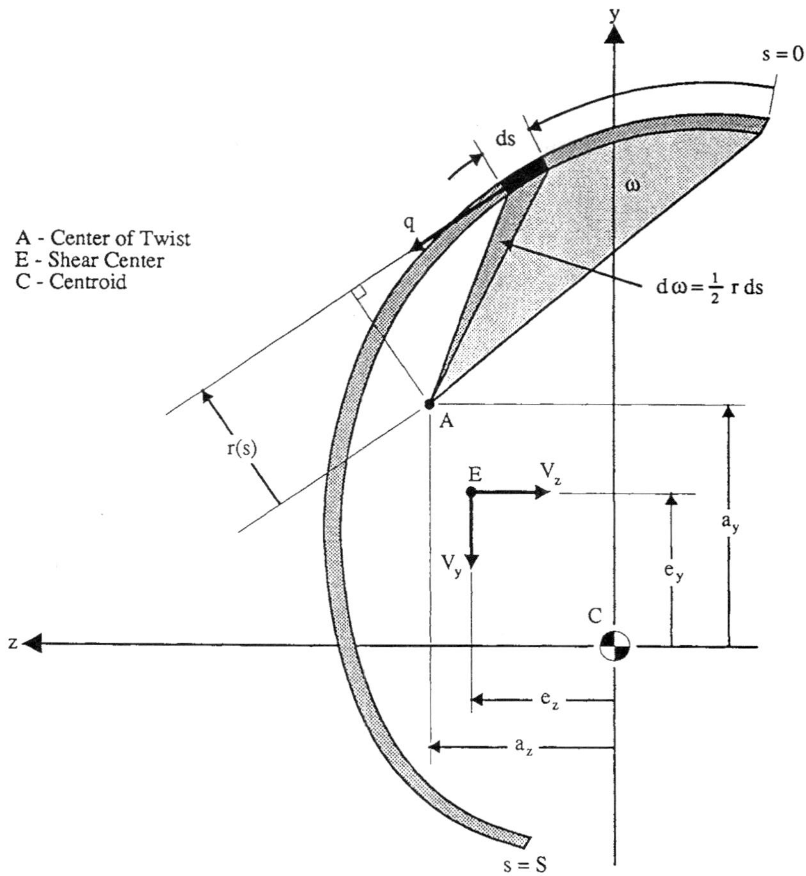

Fig. 6.11 Geometry for the location of shear center on a thin-walled cross-section.

placed at the shear center cause twisting, but our discussion here is restricted to small deflections and rotations.

We derive equations to locate the shear center by considering the thin-walled open cross section of the general shape shown in Fig. 6.11. The beam section experiences transverse shear forces V_y and V_z which induce a shear flow $q(s)$ that varies from zero at the free edge at $s = 0$ to zero at the other free edge at $s = S$. Once again, the origin of the coordinate system is at the centroid, and the y–z axes are principal axes. The shear center E is located by the coordinates e_y and e_z. The sum of the torques about any point in the plane is zero if the shear forces are applied at the shear center. To sum torques, we choose an arbitrary point A. For the section to be in equilibrium, we orient V_y and V_z in their negative directions. Taking moments about A, the total torque is

$$-V_z(e_y - a_y) + V_y(e_z - a_z) + \int_0^S qr\, ds = 0 \tag{6.43}$$

where S is the total length of the tube wall, and $r(s)$ is the perpendicular distance from point A to the line of action of $q ds$, which is tangent to the tube wall. The triangular infinitesimal element of area shown in Fig. 6.11 is

$$d\omega = \tfrac{1}{2} r \, ds \tag{6.44}$$

As s increases, the shaded area swept out by the element of area is known as the *sectorial area*, and it is denoted by ω. Substituting the shear flow from Eq. (6.42), we have

$$-V_z(e_y - a_y) + V_y(e_z - a_z) - \frac{2V_y}{I_z}\int_0^s Q_z(s)\, d\omega - \frac{2V_z}{I_y}\int_0^s Q_y(s)\, d\omega = 0$$

Equating the coefficients of V_y and V_z to zero, we find that

$$e_y = a_y - \frac{2I_{\omega y}}{I_y} \tag{6.45a}$$

$$e_z = a_z + \frac{2I_{\omega z}}{I_z} \tag{6.45b}$$

where we have introduced the sectorial products of inertia defined by

$$I_{\omega y} = \int_0^s Q_y(s)\, d\omega \tag{6.46a}$$

$$I_{\omega z} = \int_0^s Q_z(s)\, d\omega \tag{6.46b}$$

Equations (6.45) define the location of the shear center formally. For basic cross sections, textbooks[5–8] often locate the shear center by writing the equilibrium equation (6.43) for each particular cross section.

Regarding the location of the shear center, Oden[5] offers three useful observations: 1) The shear center for sections with two intersecting rectangular flanges is the point of intersection of the axes of the flanges. 2) The shear center of sections with one axis of symmetry lies on that axis. 3) The shear center of sections with two axes of symmetry occurs at the centroid of the section.

Warping displacement. In Sec. 6.3.1 as we began our development of small-deflection beam theory, we made three basic assumptions that led to the beam's displacement components given in Eqs. (6.7). The first assumption, which stated that plane sections before bending remained plane after bending, led to the representation for $u(x, y, z, t)$ given in Eq. (6.7a). Since u varies linearly with y and z through the depth of the beam, we found in Eqs. (6.8) that the shearing strains $\varepsilon_{xy} = \varepsilon_{xz} = 0$. When we considered Saint Venant torsion theory, we noted that torsion causes warping for noncircular cross sections and induces an axial displacement u that varies over the cross section. However, we assumed unrestrained uniform torsion, and no axial stresses were introduced.

In a long, slender beam of open cross section, the effect of transverse shearing distortions is also small and may be neglected. Warping distortions due to torsion, however, may be large and, if restrained (nonuniform torsion), significant axial stresses may occur. To begin our development of the warping stresses, we construct here an equation for the axial displacement u in terms of the beam slopes and angle of twist per unit length.

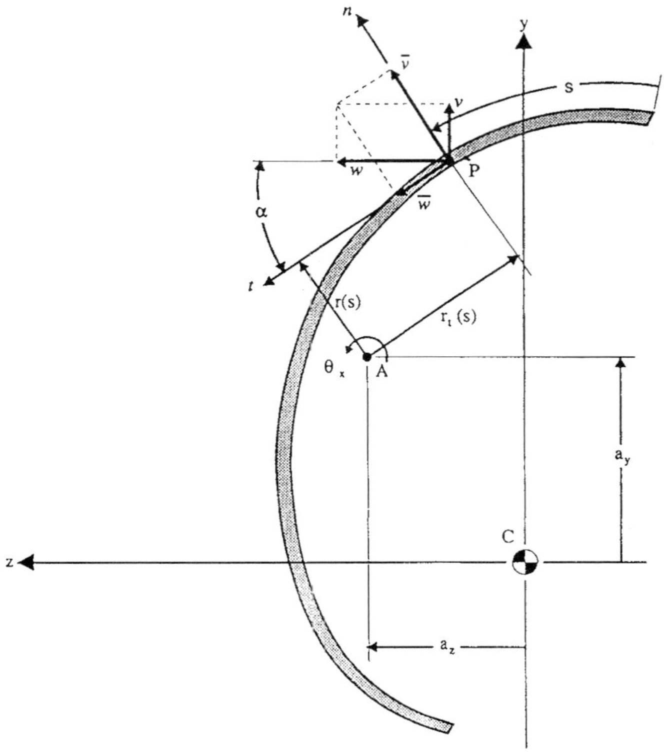

Fig. 6.12 Displacement components at a point on the thin-walled cross section.

Figure 6.12 shows transverse displacements v and w at an arbitrary point P on the wall of the cross section. At P, we resolve these displacements into components $\bar{v}$ and $\bar{w}$ in a local coordinate system with axes tangent and normal to the wall. If the beam undergoes bending and twisting simultaneously, we can represent the displacement components at P in the tangential and normal coordinates as

$$\bar{v} = v\cos\alpha + w\sin\alpha + r_t(s)\theta_x \tag{6.47}$$

$$\bar{w} = -v\sin\alpha + w\cos\alpha + r(s)\theta_x \tag{6.48}$$

where α is the angle between the t and z axes, and $r_t(s)$ and $r(s)$ are radii defined in Fig. 6.12. In particular, $r(s)$ is the perpendicular distance from point A to the tangent line at point P. If we invoke the assumption that transverse beam shearing strains are zero, then the shearing strain between the tangential and x axes $\varepsilon_{xs} = 0$. From the definition of shearing strain, we may write

$$\frac{\partial u}{\partial s} + \frac{\partial \bar{w}}{\partial x} = 0$$

Substituting Eq. (6.48) into this last result, we obtain

$$\frac{\partial u}{\partial s} = +\frac{\partial v}{\partial x}\sin\alpha - \frac{\partial w}{\partial x}\cos\alpha - r\frac{\partial \theta_x}{\partial x}$$

From geometry, $\sin\alpha = -\partial y/\partial s$ and $\cos\alpha = \partial y/\partial x$; hence,

$$\frac{\partial u}{\partial s} = -\frac{\partial v}{\partial x}\frac{\partial y}{\partial x} - \frac{\partial w}{\partial x}\frac{\partial z}{\partial s} - r\frac{\partial \theta_x}{\partial x}$$

which we can integrate from $x = 0$ to the arbitrary point P to obtain

$$u(s) - u(0) = -y\frac{\partial v}{\partial x} - z\frac{\partial w}{\partial x} - \int_0^s r\,\mathrm{d}s\,\frac{\partial \theta_x}{\partial x}$$

We write this last result as

$$u(s) = u(0) - y\frac{\partial v}{\partial x} - z\frac{\partial w}{\partial x} - 2\omega\frac{\partial \theta_x}{\partial x} \tag{6.49a}$$

where ω is the sectorial area,

$$\omega = \frac{1}{2}\int_0^s r\,\mathrm{d}s \tag{6.49b}$$

In Eq. (6.49a), $u(0)$ is the axial displacement of the edge of the open section at $s = 0$. Because it is independent of s, it represents a uniform translation of the cross section. Comparing Eq. (6.49a) with Eq. (6.7a), we see that the first three terms of the equations are the same. These terms represent uniform translation and rotation of a cross-sectional plane. The last term $-2\omega\partial\theta_x/\partial x$ is introduced by the angle of twist θ_x. This term represents the axial warping displacement introduced by torsion. Because the sectorial area $\omega(s)$ varies with s, the warping displacement varies over the cross section. If there is no torsion, so that $\theta_x = 0$, Eq. (6.49a) reduces to Eq. (6.7a), and plane sections remain plane.

Stress resultants. Now that we have an expression for the axial displacement at any point on the cross section of the thin-walled beam, we can work our way through beam theory as we did in Sec. 6.3.1. The first step is to substitute Eq. (6.49a) into Hooke's law, Eq. (6.9a), to obtain the normal stress,

$$\sigma_x = E\left[\frac{\partial u_0}{\partial x} - y\frac{\partial^2 v}{\partial x^2} - z\frac{\partial^2 w}{\partial x^2} - 2\omega\frac{\partial^2 \theta_x}{\partial x^2}\right] - E\alpha\Delta T \tag{6.50}$$

Then we substitute σ_x into the definitions of the stress resultants given in Eqs. (6.6). After evaluating the integrals over the cross section, we find that

$$N_x = EA\frac{\partial u_0}{\partial x} - ES_\omega\frac{\partial^2 \theta_x}{\partial x^2} - N_T \tag{6.51a}$$

$$M_y = -EI_y\frac{\partial^2 w}{\partial x^2} + 2EI_{\omega y}\frac{\partial^2 \theta_x}{\partial x^2} - M_{Ty} \tag{6.51b}$$

$$M_z = -EI_z\frac{\partial^2 v}{\partial x^2} + 2EI_{\omega y}\frac{\partial^2 \theta_x}{\partial x^2} - M_{Tz} \tag{6.51c}$$

where we have used

$$\int_A y\,\mathrm{d}A = \int_A z\,\mathrm{d}A = 0$$

because the origin is at the centroid. In addition, we have set

$$\int_A yz\,\mathrm{d}A = 0$$

since y and z are principal axes. Three additional integrals introduced are defined by

$$S_\omega = 2\int_0^S \omega(s)\,\mathrm{d}A \tag{6.51d}$$

$$I_{\omega y} = -\int_0^S \omega z\,\mathrm{d}A \tag{6.51e}$$

$$I_{\omega z} = -\int_0^S \omega y\,\mathrm{d}A \tag{6.51f}$$

where $\mathrm{d}A = h\,\mathrm{d}s$. The integrals defined in Eqs. (6.51e) and (6.51f) are, in fact, the same as the ones we introduced earlier as the sectorial products of inertia in Eqs. (6.46a) and (6.46b). Consider, for example, Eq. (6.51e). Using the definition of Q_y, we may write

$$I_{\omega y} = -\int_0^S \omega(z\,\mathrm{d}A) = -\int_0^S \omega\,\mathrm{d}Q_y$$

Then we integrate by parts to obtain

$$I_{\omega y} = -[\omega Q_y]_0^S + \int_0^S Q_y\,\mathrm{d}\omega$$

Given that y and z are centroidal axes, $Q_y(S) = 0$ and, by definition, $Q_y(0) = 0$. Thus, Eq. (6.51e) is the same as Eq. (6.46a). Also in Eqs. (6.51), we have introduced the thermal force N_T and the thermal moments M_{Ty} and M_{Tz}, previously defined by Eqs. (6.12).

From Eqs. (6.51), we solve for the displacement derivatives to obtain

$$\frac{\partial u_0}{\partial x} = \frac{N_x + N_T}{EA} + \frac{S_\omega}{A}\frac{\partial^2\theta_x}{\partial x^2} \tag{6.52a}$$

$$\frac{\partial^2 v}{\partial x^2} = -\frac{M_z + M_{zT}}{EI_z} + \frac{2I_{\omega z}}{I_z}\frac{\partial^2\theta_x}{\partial x^2} \tag{6.52b}$$

$$\frac{\partial^2 w}{\partial x^2} = -\frac{M_y + M_{yT}}{EI_y} + \frac{2I_{\omega y}}{I_y}\frac{\partial^2\theta_x}{\partial x^2} \tag{6.52c}$$

In comparison to the earlier results (with $I_{yz} = 0$), we see that the effect of the warping stress is to introduce a new term in each equation that is proportional to $\partial^2\theta_x/\partial x^2$. Close examination of Eqs. (6.52b) and (6.52c) shows that the coefficients of the second derivative of θ_x are terms we encountered earlier in our discussion of shear center. With reference to Eqs. (6.45), we notice that

$$\frac{2I_{\omega z}}{I_z} = e_z - a_z \qquad \frac{2I_{\omega y}}{I_y} = -(e_y - a_y)$$

To understand the significance of these equations, suppose the beam is loaded such that all mechanical and thermal bending moments are zero. Then the beam will experience curvatures given by

$$\frac{\partial^2 v}{\partial x^2} = (e_z - a_z)\frac{\partial^2 \theta_x}{\partial x^2} \qquad \frac{\partial^2 w}{\partial x^2} = -(e_y - a_y)\frac{\partial^2 \theta}{\partial x^2}$$

These last two equations show that the twist angle θ_x introduces beam curvatures. The curvatures are proportional to the distances between point A and the shear center (Fig. 6.11). So far in our discussion, the location of point A has been arbitrary. The foregoing results show that, if we choose point A to be at the shear center by selecting $a_z = e_z$ and $a_y = e_y$, the shear center will become the center of twist. In other words, pure torsion about an axis through the shear center will introduce no bending of the beam. This conclusion corroborates the statement we made earlier: *In pure torsion, the locus of the shear centers defines the elastic axis.* Selecting point A at the shear center means that we may rewrite Eqs. (6.52) as

$$\frac{\partial u_0}{\partial x} = \frac{N_x + N_T}{EA} + \frac{S_\omega}{A}\frac{\partial^2 \theta_x}{\partial x^2} \tag{6.53a}$$

$$\frac{\partial^2 v}{\partial x^2} = -\frac{M_z + M_{zT}}{EI_z} \tag{6.53b}$$

$$\frac{\partial^2 w}{\partial x^2} = -\frac{M_y + M_{yT}}{EI_y} \tag{6.53c}$$

Finally, we substitute these derivatives into Eq. (6.50) to express the stress σ_x in terms of the stress resultants,

$$\sigma_x = \frac{N_x + N_T}{A} + \frac{(M_z + M_{zT})y}{I_z} + \frac{(M_y + M_{yT})z}{I_y} - E\alpha\Delta T + E\left[\frac{S_\omega}{A} - 2\omega(s)\right]\frac{\partial^2 \theta_x}{\partial x^2} \tag{6.54}$$

The preceding equation gives the normal stress distribution in a thin-walled beam with open cross section. The coordinate axes are located at the centroid of the cross section. The elastic axis—the center of twist—is located at the shear center. The bending moments and moments of inertia are computed with respect to principal y and z axes through the centroid. The sectorial quantity S_ω is based on a radius $r(s)$ measured from the shear center.

Comparing Eq. (6.54) with Eq. (6.14) for a compact cross section shows that the equations are the same except for the last term in Eq. (6.54). This last term represents the warping stress introduced by nonuniform torsion.

Warping stresses and torque. Earlier in this section in Eq. (6.31), we resolve the normal stress σ_x in a thin-walled beam with open cross section into the sum of bending and warping constituents. We give the bending constituent $(\sigma_x)_B$ in Eq. (6.32). From Eq. (6.54), we can now identify the warping constituent as

$$(\sigma_x)_W = E\left[\frac{S_\omega}{A} - 2\omega(s)\right]\frac{\partial^2 \theta_x}{\partial x^2} \tag{6.55}$$

where the sectorial quantity S_ω is defined in Eq. (6.51a), and it is a constant for a particular cross section. The sectorial area $\omega(s)$, Eq. (6.49b), varies with s and determines the variation of the warping stress over the cross section. Alternatively, we may write the warping stress as

$$(\sigma_x)_W = 2E(\omega_0 - \omega)\frac{\partial^2\theta_x}{\partial x^2} \tag{6.56a}$$

where

$$\omega_0 = S_\sigma/2A \tag{6.56b}$$

The warping normal stress $(\sigma_x)_W$ varies with x for nonuniform torsion through the x dependence of $\partial^2\theta_x/\partial x^2$. A warping shearing stress component $(\sigma_{xs})_W$ is introduced by the variation of $(\sigma_x)_W$ with x. We saw earlier in Eq. (6.33) that the shearing stress σ_{xs} has three constituents, including a contribution $(\sigma_{xs})_{SV}$ from Saint Venant torsion, a contribution $(\sigma_{xs})_B$ from Bernoulli–Euler bending, and a warping contribution $(\sigma_{xs})_W$. We have developed in Eqs. (6.37) and (6.41) equations for $(\sigma_{xs})_B$ in terms of the shear flow q_B. Following this same procedure, we now develop an equation for $(\sigma_{xs})_W$, the warping shearing stress.

With reference to Eq. (6.38), we recall that equilibrium of an element of the thin wall provides the relationship between the shear flow q and the normal stress gradient $\partial\sigma_x/\partial x$. In integral form,

$$q(x,s) = -\int_0^s \frac{\partial\sigma_x}{\partial x}\,\mathrm{d}A$$

where the tube-wall element of area $\mathrm{d}A = h\,\mathrm{d}s$. Substituting the warping stress given in Eq. (6.56a),

$$q_W(x,s) = -2E\int_0^S (\omega_0 - \omega)\,\mathrm{d}A\frac{\partial^3\theta_x}{\partial x^3}$$

where q_W denotes the shear flow due to warping. Alternatively, we may write this last result as

$$q_W(x,s) = -EQ_\omega(s)\frac{\partial^3\theta_x}{\partial x^3} \tag{6.57a}$$

where the sectorial quantity Q_ω is defined by

$$Q_\omega(s) = 2\int_0^s (\omega_0 - \omega)\,\mathrm{d}A \tag{6.57b}$$

Equation (6.57a) shows that a shear flow q_W is induced by nonuniform torsion. The shear flow varies around the cross section through the sectorial quantity $(Q_\omega(s)$ and along the beam's length through the dependence on x of the third derivative of θ_x.

The final step in this development of nonuniform torsion is to establish the relationship between the torque developed by warping and the angle of twist θ_x. In Eq. (6.35), we express that the torque M_x had three constituents with contributions from Saint Venant torsion as well as contributions from warping and normal stress. We conclude this paragraph by developing an equation for the warping torque.

We obtain the equation for the torque developed by warping by integrating the moment of the shear flow about the shear center over the cross section. We refer to Fig. 6.12, where the center of twist, point A, is located at the shear center E. Then we write

$$(M_x)_W = \int_0^S q_W r \, \mathrm{d}s$$

where we integrate the shear flow over the entire wall from $s = 0$ to $s = S$. Substituting Eq. (6.57a),

$$(M_x)_W = -E \int_0^S Q_\omega(s) r \, \mathrm{d}s \frac{\partial^3 \theta_x}{\partial x^3}$$

or, alternately,

$$(M_x)_W = -E\Gamma \frac{\partial^3 \theta_x}{\partial x^3} \tag{6.58a}$$

where

$$\Gamma = \int_0^S Q_\omega r \, \mathrm{d}s \tag{6.58b}$$

The constant Γ is called the *warping constant* of the cross section, and the product $E\Gamma$ is called the *warping rigidity*. The warping constant may be expressed in a different form by an integration by parts. First, note, from Eq. (6.44), $2\,\mathrm{d}\omega = r\,\mathrm{d}s$. Because ω_0, Eq. (6.56b), is a constant, we may also write $2\,\mathrm{d}(\omega - \omega_0) = r\,\mathrm{d}s$. Then,

$$\Gamma = \int_0^S Q_\omega r \, \mathrm{d}s = \int_0^S Q_\omega 2 \, \mathrm{d}(\omega - \omega_0)$$

Now, integrating by parts,

$$\Gamma = [2Q_\omega(\omega - \omega_0)]_0^S - \int_0^S 2(\omega - \omega_0) \, \mathrm{d}Q_\omega$$

From the definition of Q_ω, Eq. (6.57b), we see that $\mathrm{d}Q_\omega = 2(\omega_0 - \omega)\,\mathrm{d}A$. Moreover $Q_\omega(0) = Q_\omega(S) = 0$. Finally, the warping constant is expressed as

$$\Gamma = 4 \int_0^S (\omega_0 - \omega)^2 \, \mathrm{d}A \tag{6.59}$$

We should emphasize that the warping constant Γ is a property of the beam cross section like the cross-sectional area or moments of inertia. Since the definition given by Eq. (6.59) involves the sectorial area ω, the definition appears obscure compared to the usual integral definitions of moments of inertia. But the integral has been evaluated for several thin-walled open cross sections that experience warping, and the results are tabulated in engineering texts and handbooks. Formulas for shear centers and warping constants for several cross sections are given in Refs. 18–19. We illustrate the evaluation of the warping constant in Sec. 6.4, Example 6.5.

Torque due to normal stress. In Saint Venant torsion theory, the normal stress σ_x is perpendicular to the plane containing the shearing stress, and it does not contribute to the torque. The torque is obtained by integrating the shearing stresses over the cross section of the beam. However, in beams that are more flexible in torsion, the beam cross section can rotate through a sufficiently large angle to give a component of σ_x in the direction of the shearing stress. For open cross sections that are particularly weak in torsion, the effect of the normal stress may be significant.

Engineers were made aware of the effect of axial stress on torsion from work in the 1930s on torsional-flexural buckling. In the 1950s, they began to realize that thermal stress could also affect torsion. A particular problem of early interest was the influence of aerodynamic heating on the torsional stiffness of thin wings. Papers written in 1956 by Bisplinghoff[20] and by Budiansky and Mayers[21] address the fundamental issues. The concern arose because a wing's chordwise temperature distribution induces compressive stresses that reduce the torsional rigidity. The reduction in torsional rigidity lowers torsional buckling loads and frequencies and alters flutter speeds. Refs. 22–27 describe closely related studies of the influence of temperature on torsional rigidity.

We continue our development of the formulation for thin-walled beams of open cross section by establishing the torque due to normal stress. In Eq. (6.35), we state that the beam torque has contributions from Saint Venant torsion $(M_x)_{SV}$, warping torsion $(M_x)_W$, torque from the normal stress $(M_x)_{\sigma_x}$, and torque from the temperature variation $(M_x)_T$. Equation (6.36) defines the Saint Venant contribution, and Eq. (6.58a) defines the warping torque. To continue our formulation, we establish an equation for $(M_x)_{\sigma_x}$.

We begin by restating briefly how we describe the bending and torsional deformation for the thin-walled beam. Consider again the geometry of the cross section as shown in Fig. 6.11. We have decided to locate the center of twist, point A, at the shear center, point E. The x axis through the shear center we call the *elastic axis*. Twisting of the beam takes place around the elastic axis. The transverse bending displacement components $v(x, t)$ and $w(x, t)$ are the displacements of the elastic axis. With the exception of the center of twist, all points on the cross section experience transverse displacements due to bending and twisting. For small angles of rotation, the displacements of an arbitrary point P are given by

$$v_P = v - (z - e_z)\theta_x \tag{6.60a}$$

$$w_P = w + (y - e_y)\theta_x \tag{6.60b}$$

where y and z are coordinates of point P measured in the centroidal coordinate system; e_y and e_z are coordinates of the shear center E in the same coordinate system. Using Eqs. (6.60), we may evaluate the displacements of any point on the cross section of the deformed beam. For example, after deformation, the centroid originally located at $x = y = 0$ has moved to a new position through displacements,

$$v_o = v + e_z\theta_x \tag{6.61a}$$

$$w_o = w - e_y\theta_x \tag{6.61b}$$

In contrast, in Sec. 6.3.1 in our development of small-deflection behavior of solid beams of compact cross section, we neglect the torsional displacement components, and all points on a cross section have the same transverse bending displacements.

To consider the effect of the normal stress on the torque, we need to know the orientation of the stress σ_x after deformation. From Eqs. (6.60), we may compute

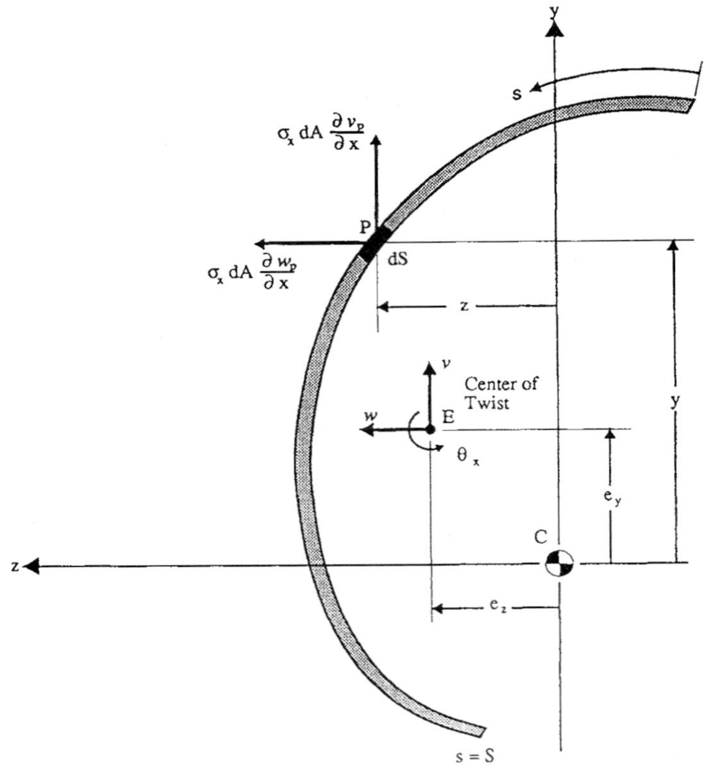

Fig. 6.13 Geometry for evaluation torque due to normal stress σ_x.

the slopes of a line through point P on the thin wall. Thus,

$$\frac{\partial v_p}{\partial x} = \frac{\partial v}{\partial x} - (z - e_z)\frac{\partial \theta_x}{\partial x} \tag{6.62a}$$

$$\frac{\partial w_p}{\partial x} = \frac{\partial w}{\partial x} + (y - e_y)\frac{\partial \theta_x}{\partial x} \tag{6.62b}$$

Because the stress component is tangent to this line, we may use these slopes to evaluate the stress components in the y–z plane.Figure 6.13 shows the y and z components of the forces obtained at point P by multiplying the force $\sigma_x \, dA$ by the appropriate slope, where we assume small angles so that $\sin(\partial v_p/\partial x) \cong \partial v_p/\partial x$ and $\sin(\partial w_p/\partial x) \cong \partial w_p/\partial x$. Summing torques about the center of twist,

$$(M_x)_{\sigma_x} = \int_o^S \sigma_x \left[(y - e_y)\frac{\partial w_p}{\partial x} - (z - e_z)\frac{\partial v_p}{\partial x}\right] dA$$

and substituting for the slopes using Eqs. (6.62), we find that

$$(M_x)_{\sigma_x} = \int_o^S \sigma_x (y - e_y) \, dA \frac{\partial w}{\partial x} - \int_o^S \sigma_x (z - e_z) \, dA \frac{\partial v}{\partial x} + \int_o^S \sigma_x \left[(y - e_y)^2 + (z - e_z)^2\right] dA \frac{\partial \theta_x}{\partial x}$$

The remaining steps are to substitute for σ_x and evaluate the integrals. For this process, the stress equation (6.54) is separated into mechanical and thermal components. In the evaluation, only linear terms are retained, and terms that involve products of bending moments and slopes are neglected. As a result of these operations, we obtain the linear equation,

$$(M_x)_{\sigma_x} = N_x e_z \frac{\partial v}{\partial x} - N_x e_y \frac{\partial v}{\partial x} + \left[\frac{I_E}{A} N_x + \int_0^S \sigma_{xT} \rho^2 \mathrm{d}A \right] \frac{\partial \theta_x}{\partial x} \tag{6.63a}$$

where

$$I_E = I_y + I_z + A(e_y^2 + e_z^2) \tag{6.63b}$$

$$\sigma_{xT} = \frac{N_T}{A} + \frac{M_{zT}\,y}{I_z} + \frac{M_{yT}\,y}{I_y} - E\alpha\Delta T \tag{6.63c}$$

$$\rho^2 = (y - e_y)^2 + (z - e_z)^2 \tag{6.63d}$$

In these equations, I_E is the polar moment of inertia with respect to the shear center E (not to be confused with the torsion constant J). The thermal portion of the stress is denoted by σ_{xT}, and the distance from the shear center E to a point P on the beam wall is denoted by ρ. Note that ρ is not the same as the sectorial radius r (see Fig. 6.12) because r is the perpendicular distance from E to a tangent line through point P.

Equation (6.63a) is the desired result, and we see that this torque constituent couples torsion with the axial force N_x and the thermal stress σ_{xT}. This coupling is a unique aspect of the bending-torsion behavior of thin-walled beams with open sections.

Torque due to axial temperature gradient. The final contribution to the total torque developed on the cross section of the thin-walled beam arises from the shear flow induced by an axial temperature gradient. Earlier, we developed Eq. (6.41) for the shear flow q_B, which we rewrite here as

$$q_B = -\frac{\partial N_T}{\partial x}\frac{A(s)}{A} - \frac{(V_y + V_{yT})Q_z}{I_z} - \frac{(V_z + V_{zT})Q_y}{I_y} + q_T$$

where the shear flow due to temperature is

$$q_T = E\alpha \int_0^S \frac{\partial(\Delta T)}{\partial x} \mathrm{d}A$$

Now, if we integrate the moment of this shear flow about the center of twist, we obtain the contribution to the torque from the temperature gradient. Thus, taking moments about the center of twist,

$$(M_x)_T = \int_0^S q_B r \,\mathrm{d}s$$

or, given that $2\,\mathrm{d}\omega = r\,\mathrm{d}s$,

$$(M_x)_T = 2 \int_0^S q_B r \,\mathrm{d}\omega \tag{6.64}$$

Substituting Eq. (6.41) in Eq. (6.64), we find that

$$(M_x)_T = \frac{\partial N_T}{\partial x}\frac{1}{A}\int_o^S A(s)\,\mathrm{d}\omega + \int_o^S q_T\,\mathrm{d}\omega \tag{6.65}$$

where the terms associated with the shear forces vanish because

$$\int_o^S Q_y\,\mathrm{d}\omega = \int_0^S Q_y\,\mathrm{d}\omega = 0$$

The preceding integrals are zero because we use the shear center as the center of twist; see Eqs. (6.45) and (6.46). The first term on the right-hand side of Eq. (6.65) arises if the resultant thermal force N_T varies with x. The second term also depends on the temperature gradient, as we can see from Eq. (6.40d). We can obtain a more convenient form for this latter term by integration by parts. We follow the procedure used in deriving Eq. (6.59). After the integration by parts,

$$(M_x)_T = -\frac{\partial N_T}{\partial x}\frac{1}{A}\int_0^S A(s)\,\mathrm{d}\omega - E\alpha\int_o^S [\omega(s) - \omega(S)]\frac{\partial(\Delta T)}{\partial x}\,\mathrm{d}A \tag{6.66}$$

Frequently in applications, the thermal force N_T is constant or zero, and the first term on the right-hand side of Eq. (6.66) vanishes. Then the contribution of the temperature gradient to the torque reduces to

$$(M_x)_T = -E\alpha\int_o^S [\omega(s) - \omega(S)]\frac{\partial(\Delta T)}{\partial x}\,\mathrm{d}A \tag{6.67}$$

With the development of Eqs. (6.67), we have all the constituents of the torque for thin-walled beam with open cross section.

Governing differential equations. We have now developed all the background theory to enable us to write the governing differential equations. As in the preceding sections, we do this by substituting equations for the stress resultants into equilibrium equations. To obtain a linear theory, we use the nonlinear equilibrium equations, and we neglect products of bending moments and torque with derivatives of the dependent variables. Referring to Eqs. (6.25), we write

$$\frac{\partial N_x}{\partial x} = 0 \qquad \frac{\partial V_y}{\partial x} + N_x\frac{\partial^2 v_0}{\partial x^2} + f_y = 0$$

$$\frac{\partial V_z}{\partial x} + N_x\frac{\partial^2 w_0}{\partial x^2} + f_z = 0 \qquad \frac{\partial M_x}{\partial x} + m_x = 0 \tag{6.68}$$

$$\frac{\partial M_y}{\partial x} = V_z \qquad \frac{\partial M_y}{\partial x} = V_y$$

In the preceding equations, v_0 and w_0 denote displacements at the centroid. For the present problem, we will introduce the displacements at the shear center by substituting Eqs. (6.61). Also note that, because $f_x = 0$, equilibrium requires that N_x be independent of x. Thus, N_x may be either a constant or a function of time; in many problems, it is constant. We may also eliminate the shear forces in the second and third equations by using the last two equations. After these manipulations, there are three equations:

$$\frac{\partial^2 M_z}{\partial x^2} + N_x \frac{\partial^2 v}{\partial x^2} + e_z N_x \frac{\partial^2 \theta_x}{\partial x^2} + f_y = 0 \tag{6.69a}$$

$$\frac{\partial^2 M_y}{\partial x^2} + N_x \frac{\partial^2 w}{\partial x^2} + e_y N_x \frac{\partial^2 \theta_x}{\partial x^2} + f_z = 0 \tag{6.69b}$$

$$\frac{\partial M_x}{\partial x} + m_x = 0 \tag{6.69c}$$

We obtain the first two governing equations by substituting for the bending moments using Eqs. (6.53b) and (6.53c). We obtain the third equation by substituting Eq. (6.35), with contributions from Eqs. (6.36), (6.58a), (6.63a), and (6.67). We find three simultaneous fourth-order equations:

$$EI_z \frac{\partial^4 v}{\partial x^4} - N_x \frac{\partial^2 v}{\partial x^2} - N_x e_z \frac{\partial^2 \theta_x}{\partial x^2} + \frac{\partial^2 M_{zT}}{\partial x^2} = f_y \tag{6.70a}$$

$$EI_y \frac{\partial^4 w}{\partial x^4} - N_x \frac{\partial^2 w}{\partial x^2} + N_x e_y \frac{\partial^2 \theta_x}{\partial x^2} + \frac{\partial^2 M_{yT}}{\partial x^2} = f_z \tag{6.70b}$$

$$E\Gamma \frac{\partial^4 \theta_x}{\partial x^4} - \left(GJ + \frac{N_x I_E}{A} + \int_0^S \sigma_{xT} \rho^2 \, \mathrm{d}A \right) \frac{\partial^2 \theta_x}{\partial x^2}$$
$$- N_x e_z \frac{\partial^2 v}{\partial x^2} + N_x e_y \frac{\partial^2 w}{\partial x^2} + \frac{\partial^2 M_{xT}}{\partial x^2} = m_x \tag{6.70c}$$

where we define the thermal torque as

$$M_{xT} = E\alpha \int_0^S [\omega(s) - \omega(S)] \Delta T \, \mathrm{d}A \tag{6.71}$$

Equations (6.70) are the culmination of the theoretical developments in the section. In these equations, we assume N_x to be a tensile force; otherwise, for compression, the signs of all terms containing N_x must be changed.

The axial force and the location of the shear center determine the coupling between the three equations. There are four cases.

Case 1: If $N_x = 0$, the three equations are independent, and we may solve each equation separately.

Case 2: If the shear center coincides with the centroid so that $e_y = e_z = 0$, the three equations are independent.

Case 3: If the shear center lies on the z axis ($e_y = 0$), then bending in the x–z plane occurs independently, and we can solve Eq. (6.70b) for w separately. The equations for v and θ_x are coupled, and we must solve them simultaneously.

Case 4: If the shear center lies on the y axis ($e_z = 0$), then bending in the y–z plane occurs independently, and we can solve Eq. (6.70a) for v separately. The equations for w and θ_x are coupled, and we must solve them simultaneously.

The bending-torsion formulation given in Eq. (6.70) is similar to the equations that appear in the Boley and Weiner text.[2] The derivation here differs considerably from their derivation, and the present derivation is more consistent with contemporary approaches used for thin-walled beams. The Boley and Weiner formulation does not contain the thermal torque that appears in Eq. (6.70c). The thermal torque was first introduced in a paper by Beam and Yagoda.[28] If the temperature does not vary with x, the thermal torque vanishes, and Eqs. (6.70) are identical to the equations of Boley and Weiner. The formulation will be complete when we discuss boundary conditions.

Boundary conditions. Equations (6.70) are a set of three fourth-order partial differential equations that require 12 boundary conditions. Six boundary conditions are applied at each end; of these six, four are related to the first two equations involving bending, Eqs. (6.70a) and (6.70b). Appropriate boundary conditions for bending are given in Eqs. (6.22), and we may employ these directly. The remaining two boundary conditions relate to the torsion equation (6.70c).

We assume implicitly that N_x is specified; therefore, no boundary conditions are required for extension. The boundary conditions required relate to twisting and warping of the cross section. The boundary conditions consist of specifying either the angle of twist or the torque *and* specifying the warping displacement or the warping force. Thus, at a typical beam end $x = a$ where $a = 0$ or $a = L$, we specify two boundary conditions from the following possibilities:

$$\begin{aligned} \theta_x(a,t) &= \theta_1(t) \quad \text{or} \quad M_x(a,t) = M_{x1}(t) \\ u_W(a,t) &= u_{W1}(t) \quad \text{or} \quad N_W(a,t) = N_{W1}(t) \end{aligned} \tag{6.72}$$

where the twisting moment, warping displacement, and warping force are defined as follows. We have defined the twisting moment in Eq. (6.35) as the sum of four constituents: Saint Venant torsion, warping torsion, a torque from the normal stress, and a torque from the axial temperature gradient. The torque from the normal stress is given in Eq. (6.63a). We see that it is specified by N_x and the temperature distribution and, hence, should be excluded from the boundary condition. Thus, in Eq. (6.72), the torque we specify is the sum of the remaining constituents, that is

$$M_x = GJ\frac{\partial \theta_x}{\partial x} - E\Gamma\frac{\partial^3 \theta_x}{\partial x^3} - \frac{\partial M_{xT}}{\partial x} \tag{6.73}$$

The warping displacement is the portion of the axial displacement $u(s)$ given in Eq. (6.49a) that is related to the angle of twist,

$$u_W = -2\omega\frac{\partial \theta_x}{\partial x} \tag{6.74}$$

The warping force is a similar component of the axial force given in Eq. (6.5la),

$$N_W = -ES_\omega\frac{\partial^2 \theta_x}{\partial x^2} \tag{6.75}$$

To illustrate common boundary conditions, consider the following possibilities.

Fixed end: The cross section can neither twist nor warp; hence, θ_x and u_W are zero:

$$\theta_x(a,t) = 0 \quad \text{and} \quad \frac{\partial \theta_x}{\partial x}(a,t) = 0$$

Free end: The cross section is free to twist and warp; hence, M_x and N_W are zero:

$$GJ\frac{\partial \theta_x}{\partial x}(a,t) - E\Gamma\frac{\partial^3 \theta_x}{\partial x^3}(a,t) - \frac{\partial M_{xT}}{\partial x} = 0 \quad \text{and} \quad \frac{\partial^2 \theta_x}{\partial x^2}(a,t) = 0$$

Simply supported end: The cross section cannot twist but it can warp; hence, θ_x and N_W are zero:

$$\theta_x(a,t) = 0 \qquad \text{and} \qquad \frac{\partial^3 \theta_x}{\partial x^2}(a,t) = 0$$

With the boundary conditions, we complete the formulation for the problem of bending-torsion of a thin-walled beam with open cross section. We illustrate some solutions in the next section. We will also use the formulation in later chapters.

6.4 Beam Problems

In this section, we illustrate quasistatic behavior of beams and the three beam theory formulations with a series of example problems.

Example 6.1: Statically Determinant Cantilever Beam

We consider a statically determinant cantilever beam (Fig. 6.14) of symmetrical cross section. The beam experiences a temperature distribution that varies through the depth y of the beam, but it is independent of x and z as well as time; i.e., $T = T(y)$. We seek equations for the elongation, transverse bending deflections, and normal stress.

From Eqs. (6.12), we may compute the thermal force and thermal moments. Because the temperature depends only on y, these quantities are constant and are given by

$$N_T = E\alpha\int_A [T(y) - T_{\text{ref}}]\,dA \qquad M_{yT} = 0 \qquad M_{zT} = E\alpha\int_A [T(y) - T_{\text{ref}}]y\,dA$$

The thermal bending moment M_{yT} is zero because the temperature is independent of z, and it is symmetrical about the y axis. The axial force N_x and bending moments M_y and M_z are zero at any cross section since the beam experiences no external forces. Because the cross-sectional area is symmetric about the y axis, the product of inertia $I_{yz} = 0$. From Eq. (6.13), we find that

$$\frac{\partial u_0}{\partial x} = \frac{N_T}{EA} \qquad \frac{\partial^2 v}{\partial x^2} = -\frac{M_{zT}}{EI_z} \qquad \frac{\partial^2 w}{\partial x^2} = 0$$

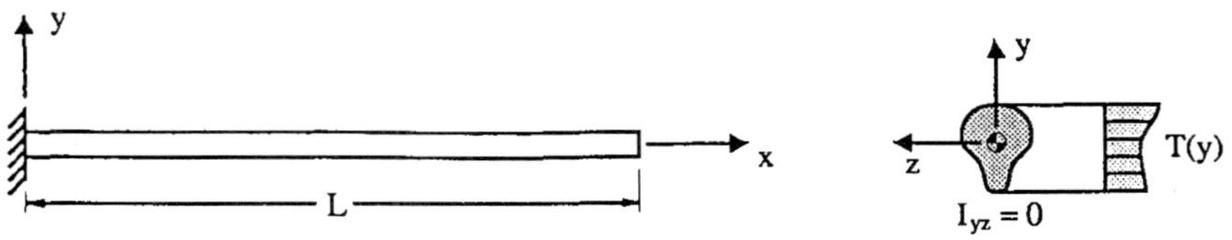

Fig. 6.14 Statically determinant cantilever beam.

The boundary conditions at the fixed end are

$$u_0(0) = 0 \qquad v(0) = \frac{dv}{dx}(0) = 0 \qquad w(0) = \frac{dw}{dx}(0) = 0$$

Integrating the differential equations and imposing the boundary conditions yields

$$u_0(x) = \frac{N_T x}{EA} \qquad v(x) = -\frac{M_{zT} x^2}{2EI_z} \qquad w(x) = 0$$

In this case, because of the symmetrical cross section and temperature distribution, the beam deforms in the x–y plane. Unless we know the temperature distribution, we cannot tell the sense of the deflection. Intuitively, we expect that, if the top of the beam is hotter than the bottom, the top fibers of the beam will expand more than the bottom fibers, and the beam will deflect down. Thus, for this situation, we expect $M_{zT} > 0$ and $v < 0$.

The normal stress is given by Eq. (6.14) and, after substitutions, we find that

$$\sigma_x = \frac{N_T}{A} + \frac{M_{zT} y}{I_z} - E\alpha[T(y) - T_{\text{ref}}]$$

If the beam were subjected, to mechanical loads, we could include their effects in the preceding analysis, or we could perform the analyses separately and superimpose the results because the problem is linear.

Example 6.2: Statically Indeterminant Beam

Consider now the statically indeterminant beam problem shown in Fig. 6.15. Again, we assume that the cross section is symmetrical and that the temperature is a function only of y; $T = T(y)$. The beam is built in at $x = 0$ and $x = L$ so that the transverse displacements and slopes at the two supports are zero. The horizontal displacement is zero at the left end, but the right end is free to move as the beam elongates. The problem is statically indeterminant because there are four unknown reactions—two bending moments and two shear forces at the supports—and there are only three equations of static equilibrium.

As in Example 6.1, we know that N_x and M_{zT} are constant, and that M_{yT} is zero. Because the cross section is symmetric, $I_{yz} = 0$ and bending may occur only

Because the right end of the beam is free to move, we know that $N_x(L) = 0$ Moreover, for beam equilibrium. Eq. (6.15a) requires (for zero body force) that $\partial N_x/\partial x = 0$ or $N_x = 0$ constant. But, since $N_x(L) = 0$, then $N_x = 0$ at all points. Thus, from Eq. (6.13a),

$$\frac{\partial u_0}{\partial x} = \frac{N_T}{EA}$$

and, with $u_0(0) = 0$, we find that

$$u_0 = \frac{N_T x}{EA}$$

For bending in the x–y plane, Eqs. (6.18a–6.18c) give, for zero body forces,

$$M_z = -EI_z \frac{\mathrm{d}^2 v}{\mathrm{d}x^2} - M_{zT}$$

$$V_y = -EI_z \frac{\mathrm{d}^3 v}{\mathrm{d}x^3}$$

$$EI_z \frac{\mathrm{d}^4 v}{\mathrm{d}x^4} = 0$$

where we have used $I_{yz} = 0$, and the derivatives of the thermal moment are zero since M_{zT} is constant. The boundary conditions require that

$$v(0) = \frac{\mathrm{d}v}{\mathrm{d}x}(0) = v(L) = \frac{\mathrm{d}v}{\mathrm{d}x}(L) = 0$$

Integrating the differential equation for v, we obtain four constants of integration. Imposing the four boundary conditions, we find that the constants are zero. (We should not be surprised since we have a homogeneous equation and homogeneous boundary conditions.) Thus, the deflection is identically zero,

$$v(x) \equiv 0$$

Hence, the beam does not deflect but, rather, it remains perfectly straight. From Eq. (6.18a), we find that the bending moment is constant,

$$M_x(x) = -M_{zT}$$

and, from Eq. (6.18b), the transverse shear force is zero,

$$V_y(x) \equiv 0$$

Finally, from Eq. (6.14), the normal stress is

$$\sigma_x = \frac{N_T}{A} - E\alpha[T(y) - T_{\mathrm{ref}}]$$

where the bending term vanishes since the beam remains straight. Hence, in this case, the beam elongates, but the transverse deflection is zero. A normal stress occurs as a result of the temperature variation through the beam depth.

Suppose that the right support was restrained against expansion so that $u_0(L) = 0$. As an exercise show that

$$u_0(x) \equiv 0 \qquad v(x) \equiv 0 \qquad \sigma_x = -E\alpha[T(y) - T_{\text{ref}}]$$

These two cases are another illustration of how thermal stresses may occur even though displacements are zero.

Example 6.3: Wing Spar

Example 4.13 describes the thermal cycle experienced by a titanium wing spar during the flight of a supersonic transport. The evolution of the spar temperature distributions are presented in Fig. 4.29. During the initial portion of the flight, the aerodynamic skin temperature first drops during ascent into the cold upper atmosphere but then, as aerodynamic heating increases, the skin warms. During cruise, the skin maintains a temperature of 330°F. For the early portion of the flight, the maximum temperature gradient from the skin to the center of the web develops at about 50 min, and the gradient decreases thereafter as heat is conducted into the web. The gradient reverses during descent as the aerodynamic heating diminishes rapidly, and the skin cools. The maximum reversed gradient occurs at the end of the flight at about 290 min when the aircraft has landed.

We wish to compute the corresponding spar normal stress $\sigma_x(y)$ for the mission. The geometry of the spar cross section is shown in Fig. 6.16a, and a typical temperature distribution is sketched for an early portion of the flight in Fig. 6.16b. Note that temperatures are assumed constant in the aerodynamic skin/spar cap and that the geometry and temperatures are symmetric about the z axis through the centroid.

We evaluate the normal stress at any time by using Eq. (6.14). We consider only thermal effects, and so $N_x = M_y = M_z = 0$ and, since the spar cross section is symmetric, $I_{yz} = 0$. In addition, the temperature is a function only of y, and so $M_{yT} = 0$. With these simplifications, the stress equation reduces to

$$\sigma_x = \frac{N_T}{A} + \frac{M_{zT}\, y}{I_z} - E\alpha\Delta T$$

The thermal moment M_{zT} is also zero because of the symmetry of the temperature distribution. Thus, we are left with

$$\sigma_x = \frac{N_T}{A} - E\alpha\Delta T$$

The thermal force is given by Eq. (6.12a),

$$N_T = E\alpha \int_A (T - T_{\text{ref}})\, \mathrm{d}A$$

Because the temperature is constant in the skin/spar cap (Fig. 6.16), we evaluate the integral as the sum,

$$N_T = E\alpha d \int_{-h}^{h} T(y)\, \mathrm{d}y + 2E\alpha T_s b \int_{h}^{h+H} \mathrm{d}y$$

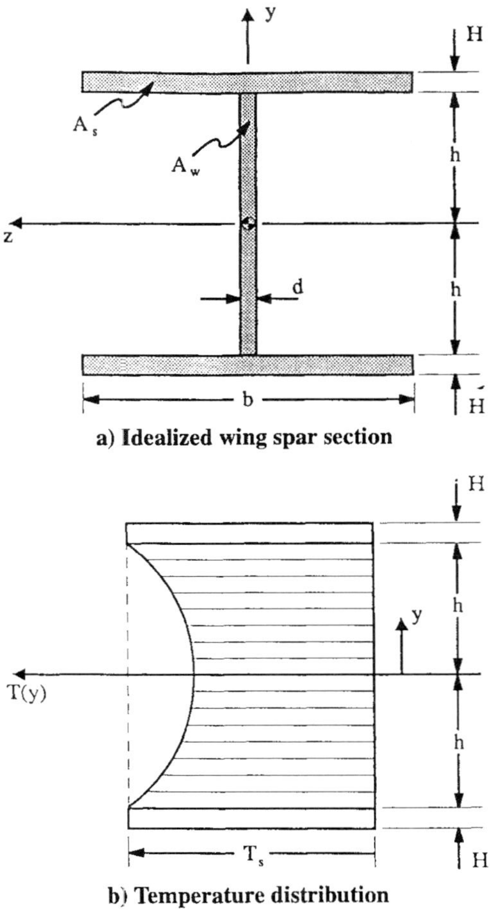

Fig. 6.16 Wing spar geometry and temperature distribution.

hence,

$$N_T = 2E\alpha T_s A_s + E\alpha d \int_{-h}^{h} T(y)\,\mathrm{d}y$$

where T_s is the skin temperature and $A_s = bH$ is the skin/spar cap area. Finally, substituting this last result into the equation for σ_x, we obtain

$$\sigma_x(y) = \frac{2E\alpha T_s A_s}{A_w + 2A_s} + \frac{E\alpha d}{A_w + 2A_s} \int_{-h}^{h} T(y)\,\mathrm{d}y - E\alpha T(y)$$

Values for the temperature distributions computed in Chapter 4 were used to evaluate the spar stresses. The integral was evaluated numerically. For the computations, $A_s/A_w = 12$.

The time histories of the skin normal stress and the temperature difference between the skin and the center of the spar web are shown in Fig. 6.17. During the initial ascent, the skin experiences a small tensile stress as the skin temperature drops to give a negative temperature difference. For most of the remainder of the

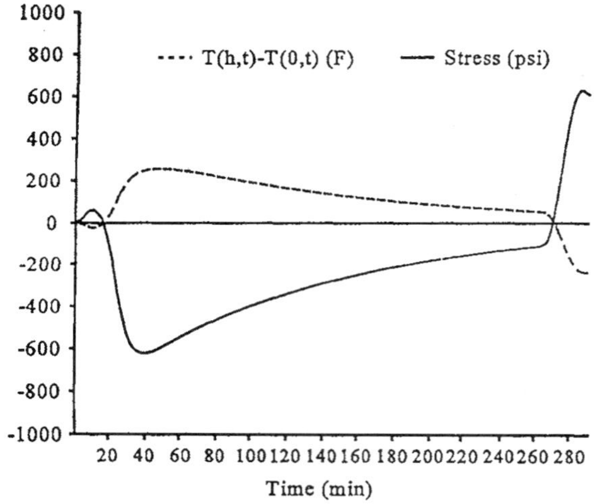

Fig. 6.17 History of wing spar skin response.

flight, skin stress is compressive as the aerodynamic heating induces a positive temperature difference. The maximum compressive stress of about 600 psi occurs at about 40 min; thereafter, the compressive stress diminishes as the temperature difference lessens. On descent, the skin stress becomes tensile again as the temperature difference reverses sign. The direct correspondence of the skin stress with the temperature difference is another demonstration of the important role of temperature gradients. Skin stresses are small because of the large skin/spar cap area.

The web stress distributions (Fig. 6.18) were computed at various times for the ascent, cruise, and descent portions of the flight. These stresses show a direct correlation with the web temperature distribution. During the ascent phase, the web stresses are compressive but quickly change to tensile as skin temperatures rise. The stresses remain tensile during most of the flight until descent, when they once again become compressive. The maximum tensile stress of 20,000 psi occurs at 50 min into the flight, and the maximum compressive stress of 18,500 psi occurs at the end of the flight.

Two additional points should be noted. First, the stresses in the skin and web are self-equilibrating. The magnitudes differ significantly because of the large difference in areas. At any selected time, the stresses have opposite sense, and an integration of the stresses over their respective areas must produce equal and opposite force resultants. The second point is that the reversal in sign of the stresses during each flight raises the possibility of thermally induced fatigue.

Example 6.4: Simply Supported Beam with Immovable Supports

We consider now a nonlinear problem of a simply supported beam with immovable supports (Fig. 6.19). By *simply supported* we mean that the transverse displacement and bending moment are zero at $x = 0$ and $x = L$. By *immovable* we mean that horizontal displacements at the supports are fully restrained. The beam is subjected to a temperature distribution such that

$$N_T(x) \equiv 0 \qquad M_{yT}(x) \equiv 0$$

$$M_{zT} = M_T \sin(\pi x/L)$$

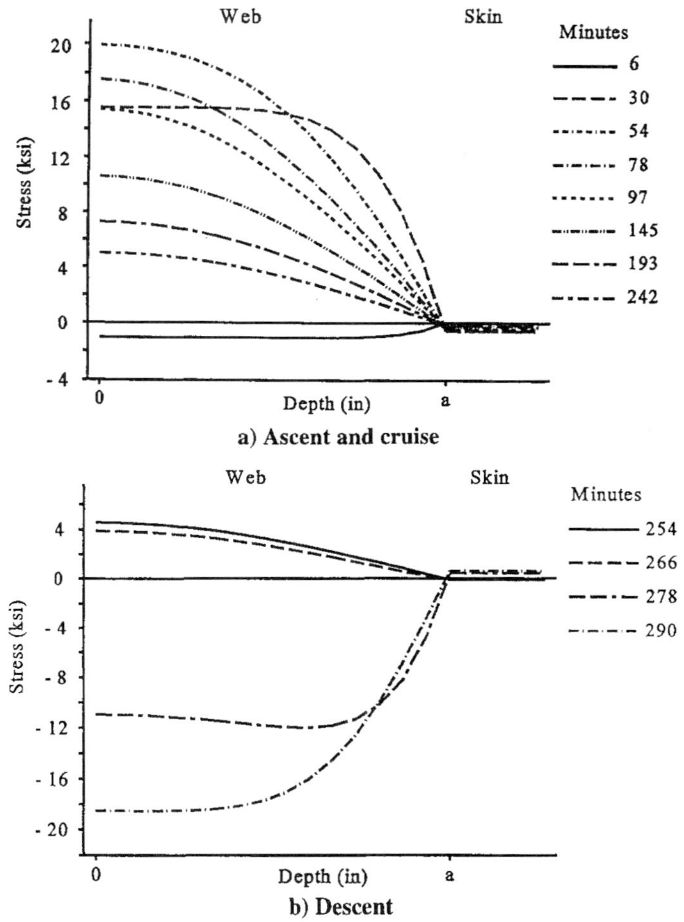

Fig. 6.18 History of wing spar web stress distribution.

where the maximum value of the thermal moment M_T is a constant. The idea of the problem follows a solution for an isothermal beam with pressure loading given by Rivello[6] and Donnell.[29]

The source of the nonlinearity is the immovable supports; for, as the beam deflects beyond the small-deflection range, the centroidal axis must elongate. Thus, bending and extension are coupled, and we must use a nonlinear formulation. Our objective is to establish the nonlinear relations between the thermal moment and extension and bending displacements. We wish to understand how stretching the

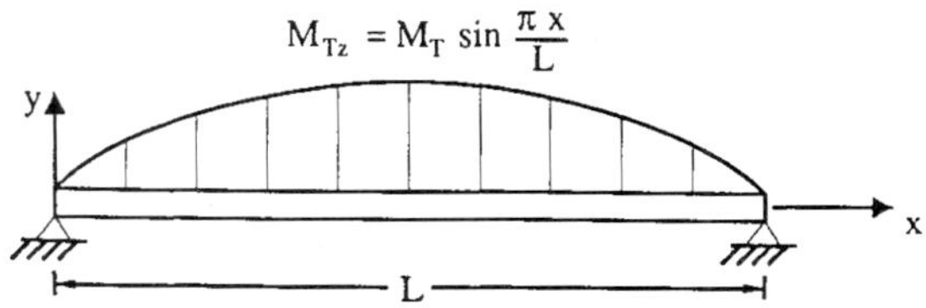

Fig. 6.19 Simply supported beam with immovable supports.

centroidal axis affects the beam response. The thermal-moment distribution and the beam boundary conditions permit us to obtain a simple analytical solution.

For bending in the x–y plane, the nonlinear equation of motion (6.28c) reduces to

$$EI_z \frac{\mathrm{d}^4 v}{\mathrm{d}x^4} - N_x \frac{\mathrm{d}^2 v}{\mathrm{d}x^2} = -\frac{\mathrm{d}^2 M_{zT}}{\mathrm{d}x^2}$$

and the boundary conditions are

$$v(0) = v(L) = 0 \qquad M_x(0) = M_x(L) = 0$$

Because the thermal moment is zero at $x = 0$ and $x = L$, the moment boundary simplifies to

$$\frac{\mathrm{d}^2 v}{\mathrm{d}x^2}(0) = \frac{\mathrm{d}^2 v}{\mathrm{d}x^2}(L) = 0$$

With $N_T = 0$ and assuming $f_x = 0$, the first equilibrium equation, Eq. (6.25a), shows that

$$\frac{\partial N_x}{\partial x} = 0 \quad \text{or} \quad N_x = \text{const}$$

For large deflections with $w = 0$, Eq. (6.26) expresses N_x by

$$N_x = EA\left[\frac{\mathrm{d}u_0}{\mathrm{d}x} + \frac{1}{2}\left(\frac{\mathrm{d}v}{\mathrm{d}x}\right)^2\right]$$

The remaining boundary conditions require that

$$u_0(0) = u_0(L) = 0$$

We begin the solution by assuming a form for $v(x)$ that satisfies the boundary conditions. Thus, we take

$$v(x) = v_0 \sin\frac{\pi x}{L}$$

where v_0 is to be determined. Substituting into the differential equation for $v(x)$ and matching coefficients on the left- and right-hand sides, we find that

$$\left(\frac{\pi^2}{L^2} EI_z + N_x\right) v_0 = M_T$$

Next, we may substitute $v(x)$ into the equation for N_x, and solve for $\mathrm{d}u_0/\mathrm{d}x$. After integration and imposing boundary conditions for u_0, we obtain

$$N_x = \frac{\pi^2}{4}\frac{EA}{L^2} v_0^2$$

The axial force N_x is independent of x as required, and it varies as the square of the displacement amplitude v_0. Combining the last two equations, we obtain a cubic algebraic equation for the amplitude v_0,

$$\frac{EA}{4} v_0^3 + EI_z v_0 = \frac{1}{\pi^2} M_T L^2$$

If we assume a rectangular cross section with $A = bh$ and $I_z = bh^3/12$, we may

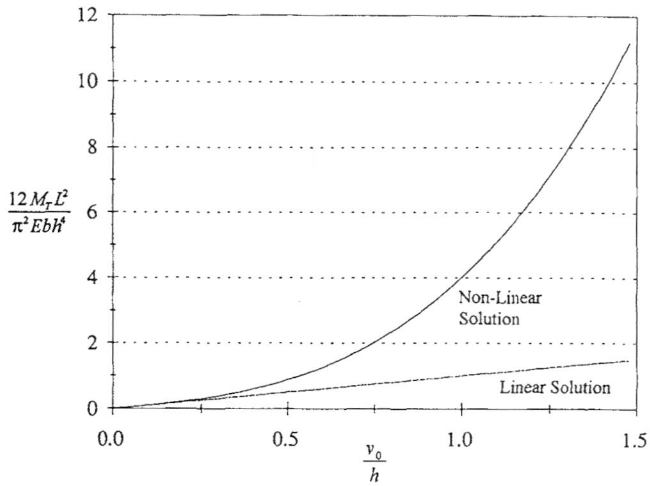

Fig. 6.20 Load-deflection solutions for a beam with immovable supports.

write this last result in nondimensional form as

$$3\left(\frac{v_0}{h}\right)^3 + \frac{v_0}{h} = \frac{12}{\pi^2}\frac{M_T L^2}{E b h^4}$$

where h is the beam depth and b the beam width. This nondimensional equation expresses the ratio of the beam deflection to beam depth as a function of a nondimensional load.

Figure 6.20 plots the nondimensional load deflection. Actually, two curves are plotted in Fig. 6.20. The first is the nonlinear solution; the second is a linear solution obtained by neglecting the nonlinear term $3(v_0/h)^3$. The figure shows that the load–deflection relation is linear only for beam deflections $v_0/h < 0.25$. After the beam deflection exceeds about one-fourth of the beam depth, the nonlinear solution begins to differ significantly from the linear solution. As the deflection enters this range, the axial force begins to develop, and the beam stiffens considerably. As the beam deflects further in the nonlinear range, the in-plane stretching becomes predominant. For deflections of the order of the beam depth, the linear solution would have substantial error. In laboratory experiments, the immovable boundary conditions are very difficult to achieve and, hence, in practical problems, nonlinear beam deflection data may be expected to fall between the two curves shown.

Example 6.5: Properties of a Segment of a Circular Tube

We consider here the evaluation of the cross-sectional properties of a segment of a circular tube (Fig. 6.21a). The tube is thin-walled, so that the thickness h is much smaller than the tube radius R. We wish to derive formulas for the cross-sectional properties used in the analysis of a beam with combined flexure and torsion. These properties include location of the centroid, the moments of inertia about centroidal axes, the shear center, and the warping constant.

The coordinate system used in the evaluation of the properties is shown in Fig. 6.21b. For evaluation of the various integrals, we use a polar coordinate system

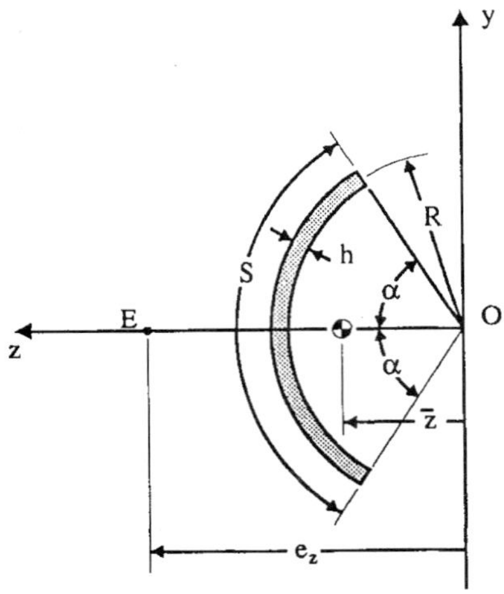

a) Cross-sectional geometry

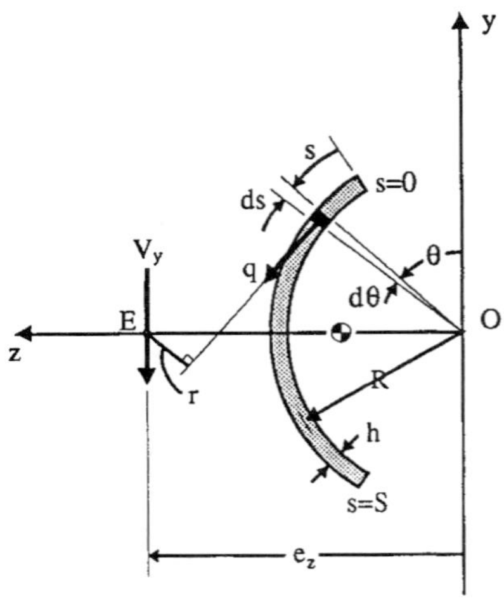

b) Integration geometry

Fig. 6.21 Segment of a circular tube.

centered at point O and a Cartesian system with axes y and z with origin at point O. For analyses of beam behavior, we would locate the Cartesian system at the centroid. For the polar coordinate system shown, the element of arc length is

$$\mathrm{d}s = R\,\mathrm{d}\theta$$

and the element of area is

$$\mathrm{d}A = Rh\,\mathrm{d}\theta$$

Centroid. By symmetry, the centroid lies in the z axis, and we may write

$$A\bar{z} = \int_A z\,\mathrm{d}A = \int_{(\pi/2)-\alpha}^{(\pi/2)+\alpha} (R\sin\theta)Rh\,\mathrm{d}\theta$$

Because $A = 2Rh\alpha$, we evaluate the integral and find that

$$\bar{z} = R\frac{\sin\alpha}{\alpha}$$

Moments of inertia. We first evaluate the moment of inertia about the y axis and then use the parallel axis theorem to transfer it to a y axis through the centroid. Thus,

$$I_y = \int_A z^2 \,\mathrm{d}\theta = \int_{(\pi/2)-\alpha}^{(\pi/2)+\alpha} (R\sin\theta)^2 Rh \,\mathrm{d}\theta$$

hence,

$$I_y = R^3 h(\alpha + \sin\alpha\cos\alpha)$$

At the centroid,

$$\begin{aligned}\bar{I}_y &= I - A\,\mathrm{d}^2 \\ &= R^3 h(\alpha + \sin\alpha\cos\alpha) - (2Rh\alpha)\left(\frac{R\sin\alpha}{\alpha}\right)^2\end{aligned}$$

Thus,

$$\bar{I}_y = R^3 h\left(\alpha + \sin\alpha\cos\alpha - \frac{R\sin^2\alpha}{\alpha}\right)$$

Because the centroid lies on the z axis, we may write directly

$$\bar{I}_z = \int_A y^2 \,\mathrm{d}\theta = \int_{(\pi/2)-\alpha}^{(\pi/2)+\alpha} (R\cos\theta)^2 Rh \,\mathrm{d}\theta$$

and we find that

$$\bar{I}_z = R^3 h(\alpha - \sin\alpha\cos\alpha)$$

Shear center. By symmetry, the shear center lies on the z axis so that $e_y = 0$, and we need to determine only e_z. To determine e_z, we write an equilibrium equation that balances the torque resulting from integrating the shear flow q over the tube and the torque resulting from a vertical shear force V_y. By definition, if the shear force acts at the shear center, the resultant torque on the cross section is zero. Taking moments about the origin,

$$\int_0^S (q\,\mathrm{d}s)R + V_y e_z = 0$$

Thus,

$$e_z = -\frac{R\int_0^S (q\,\mathrm{d}s)}{V_y}$$

Substituting for the shear flow from Eq. (6.42), we find that

$$e_z = \frac{R}{\bar{I}_z}\int_0^S (Q_z\,\mathrm{d}s)$$

Table 6.1 Values of e_z for various angles

α	30 deg	60 deg	90 deg	120 deg	150 deg	180 deg
e_z	1.028 R	1.115 R	1.273 R	1.514 R	1.814 R	2 R

The first moment of area Q_z is defined in Eq. (6.40a) as

$$Q_z = \int_0^s y\,\mathrm{d}A = \int_{(\pi/2)-\alpha}^{\theta} (R\cos\theta)Rh\,\mathrm{d}\theta$$

Hence,

$$Q_z = R^2 h(\sin\theta - \cos\alpha)$$

and

$$e_z = \frac{R}{\bar{I}_z}\int_{(\pi/2)-\alpha}^{(\pi/2)+\alpha} R^2 h(\sin\theta - \cos\alpha)R\,\mathrm{d}\theta$$

After evaluating the integral and substituting for $\bar{I}_z$, we find that

$$e_z = 2R\frac{\sin\alpha - \alpha\cos\alpha}{\alpha - \sin\alpha\cos\alpha}$$

The equation for e_z locates the shear center along the positive z axis from the origin O shown in Fig. 6.21. The location of the shear center depends on the tube radius R and the angle α. The values of e_z for various angles α are given in Table 6.1. The interesting aspect of the result is that e_z is always greater than R, so that the shear center is always to the left of the tube wall; that is, it is always outside the tube.

Warping constant. The evaluation of the warping constant is a multistep calculation, and we begin by writing the equation for the sectorial radius r. By definition, r is the perpendicular distance from the shear center to the line of action or tangent line of $q\,\mathrm{d}s$. From Fig. 6.21b, we obtain

$$r = R - e_z\sin\theta$$

Next, we evaluate the sectorial area. From Eq. (6.49b),

$$\omega = \frac{1}{2}\int_0^s r\,\mathrm{d}s = \frac{1}{2}\int_{(\pi/2)-\alpha}^{\theta} (R - e_z\sin\theta)R\,\mathrm{d}\theta$$

After integration, we obtain

$$\omega = \frac{R}{2}\left[R\left(\theta - \frac{\pi}{2} + \alpha\right) + e_z(\cos\theta - \sin\alpha)\right]$$

Next, we evaluate

$$S_\omega = 2\int_0^S \omega(s)\,\mathrm{d}A$$

which is given in Eq. (6.51d). Substituting for ω,

$$S_\omega = R\int_{(\pi/2)-\alpha}^{(\pi/2)+\alpha}\left[R\left(\theta - \frac{\pi}{2} + \alpha\right) + e_z(\cos\theta - \sin\alpha)\right] Rh\,\mathrm{d}\theta$$

which yields

$$S_\omega = 2\alpha R^2 h(\alpha R - e_z \sin\alpha)$$

From Eq. (6.56b),

$$\omega_0 = \frac{S_\omega}{2A} = \frac{R}{2}(\alpha R - e_z \sin\alpha)$$

Finally, the warping constant is defined by Eq. (6.59),

$$\Gamma = 4\int_0^S (\omega_0 - \omega)^2\,\mathrm{d}A$$

Substituting for ω and ω_0, the integral becomes

$$\Gamma = R^4\int_{(\pi/2)-d}^{(\pi/2)+d}\left(\frac{\pi}{2} - \theta - \frac{e_z}{R}\cos\theta\right)^2 Rh\,\mathrm{d}\theta$$

After evaluation,

$$\Gamma = R^5 h\left[\frac{2}{3}\alpha^3 + \frac{e_z^2}{R^2}(\alpha - \sin\alpha\cos\alpha) + 4\frac{e_z}{R}(\alpha\cos\alpha - \sin\alpha)\right]$$

Substituting for the shear center coordinate e_z gives the final result,

$$\Gamma = R^5 h\left[\frac{2}{3}\alpha^3 - \frac{4(\sin\alpha - \alpha\cos\alpha)^2}{(\alpha - \sin\alpha\cos\alpha)}\right]$$

The preceding equation gives the warping constant for the segment of a circular tube. Note that Γ has dimensions of (length)^6.

Saint Venant torsion constant. As mentioned earlier, the torsion constant $J = Sh^3/3$, where S is length of the centerline of the tube. For the tube segment, $S = 2\alpha R$; hence,

$$J = \tfrac{2}{3}\alpha Rh^3$$

Since h is small, this result shows that the torsion constant J is quite small for a thin-walled tube.

Slit tube. A slit tube is a special case of the preceding formulas when $\alpha = \pi$. For the slit tube, the foregoing results may be summarized as follows.

$$\bar{y} = \bar{z} = 0 \qquad \bar{I}_y = \bar{I}_z = \pi R^3 h \qquad e_y = 0$$

$$e_z = 2R \qquad \Gamma = \tfrac{2}{3}\pi(\pi^2 - 6)R^5 h \qquad J = \tfrac{2}{3}\pi Rh^3$$

Segments of open tubes are, of course, much more flexible in torsion than closed tubes. However, tube segments have been used in spacecraft applications as deployable booms. Overlapping slit tubes, where $\alpha > \pi$, have been used in several deployable booms.

Slit, overlapped tubes can be stored by rolling them into a drum or cassette similar to a carpenter's tape. A boom made with a single tube is known as a STEM (storable tubular extendible member). Two concentric STEMs may also be deployed to form a boom, and these designs are called Bi-STEMs. Bi-STEMs are used as booms on the Hubble Space Telescope solar arrays. Some design considerations for STEMs are described by Rimrott in Ref. 30. Torsional buckling of slit, overlapped, thin-walled tubes is discussed by Rimrott and Elliott.[31]

Example 6.6: Simply Supported Beam with Axial Load

We analyze the flexural-torsional behavior of a simply supported beam with a compressive axial load P (Fig. 6.22). Thus, we wish to solve the set of equations given in Eq. (6.70). In these equations, we take $N_x = -P$, and we neglect the effect of the thermal stress on the torsional stiffness. To further simplify the problem, we assume a cross section that has $e_y = 0$ and $e_z > 0$ and $I_y = I_z = I$. The slit tube of Example 6.5 is such a cross section. The beam is subjected to a temperature distribution such that

$$N_T = 0 \qquad M_{xT} = M_x^T \sin \pi x/L$$

$$M_{zT} = 0 \qquad M_{yT} = M_y^T \sin \pi x/L$$

All other distributed forces and moments are zero. With these assumptions, Eqs. (6.70) reduce to

$$EI\frac{d^4 v}{dx^4} + P\frac{d^2 v}{dx^2} + Pe_z\frac{d^2\theta_x}{dx^2} = 0$$

$$EI\frac{d^4 w}{dx^4} + P\frac{d^2 w}{dx^2} - \left(\frac{\pi}{L}\right)^2 M_y^T \sin\frac{\pi x}{L} = 0$$

$$E\Gamma\frac{d^4\theta_x}{dx^4} - \left(GJ - \frac{I_E}{A}P\right)\frac{d^2\theta_x}{dx^2} + Pe_z\frac{d^2 v}{dx^2} - \left(\frac{\pi}{L}\right)^2 M_x^T \sin\frac{\pi x}{L} = 0$$

For simple supports, the transverse deflections and bending moments are zero at $x = 0$ and $x = L$. In addition, the angle of twist is zero, and we permit the end

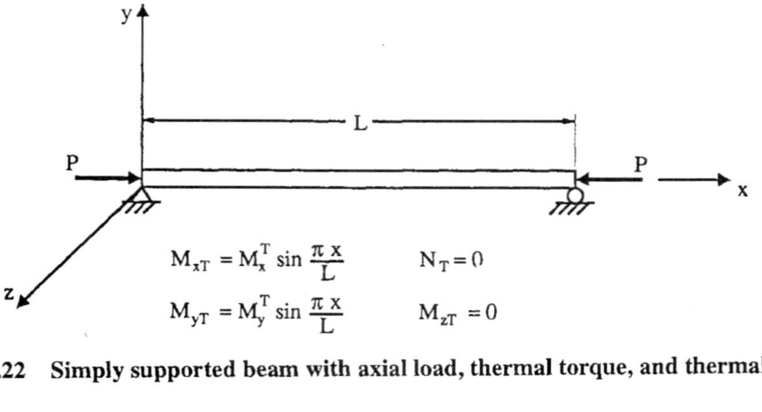

Fig. 6.22 Simply supported beam with axial load, thermal torque, and thermal moment.

cross sections to warp. Noting that the thermal moments are zero at the ends, we find that the boundary conditions are

$$v = w = \theta_x = 0 \qquad \frac{d^2 v}{dx^2} = \frac{d^2 w}{dx^2} = \frac{d^2 \theta_x}{dx^2} = 0$$

at each end of the beam.

With this problem statement, an examination of the differential equations shows that the beam response consists of bending in the x–y plane and bending in the x–z plane coupled with torsion. Thus, we may solve the differential equation for $w(x)$ independently of the first and third equations. These equations must be solved simultaneously because the displacement v and angle of twist θ_x are coupled. To solve the equations, we assume a solution of the form

$$v = v_0 \sin\frac{\pi x}{L} \qquad w = w_0 \sin\frac{\pi x}{L} \qquad \theta_x = \theta_0 \sin\frac{\pi x}{L}$$

where v_0, w_0, and θ_0 are amplitudes to be determined. These functions automatically satisfy all boundary conditions. The remaining steps determine these amplitudes so that the governing differential equations are satisfied.

Solving the uncoupled flexural problem first, we substitute the equation for w into the second differential equation. We may then solve directly for the amplitude w_0, to obtain

$$w_0 = -\frac{M_y^T}{P_{cr} - P}$$

where $P_{cr} = \pi^2 EI/L^2$. We recognize P_{cr} as the familiar Euler buckling load for a simply supported column. The minus arises because a positive thermal moment causes a negative w displacement. The final solution for the uncoupled flexural portion of the problem is

$$w = -\frac{M_y^T}{P_{cr} - P} \sin\frac{\pi x}{L}$$

This solution is valid only for $P < P_{cr}$. As P approaches P_{cr}, w becomes infinitely large, indicating buckling in the x–z plane.

The solution to the coupled flexure-torsion portion of the problem is obtained by substituting the forms for v and θ_x in the first and third differential equations. This yields two simultaneous equations:

$$(P_{cr} - P)v_0 - Pe_z\theta_0 = 0$$

$$\left(\frac{\pi^2 E\Gamma}{L^2} + GJ - \frac{I_E}{A}P\right)\theta_0 - Pe_z v_0 = \frac{\pi^2}{L^2} M_x^T$$

Solving for v_0 and θ_0, we obtain

$$v_0 = \frac{Pe_z}{D} M_x^T$$

$$\theta_0 = \frac{P_{cr} - P}{D} M_x^T$$

where

$$D = \left[E\Gamma\frac{\pi^2}{L^2} + GJ - \frac{I_E}{A}P\right](P_{cr} - P) - P^2 e_z^2$$

The final solution is

$$v = \frac{Pe_z}{D} M_x^T \sin\frac{\pi x}{L}$$

$$\theta_x = \frac{P_{cr} - P}{D} M_x^T \sin\frac{\pi x}{L}$$

We have already seen that, because P is compressive, flexural buckling may occur in the x–z plane as P approaches P_{cr}, the Euler buckling load. The solution for v and θ shows that there is also the possibility of flexural-torsional buckling when the denominator D approaches zero. Thus, for the value of P, say P_θ, that satisfies

$$D = \left[E\Gamma\frac{\pi^2}{L^2} + GJ - \frac{I_E}{A}P_\theta\right](P_{cr} - P_\theta) - P_\theta^2 e_z^2 = 0$$

the displacement v_0 and angle of twist θ_x will simultaneously become infinitely large. Hence, there is a second critical buckling load for the problem. This critical buckling load P_θ may be smaller or larger than P_{cr}. Hence, for a given material and cross-sectional geometry, we must calculate P_{cr} and P_θ to determine the lowest buckling load for the beam. For the constant material properties that we have assumed, the buckling loads are independent of temperature. The role of temperature in the problem is to cause the beam to respond as a beam column. For a constant force P less than the buckling load, the beam bends and twists, and the transverse displacements and angle of twist increase linearly with the thermal moment and torque.

6.5 Closing Remarks

This chapter presents engineering theories for thermal-stress analysis of rods and beams. Rods are considered first, and basic equations are presented for quasistatic as well as dynamic initial/boundary-value problems. Only quasistatic rod examples are presented. For beams, only quasistatic formulations are presented, but appropriate body forces and moments are retained, and so, by using D'Alembert's principle, appropriate equations of motion may be written easily. Three thermoelastic beam formulations are presented: 1) classical Bernoulli–Euler theory for beams with compact cross sections, 2) a large-deflection approach that assumes moderate rotations, and 3) the theory for thin-walled beams with open cross sections. The formulations permit three-dimensional flexure, extension, and torsion. A series of examples illustrate salient features of the formulations and solution approaches for practical problems. The chapter provides the basic equations for more complicated beam problems, including thermal buckling and thermally induced vibrations, that will be studied in later chapters.

In the next chapter, we provide a similar development of the basic thermomechanics formulations for flat plates.

References

[1]Gatewood, B. E., *Thermal Stresses*, McGraw–Hill, New York, 1957.

[2]Boley, B. A., and Weiner, J. H., *Theory of Thermal Stresses*, Wiley, 1960; reprinted ed. available from Krieger, Malabar, FL.

[3]Johns, D. J., *Thermal Stress Analysis*, Pergamon, Oxford, England, UK, 1965.

[4]Burgreen, D., *Element of Thermal Stress Analysis*, C. P. Press, Jamaica, NY, 1971.

[5]Oden, J. T., *Mechanics of Elastic Structures*, McGraw–Hill, New York, 1967.

[6]Rivello, R. M., *Theory and Analysis of Flight Structures*, McGraw–Hill, New York, 1969.

[7]Allen, D. H., and Haisler, W. E., *Introduction to Aerospace Structural Analysis*, Wiley, New York, 1985.

[8]Donaldson, B. K., *Analysis of Aircraft Structures: An Introduction*, McGraw–Hill, New York, 1993.

[9]Clarke, A. C., *Fountains of Paradise*, Harcourt, Brace, Jovanovich, New York, 1979.

[10]Pearson, J., "The Orbital Tower: A Spacecraft Launcher Using the Earth's Rotational Energy," *Acta Aeronautica*, Vol. 2, 1975, pp. 785–799.

[11]Colombo, G., Gaposchkin, E. M., Grossi, M. D., and Weiffenback, G. C., "The Skyhook: A Shuttle-Borne Tool for Low-Orbital-Altitude Research," *Meccanica*, Vol. 10, No. 1, 1975, pp. 3–20.

[12]Hurlbut, F. C., and Potter, J. L., "Tethered Aerothermodynamic Research Needs," *Journal of Spacecraft and Rockets*, Vol. 28, No. 1, 1991, pp. 50–57.

[13]Greenwood, T., *Principles of Dynamics*, Prentice–Hall, Englewood Cliffs, NJ, 1965, pp. 328–330.

[14]Brush, D. O., and Almroth, B. O., *Buckling of Bars, Plates and Shells*, McGraw–Hill, New York, 1975, p. 8.

[15]Vlasov, V. Z., *Thin-Wall Elastic Beams*, Israel Program for Scientific Translations, Jerusalem, 1961.

[16]Galambos, T. V., *Structural Members and Frames*, Prentice–Hall, Englewood Cliffs, NJ, 1968.

[17]Murray, N. W., *Introduction to the Theory of Thin-Walled Structures*, Clarendon Press, Oxford, England, UK, 1984.

[18]Timoshenko, S. P., and Gere, J. M., *Theory of Elastic Stability*, 2nd ed., McGraw–Hill, New York, 1961, p. 530.

[19]Roark, R. J., and Young, W. C., *Roark's Formulas for Stress and Strain*, 6th ed., McGraw–Hill, New York, 1989, pp. 366–369.

[20]Bisplinghoff, R. L., "Some Structural and Aeroelastic Considerations of High Speed Flight," *Journal of the Aeronautical Sciences*, Vol. 23, No. 4, 1956, pp. 289–329.

[21]Budiansky, B., and Mayers, J., "Influence of Aerodynamic Heating on the Effective Torsional Stiffness of Thin Wings," *Journal of the Aeronautical Sciences*, Vol. 23, No. 12, 1956, pp. 1081–1108.

[22]Hoff, N. J., "Approximate Analysis of the Reduction in Torsional Rigidity and of the Torsional Buckling of Solid Wings Under Thermal Stress," *Journal of the Aeronautical Sciences*, Vol. 23, No. 6, 1956, pp. 603–604.

[23]Singer, J., and Hoff, N. J., "Effect of the Change in Thermal Stresses Due to Large Deflections on the Torsional Rigidity of Wings," *Journal of the Aeronautical Sciences*, Vol. 24, No. 4, 1957, pp. 310–311.

[24]Kochanski, S. L., and Argyris, J. H., "Some Effects of Kinetic Heating on Stiffness of Thin Wings," *Aircraft Engineering*, Vol. 29, No. 344, 1957, pp. 310–318.

[25]Bisplinghoff, R. L., "Further Remarks on the Torsional Rigidity of Thermally Stressed Wings," *Journal of the Aeronautical Science*, Vol. 25, No. 10, 1958, pp. 657–658.

[26]van der Neut, A., "Buckling Caused by Thermal Stresses," *High Temperature Effects in Aircraft Structures*, edited by N. J. Hoff, Pergamon, New York, 1958, pp. 215–247.

[27]Bisplinghoff, R. L., and Dugundji, J., "Influence of Aerodynamic Heating on Aeroelastic Phenomena," *High Temperature Effects in Aircraft Structures*, edited by N. J. Hoff, Pergamon, New York, 1958, pp. 288–312.

[28]Beam, R. M., and Yagoda, H. P., "On the Torsional Static Stability and Response of Open Section Tubes Subjected to Thermal Radiation Loading," *International Journal of Solids and Structures*, Vol. 9, No. 1, 1973, pp. 151–175.

[29]Donnell, L. H., "Shell Theory," *Proceedings of Fourth Midwest Conference on Solid Mechanics* (Austin, TX), Univ. of Texas, Austin, 1959, pp. 1–17.

[30]Rimrott, F.P.J., "Storable Tubular Extendible Member—A Unique Machine Element," *Machine Design*, Dec. 1965, pp. 157–165.

[31]Rimrott, F. P. J., and Elliott, T., "Torsion of Slit, Overlapped, Thin-Walled Tubes," *Journal of Spacecraft and Rockets*, Vol. 3, No. 6, 1966, pp. 873–876.

Problems

6.1. A steel rod 20 in. long is fastened between two walls. The rod has an initial temperature of 90°F, and the initial stress is zero. $E = 29 \times 10^6$ psi, and $\alpha = 7 \times 10^{-6}$ 1/°F.

(a) What is the stress in the rod when the rod temperature is −30°F?

(b) If the walls move inward 0.010 in. at constant temperature, what is the rod stress?

6.2. An aluminum tube 6 ft long is restrained from thermal expansion by two rigid walls. The tube has an outside diameter of 2 in. and a wall thickness of 0.125

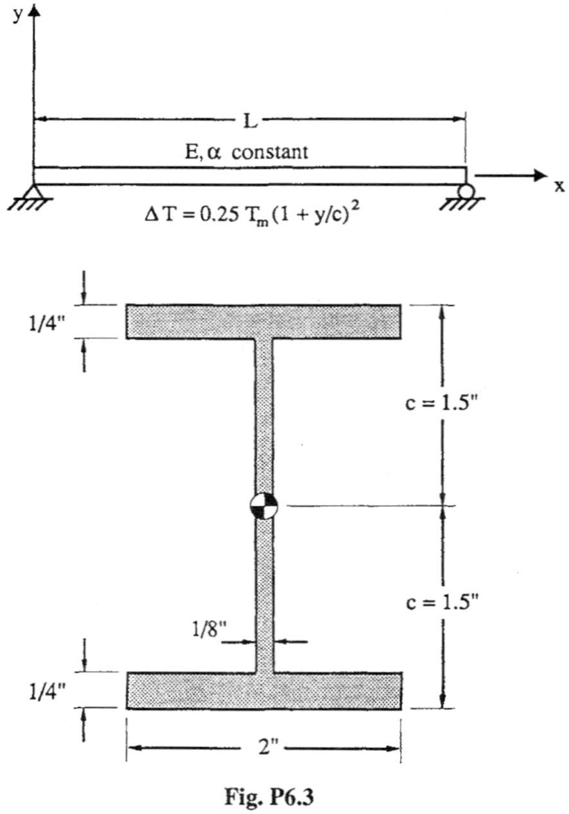

Fig. P6.3

in. Determine the temperature change that will cause Euler buckling of the tube. Assume simply supported ends.

6.3. A simply supported beam with the cross section shown (Fig. P6.3) experiences a steady-state temperature change $\Delta T = 0.25T_m(1 + y/c)^2$ where T_m is a constant and $2c$ is the beam depth. After first sketching the temperature change vs y, determine the following quantities for the beam in terms of E, α, T_m, and L:

(a) Thermal force N_T.
(b) Thermal moment M_{Tz}.
(c) Beam elongation $u_0(L)$.
(d) Maximum deflection $v(L/2)$.
(e) Maximum normal stress σ_x. Sketch the stress distribution through the beam depth.

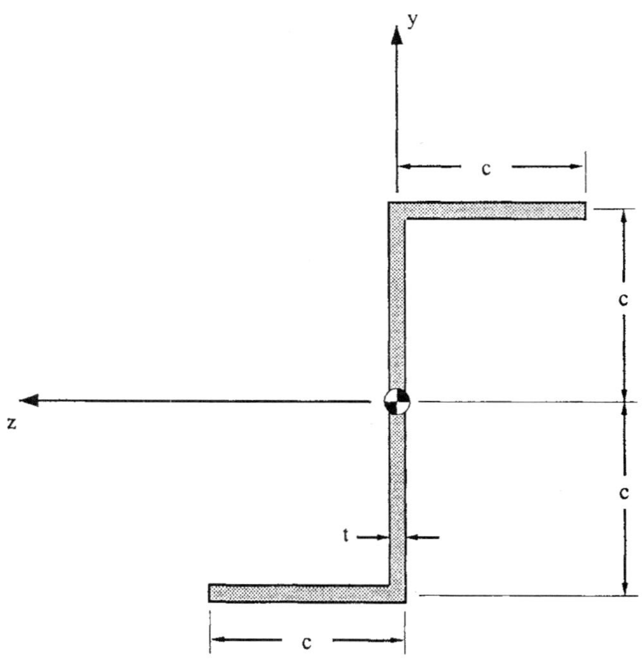

Fig. P6.4

6.4. A cantilever beam of length L with the asymmetric cross section shown in Fig. P6.4 experiences a temperature change $\Delta T = T_0(y/c)^2$, where T_0 is a constant and $2c$ is the beam depth. Determine:

(a) The distribution of the normal stress σ_x.
(b) The deflection components $v(L)$ and $w(L)$.

6.5. Repeat Problem 6.4 for the temperature change $\Delta T = T_0(y/c)$.

6.6. A fixed simply supported beam of rectangular cross section with depth h and width b has a steady-state temperature change $\Delta T = 0.25T_m(1 + y/h)^2$.

(a) Determine the maximum transverse deflection.
(b) Determine the maximum value of the strain on the top surface of the beam.

6.7. A fixed-fixed rectangular beam has a temperature distribution

$$T(x, y) = \frac{T_0 y}{h}\left(1 - \sin\frac{\pi x}{L}\right)$$

where T_0 is a constant and the reference temperature zero. Derive equations for:

(a) Thermal force N_T.
(b) Force N_x.
(c) Thermal moment M_{Tz}.
(d) In-plane displacement $u(x)$.
(e) Transverse displacement $v(x)$.

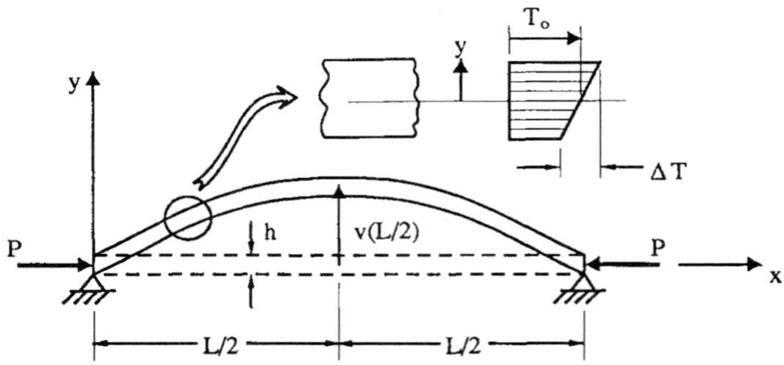

Fig. P6.8

6.8. A thin beam of rectangular cross-section (Fig. P6.8) is fully restrained against in-plane expansion and may bend in the x–y plane. The beam experiences a linear temperature distribution $T(y) = T_0 + \Delta T y/h$, where T_0 is a uniform temperature, ΔT denotes the difference between the top and bottom surface temperatures, and h is the beam depth. The reference temperature is zero.

(a) Derive an equation for the transverse deflection $v(x)$ in terms of the thermal moment M_{Tz} and the unknown in-plane restraint force P.

(b) Derive an equation for determining the restraint force P in terms of T_0, ΔT, and other beam parameters.

6.9. For the beam problem described in Problem 6.8, construct graphs of P vs $v(L/2)$ for $\Delta T = 1, 5, 10$, and 50°F. Use the following data:

$$E = 28.4 \times 10^6 \text{ lb/in.}^2$$
$$\alpha = 6.52 \times 10^{-6}\ 1/^\circ\text{F}$$
$$b = 1.0 \text{ in.}$$
$$h = 0.1 \text{ in.}$$
$$L = 4.0 \text{ in.}$$

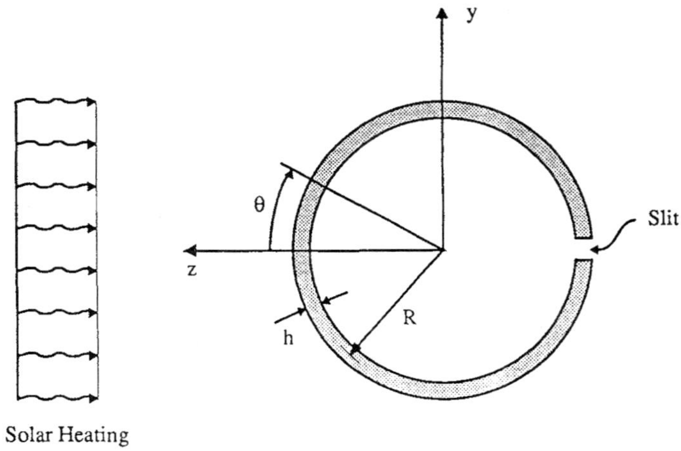

Fig. P6.10

6.10. A cantilever thin-walled slit tube (Fig. P6.10) used as a satellite boom experiences solar heating. Assume that the temperature distribution is given by $T = \bar{T} + T_m \cos\theta$, where $\bar{T}$ is the average tube temperature, and T_m is a perturbation from the mean temperature. (See Sec. 4.8 for further details.) Derive an equation for the deflection of the end of the boom if the boom length is L.

This page intentionally left blank

7
Thermal Structures Problems: Plates

7.1 Introduction

In Chapter 6, we began our treatment of thermal structures problems by developing formulations and solving basic thermal-stress problems for rods and beams. We continue our discussion here for plates. As before, we limit our discussion to fundamental engineering theories for thermal-stress analyses. We consider only homogeneous, isotropic plates, and we limit our solutions to rectangular shapes. For plate applications with nonrectangular shapes, we will abandon the analytical approach and turn to finite element analysis. Some of the theory developed here will be used in our development of finite elements.

There is a rich literature available for isothermal plate theory as well as plate theory, including thermal effects. Thermal-stress texts describing plate theory include the books by Boley and Weiner,[1] Nowacki,[2] Johns,[3] Parkus,[4] and Nowinski.[5] The first four books appeared in the 1960s, and the fifth in the 1970s. In the 1980s, Tauchert[6,7] presented a more recent description of statical and dynamical thermal-stress problems for plates. In 1991, Tauchert[8] also presented a survey of recent literature on thermally induced flexure, buckling, and vibrations of plates.

In formulating engineering mechanics theories for plates, we make simplifying assumptions guided by the thermoelasticity foundation established in Chapter 5. Actually, in Sec. 5.5.2, we present a thermoelasticity solution for an unrestrained thin plate. The problem is solved using the three-dimensional equations of thermoelasticity, and the solution illustrates basic features of plate bending that we reconsider in more detail in this chapter. In the treatment presented here, the approach is to use engineering assumptions to derive an approximate theory including partial differential equations and boundary conditions for a boundary/initial-value problem formulation. As with beams, we assume quasistatic behavior but retain appropriate body forces so that the equations of motion may be derived directly by D'Alembert's principle.

We begin with classical small-deflection plate theory and then extend the results to large deflections in the sense of von Kármán's plate theory.

7.2 Small-Deflection Formulation

In Sec. 5.5.2, we presented a thermoelasticity solution for an unrestrained thin flat plate. We use the solution obtained there to guide us in our classical small-deflection plate bending formulation. We have in mind a rectangular plate (Fig. 7.1) of plane dimensions a and b with thickness h. The quotient b/a is called the *aspect ratio*. We assume a thin plate so that the thickness $h \ll a$, where a is a typical plate dimension. We assume that, before loads are applied, the plate is perfectly flat. The plate is subjected to a transverse surface load $q(x, y, t)$ with units of force per unit area. In addition, the plate experiences a temperature change $\Delta T = T(x, y, z, t) - T_{\text{ref}}$, where T denotes the temperature and T_{ref} is the reference temperature. Thus, we permit the temperature to vary with the plate spatial dimensions and through

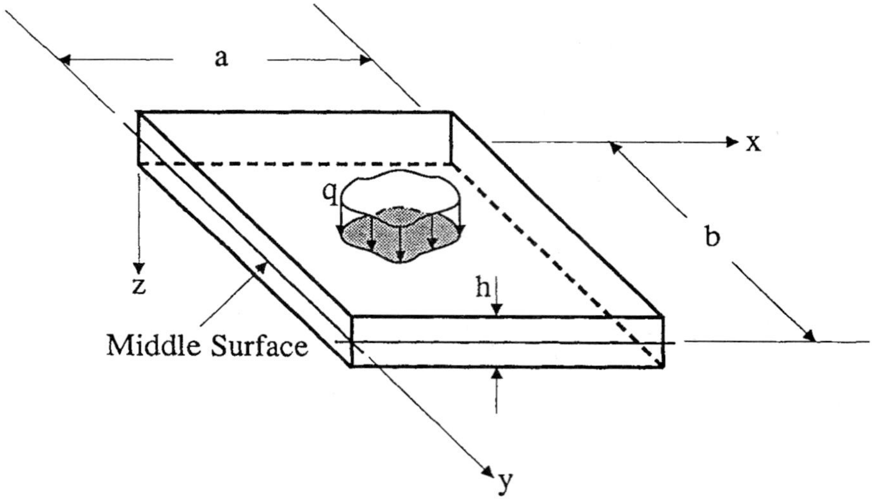

Fig. 7.1 Rectangular plate

the thickness h. See Chapter 4 for a discussion of temperature variations in thin-walled structures. Roughly speaking, for metallic plates, we may expect the spatial temperature variations to be large but the thickness temperature gradients to be small.

7.2.1 Displacements and Strains

We make three basic assumptions:

1) Line elements normal to the middle surface before deformation remain normal during deformation (Fig. 7.2).

2) The transverse displacement w results from bending alone; line elements normal to the middle surface do not extend or contract, and shearing deformations are negligible.

3) Each layer of the plate is in a state of plane stress with stress components σ_x, σ_y, and σ_{xy}.

With these assumptions, we write the plate's displacement components as

$$u(x, y, z, t) = u_0(x, y, t) - z\frac{\partial w}{\partial x}(x, y, t) \tag{7.1a}$$

$$v(x, y, z, t) = v_0(x, y, t) - z\frac{\partial w}{\partial y}(x, y, t) \tag{7.1b}$$

$$w(x, y, z, t) = w_0(x, y, t) \tag{7.1c}$$

where u_0, v_0 denote the in-plane displacements of a point on the middle surface. The derivatives $\partial w/\partial x$ and $\partial w/\partial y$ denote the plate's slopes in the x and y directions, respectively. Equations (7.1a) and (7.1b) follow directly from assumption 1. The first two equations express the assumption that a line element normal to the middle surface before deformation translates and rotates during deformation. The last equation expresses assumption 2: at a position on the plate located by x, y, all points through the plate thickness experience the same vertical displacement.

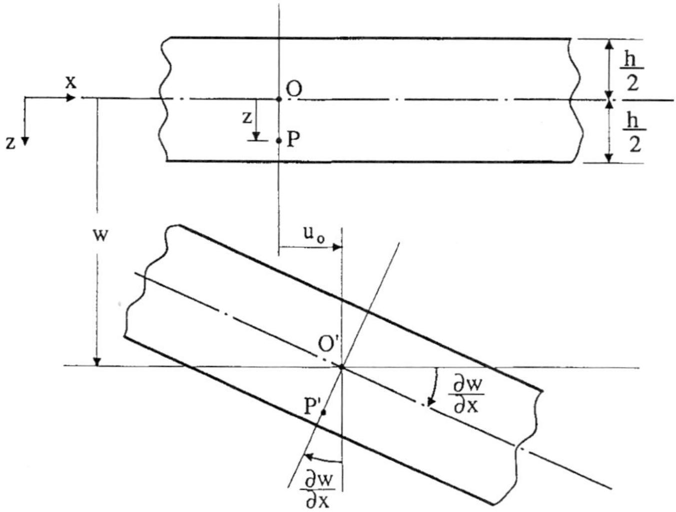

Fig. 7.2 Plate deformations (greatly exaggerated).

For the assumed displacements of Eqs. (7.1), the linear strain-displacement relations (5.2) yield

$$\varepsilon_x = \frac{\partial u}{\partial x} = \frac{\partial u_0}{\partial x} - z\frac{\partial^2 w}{\partial x^2} \tag{7.2a}$$

$$\varepsilon_y = \frac{\partial v}{\partial y} = \frac{\partial v_0}{\partial y} - z\frac{\partial^2 w}{\partial y^2} \tag{7.2b}$$

$$\varepsilon_{xy} = \frac{1}{2}\left(\frac{\partial u}{\partial y} + \frac{\partial v}{\partial x}\right) = \frac{1}{2}\left(\frac{\partial u_0}{\partial y} + \frac{\partial v_0}{\partial x}\right) - z\frac{\partial^2 w}{\partial x \partial y} \tag{7.2c}$$

where all other strain components are zero. In the preceding strain expressions, the first term on the right-hand side of each equation represents strain in the middle surface. The second term represents strain induced by bending. The bending strains occur in a plane parallel to the middle surface given by the coordinate z. The second derivatives in the latter terms are the plate curvatures, which we denote as

$$\kappa_x = -\frac{\partial^2 w}{\partial x^2} \tag{7.3a}$$

$$\kappa_y = -\frac{\partial^2 w}{\partial y^2} \tag{7.3b}$$

$$\kappa_{xy} = -\frac{\partial^2 w}{\partial x \partial y} \tag{7.3c}$$

The first two curvatures κ_x and κ_y are the bending curvatures, and κ_{xy} is the twisting curvature.

7.2.2 Stresses

According to assumption 3, each plate layer is in a state of plane stress. From Eq. (5.13), we may write Hooke's law for an isotropic, homogeneous material as

$$\sigma_x = \frac{E}{1-\upsilon^2}\left(\varepsilon_x + \upsilon\varepsilon_y\right) - \frac{E\alpha}{1-\upsilon}\Delta T$$

$$\sigma_y = \frac{E}{1-\upsilon^2}\left(\varepsilon_y + \upsilon\varepsilon_x\right) - \frac{E\alpha}{1-\upsilon}\Delta T$$

$$\sigma_{xy} = \frac{E}{1+\upsilon}\varepsilon_{xy}$$

where ΔT denotes the temperature change. Substituting the strain components from Eqs. (7.2), we obtain

$$\sigma_x = \frac{E}{1-\upsilon^2}\left(\frac{\partial u_0}{\partial x} + \upsilon\frac{\partial v_0}{\partial y}\right) - \frac{E}{1-\upsilon^2}\left(\frac{\partial^2 w}{\partial x^2} + \upsilon\frac{\partial^2 w}{\partial y^2}\right)z - \frac{E\alpha}{1-\upsilon}\Delta T \quad (7.4a)$$

$$\sigma_y = \frac{E}{1-\upsilon^2}\left(\frac{\partial v_0}{\partial y} + \upsilon\frac{\partial u_0}{\partial x}\right) - \frac{E}{1-\upsilon^2}\left(\frac{\partial^2 w}{\partial y^2} + \upsilon\frac{\partial^2 w}{\partial x^2}\right)z - \frac{E\alpha}{1-\upsilon}\Delta T \quad (7.4b)$$

$$\sigma_{xy} = \frac{E}{2(1+\upsilon)}\left(\frac{\partial u_0}{\partial y} + \frac{\partial v_0}{\partial x}\right) - \frac{E}{1+\upsilon}\frac{\partial^2 w}{\partial x \partial y}z \quad (7.4c)$$

From these equations, we can see how the stress components vary with deformation. The first terms on the right-hand sides of Eqs. (7.4) are the normal stress components and a shearing stress component associated with deformation of the middle surface of the plate. Superimposed on the middle surface stresses are stresses that vary linearly with z through the plate thickness. These are the stresses induced by plate bending and twisting. Finally, in the normal stress components, there is a thermal-stress component proportional to ΔT that is induced by restrained local thermal expansion or contraction.

7.2.3 Stress Resultants

Figure 7.3 shows the normal and shearing stresses acting on a plate element. The stresses act on a rectangular lamina of the plate of thickness dz. The lamina is located at a distance z from the middle surface of the plate. The lamina has unit dimensions in the x and y directions. Each lamina is in a state of plane stress, with normal stresses σ_x and σ_y and a shearing stress σ_{xy}. Let us integrate the forces and moments of these stresses through the plate thickness h. Thus, the stress resultants are defined by

$$N_x = \int_{-(h/2)}^{h/2} \sigma_x \, \mathrm{d}z \quad (7.5a)$$

$$N_y = \int_{-(h/2)}^{h/2} \sigma_y \, \mathrm{d}z \quad (7.5b)$$

$$N_{xy} = N_{yx} = \int_{-(h/2)}^{h/2} \sigma_{xy} \, \mathrm{d}z \quad (7.5c)$$

$$M_x = \int_{-(h/2)}^{h/2} \sigma_x \, z \, \mathrm{d}z \quad (7.5d)$$

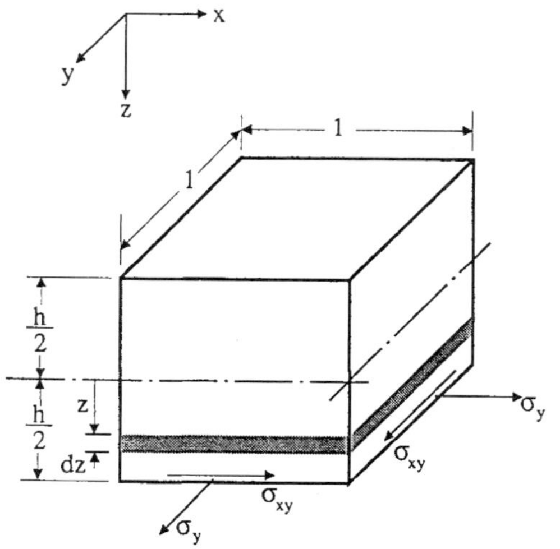

Fig. 7.3 Stresses on an element of a flat plate.

$$M_y = \int_{-(h/2)}^{h/2} \sigma_y \, z \, dz \tag{7.5e}$$

$$M_{xy} = M_{yx} = -\int_{-(h/2)}^{h/2} \sigma_{xy} \, z \, dz \tag{7.5f}$$

The stress resultants are shown in their positive senses in Fig. 7.4. Figure 7.4a shows the first three stress resultants—the membrane forces. Figure 7.4b shows the last three stress resultants—the bending and twisting moments. Note that Figs. 7.3 and 7.4 define the positive senses of the stress resultants. The positive senses for N_x, N_y, and N_{xy}, as well as M_x and M_y, follow directly from the directions of the stresses shown in Fig. 7.3. The twisting moments $M_{xy} = M_{yx}$ are defined as the negative of the moments of the shearing stresses. The stress resultants are defined per unit length.

The stress resultants can now be expressed in terms of the displacement derivatives. To do this, we substitute the stresses from Eqs. (7.4) into Eqs. (7.5) and evaluate the integrals through the plate thickness:

$$N_x = \int_{-(h/2)}^{h/2} \sigma_x \, dz = \frac{Eh}{1-\upsilon^2}\left(\frac{\partial u_0}{\partial x} + \upsilon \frac{\partial v_0}{\partial y}\right) - N_T \tag{7.6a}$$

$$N_y = \int_{-(h/2)}^{h/2} \sigma_y \, dz = \frac{Eh}{1-\upsilon^2}\left(\frac{\partial v_0}{\partial y} + \upsilon \frac{\partial u_0}{\partial x}\right) - N_T \tag{7.6b}$$

$$N_{xy} = N_{yx} = \int_{-(h/2)}^{h/2} \sigma_{xy} \, dz = \frac{Eh}{2(1+\upsilon)}\left(\frac{\partial u_0}{\partial y} + \frac{\partial v_o}{\partial x}\right) \tag{7.6c}$$

$$M_x = \int_{-(h/2)}^{t/2} \sigma_x z \, dz = -D\left(\frac{\partial^2 w}{\partial x^2} + \upsilon \frac{\partial^2 w}{\partial y^2}\right) - M_T \tag{7.6d}$$

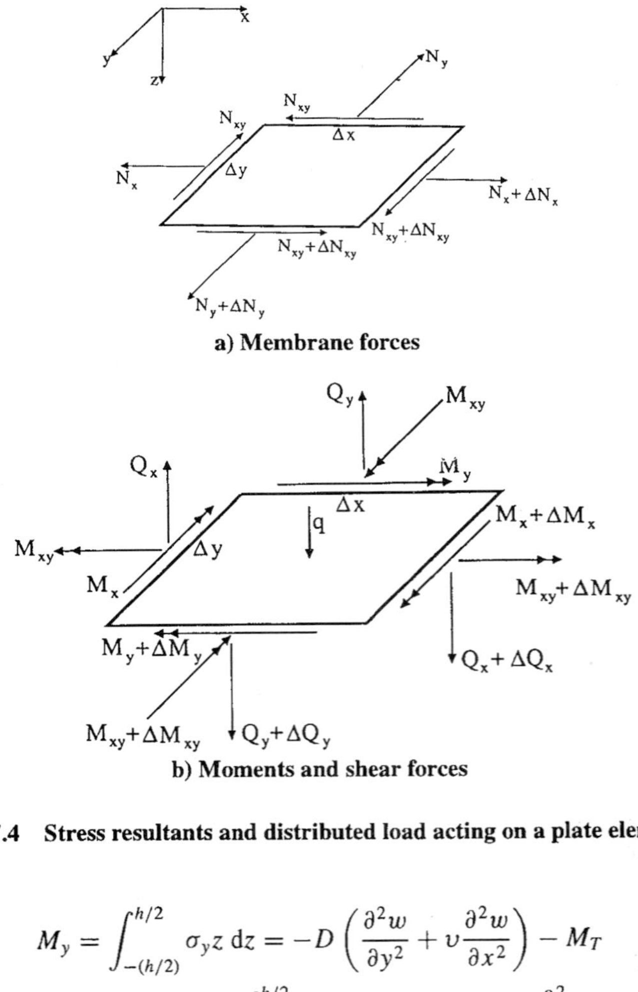

Fig. 7.4 Stress resultants and distributed load acting on a plate element.

$$M_y = \int_{-(h/2)}^{h/2} \sigma_y z \, \mathrm{d}z = -D\left(\frac{\partial^2 w}{\partial y^2} + \upsilon \frac{\partial^2 w}{\partial x^2}\right) - M_T \tag{7.6e}$$

$$M_{xy} = M_{yx} = -\int_{-(h/2)}^{h/2} \sigma_{xy} z \, \mathrm{d}z = (1-\upsilon) D \frac{\partial^2 w}{\partial x \partial y} \tag{7.6f}$$

where D is the plate flexural rigidity defined by

$$D = \frac{Eh^3}{12(1-\upsilon^2)} \tag{7.7}$$

The thermal force per unit length N_T is given by

$$N_T = \frac{E\alpha}{1-\upsilon} \int_{-(h/2)}^{h/2} \Delta T \, \mathrm{d}z \tag{7.8}$$

and the thermal moment per unit length M_T is given by

$$M_T = \frac{E\alpha}{1-\upsilon} \int_{-(h/2)}^{h/2} \Delta T z \, \mathrm{d}z \tag{7.9}$$

In the definitions of the thermal force and thermal moment, we have assumed that the elastic modulus E and coefficient of thermal expansion α are constant.

By combining Eqs. (7.6) and (7.4), we can express the plate stresses in terms of the stress resultants:

$$\sigma_x = \frac{N_x + N_T}{h} + \frac{12(M_x + M_T)}{h^3}z - \frac{E\alpha}{1-v}\Delta T \tag{7.10a}$$

$$\sigma_y = \frac{N_y + N_T}{h} + \frac{12(M_y + M_T)}{h^3}z - \frac{E\alpha}{1-v}\Delta T \tag{7.10b}$$

$$\sigma_{xy} = \frac{N_{xy}}{h} - \frac{12M_{xy}}{h^3}z \tag{7.10c}$$

Equations (7.10a) and (7.10b) are similar to the equation for the normal stress in a beam, Eq. (6.14). They express the plate normal stresses as the sum of three terms. The first term on the right-hand side is a uniform stress determined by the membrane and thermal forces. The second term is a bending stress determined by the bending and thermal moments; it varies linearly through the plate thickness. The third term is proportional to ΔT, and so it is a thermal stress that varies with the temperature distribution.

In plate analysis, we cannot determine the internal stress resultants from equilibrium of plate sections. Instead, the displacements u_0, v_0, and w must be determined first. Then, from differentiation, we obtain the stress resultants using Eqs. (7.6). Finally, we obtain the plate stress from Eqs. (7.10).

7.2.4 Equilibrium Equations

To develop the equilibrium equations, we consider the plate elements shown in Fig. 7.4. For small-deflection plate theory, we write the equilibrium equations for the undeformed configuration. Consider first the membrane forces in Fig. 7.4a; we write $\Sigma F_x = 0$ and $\Sigma F_y = 0$ to obtain

$$\frac{\partial N_x}{\partial x} + \frac{\partial N_{xy}}{\partial y} = 0 \tag{7.11a}$$

$$\frac{\partial N_{xy}}{\partial x} + \frac{\partial N_y}{\partial y} = 0 \tag{7.11b}$$

In these equations, we have neglected in-plane body forces. Next, we consider the moments and shear forces in Fig. 7.4b; we write $\Sigma F_z = 0$, $\Sigma M_x = 0$, and $\Sigma M_y = 0$ to obtain

$$\frac{\partial Q_x}{\partial x} + \frac{\partial Q_y}{\partial y} + q = 0 \tag{7.12a}$$

$$\frac{\partial M_{xy}}{\partial x} - \frac{\partial M_y}{\partial y} + Q_y = 0 \tag{7.12b}$$

$$\frac{\partial M_{xy}}{\partial y} - \frac{\partial M_x}{\partial x} + Q_x = 0 \tag{7.12c}$$

In Eq. (7.12a), we have included the transverse loading per unit area, q. If we wish to convert the equation into an equation of motion, we may use this term to introduce an inertia force. Otherwise, body forces and moments are neglected.

7.2.5 Governing Differential Equations

For our small-deflection formulation, the in-plane membrane behavior is independent of the bending response. Thus, we may formulate two independent boundary/initial-value problems.

Membrane problem. The membrane problem requires us to determine the membrane stress resultants N_x, N_y, and N_{xy} so that the equilibrium equations (7.11) are satisfied. This problem, in fact, is identical to the plane stress problem of thermoelasticity discussed in Sec. 5.2.4. The approach there is to introduce a stress function F and reduce the problem to a nonhomogeneous biharmonic equation with the stress function as the single unknown. We may refer to Sec. 5.2.4 and follow exactly the same approach here. First, we introduce a stress function so that the membrane forces are defined by

$$N_x = \frac{\partial^2 F}{\partial y^2} \qquad N_{xy} = -\frac{\partial^2 F}{\partial x \partial y} \qquad N_y = \frac{\partial^2 F}{\partial x^2} \tag{7.13}$$

With these definitions, the equilibrium equations (7.11) are satisfied identically. Because we use a stress-function approach, the strain-compatibility equation of elasticity must also be satisfied. The details of the remaining steps are given in Chapter 5. Using our current notation, we obtain

$$\nabla^4 F = -(1 - \upsilon)\nabla^2 N_T \tag{7.14a}$$

where

$$\nabla^4 F = \nabla^2(\nabla^2 F) = \frac{\partial^4 F}{\partial x^4} + 2\frac{\partial^4 F}{\partial x^2 \partial y^2} + \frac{\partial^4 F}{\partial y^4} \tag{7.14b}$$

and the thermal force is defined in Eq. (7.8).

For a given temperature distribution, the solution procedure is first to solve Eq. (7.14) for the stress function F and then, by differentiation, to determine the membrane forces using Eqs. (7.13). Finally, we obtain the membrane displacements u_0 and v_0 by integration. To do this, we may use Eqs. (7.6a–7.6c), where the left-hand sides, the membrane forces, are known; hence, we may solve for the displacements on the right-hand side. Although we have three equations and two unknowns, we are assured that u_0 and v_0 are unique because we have satisfied the strain-compatibility equation.

Bending problem. For the bending problem, we use a displacement formulation and develop a single partial differential equation with the displacement w as the unknown. As the first step, we solve Eqs. (7.12c) and (7.12b) for Q_x and Q_y, respectively, and then substitute the results into Eq. (7.12a). Thus, we have

$$\frac{\partial^2 M_x}{\partial x^2} - 2\frac{\partial^2 M_{xy}}{\partial x \partial y} + \frac{\partial^2 M_y}{\partial y^2} + q = 0$$

Next, we use the definitions of the moments given in Eqs. (7.6d–7.6f) to introduce the displacement w. Consequently, the governing equation is found to be

$$D\left(\frac{\partial^4 w}{\partial x^4} + 2\frac{\partial^4 w}{\partial x^2 \partial y^2} + \frac{\partial^4 w}{\partial y^4}\right) = q - \nabla^2 M_T \tag{7.15a}$$

or

$$D\nabla^4 w = q - \nabla^2 M_T \tag{7.15b}$$

where D is the flexural rigidity defined in Eq. (7.7), and M_T is the thermal moment defined in Eq. (7.9).

For a given temperature distribution, we solve the plate bending equation (7.15) for the displacement w. Then, by differentiation, we evaluate the bending and twisting moments. The stresses in the plate are calculated by superimposing the membrane and bending contributions using Eqs. (7.10).

7.2.6 Boundary Conditions

To obtain solutions to the membrane and bending problems, we need four boundary conditions on each plate edge. We may separate the boundary conditions into cases of membrane and bending.

Membrane. Typically, on an edge, we must specify either the displacement or the corresponding surface traction. Thus, on the edges $x = 0, a$, we specify

$$\begin{aligned} u_0(0,t) &= u_1(t) \quad \text{or} \quad N_x(0,t) = F_1(t) \\ v_0(0,t) &= v_1(t) \quad \text{or} \quad N_{xy}(0,t) = G_1(t) \\ u_0(a,t) &= u_2(t) \quad \text{or} \quad N_x(a,t) = F_2(t) \\ v_0(a,t) &= v_2(t) \quad \text{or} \quad N_{xy}(a,t) = G_2(t) \end{aligned} \tag{7.16}$$

and, on the edges $y = 0, b$, we specify a similar set of conditions with the normal membrane force N_y and shear force N_{xy}.

Bending. The boundary conditions for plate bending are slightly more complicated than those encountered in the membrane problem, but the boundary conditions are described thoroughly in the literature, including the discussions by Boley and Weiner[1] and Tauchert.[6] A discussion of plate boundary conditions for classical isothermal problems appears in the book by Timoshenko and Woinowsky-Krieger.[9] The unusual aspects of plate boundary conditions concern the edge shear forces and concentrated forces that may occur at the corners of rectangular plates. The most consistent approach for developing plate boundary conditions is to use the principle of minimum potential energy.[6] On a typical edge, the shear force and twisting moment are combined to form an equivalent shear force called the Kirchhoff force. On a typical edge, we must specify the displacement or the Kirchhoff shear force *and* the slope or the bending moment normal to the edge. Thus, on the edges $x = 0, a$, we specify

$$\begin{aligned} w(0,t) &= w_1(t) \quad \text{or} \quad V_x(0,t) = V_1(t) \\ \frac{\partial w}{\partial x}(0,t) &= \theta_1(t) \quad \text{or} \quad M_x(0,t) = M_1(t) \\ w(a,t) &= w_2(t) \quad \text{or} \quad V_x(a,t) = V_2(t) \\ \frac{\partial w}{\partial x}(a,t) &= \theta_2(t) \quad \text{or} \quad M_x(a,t) = M_2(t) \end{aligned} \tag{7.17}$$

In the preceding equations, the Kirchhoff shear force V_x has been introduced; it is defined by

$$V_x = Q_x - \frac{\partial M_{xy}}{\partial y} \tag{7.18}$$

Similar boundary conditions are specified on the edges $y = 0, b$, with the Kirchhoff shear force V_y and the normal moment M_y.

The practical boundary conditions encountered most frequently are the cases of clamped, simply supported, and free edges. Let n denote the direction normal to a plate edge, and s the direction tangential to an edge. Then the boundary conditions on a typical edge are
clamped:

$$w = 0 \quad \text{and} \quad \frac{\partial w}{\partial n} = 0 \tag{7.19a}$$

simply supported:

$$w = 0 \quad \text{and} \quad M_n = 0 \tag{7.19b}$$

free:

$$M_n = 0 \quad \text{and} \quad V_n = 0 \tag{7.19c}$$

where

$$V_n = Q_n - \frac{\partial M_{ns}}{\partial s} \tag{7.19d}$$

Note that, if we prescribe a moment on the boundary, temperature is included in the boundary condition through the thermal moment; see Eq. (7.6d) or (7.6e).

In this section, we present the basic equations for the classical formulation of thin, homogeneous, isotropic plates with small displacements and rotations. The membrane and bending problems are uncoupled and may be solved separately. The resulting normal stresses may then be superimposed to produce the total stresses. In Sec. 5.5, we illustrated the solution of the membrane problem in our discussion of plane stress problems. Later in this chapter, in Sec. 7.4, we present classical solutions for small-deflection plate bending problems.

7.3 Large-Deflection Formulation

In the preceding section, we developed the classical linear formulation for small deflections for thin plates. The theory predicts that the in-plane, i.e., membrane, response is uncoupled from bending. Thus, in-plane forces, either tension or compression, do not alter transverse bending. Furthermore, transverse mechanical forces do not induce in-plane deformations. Similar behavior occurs for the corresponding thermal loads. The theory may be expected to represent accurately the response of plates only when transverse displacements are small compared to the plate thickness.

Once transverse displacements become of the same order of magnitude as the plate thickness, the in-plane membrane response becomes coupled to the transverse bending. As the plate bends, the middle surface stretches and significant membrane forces develop. The load–transverse deflection response becomes nonlinear. As the transverse load increases, the plate's bending stiffness increases. If we assume linear behavior, we will overestimate the transverse bending displacements by a large margin.

The large-deflection plate theory we seek was first presented by Theodore von Kármán in 1910, and the equations are known as the von Kármán plate equations. A classical development of the isothermal form of the von Kármán equations appears in the text by Timoshenko and Woinowsky-Krieger.[9] More modern development of the equations is given by Langhaar[10] and Fung.[11] Langhaar derives the equations, including temperature effects, using the principle of minimum potential energy. Shames and Dym,[12] in a more recent book, use the energy approach to develop the isothermal equations. We will develop the thermoelastic von Kármán plate equations using the equilibrium of forces approach that we used for small deflections.

The preceding development of small-deflection plate theory is based on 1) using the linear strain-displacement equations and 2) writing the equilibrium equations for an undeformed small element of the plate. To develop the von Kármán plate equations, we introduce nonlinear terms in the strain-displacement equations and we write the equilibrium equations for a deformed element of the plate. Otherwise, the basic assumptions of classical plate theory apply.

7.3.1 Strains

For large deflections, we use the nonlinear strain-displacement equations from continuum mechanics. From Eq. (3.8), we have

$$\varepsilon_x = \frac{\partial u}{\partial x} + \frac{1}{2}\left[\left(\frac{\partial u}{\partial x}\right)^2 + \left(\frac{\partial v}{\partial x}\right)^2 + \left(\frac{\partial w}{\partial x}\right)^2\right]$$

$$\varepsilon_y = \frac{\partial v}{\partial y} + \frac{1}{2}\left[\left(\frac{\partial u}{\partial y}\right)^2 + \left(\frac{\partial v}{\partial y}\right)^2 + \left(\frac{\partial w}{\partial y}\right)^2\right]$$

$$\varepsilon_{xy} = \frac{1}{2}\left[\frac{\partial u}{\partial y} + \frac{\partial v}{\partial x} + \frac{\partial u}{\partial x}\frac{\partial u}{\partial y} + \frac{\partial v}{\partial x}\frac{\partial v}{\partial y} + \frac{\partial w}{\partial x}\frac{\partial w}{\partial y}\right]$$

One of von Kármán's contributions was to suggest that the quadratic terms in $\partial w/\partial x$ and $\partial w/\partial y$ be retained in the strain components but that the other quadratic terms be neglected. With this assumption,

$$\varepsilon_x = \frac{\partial u}{\partial x} + \frac{1}{2}\left(\frac{\partial w}{\partial x}\right)^2 \tag{7.20a}$$

$$\varepsilon_y = \frac{\partial v}{\partial y} + \frac{1}{2}\left(\frac{\partial w}{\partial y}\right)^2 \tag{7.20b}$$

$$\varepsilon_{xy} = \frac{1}{2}\left[\frac{\partial u}{\partial y} + \frac{\partial v}{\partial x} + \frac{\partial w}{\partial x}\frac{\partial w}{\partial y}\right] \tag{7.20c}$$

Substituting the displacements from Eqs. (7.1a) and (7.1b), we obtain

$$\varepsilon_x = \frac{\partial u_0}{\partial x} + \underline{\frac{1}{2}\left(\frac{\partial w}{\partial x}\right)^2} - z\frac{\partial^2 w}{\partial x^2} \tag{7.21a}$$

$$\varepsilon_y = \frac{\partial v_0}{\partial y} + \underline{\frac{1}{2}\left(\frac{\partial w}{\partial y}\right)^2} - z\frac{\partial^2 w}{\partial y^2} \tag{7.21b}$$

$$\varepsilon_{xy} = \frac{1}{2}\left[\frac{\partial u_0}{\partial y} + \frac{\partial v_0}{\partial x} + \underline{\frac{\partial w}{\partial x}\frac{\partial w}{\partial y}}\right] - z\frac{\partial^2 w}{\partial x \partial y} \tag{7.21c}$$

In Eqs. (7.21), the underlined nonlinear terms are uniform through the plate thickness; hence, they represent contributions to the membrane strain.

7.3.2 Stresses

Substituting the strain components from Eqs. (7.21) into Hooke's law for plane stress, Eq. (5.13), we obtain

$$\sigma_x = \frac{E}{1-\upsilon^2}\left[\frac{\partial u_0}{\partial x} + \upsilon\frac{\partial v_0}{\partial y} + \underline{\frac{1}{2}\left(\frac{\partial w}{\partial x}\right)^2 + \frac{\upsilon}{2}\left(\frac{\partial w}{\partial y}\right)^2}\right]$$
$$- \frac{E}{1-\upsilon^2}\left(\frac{\partial^2 w}{\partial x^2} + \upsilon\frac{\partial^2 w}{\partial y^2}\right)z - \frac{E\alpha}{1-\upsilon}\Delta T \tag{7.22a}$$

$$\sigma_y = \frac{E}{1-\upsilon^2}\left[\frac{\partial v_0}{\partial y} + \upsilon\frac{\partial u_0}{\partial x} + \underline{\frac{1}{2}\left(\frac{\partial w}{\partial y}\right)^2 + \frac{\upsilon}{2}\left(\frac{\partial w}{\partial x}\right)^2}\right]$$
$$- \frac{E}{1-\upsilon^2}\left(\frac{\partial^2 w}{\partial y^2} + \upsilon\frac{\partial^2 w}{\partial x^2}\right)z - \frac{E\alpha}{1-\upsilon}\Delta T \tag{7.22b}$$

$$\sigma_{xy} = \frac{E}{2(1+\upsilon)}\left(\frac{\partial u_0}{\partial y} + \frac{\partial v_0}{\partial x} + \underline{\frac{\partial w}{\partial x}\frac{\partial w}{\partial y}}\right) - \frac{E}{1+\upsilon}\frac{\partial^2 w}{\partial x \partial y}z \tag{7.22c}$$

From the preceding equations, we see that the underlined nonlinear terms also contribute to the membrane stress.

7.3.3 Stress Resultants

Following the definitions of the stress resultants given in Eqs. (7.5), we substitute Eqs. (7.22) and perform the integrations to obtain new equations for the membrane forces:

$$N_x = \frac{Eh}{1-\upsilon^2}\left[\frac{\partial u_0}{\partial x} + \upsilon\frac{\partial v_0}{\partial y} + \underline{\frac{1}{2}\left(\frac{\partial w}{\partial x}\right)^2 + \frac{\upsilon}{2}\left(\frac{\partial w}{\partial x}\right)^2}\right] - N_T \tag{7.23a}$$

$$N_y = \frac{Eh}{1-\upsilon^2}\left[\frac{\partial v_0}{\partial y} + \upsilon\frac{\partial u_0}{\partial x} + \underline{\frac{1}{2}\left(\frac{\partial w}{\partial y}\right)^2 + \frac{\upsilon}{2}\left(\frac{\partial w}{\partial x}\right)^2}\right] - N_T \tag{7.23b}$$

$$N_{xy} = N_{yx} = \frac{Eh}{(1+\upsilon)}\left[\frac{\partial u_0}{\partial y} + \frac{\partial v_0}{\partial x} + \underline{\frac{\partial w}{\partial x}\frac{\partial w}{\partial y}}\right] \tag{7.23c}$$

The bending and twisting moments are the same as those in Eqs. (7.6d–7.6f).

7.3.4 Equilibrium Equations

We write the equilibrium equations for the deformed element of the plate shown in Fig. 7.5. For clarity, only the membrane forces are shown, but the shear forces, the bending and twisting moments, and the transverse surface load exist as well (see Fig. 7.4b). Figure 7.5 shows that the edges of the element make small angles θ_x and θ_y with the x–y plane and that these angles vary across the element. The membrane forces are thus inclined to the x–y plane. When we sum forces in the x and y directions, we multiply the membrane forces by the cosines of these angles. In the limit, as we shrink the element to zero, the cosines approach 1; hence, the x and y equilibrium equations are the same as those derived in Eqs. (7.11). However, when we sum forces in the z direction, we multiply the membrane forces by the sines of these angles; consequently, the membrane forces introduce additional terms into the z equilibrium equation. From $\Sigma F_z = 0$, we obtain the new equation,

$$\frac{\partial}{\partial x}\left(Q_x + \underline{N_x\frac{\partial w}{\partial x} + N_{xy}\frac{\partial w}{\partial y}}\right) + \frac{\partial}{\partial y}\left(Q_y + \underline{N_y\frac{\partial w}{\partial y} + N_{xy}\frac{\partial w}{\partial x}}\right) + q = 0 \tag{7.24}$$

The underlined terms are the contributions introduced by considering the plate element in the deformed configuration. The new terms are nonlinear because the membrane forces are not constant; the membrane forces change as the plate deforms and the middle surface stretches.

The equilibrium equations given in Eqs. (7.12b) and (7.12c) resulting from $\Sigma M_x = 0$ and $\Sigma M_y = 0$ are unchanged.

7.3.5 Governing Differential Equations

Unlike the small-deflection formulation, the membrane and bending problems are now fully coupled. Thus, we can no longer formulate two independent problems; rather, we must formulate two simultaneous partial differential equations.

Membrane equation. The membrane problem consists of determining in-plane forces so that the equilibrium equations given by Eqs. (7.11) are satisfied. As before, we can do this by using a stress function F. Thus, if we define the in-plane forces by Eqs. (7.13), the equilibrium equations are satisfied identically. The remaining steps involve deriving the partial differential equation that F satisfies. Since we use a stress function, we must satisfy a compatibility equation so that the

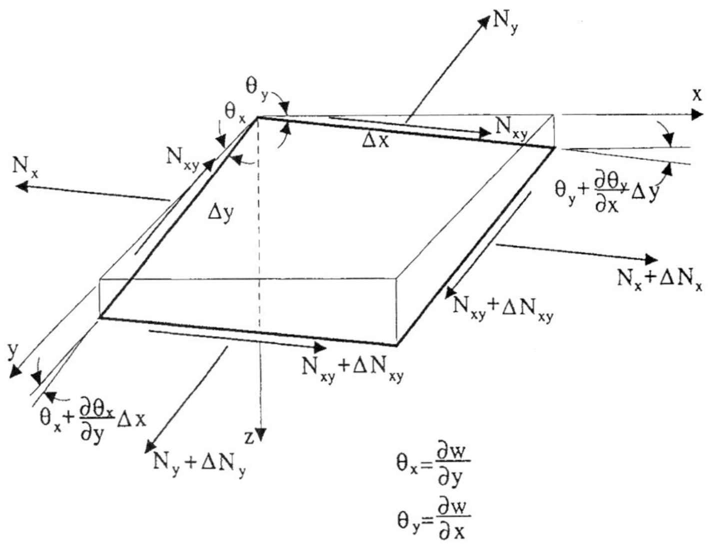

Fig. 7.5 Membrane stress resultants on a deformed plate element.

displacements will be unique. We return to the compatibility equation (5.18) for the plane stress problem in Chapter 5:

$$\frac{\partial^2 \varepsilon_x}{\partial y^2} + \frac{\partial^2 \varepsilon_y}{\partial x^2} = 2\frac{\partial^2 \varepsilon_{xy}}{\partial x \partial y}$$

We recall that this equation holds for the linear strain-displacement equations, but we want a compatibility equation that applies for the nonlinear membrane strains that occur on the plate's middle surface. Let us denote these components as ε_x^0, ε_y^0, and ε_{xy}^0. From Eqs. (7.21), we can express these middle surface strains as

$$\varepsilon_x^0 = \frac{\partial u_0}{\partial x} + \frac{1}{2}\left(\frac{\partial w}{\partial x}\right)^2 \tag{7.25a}$$

$$\varepsilon_y^0 = \frac{\partial v_0}{\partial y} + \frac{1}{2}\left(\frac{\partial w}{\partial y}\right)^2 \tag{7.25b}$$

$$\varepsilon_{xy}^0 = \frac{1}{2}\left(\frac{\partial u_0}{\partial y} + \frac{\partial v_0}{\partial x} + \frac{\partial w}{\partial x}\frac{\partial w}{\partial y}\right) \tag{7.25c}$$

If we substitute these components into Eq. (5.18), we obtain the desired compatibility equation,

$$\frac{\partial^2 \varepsilon_x^0}{\partial x^2} + \frac{\partial^2 \varepsilon_y^0}{\partial x^2} - 2\frac{\partial^2 \varepsilon_{xy}^0}{\partial x \partial y} = \left(\frac{\partial^2 w}{\partial x \partial y}\right)^2 - \frac{\partial^2 w}{\partial x^2}\frac{\partial^2 w}{\partial y^2} \tag{7.26}$$

For small deflections, the nonlinear terms on the right-hand side are neglected, and Eq. (7.26) reduces to Eq. (5.18).

If we introduce Eqs. (7.25) into Eqs. (7.23), we find the relationship between the middle surface forces and middle surface strains,

$$N_x = \frac{Eh}{1-\upsilon^2}\left(\varepsilon_x^0 + \upsilon\varepsilon_y^0\right) - N_T \tag{7.27a}$$

$$N_y = \frac{Eh}{1-\upsilon^2}\left(\varepsilon_y^0 + \upsilon\varepsilon_x^0\right) - N_T \tag{7.27b}$$

$$N_{xy} = N_{yx} = \frac{Eh}{1+\upsilon}\varepsilon_{xy}^0 \tag{7.27c}$$

Introducing the definitions of the stress function from Eq. (7.13) on the left-hand side, we solve for the strain components,

$$\varepsilon_x^0 = \frac{1}{Eh}\left(\frac{\partial^2 F}{\partial y^2} - \upsilon\frac{\partial^2 F}{\partial x^2}\right) + \frac{(1-\upsilon)}{Eh}N_T \tag{7.28a}$$

$$\varepsilon_y^0 = \frac{1}{Eh}\left(\frac{\partial^2 F}{\partial x^2} - \upsilon\frac{\partial^2 F}{\partial y^2}\right) + \frac{(1-\upsilon)}{Eh}N_T \tag{7.28b}$$

$$\varepsilon_{xy}^0 = -\frac{1+\upsilon}{Eh}\frac{\partial^2 F}{\partial x \partial y} \tag{7.28c}$$

Eliminating the strain components between the compatibility equation (7.26) and Eqs. (7.28), we obtain

$$\nabla^4 F + (1-\upsilon)\nabla^2 N_T + Eh\left[\frac{\partial^2 w}{\partial x^2}\frac{\partial^2 w}{\partial y^2} - \left(\frac{\partial^2 w}{\partial x \partial y}\right)^2\right] = 0 \tag{7.29}$$

Equation (7.29) is the partial differential equation that describes the membrane behavior. For small deflections, we neglect the nonlinear terms, and it reduces the linear equation (7.14a). Equation (7.29) is the first of the two equations that make up von Kármán's plate theory.

Bending equation. For the bending equation, we return to the $\Sigma F_z = 0$ equilibrium equation (7.24). Performing the indicated differentiations, we find

$$\frac{\partial Q_x}{\partial x} + \frac{\partial Q_y}{\partial y} + N_x\frac{\partial^2 w}{\partial x^2} + 2N_{xy}\frac{\partial^2 w}{\partial x \partial y} + N_y\frac{\partial^2 w}{\partial y^2} + q$$
$$+\left(\frac{\partial N_x}{\partial x} + \frac{\partial N_{xy}}{\partial y}\right)\frac{\partial w}{\partial x} + \left(\frac{\partial N_{xy}}{\partial x} + \frac{\partial N_y}{\partial y}\right)\frac{\partial w}{\partial y} = 0$$

First, we note that the terms in parentheses are zero because of the in-plane equilibrium equations (7.11). Then, if we eliminate the shear forces via Eqs. (7.12b) and (7.12c), we obtain

$$\frac{\partial^2 M_x}{\partial x^2} - 2\frac{\partial^2 M_{xy}}{\partial x \partial y} + \frac{\partial^2 M_y}{\partial y^2} + \underline{N_x\frac{\partial^2 w}{\partial x^2} + 2N_{xy}\frac{\partial^2 w}{\partial x \partial y} + N_y\frac{\partial^2 w}{\partial y^2}} + q = 0 \tag{7.30}$$

To obtain the second von Kármán equation, we introduce the definitions of the moments from Eqs. (7.6d–7.6f) and the membrane forces from Eqs. (7.13). Then we obtain

$$D\left(\frac{\partial^4 w}{\partial x^4} + 2\frac{\partial^4 w}{\partial x^2 \partial y^2} + \frac{\partial^4 w}{\partial y^4}\right)$$
$$= \underline{\frac{\partial^2 F}{\partial y^2}\frac{\partial^2 w}{\partial x^2} - 2\frac{\partial^2 F}{\partial x \partial y}\frac{\partial^2 w}{\partial x \partial y} + \frac{\partial^2 F}{\partial x^2}\frac{\partial^2 w}{\partial y^2}} + q - \nabla^2 M_T$$

or

$$D\nabla^4 w = \underline{\frac{\partial^2 F}{\partial y^2}\frac{\partial^2 w}{\partial x^2} - 2\frac{\partial^2 F}{\partial x \partial y}\frac{\partial^2 w}{\partial x \partial y} + \frac{\partial^2 F}{\partial x^2}\frac{\partial^2 w}{\partial y^2}} + q - \nabla^2 M_T \tag{7.31}$$

Equations (7.29) and (7.31) are the von Kármán equations for large deflections of plates. Rewriting the equations slightly, we have

$$\nabla^4 w = \frac{1}{D}\left[q + \frac{\partial^2 F}{\partial y^2}\frac{\partial^2 w}{\partial x^2} - 2\frac{\partial^2 F}{\partial x \partial y}\frac{\partial^2 w}{\partial x \partial y} + \frac{\partial^2 F}{\partial x^2}\frac{\partial^2 w}{\partial y^2} - \nabla^2 M_T\right] \tag{7.32a}$$

$$\nabla^4 F = Eh\left[\left(\frac{\partial^2 w}{\partial x \partial y}\right)^2 - \frac{\partial^2 w}{\partial x^2}\frac{\partial^2 w}{\partial y^2}\right] - (1-\upsilon)\nabla^2 N_T \tag{7.32b}$$

The most important feature of the von Kármán plate equations is that the equations are coupled and nonlinear. Fung[11] points out that the quantity

$$\frac{\partial^2 w}{\partial x^2}\frac{\partial^2 w}{\partial y^2} - \left(\frac{\partial^2 w}{\partial x \partial y}\right)^2 \tag{7.33}$$

is the Gaussian curvature of the plate. The Gaussian curvature given by Eq. (7.33) is a mathematical quantity that is invariant when we rotate the x–y axes through an arbitrary angle. The Gaussian curvature is zero if a surface is "developable." A developable surface may be deformed without extension. If a plate deforms into a developable surface, the right-hand side of Eq. (7.32b) is independent of w. Then the midplane forces arise because of forces on the boundary. On the other hand, if the middle surface deforms into a nondevelopable surface, the Gaussian curvature on the right-hand side of Eq. (7.32b) is not zero. Then the middle surface stretches as the plate deforms, and membrane forces depend on w.

7.3.6 Boundary Conditions

The boundary conditions for the von Kármán equations are generally the same as those we discussed in Sec. 7.2.6 for classical small-deflection plate theory. The one difference pertains to the definition of the Kirchhoff shear force. When we wrote the $\Sigma F_z = 0$ for the deformed plate element, we introduced vertical components into the equilibrium equation (7.24). A derivation[12] of the equilibrium equations and boundary conditions by the principle of minimum potential energy shows that, for large deflections, the Kirchhoff shear forces must be modified to include appropriate components of the membrane forces. Tauchert[6] also discusses the Kirchhoff shear boundary condition for the von Kármán equations. Thus, on a typical edge,

$$V_n = Q_n - \frac{\partial M_{ns}}{\partial s} + N_n \frac{\partial w}{\partial n} + N_{ns} \frac{\partial w}{\partial s} \tag{7.34}$$

where n denotes the direction normal to the edge, and s denotes the direction tangent to the edge.

In this section, we present the basic equations for the von Karman formulation for thin, homogeneous isotropic plates with comparatively large deflections and rotations. The membrane and bending equations are coupled and must be solved simultaneously. The formulation is used widely for the study of large-deflection and plate buckling problems.

Chuen-Yuan Chia[13] describes nonlinear analysis of isothermal plates, presenting analytical solutions to the von Kármán equations for a variety of plates with mechanical loads. There are, in comparison, relatively few analytical solutions to the von Kármán equations with temperature distributions. We describe one solution in Example 7.5.

7.4 Plate Problems

Given that a large number of textbooks, Ref. 9 for example, present solutions for isothermal plate problems, we will concentrate here on solutions for deformations and stresses due to temperature distributions. Some solutions for thermal stress problems in plates can be found in the books cited earlier[1–6]; here we describe a few quasistatic solutions for rectangular plates.

Example 7.1: Simply Supported Plate with *T(x, y, z)*

We consider a rectangular plate (Fig. 7.6) with all edges simply supported. The plate experiences an arbitrary temperature distribution $T(x, y, z)$ and the transverse loading $q = 0$. We assume small deflections; hence, the problem is linear, and solutions due to temperature and the transverse loading may be developed separately and superimposed. Our objective is to determine the bending deflection $w(x, y)$; once w is known, bending moments and stresses may be derived directly. The boundary-value problem consists of solving Eq. (7.15b),

$$D\nabla^4 w + \nabla^2 M_T = 0$$

subject to two boundary conditions on each of the four edges of the plate. In Eq. (7.15b), M_T is the thermal moment defined as

$$M_T = \frac{E\alpha}{1-\upsilon}\int_{-(h/2)}^{h/2} \Delta T z \, \mathrm{d}z$$

where $\Delta T = T(x, y, z) - T_{\text{ref}}$. For an arbitrary temperature distribution, the thermal moment varies over the plate, i.e., $M_T = M_T(x, y)$. The boundary conditions are

$$w(0, y) = 0 \qquad w(x, 0) = 0$$
$$w(a, y) = 0 \qquad w(x, b) = 0$$

and

$$M_x(0, y) = 0 \qquad M_y(x, 0) = 0$$
$$M_x(a, y) = 0 \qquad M_y(x, b) = 0$$

Let us consider the moment boundary conditions in more detail. On the edges parallel to the y axes,

$$M_x = -D\left(\frac{\partial^2 w}{\partial x^2} + \upsilon\frac{\partial^2 w}{\partial y^2}\right) - M_T$$

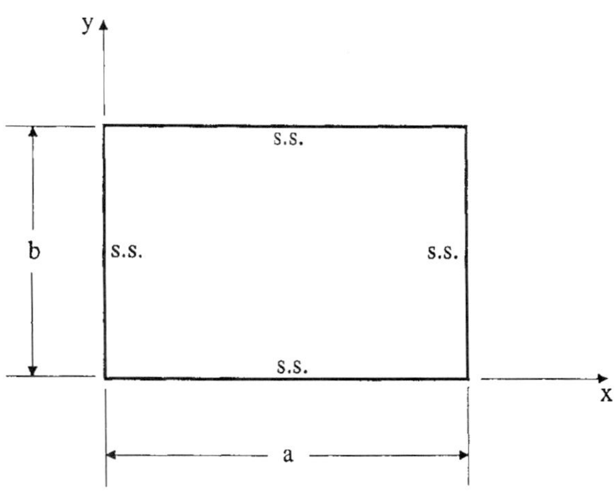

Fig. 7.6 Simply supported rectangular plate.

From the displacement boundary conditions on these edges, we have $\partial^2 w/\partial y^2 = 0$, assuming, as we do, that the edge is perfectly straight. Thus, the moment boundary condition reduces to

$$M_x = -D\frac{\partial^2 w}{\partial x^2} - M_T = 0 \qquad \text{on } x = 0 \text{ and } x = a$$

Similarly,

$$M_y = -D\frac{\partial^2 w}{\partial y^2} - M_T = 0 \qquad \text{on } y = 0 \text{ and } y = b$$

Alternatively, we may say that the moment boundary condition on a typical edge is given by

$$D\nabla^2 w + M_T = 0$$

where we recognize that one term in ∇^2 is always zero.

An important point that we should observe immediately is that the moment boundary conditions are nonhomogeneous. The nonhomogeneous boundary conditions are a signal of mathematical difficulty because separation of variables cannot be used unless the boundary conditions are homogeneous. This difficulty is often resolved by forming the solution as the sum of two or more parts such that a new nonhomogeneous partial differential equation with homogeneous boundary conditions is introduced. Then the new problem is solved by separation of variables. Parkus,[4] Tauchert,[6] and Vinson[14] use this approach to solve rectangular-plate problems with nonhomogeneous thermal-moment boundary conditions.

We use an approach[1,9] that reduces the plate bending problem to that of the deflection of a membrane. The fourth-order plate equation is replaced by two second-order equations. We begin by writing Eq. (7.15b) as

$$\nabla^2\left(D\nabla^2 w + M_T\right) = 0$$

which is equivalent to two equations:

$$D\nabla^2 w + M_T = f(x, y) \qquad \nabla^2 f = 0$$

where $f(x, y)$ is a function to be determined. Each of these equations is second-order and requires only one boundary condition on each plate edge. On a plate edge, the moment boundary condition shows that the left-hand side of the first of the two preceding equations is zero. Thus, on the plate edges, the function f is zero. Now we realize from the second of the two preceding equations that we have a homogeneous partial differential equation with homogeneous boundary conditions; therefore, the solution is $f(x, y) = 0$. Thus, the problem reduces to the single second order nonhomogeneous equation,

$$D^2 w = -M_T(x, y)$$

with the homogeneous boundary condition $w = 0$ on each edge. This problem is easily solved by separation of variables using a double Fourier sine series,

$$w(x, y) = \sum_{m=1}^{\infty}\sum_{n=1}^{\infty} w_{mn} \sin\frac{m\pi x}{a} \sin\frac{n\pi y}{b}$$

where w_{mn} are Fourier coefficients to be determined. The Fourier series automatically satisfies the displacement boundary conditions. To determine w_{mn}, we expand the thermal moment in a similar series,

$$M_T(x, y) = \sum_{m=1}^{\infty}\sum_{n=1}^{\infty} a_{mn} \sin\frac{m\pi x}{a} \sin\frac{n\pi y}{b}$$

where

$$a_{mn} = \frac{4}{ab}\int_0^b\int_0^a M_T(x, y) \sin\frac{m\pi x}{a} \sin\frac{n\pi y}{b}$$

Substituting the infinite series for $W(x, y)$ and $M_T(x, y)$ into the second order partial differential equation, we match coefficients to find

$$w_{mn} = \frac{a_{mn}}{\pi^2 D(m^2/a^2 + n^2/b^2)}$$

and, consequently,

$$w(x, y) = \frac{1}{\pi^2 D}\sum_{m=1}^{\infty}\sum_{n=1}^{\infty} \frac{a_{mn}}{(m^2/a^2 + n^2/b^2)} \sin\frac{m\pi x}{a} \sin\frac{n\pi y}{b}$$

A solution like this one in the form of a double Fourier sine series is called a Navier solution. For a given thermal moment, we evaluate the Fourier coefficients a_{mn} and then sum the infinite series of the preceding equation to obtain $w(x, y)$. The series converge satisfactorily for the deflection w, but we encounter computational difficulty when we seek bending moments. From Eqs. (7.6d) and (7.6e), we see that the moments depend on the second derivatives of the preceding equation. If we compute these derivatives, the series may fail to converge because we will encounter coefficients of the order $a_{mn}m^2/(m^2+n^2)$. Therefore, this solution is not practical for the determination of plate moments, and we must find an alternative solution. We develop such a solution in Example 7.3.

Example 7.2: Simply Supported Plate with *T(z)*

For the special case of a rectangular simply supported plate with a temperature varying only through the plate thickness, we can use the preceding solution to derive a series solution for $w(x, y)$. From Example 7.1, we can evaluate the coefficients a_{mn}. Since $T = T(z)$, the thermal moment is constant, and we have

$$a_{mn} = \frac{4M_T}{ab}\int_0^b \sin\frac{n\pi y}{b}\,dy \int_0^a \sin\frac{m\pi x}{a}\,dx$$

Hence,

$$\begin{aligned} a_{mn} &= \frac{16M_T}{\pi^2}\frac{1}{mn} \qquad m, n \text{ odd} \\ &= 0 \qquad m, n \text{ even} \end{aligned}$$

Thus, the solution is

$$w(x, y) = \frac{16}{\pi^4} \frac{M_T}{D} \sum_{m=1,3,5}^{\infty} \sum_{n=1,3,5}^{\infty} \frac{1}{mn(m^2/a^2 + n^2/b^2)} \sin \frac{m\pi x}{a} \sin \frac{n\pi y}{b}$$

As in Example 7.1, this series solution works well for the deflections, but bending moments derived from it are known to converge poorly.

Example 7.3: Levy Solution for Simply Supported Plate with *T(z)*

The Levy solution addresses the problem of a plate (Fig. 7.6) with simply supported edges by using a different series expansion. We assume that the temperature varies only through the plate thickness, so that the thermal moment is constant. We follow the approach used in Example 7.1, where the plate bending equation is separated into two second-order membrane equations. One difference here is that $\nabla^2 M_T = 0$, given that M_T is constant throughout the plate. Then the plate equation (7.15b) becomes

$$D\nabla^4 w = 0$$

which is equivalent to the two equations,

$$D\nabla^2 w = f \qquad \nabla^2 f = 0$$

As before, the bending-moment boundary condition on a typical edge can be written as

$$D\nabla^2 w = -M_T$$

Thus, in this case, we can say that on the plate edges, $f = -M_T$. Then we have the boundary-value problem

$$\nabla^2 f = 0$$

with the boundary condition $f = -M_T$ on all plate edges. A solution to this problem is $f(x, y) = -M_T$. Hence, our original fourth-order problem reduces to the second-order nonhomogeneous equation

$$D\nabla^2 w = -M_T$$

with the boundary condition $w = 0$ on the edges $x = 0$ and $x = a$.

In the Levy approach, we expand the solution in the series

$$w(x, y) = \sum_{m=1}^{\infty} Y_m(y) \sin \frac{m\pi x}{a}$$

which satisfies the boundary conditions $w = 0$ at $x = 0$ and $x = a$. Then we expand the thermal moment as a Fourier sine series,

$$M_T(x, y) = \sum_{m=1}^{\infty} a_m \sin \frac{m\pi x}{a}$$

where

$$a_m = \frac{2}{a}\int_0^a M_T \sin\frac{m\pi x}{a}\,\mathrm{d}x$$

Because M_T is constant, we can evaluate the integral, which yields

$$\begin{aligned} a_m &= \frac{4M_T}{\pi}\frac{1}{m} && m \text{ odd} \\ &= 0 && m \text{ even} \end{aligned}$$

Thus,

$$M_T(x) = \frac{4}{\pi}M_T \sum_{m=1,3,5}^{\infty} \frac{1}{m}\sin\frac{m\pi x}{a}$$

Substituting the two infinite series equations into the $D\nabla^2 w$ equation, we equate coefficients to obtain the ordinary differential equations,

$$\begin{aligned} \frac{\mathrm{d}^2 Y_m}{\mathrm{d}x^2} - \beta_m^2 Y_m &= -\frac{4}{\pi}\frac{M_T}{D}\frac{1}{m} && m \text{ odd} \\ \frac{\mathrm{d}^2 Y_m}{\mathrm{d}x^2} - \beta_m^2 Y_m &= 0 && m \text{ even} \end{aligned}$$

where $\beta_m = m\pi/a$. These equations are solved subject to the boundary conditions $Y_m(0) = Y_m(b) = 0$. A nontrivial solution exists only for m odd, and it is given by

$$Y_m = \frac{4}{\pi}\frac{M_T}{D}\frac{1}{m\beta_m^2}\left[1 - \frac{\sinh\beta_m(b-y) + \sinh\beta_m y}{\sinh\beta_m b}\right] \qquad m \text{ odd}$$

Then the final solution for $w(x, y)$ is obtained by substituting Y_m into the Levy sine series expansion to yield

$$w(x, y) = \frac{4}{\pi}\frac{M_T}{D}\sum_{m=1,3,5}^{\infty}\frac{1}{m\beta_m^2}\left[1 - \frac{\sinh\beta_m(b-y) + \sinh\beta_m y}{\sinh\beta_m b}\right]\sin\frac{m\pi x}{a}$$

From this equation, the bending and twisting moments may be computed using Eqs. (7.6d–7.6f). The stress components can then be evaluated using Eqs. (7.10).

Boley and Weiner[1] note that more convenient equations result if the coordinate axes in Fig. 7.6 are shifted so that the new x axis lies on the horizontal centerline of the plate. Thus, we introduce

$$y' = y - b/2$$

and the displacement w takes the form

$$w(x, y) = \frac{4}{\pi}\frac{M_T}{D}\sum_{m=1,3,5}^{\infty}\frac{1}{m\beta_m^2}\left[1 - \frac{\cosh\beta_m y'}{\cosh(\beta_m b/2)}\right]\sinh\frac{m\pi x}{a}$$

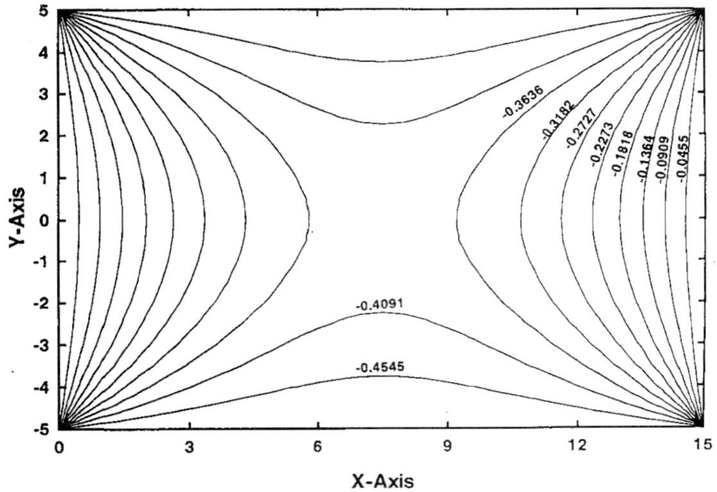

a) $\sigma_x/E\alpha\Delta T_h$

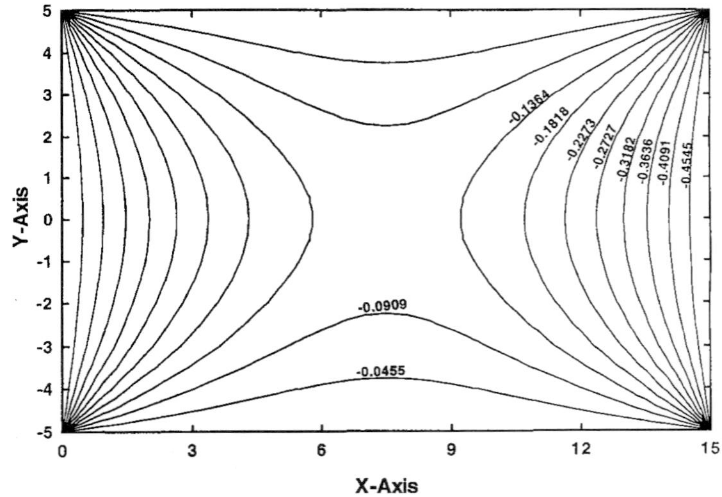

b) $\sigma_y/E\alpha\Delta T_h$

Fig. 7.7 Thermal stresses on the upper surface of a simply supported plate due to a linear temperature gradient through the thickness.

The corresponding expressions for the bending and twisting moments may be derived from Eqs. (7.6d–7.6e) to obtain

$$M_x = -\frac{4}{\pi}(1-\upsilon)M_T \sum_{m=1,3,5}^{\infty} \frac{1}{m}\frac{\cosh(m\pi y'/a)}{\cosh(m\pi b/2a)} \sin\frac{m\pi x}{a}$$

$$M_y = -\frac{4}{\pi}(1-\upsilon)M_T \sum_{m=1,3,5}^{\infty} \frac{1}{m}\left[1-\frac{\cosh(m\pi y'/a)}{\cosh(m\pi b/2a)}\right] \sin\frac{m\pi x}{a}$$

$$M_{xy} = -\frac{4}{\pi}(1-\upsilon)M_T \sum_{m=1,3,5}^{\infty} \frac{1}{m}\frac{\cosh(m\pi y'/a)}{\cosh(m\pi b/2a)} \cos\frac{m\pi x}{a}$$

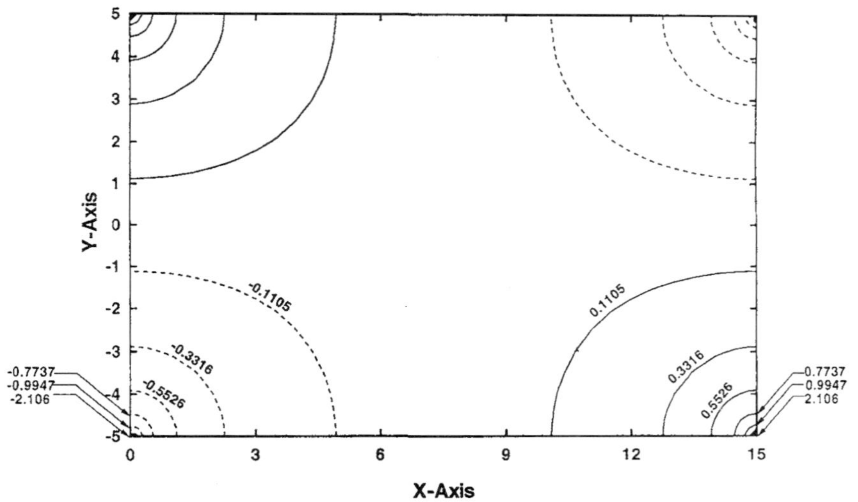

c) $\sigma_{xy}/E\alpha\Delta T_h$

Fig. 7.7 (Cont.) Thermal stresses on the upper surface of a simply supported plate due to a linear temperature gradient through the thickness.

To illustrate the thermal-stress distributions, we assume a linear temperature through the plate thickness, $T(z) = \Delta T_h z/h$, where ΔT_h is the temperature change from $z = -h/2$ to $z = +h/2$. For this temperature distribution, we find, from Eq. (7.8), that the thermal force $N_T = 0$ and, from Eq. (7.9), that the thermal moment is given by

$$M_T = \frac{1}{12}\frac{E\alpha}{1-\upsilon}h^2\Delta T_h$$

Using Eqs. (7.10), we may evaluate the plate stresses. The plate has no in-plane forces, so that $N_x = N_y = N_{xy} = 0$. Substituting for the thermal moment and temperature, we obtain

$$\sigma_x = \frac{12M_x z}{h^3} \qquad \sigma_y = \frac{12M_y z}{h^3} \qquad \sigma_{xy} = -\frac{12M_{xy}z}{h}$$

Using the series representations above for the moments, we may now evaluate the thermal stresses at any point on the plate. For example, the σ_x stress is given by

$$\sigma_x = -\frac{4}{\pi}\frac{z}{h}E\alpha\Delta T_h \sum_{m=1,3,5}^{\infty}\frac{1}{m}\frac{\cosh(m\pi y'/a)}{\cosh(m\pi b/2a)}\sin\frac{m\pi x}{a}$$

This result illustrates that a typical thermal stress 1) is proportional to $E\alpha\Delta T_h$, 2) is independent of Poisson's ratio, 3) varies linearly through the plates thickness, and 4) depends on the aspect ratio a/b.

The results show that the maximum values of σ_x and σ_y occur on the edges of the plate and are independent of the aspect ratio:

$$(\sigma_x)_{\max} = \pm\tfrac{1}{2}E\alpha\Delta T_h \qquad \text{on the edges } y' = \pm b/2$$

$$(\sigma_y)_{\max} = \pm\tfrac{1}{2}E\alpha\Delta T_h \qquad \text{on the edges } x = 0, a$$

Contour plots of nondimensional stresses $\sigma/E\alpha\Delta T_h$ on the upper plate surface are presented in Figs. 7.7a–7.7c for a rectangular plate of aspect ratio $a/b = 1.5$. It can be seen that the contours approach the maximum values given by the preceding equation. The maximum values of the shearing stress occur at the corners of the plate corresponding to concentrated corner reactions. For this aspect ratio, the stresses at the center of the plate are

$$\sigma_x(a/2, 0) = -0.381E\alpha\Delta T_h$$
$$\sigma_y(a/2, 0) = -0.119E\alpha\Delta T_h$$
$$\sigma_y(a/2, 0) = 0$$

Johns[3] presents a graph for σ_x and σ_y as a function of aspect ratio showing that, for a square plate, the stresses at the center of the plate are $\sigma_x = \sigma_y = 0.25E\alpha\Delta T_h$ and that, when $a/b \to \infty$, the stress $\sigma_x = 0.5E\alpha\Delta T_h$ and $\sigma_y = 0$.

Example 7.4: Membrane Stresses in an Unrestrained Plate

In our discussion of von Kármán plate theory, we saw that the membrane stresses have an important role in the nonlinear plate response because, for large deflections, the membrane forces appear in the plate bending equation. As the plate bends, the membrane forces change as the plate middle surface stretches. To gain further appreciation of membrane stresses, we consider an example of a perfectly flat rectangular plate that experiences the "tentlike" temperature distribution shown in Fig. 7.8. The temperature is uniform through the plate thickness. The problem was considered by Heldenfels and Roberts[15] in a 1952 evaluation study of strain gauges for thermal applications. These authors present an approximate analytical solution and a comparison of analytical and experimental results. The experimental results were found to be in good agreement with the approximate analytical solution. We center our discussion on the behavior of the membrane stresses and omit details of the solution procedure.

In Sec. 5.5.1, we present a thermoelasticity solution for a thin rectangular beam with a temperature distribution through the beam depth. The beam is infinitely long, and so the resulting stress distribution is independent of x. The thin rectangular beam solution gives an approximate representation for the σ_x stress in the center of the Heldenfels plate. For the tentlike temperature, the thermoelasticity solution from Chapter 5 predicts the σ_x temperature distribution shown in Fig. 7.8 and yields $\sigma_y = \sigma_{xy} = 0$. The stress is proportional to the temperature difference T_0 between the center of the plate and the plate edge. The stress distribution is self-equilibrating, with compressive stresses in the plate center and tensile stresses on the outer edges.

Heldenfels' solution provides insight into the end effects introduced by a finite-length plate. Heldenfels' approximate analytical solution used the stress-function approach with the principle of minimum complementary energy. The stress distributions in the plate were predicted as

$$\sigma_x = E\alpha T_0\left(\frac{y}{b} - 1\right)\left(C_1 \sinh\frac{\alpha x}{a}\sin\frac{\beta x}{a} + C_2 \cosh\frac{\alpha x}{a}\cos\frac{\beta x}{a} + 1\right)$$

$$\sigma_y = \frac{E\alpha T_0}{12}\left(1 - 3\frac{y^2}{b^2} + 2\frac{y^3}{b^3}\right)\left(C_3 \sinh\frac{\alpha x}{a}\sin\frac{\beta x}{a} + C_4 \cosh\frac{\alpha x}{a}\cos\frac{\beta x}{a}\right)$$

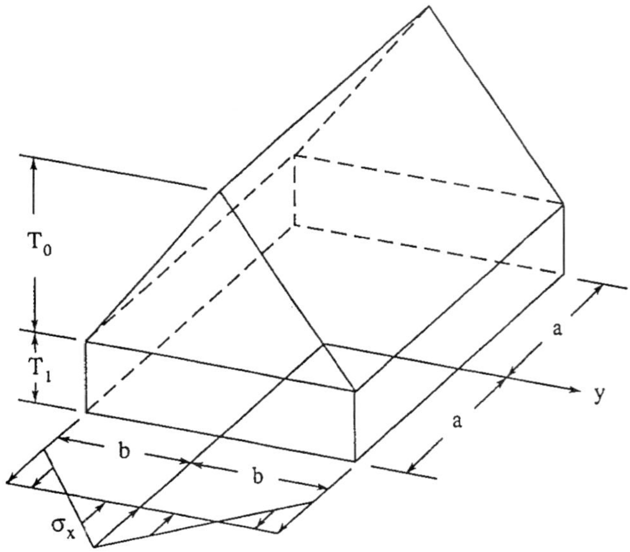

Fig. 7.8 Unrestrained rectangular plate with tentlike temperature distribution.

$$\sigma_{xy} = \frac{E\alpha T_0}{2}\left(1 - \frac{y}{b}\right)\frac{y}{b}\left(C_5 \sinh\frac{\alpha x}{a}\cos\frac{\beta x}{a} + C_6 \cosh\frac{\alpha x}{a}\sin\frac{\beta x}{a}\right)$$

where $C_1, C_2, \ldots, C_6$, as well as α, and β are constants given in Ref. 15. The preceding equations are valid in the first quadrant, where $0 \leq x \leq a$ and $0 \leq y \leq b$. For the finite-length plate, we have a biaxial stress state with all three components of stress, σ_x, σ_y, and σ_{xy}. Moreover, the stresses vary with x and y to satisfy the stress-free boundary conditions of the free edges.

Figure 7.9 shows contours of the normal stresses for an aluminum plate with an aspect ratio of 1.57. The plot of the σ_x contours in Fig. 7.9a shows that there is a large central core of the plate that experiences compressive stress. The σ_x stress distribution along the y axis agrees well with the thermoelasticity solution. In addition, as x approaches a, σ_x goes to zero to meet the free-edge condition. The plot of the σ_y contours in Fig. 7.9b shows that the plate experiences compressive σ_y stresses near the center. Interestingly, the largest σ_y stress occurs at $x = a$, $y = 0$ and is tensile. The σ_{xy} shearing stresses (not shown) are generally much smaller than the normal stresses.

The Heldenfels solution illustrates, for a rectangular plate, how a relatively simple spatial temperature distribution can induce a complex biaxial stress state. An important feature of the stress distribution is the biaxial compressive stresses in the hotter central region of the plate. We discuss the significance of these stresses in the next example.

Example 7.5: Large-Deflection Solution of a Simply Supported Plate

In a companion study to the Heldenfels research, Gossard et al.[16] evaluated the thermal-buckling behavior of the same rectangular plate. We describe details of

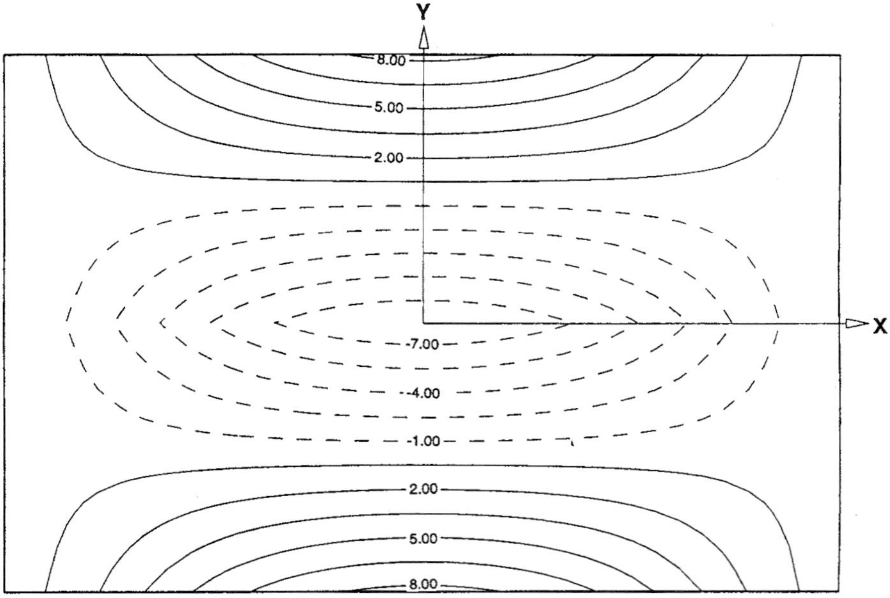

a) Contours of σ_x

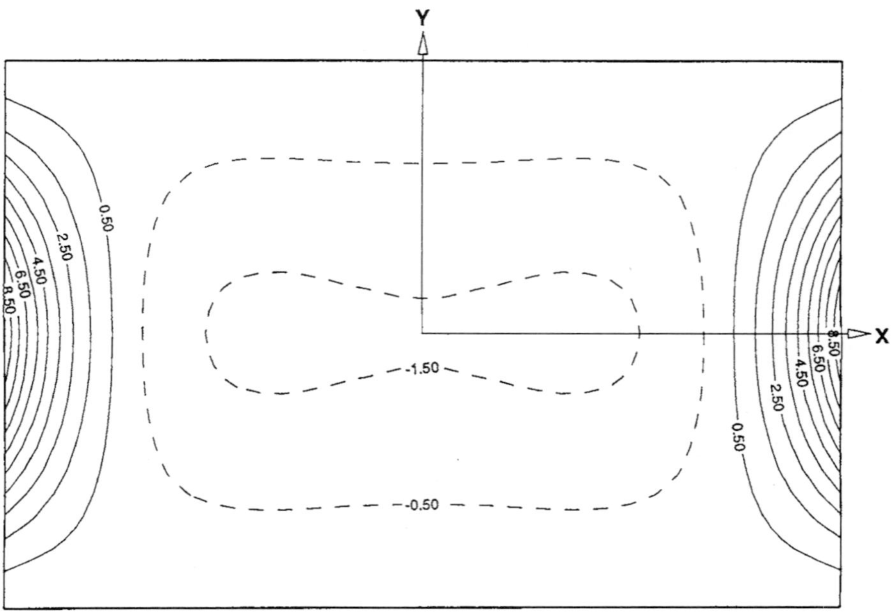

b) Contours of σ_y

Fig. 7.9 Membrane stresses in an unrestrained rectangular plate with tentlike temperature distribution (Ref. 15).

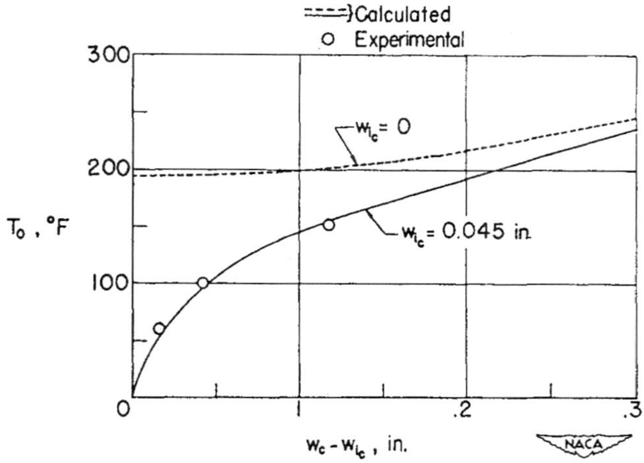

Fig. 7.10 Temperature-transverse deflection for unrestrained plate with tentlike temperature distribution (from Ref. 16).

the thermal buckling in Chapter 10, and here we discuss some qualitative features of the plate behavior.

The results presented in Example 7.4 show that, for the tentlike temperature distribution, the central region of the plate experiences biaxial compressive stresses. The stresses are proportional to the temperature difference T_0 between the temperatures on the centerline and the plate edge. Analysis of a *perfectly flat* plate, according to small-deflection theory, predicts that, as the plate temperatures increase, the unrestrained flat plate expends with u and v displacements but the plate remains flat with $w \equiv 0$. The membrane stresses increase linearly with T_0 until yielding occurs.

The experiment described by Gossard et al.[16] and a more recent experiment by Thornton et al.[17] show that an actual plate behaves quite differently. A plate in a laboratory experiment always has some small initial imperfection so that it is not perfectly flat. The interaction of the compressive thermal stresses with the initial imperfections causes the plate to bend immediately as the temperature increases. Consequently, pure membrane behavior is not attained, and the plate response consists of simultaneous bending and stretching. This behavior cannot be described by the small-deflection theory; instead, we must use von Kármán plate theory. In the Gossard paper, an approximate solution to the large-deflection problem is given assuming that the plate is free to expand in the plane and that it has simple supports. The plate considered is the same one studied by Heldenfels, and it is aluminum with a thickness of 0.25 in.

Figure 7.10 compares the calculated and experimental deflections from Ref. 16. The temperature difference T_0 is plotted vs w_c–w_{i_c}, where w_c denotes the deflection at the center of the plate, and w_{i_c} denotes the initial center deflection. The results demonstrate how the temperature-deflection curve becomes nonlinear for very small deflections: Note that when the curve begins to "bend over," the deflection are much smaller than the plate thickness of 0.25 in. For the relatively small temperature range considered (about 150°F), the agreement between the calculated and experimental deflections is very good. The dashed line on the figure for $w_{i_c} = 0$ intersects the vertical axis at about $T_0 = 190$°F, which is the plate bifurcation buckling temperature.The dashed line is an approximate post-buckling solution to the von Kármán equations. In Chapter 10, we discuss the bending/buckling behavior in further detail and present more recent results from finite element analyses.

7.5 Closing Remarks

This chapter presents engineering theories for thermal-stress analysis of flat plates. Classical theories are presented for thin rectangular plates, with homogeneous, isotropic materials assumed. Basic formulations are developed for small-deflection and von Kármán plate theories. A series of example problems illustrate some classical solutions and discuss qualitative features of the thermal stresses. A notable omission is a discussion of plate theories for composite plates, which are finding increased applications in flight vehicles. A modern treatment of anisotropic and heterogeneous plates is given by Tauchert.[6] In addition, several texts—such as that of Whitney[18]—are devoted exclusively to the study of laminated anisotropic plates. This chapter provides basic equations of plate theory that we use in later chapters as we develop finite element thermal-structural analysis methods and describe more complex thermal-structural problems such as thermal buckling.

References

[1]Boley, B. A., and Weiner, J. H., *Theory of Thermal Stresses*, Wiley, New York, 1960; reprinted ed. available from Krieger Publishing, Malabar, FL.

[2]Nowacki, W., *Thermoelasticity*, Addison–Wesley, Reading, MA, 1962.

[3]Johns, D. J., *Thermal Stress Analysis*, Pergamon, Oxford, England, UK, 1965.

[4]Parkus, H., *Thermoelasticity*, Blaisdell, Waltham, MA, 1968.

[5]Nowinski, J. L., *Theory of Thermoelasticity with Applications*, Sijhoff & Noordhoff, Alphen aan den Rijn, The Netherlands, 1978.

[6]Tauchert, T. R., "Thermal Stresses in Plates—Statical Problems," *Thermal Stresses I*, edited by R. B. Hetnarski, North-Holland, Amsterdam, The Netherlands, 1986, Chap. 2.

[7]Tauchert, T. R., "Thermal Stresses in Plates—Dynamical Problems," *Thermal Stresses II*, edited by R. B. Hetnarski, North-Holland, Amsterdam, The Netherlands, 1987, Chap. 1.

[8]Tauchert, T. R., "Thermally Induced Flexure, Buckling and Vibration of Plates," *Applied Mechanics Reviews*, Vol. 44, No. 8, 1991, pp. 347–360.

[9]Timoshenko, S. P., and Woinowsky-Krieger, S., *Theory of Plates and Shells*, 2nd ed., McGraw–Hill, New York, 1959.

[10]Langhaar, H. L., *Energy Methods in Applied Mechanics*, Wiley, New York, 1962.

[11]Fung, Y. C., *Foundations of Solid Mechanics*, Prentice–Hall, Englewood Cliffs, NJ, 1965.

[12]Shames, I. H., and Dym, C. L., *Energy and Finite Element Methods in Structural Mechanics*, Hemisphere, New York, 1985.

[13]Chia, C.-Y., *Nonlinear Analysis of Plates*, McGraw–Hill, New York, 1980.

[14]Vinson, J., *The Behavior of Thin Walled Structures: Beams, Plates and Shells*, Kluwer, Boston, 1989.

[15]Heldenfels, R. R., and Roberts, W. M., "Experimental and Theoretical Determination of Thermal Stresses in a Flat Plate," NACA TN-2769, Aug. 1952.

[16]Gossard, M. L., Seide, P., and Roberts, W. M., "Thermal Buckling of Plates," NACA TN-2771, Aug. 1952.

[17]Thornton, E. A., Coyle, M. F., and McLeod, R. N., "Experimental Study of Plate Buckling Induced by Spatial Temperature Gradients," *Journal of Thermal Stresses*, Vol. 17, No. 2, 1994, pp. 191–212.

[18]Whitney, J. M., *Structural Analysis of Laminated Anisotropic Plates*, Technomic, Lancaster, PA, 1987.

[19]Washizu, K., *Variational Methods in Elasticity and Plasticity*, Pergamon, Oxford, England, UK, 1968, pp. 175–176.

[20]Dowell, E. H., *Aeroelasticity of Plates and Shells*, Noordhoff, Leyden, The Netherlands, 1975, pp. 109–119.

Problems

7.1. Derive the equilibrium equations (7.11) and (7.12) for the stress resultants for a classical thin plate undergoing small deflections.

7.2. A thin rectangular plate (Fig. P7.2) is subjected to a uniform stress σ_0 and a temperature distribution

$$T = T_0 - T_1 \cos(2\pi y/b)$$

where T_0 and T_1 are constants. The boundary conditions are

$$u(0, y) = 0 \qquad \sigma_x(a, y) = \sigma_0 \qquad v(x, 0) = 0 \qquad \sigma_y(x, b) = 0$$

A solution is proposed:

$$u = (\sigma_0 + E\alpha T_0)x/E$$

$$v = (-\nu\sigma_0 + E\alpha T_0)y/E - (1 + \nu)\alpha T_1 b \sin(2\pi y/b)/2\pi$$

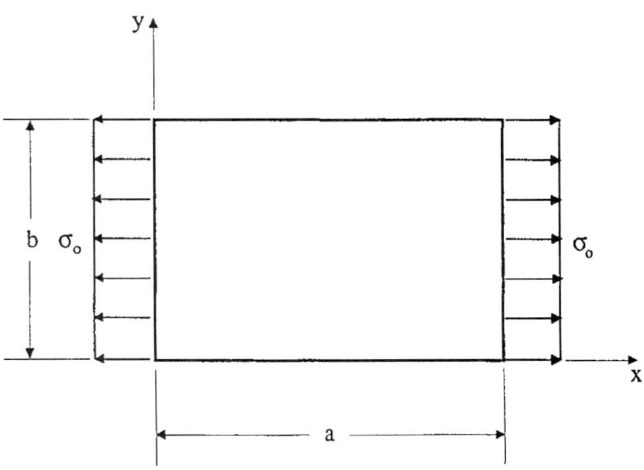

Fig. P7.2

(a) Is this an exact solution to the membrane plane stress problem?
(b) Derive equations for the strains, stresses, and membrane forces.
(c) Plot the temperature distribution $T(y)$ and stress distribution $\sigma_x(y)$.
(d) Using the plots, discuss the relationship between the temperature distribution and stress distribution.

7.3. A thin plate is heated by a specified heat flux q on its upper surface $z = h/2$ and is insulated on its lower surface $z = -h/2$. The transient thermal response of the plate is given by Eq. (4.48).

(a) Show that, after an initial thermal transient, the thickness temperature distribution is given by

$$T(z, t) = \frac{q_s h}{k}\left[\frac{\kappa t}{h^2} + \frac{1}{2}\left(\frac{z}{h} + \frac{1}{2}\right)^2 - \frac{1}{6}\right]$$

where k is the thermal conductivity and κ is the thermal diffusivity.
(b) Derive an equation for the thermal force per unit length, $N_T(t)$.
(c) Derive an equation for the thermal moment per unit length, $M_T(t)$.

7.4. The plate of the preceding problem is rectangular and simply supported. Assume the following data:

$a = 15$ in. $\qquad \alpha = 8.2 \times 10^{-6}$ 1/°F
$b = 10$ in. $\qquad k = 2.6 \times 10^{-4}$ Btu/in.-s-°F
$h = 0.125$ in. $\qquad \kappa = 6.7 \times 10^{-3}$ in.^{2}/s
$E = 29 \times 10^6$ lb/in.^{2} $\qquad T_{\text{ref}} = 70$°F
$\upsilon = 0.32$

(a) Calculate the temperature difference ΔT_h between the top and bottom surfaces of the plate for $q = 1,\ 10,\ 100$ Btu/ft^2-s.
(b) Calculate the thermal force and thermal moment for the above heat fluxes.
(c) Compute the deflection at the center of the plate for the above heat fluxes.
(d) Compute the stress components σ_x and σ_y for the above heat fluxes.
(e) Present your results in a table, and discuss how the plate response varies with the heat flux.

7.5. A rectangular plate (Fig. P7.5) is simply supported on two opposite edges and clamped on the other two edges. The plate temperature distribution gives a constant thermal moment M_T. We propose a solution, based on Ref. 6, of the form

$$w(x, y) = W(x, y) + \frac{M_T a^2}{2D}\left(\frac{x}{a} - \frac{x^2}{a^2}\right)$$

and argue that

$$\nabla^4 W = 0$$

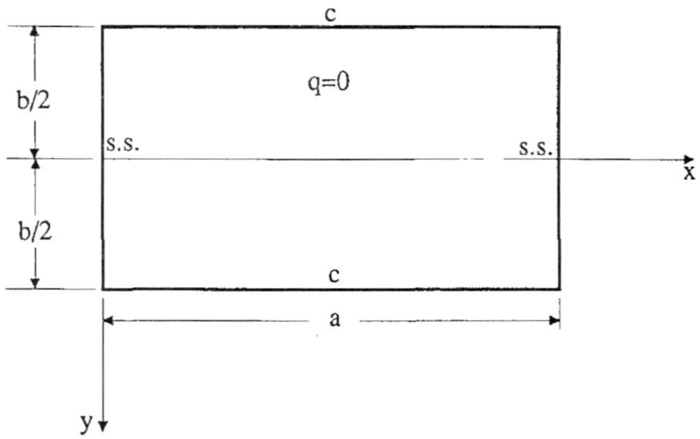

Fig. P7.5

and

$$W = 0 \qquad \frac{\partial^2 W}{\partial x^2} = 0 \qquad \text{at } x = 0 \text{ and } x = a$$

(a) Derive the preceding equation and boundary conditions, and determine if they are correct.

(b) Assume a Levy solution,

$$W(x, y) = \sum_{m=1}^{\infty} Y_m(y) \sin \frac{m\pi x}{a}$$

and derive a general solution for $Y_m(y)$.

(c) Impose the appropriate boundary conditions on $Y_m(y)$, and solve for the constants of integration. Write the final solution for $w(x, y)$.

(d) Sketch the deflected shape of the plate.

7.6. Derive the $\Sigma F_z = 0$ equilibrium equation (7.24) for a plate element in a deformed configuration.

7.7. Consider a shallow, curved plate where $w_0(x, y)$ denotes the initial position of the middle surface relative to the x–y plane. The basic equations of shallow, curved plates are developed by Chia,[13] Washizu,[19] and Dowell.[20] The strain-displacement relations are

$$\varepsilon_x = \frac{\partial u_0}{\partial x} + \frac{1}{2}\left(\frac{\partial w}{\partial x}\right)^2 + \frac{\partial w_0}{\partial x}\frac{\partial w}{\partial x} - z\frac{\partial^2 w}{\partial x^2}$$

$$\varepsilon_y = \frac{\partial v_0}{\partial y} + \frac{1}{2}\left(\frac{\partial w}{\partial y}\right)^2 + \frac{\partial w_0}{\partial y}\frac{\partial w}{\partial y} - z\frac{\partial^2 w}{\partial y^2}$$

$$\varepsilon_{xy} = \frac{1}{2}\left[\frac{\partial u_0}{\partial y} + \frac{\partial v_0}{\partial x} + \frac{\partial w}{\partial x}\frac{\partial w}{\partial y} + \frac{\partial w_0}{\partial x}\frac{\partial w}{\partial y} + \frac{\partial w_0}{\partial y}\frac{\partial w}{\partial x}\right] - z\frac{\partial^2 w}{\partial x \partial y}$$

Derive the strain-compatibility equation equivalent to Eq. (7.26). Then, show that the stress function satisfies

$$\nabla^4 F + (1-\upsilon)\nabla^2 N_T + Eh\left[\frac{\partial^2 w}{\partial x^2}\frac{\partial^2 w}{\partial y^2} - \left(\frac{\partial^2 w}{\partial x \partial y}\right)^2 - 2\frac{\partial^2 w_0}{\partial x \partial y}\frac{\partial^2 w}{\partial x \partial y}\right.$$
$$\left. + \frac{\partial^2 w_0}{\partial x^2}\frac{\partial^2 w}{\partial y^2} + \frac{\partial^2 w_0}{\partial y^2}\frac{\partial^2 w}{\partial x^2}\right] = 0$$

7.8. The second equation of the pair of equations that describes the response of shallow, curved plates with initial deflection $w_0(x, y)$ comes from the equilibrium equation $\Sigma F_z = 0$. Show that the appropriate equation is

$$\nabla^4 w = \frac{1}{D}\left[q + \frac{\partial^2 F}{\partial y^2}\left(\frac{\partial^2 w_0}{\partial x^2} + \frac{\partial^2 w}{\partial x^2}\right) - 2\frac{\partial^2 F}{\partial x \partial y}\left(\frac{\partial^2 w_0}{\partial x \partial y} + \frac{\partial^2 w}{\partial x \partial y}\right)\right.$$
$$\left. + \frac{\partial^2 F}{\partial x^2}\left(\frac{\partial^2 w_0}{\partial y^2} + \frac{\partial^2 w}{\partial y^2}\right) - \nabla^2 M_T\right]$$

Note that the equations developed in the last two problems are a generalization of von Kármán's plate equations (7.32).

7.9. For the shallow, curved plate described in the two preceding problems, show how the initial deflection $w_0(x, y)$ affects the equations for the membrane forces and bending and twisting moments.

7.10. For the shallow, curved plate described in the three preceding problems, write appropriate boundary conditions for in-plane and out-of-plane supports.

8
Finite Element Thermal-Structural Analysis

8.1 Introduction

In the last two chapters, we developed analytical formulations for thermal-structural analysis of rods, beams, and plates. The development emphasizes formulation of the governing partial differential equations. Analytical solutions are given for boundary-value problems to illustrate solution procedures and qualitative aspects of the thermal-structural behavior. Analytical solutions are usually available only for relatively simple geometrical shapes and are typically limited to linear problems. For these reasons, when we address practical aerospace thermal-structural applications, we turn to computational methods. In Chapter 4, we discussed computational methods for heat-transfer problems, which include finite difference, finite volume, and finite element methods. All three methods are used in industry to solve heat-transfer problems although the finite difference and finite volume methods are the most popular. For structural applications, the preeminent computational method is finite elements. In this chapter, we develop the finite element method for both thermal and structural analyses.

Structural finite element analysis began with the classic 1956 paper by Turner et. al.[1] in which the concept of the triangular plane stress element was introduced. In a recent paper,[2] Clough describes early developments in finite elements and explains how he introduced the label "finite elements" in a 1960 paper. Since then, there has been a virtual explosion of finite element literature, with literally thousands of papers. In a survey paper, Noor[3] identifies over 200 international finite element symposia, conferences, and short courses that took place between 1964 and 1991. The paper also identifies over 400 finite element books, the vast majority of which describe isothermal-structural applications, although some books deal with applications to other disciplines as well. References 4–7 are recent editions of finite element books that include thermal and structural applications. The work by Hsu[8] is perhaps the only book devoted solely to thermomechanical finite element applications. It is written from a mechanical engineering perspective, with applications to energy conversion problems. Our treatment here follows the approach and notation of the text by Huebner et al.[7]

We center our discussion on the special aspects of finite elements that apply to aerospace thermal structures. We presume that the reader has some knowledge of finite element fundamentals. And so, even though we develop the theory from basic principles, our discussion is more concise than finite element books. Many structural finite element books use the principles of virtual work or minimum potential energy to derive finite element equations. These are powerful methods, but they require more background in structural mechanics than we need. Our development of finite element theory relies on the method of weighted residuals to derive the equations because this approach applies to both thermal and structural problems. Moreover, the method of weighted residuals is equivalent mathematically to the method of virtual work, and so we will derive exactly the same equations. The only differences are that we will use the same method to derive both thermal and structural equations and that we require no background in energy methods. We first develop finite element methodology for transient heat-transfer problems and then consider problems quasistatic structures. Examples illustrate thermal and structural applications.

8.2 Thermal Analysis

We consider aerospace applications dominated by conduction heat transfer in the structure. We permit general boundary conditions so that a variety of heat sources and losses may be simulated. Temperature-dependent properties and radiation exchanges are permitted, and so the problem is nonlinear. Our objective is to compute the temperature $T(x, y, z, t)$ throughout the structure so that, in a following structural analysis, we may compute detailed deformations and stresses. Our finite element formulation begins with the structural heat-transfer problem posed in Sec. 4.2.

8.2.1 Finite Element Formulation

Problem statement. Consider heat transfer in a three-dimensional anisotropic structure Ω bounded by surface Γ (Fig. 8.1). The problem is governed by the conservation of energy equation from Chapter 3:

$$-\left(\frac{\partial q_x}{\partial x} + \frac{\partial q_y}{\partial y} + \frac{\partial q_z}{\partial z}\right) + Q = \rho c \frac{\partial T}{\partial t} \tag{8.1}$$

where q_x, q_y, q_z are the conduction heat fluxes, $Q(x, y, z, t)$ is the internal heat-generation rate per unit volume, ρ is the density, and c is the specific heat. Fourier's law, Eq. (3.31), for an anisotropic material is

$$\begin{aligned} q_x &= -\left(k_{11}\frac{\partial T}{\partial x} + k_{12}\frac{\partial T}{\partial y} + k_{13}\frac{\partial T}{\partial z}\right) \\ q_y &= -\left(k_{21}\frac{\partial T}{\partial x} + k_{22}\frac{\partial T}{\partial y} + k_{23}\frac{\partial T}{\partial z}\right) \\ q_z &= -\left(k_{31}\frac{\partial T}{\partial x} + k_{32}\frac{\partial T}{\partial y} + k_{33}\frac{\partial T}{\partial z}\right) \end{aligned} \tag{8.2}$$

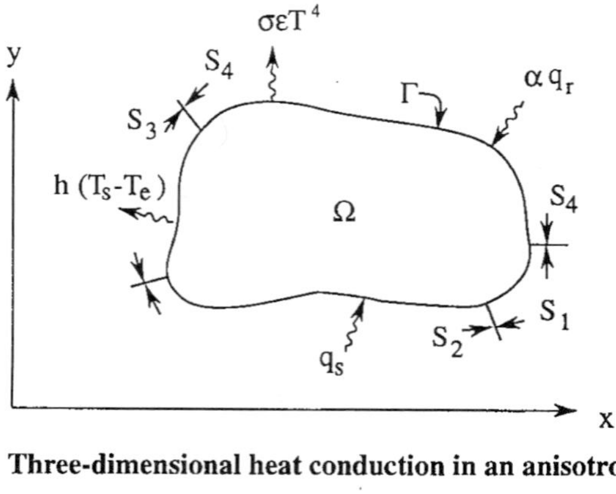

Fig. 8.1 Three-dimensional heat conduction in an anisotropic solid.

where k_{ij} is the symmetric thermal-conductivity tensor. The material properties ρ, c and k_{ij} may be temperature-dependent. The heat-conduction equation is solved subject to boundary conditions on all portions of the surface Γ and an initial condition. Heat-conduction boundary conditions take several forms; consider the frequently encountered conditions of specified surface temperature, specified surface heat flux, convective heat exchange, and radiation exchange. The boundary conditions (Fig. 8.1) are

$$T_s = T_s(x, y, z, t) \qquad \text{on } S_1 \tag{8.3a}$$

$$q_n = -q_s \qquad \text{on } S_2 \tag{8.3b}$$

$$q_n = h(T_s - T_e) \qquad \text{on } S_3 \tag{8.3c}$$

$$q_n = \sigma \varepsilon T_s^4 - \alpha q_R \qquad \text{on } S_4 \tag{8.3d}$$

where

T_s = specified surface temperature
q_n = heat flux normal to the surface
q_s = specified surface heat flux (positive into the surface)
h = convective heat-transfer coefficient
T_e = convective exchange temperature
σ = Stefan–Boltzmann constant
ε = surface emissivity
α = surface absorptivity
q_R = incident radiation heat flux

The surface emissivity and absorptivity may be temperature-dependent.

Finite element equations. The solution domain Ω is divided into M elements of r nodes each. By the usual procedure, we express the temperature and temperature gradients within each element as

$$T(x, y, z, t) = [N(x, y, z)]\{T(t)\} \tag{8.4a}$$

and

$$\begin{Bmatrix} \dfrac{\partial T}{\partial x}(x, y, z, t) \\ \dfrac{\partial T}{\partial y}(x, y, z, t) \\ \dfrac{\partial T}{\partial z}(x, y, z, t) \end{Bmatrix} = [B(x, y, z)]\{T(t)\} \tag{8.4b}$$

where $[N]$ is the temperature interpolation matrix, $[B]$ the temperature gradient interpolation matrix, and $\{T\}$ the vector of nodal temperatures. The temperature interpolation matrix is a row matrix of the interpolation functions,

$$[N] = [N_1 \quad N_2 \quad \ldots \quad N_r] \tag{8.4c}$$

and the temperature gradient interpolation matrix is a rectangular array of interpolation function derivatives,

$$[B] = \begin{bmatrix} \dfrac{\partial N_1}{\partial x} & \dfrac{\partial N_2}{\partial x} & \cdots & \dfrac{\partial N_r}{\partial x} \\ \dfrac{\partial N_1}{\partial y} & \dfrac{\partial N_2}{\partial y} & \cdots & \dfrac{\partial N_r}{\partial y} \\ \dfrac{\partial N_1}{\partial z} & \dfrac{\partial N_2}{\partial z} & \cdots & \dfrac{\partial N_r}{\partial z} \end{bmatrix} \tag{8.4d}$$

We focus our attention now on a single finite element of domain Ω_e, and we use the method of weighted residuals (MWR) to derive the element equations. We satisfy the conservation of energy equation for each element; consequently, the finite element equations represent an element energy balance. We first substitute the temperature representation, Eq. (8.4a), into Eq. (8.1). Because this approximate temperature does not satisfy conservation of energy exactly, the result is a residual function. In the MWR, we multiply the residual by a weighting function and integrate the weighted residual over the solution domain. Here we choose as weighting functions the interpolation functions N_i, $i = 1, 2, \ldots, r$, and the solution domain is the element domain Ω_e. Hence, the MWR requires

$$\int_{\Omega_e} \left(\frac{\partial q_x}{\partial x} + \frac{\partial q_y}{\partial y} + \frac{\partial q_z}{\partial z} - Q + \rho c \frac{\partial T}{\partial t} \right) N_i \, \mathrm{d}\Omega = 0 \qquad i = 1, 2, \ldots, r \tag{8.5}$$

Note that, for an element, Eq. (8.5) represents r equations for r nodal temperatures. Following standard procedures, we integrate the derivatives of the heat fluxes by parts using Gauss's theorem. The integration introduces a surface integral on the element boundary, which we break into the sum of integrals over S_1, S_2, S_3, and S_4 of the surface, where the boundary conditions are specified. After the integration by parts, we can write the results as

$$\begin{aligned} &\int_{\Omega_e} \rho c \frac{\partial T}{\partial t} N_i \, \mathrm{d}\Omega - \int_{\Omega_e} \begin{bmatrix} \dfrac{\partial N_i}{\partial x} & \dfrac{\partial N_i}{\partial y} & \dfrac{\partial N_i}{\partial z} \end{bmatrix} \begin{Bmatrix} q_x \\ q_y \\ q_z \end{Bmatrix} \mathrm{d}\Omega = \int_{\Omega_e} Q N_i \, \mathrm{d}\Omega \\ &\quad - \int_{S_1} (\vec{q} \cdot \hat{n}) N_i \, \mathrm{d}S + \int_{S_2} q_s N_i \, \mathrm{d}S - \int_{S_3} h(T - T_e) N_i \, \mathrm{d}S \\ &\quad - \int_{S_4} (\sigma \varepsilon T^4 - \alpha q_r) N_i \, \mathrm{d}S \qquad i = 1, 2, \ldots, r \end{aligned} \tag{8.6}$$

where $\vec{q} = q_x \hat{i} + q_y \hat{j} + q_z \hat{k}$ is the heat-flux vector, and $\hat{n}$ is a unit normal vector to the surface. As the last step, we introduce the interpolations for the element temperature and heat-flux components. For convenience, we write the heat fluxes from Fourier's law, Eqs. (8.2), in matrix form as

$$\begin{Bmatrix} q_x \\ q_y \\ q_z \end{Bmatrix} = -[k] \begin{Bmatrix} \dfrac{\partial T}{\partial x} \\ \dfrac{\partial T}{\partial y} \\ \dfrac{\partial T}{\partial z} \end{Bmatrix}$$

and then eliminate the temperature gradients using Eq. (8.4b). Thus,

$$\begin{Bmatrix} q_x \\ q_y \\ q_z \end{Bmatrix} = -[k][B]\{T\} \tag{8.7}$$

Substituting Eqs. (8.4a) and (8.7) into Eq. (8.6), we write the results in the matrix form

$$\begin{aligned}&[C]\{\dot{T}\} + [[K_c] + [K_h]]\{T\} \\ &\quad = \{R_T\} + \{R_Q\} + \{R_q\} + \{R_h\} + \{R_\sigma\} + \{R_R\}\end{aligned} \tag{8.8}$$

where the element matrices and vectors are defined by the integrals:

$$[C] = \int_{\Omega_e} \rho c \{N\}\lfloor N \rfloor \, \mathrm{d}\Omega \tag{8.9a}$$

$$[K_c] = \int_{\Omega_e} [B]^T [k][B] \, \mathrm{d}\Omega \tag{8.9b}$$

$$[K_h] = \int_{S_3} h\{N\}\lfloor N \rfloor \, \mathrm{d}S \tag{8.9c}$$

$$\{R_T\} = -\int_{S_1} (\vec{q} \cdot \hat{n})\{N\} \, \mathrm{d}S \tag{8.9d}$$

$$\{R_Q\} = \int_{\Omega_e} Q\{N\} \, \mathrm{d}\Omega \tag{8.9e}$$

$$\{R_q\} = \int_{S_2} q_s\{N\} \, \mathrm{d}S \tag{8.9f}$$

$$\{R_h\} = \int_{S_3} hT_e\{N\} \, \mathrm{d}S \tag{8.9g}$$

$$\{R_\sigma\} = -\int_{S_4} \sigma\varepsilon T^4\{N\} \, \mathrm{d}S \tag{8.9h}$$

$$\{R_R\} = \int_{S_4} \alpha q_R\{N\} \, \mathrm{d}S \tag{8.9i}$$

The coefficient matrix $[C]$ is the element capacitance matrix. The coefficient matrices $[K_c]$ and $[K_h]$ are element conductance matrices and relate, respectively, to conduction and surface convection. The vector R_T is the vector of nodal heat flows that are necessary to keep element nodes at specified temperatures. Normally, this integral is not evaluated because specified nodal temperatures are imposed when we solve the global equations. The vectors $\{R_Q\}$, $\{R_q\}$, and $\{R_h\}$ are element heat load vectors relating to specified internal heat generation, surface heating, and surface convection. The last two integrals, $\{R_\sigma\}$ and $\{R_R\}$, apply to emitted and absorbed radiation, respectively.

The element matrices defined in Eqs. (8.9) are assembled in the usual way to form the global equations, which have the form

$$[C(T,t)]\{\dot{T}\} + [K(T,t)]\{T\} = \{R(T,t)\} \tag{8.10}$$

where we indicate that the global capacitance matrix $[C(T,t)]$, the global conductance matrix $[K(T,t)]$, and the global load vector $\{R(T,t)\}$ may be temperature- and time-dependent. Equation (8.10) represents a set of simultaneous, first-order, nonlinear ordinary differential equations. Before we specify nodal temperature boundary conditions, there are as many equations as system nodes since each node has a single unknown temperature.

Element matrices. The element matrices defined in Eqs. (8.9) are the basis for finite element heat-transfer analysis. The form of the matrices depends on the element shape, whether the thermal properties are constant or vary with temperature, and the order of the temperature interpolation functions. Element shapes may be one-, two-, or three-dimensional in the form of lines, triangles, quadrilaterals, tetrahedrons, hexahedrons, and so forth. If thermal properties such as specific heat or thermal conductivity are constant, the element matrices are constant and do not vary during the solution process. If thermal properties are temperature-dependent, then element matrices require periodic updates during the solution as the properties vary with temperature. Temperature interpolation functions may be linear, quadratic, or higher-order functions of x, y, z within an element; the order determines the number of element nodes. Dependent on the element shape and the interpolation functions, the evaluation of the integrals in Eqs. (8.9) may be done analytically in closed form or may be performed numerically within the computer program.

To illustrate the application of the foregoing element equations, we consider a one-dimensional linear element with conduction and convection. We discuss the contributions of radiation heat transfer in Sec. 8.2.3. Reference 7 describes evaluation of heat-transfer matrices for more complicated elements.

A one-dimensional linear element (Fig. 8.2) is often called a *rod element*. The element resembles a structural truss member, but it is much more versatile than a truss element because it models a variety of one-dimensional heat-conduction modes. We assume that the element has constant cross-sectional area A and perimeter p. The perimeter p is used to define the appropriate area for surface heat transfer. The element has two nodes, and the assumed temperature varies linearly between nodes. We write the interpolation functions with a local x coordinate as

$$N_1(x) = 1 - x/L \qquad N_2(x) = x/L$$

where L is the element length. By Eq. (8.4d),

$$B_1(x) = \frac{\partial N_1}{\partial x} = -1/L \qquad B_2(x) = \frac{\partial N_2}{\partial x} = 1/L$$

Hence, the temperature gradient and heat flux are constant within an element. Element matrices are readily evaluated using $[k] = k$, $d\Omega = A dx$, and $dS = p dx$. From Eqs. (8.9),

$$[C] = \rho c \int_0^L \{N\}[N] A \, dx = \frac{\rho c A L}{6}\begin{bmatrix} 2 & 1 \\ 1 & 2 \end{bmatrix} \tag{8.11a}$$

$$[K_c] = k \int_0^L \{B\}[B] A \, dx = \frac{kA}{L}\begin{bmatrix} 1 & -1 \\ -1 & 1 \end{bmatrix} \tag{8.11b}$$

$$[K_h] = h \int_0^L \{N\}[N] p \, dx = \frac{hpL}{6}\begin{bmatrix} 2 & 1 \\ 1 & 2 \end{bmatrix} \tag{8.11c}$$

$$[R_Q] = Q \int_0^L \{N\} A \, dx = \frac{QAL}{2}\begin{Bmatrix} 1 \\ 1 \end{Bmatrix} \tag{8.11d}$$

$$[R_Q] = q_s \int_0^L \{N\} p \, dx = \frac{q_s p L}{2}\begin{Bmatrix} 1 \\ 1 \end{Bmatrix} \tag{8.11e}$$

$$[R_h] = hT_e \int_0^L \{N\} p \, dx = \frac{hT_e pL}{2}\begin{Bmatrix} 1 \\ 1 \end{Bmatrix} \tag{8.11f}$$

In finite element terminology, the capacitance matrix, Eq. (8.11a), is called a consistent capacitance matrix because we use the same interpolation functions to derive it as we use for the other matrices. Note that the coefficient of the matrix is the product of ρc and the element volume AL; hence, it represents the element's capability to store energy. Another form of the capacitance matrix is also used:

$$[C] = \frac{\rho c A L}{2}\begin{bmatrix} 1 & 0 \\ 0 & 1 \end{bmatrix} \quad \text{(lumped)} \tag{8.11g}$$

where the matrix is diagonal. This matrix is called a *lumped capacitance matrix* because one-half of the element's capacitance is placed (or "lumped") at each

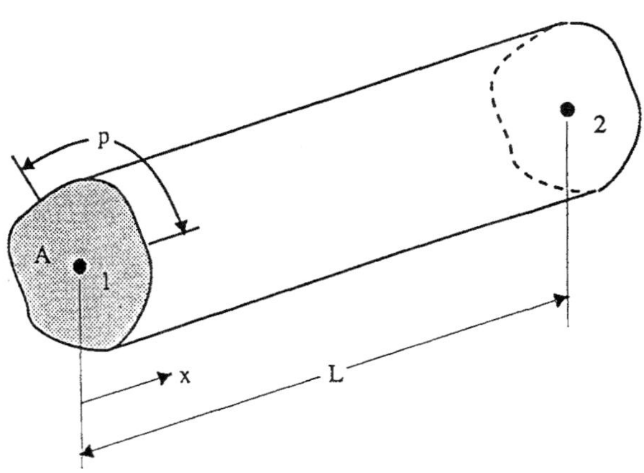

Fig. 8.2 One-dimensional heat-transfer element.

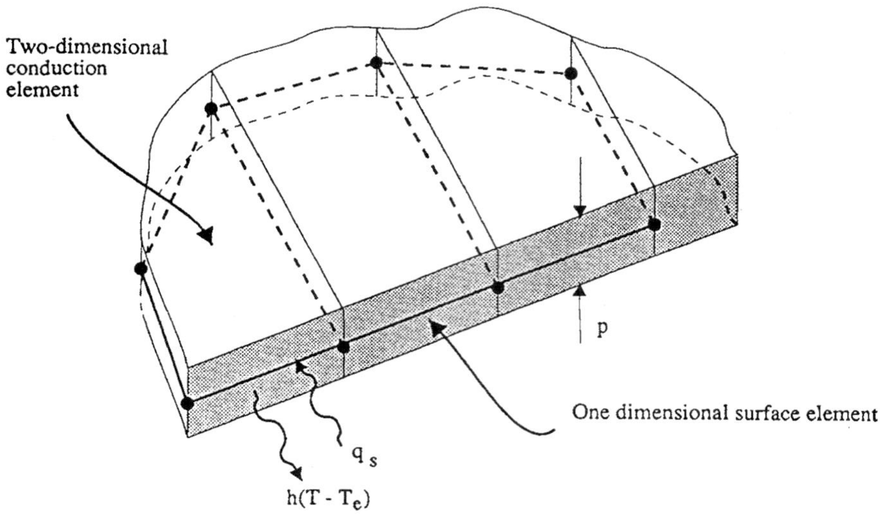

Fig. 8.3 Two-dimensional structure with surface heat transfer.

element node. In solution algorithms, a lumped capacitance matrix can provide computational advantages.

The foregoing heat-flux and convection matrices may be used in combination with two-dimensional conduction elements (Fig. 8.3) to model surface heat exchanges on an edge of a two-dimensional structure. The perimeter p is the thickness of the conduction element.

8.2.2 Solution Methods

Solution algorithms depend on whether a problem is steady-state or transient, linear or nonlinear. We describe popular algorithms for nonlinear steady-state problems and linear transient problems. Algorithm performance is an important aspect of the effectiveness of finite element software. We only summarize the highlights of these basic algorithms. For further discussion readers are encouraged to consult Refs. 4–7.

Nonlinear steady-state solutions. For steady-state problems, we may write the nonlinear equilibrium equations from Eq. (8.10) as

$$\{F(T)\} = [K(T)]\{T\} - \{R(T)\} \tag{8.12}$$

where $\{F(T)\}$ is a vector that represents the unbalance, or residual, in the equations for an approximate solution vector $\{T\}$. To find the solution to the nonlinear equations, computational algorithms use iterative schemes to drive the residual vector toward zero. In finite elements, a popular procedure is the Newton–Raphson iteration. The algorithm is

$$[J]^m\{\Delta T\}^{m+1} = -\{F\}^m \tag{8.13a}$$

$$\{T\}^{m+1} = \{T\}^m + \{\Delta T\}^{m+1} \tag{8.13b}$$

where the superscript m denotes the mth iteration, $[J]^m$ is the Jacobian, and $\{\Delta T\}^{m+1}$ is a vector of nodal temperature increments. Convergence may be specified by requiring the temperature increments to meet a convergence criterion. The elements of the Jacobian matrix are defined as

$$J_{ij} = \frac{\partial F_i}{\partial T_j} \tag{8.14}$$

In computer implementations of the procedure, the Jacobians and residual vectors are evaluated for each element and then assembled into global matrices. Reference 9 describes a procedure for evaluation of the Jacobians and residual heat load vectors for elements with temperature-dependent thermal properties. In Sec. 8.2.3, we consider conduction with radiation boundary conditions and develop radiation matrices.

Transient solutions. The most popular method for computing transient solutions to Eq. (8.10) is by recursive time-marching schemes. In this approach, recursion formulas relate the values of the temperature vector at time t to the value of the temperature vector of a later time $t + \Delta t$, where Δt is the time step. The recursion formula makes it possible to march out the solution in time, starting from an initial condition at time $t = 0$ and continuing step by step until the desired duration is reached.

A popular time-marching approach is based on a family of algorithms known as the θ algorithms. Consider Eq. (8.10) for the special case of linear problems where the coefficient matrices are constant and the heat load vector is a function only of time. Then we may write Eq. (8.10) as

$$[C]\{\dot{T}\} + [K]\{T\} = \{R(t)\} \tag{8.15}$$

Let t_n denote a typical time in the response so that $t_{n+1} = t_n + \Delta t$, where $n = 1, 2, \ldots, N$. The algorithms represent the first-order matrix equations by finite difference approximations to $\{dT/dt\}$ at an intermediate time within each time step. A family of algorithms results by introducing a parameter θ such that $t_\theta = t_n + \theta \Delta t$, where $0 \leq \theta \leq 1$. To develop the recursion algorithms, we write Eq. (8.15) at time t_θ as

$$[C]\{\dot{T}\}_\theta + [K]\{T\}_\theta = \{R(t_\theta)\}$$

and introduce the approximations

$$\{\dot{T}\}_\theta = \frac{\{T\}_{n+1} - \{T\}_n}{\Delta t}$$

$$\{T\}_\theta = (1-\theta)\{T\}_n + \theta\{T\}_{n+1}$$

$$\{R(t_\theta)\} = (1-\theta)\{R\}_n + \theta\{R\}_{n+1}$$

Combining these last four equations, we may write the result as

$$\big[[C] + \theta\Delta t[K]\big]\{T\}_{n+1} = \{\bar{R}\}_{n+1} \tag{8.16a}$$

where

$$\{\bar{R}\}_{n+1} = [[C] - (1-\theta)\Delta t[K]]\{T\}_n + (1-\theta)\Delta t\{R\}_n + \theta\Delta t\{R\}_{n+1} \tag{8.16b}$$

The temperature vector $\{T\}_{n+1}$ on the left-hand side of Eq. (8.16a) is unknown, and the equivalent load vector $\{\bar{R}\}_{n+1}$ on the right-hand side is known. Thus, by solving Eq. (8.16a) repeatedly, we may advance the solution in time. Equation (8.16) represents a general family of algorithms; a particular algorithm depends on the value of θ selected. Two particular variants of the θ family of algorithms are popular.

$\theta = 0$. If we substitute $\theta = 0$ in Eq. (8.16a), we see that the coefficient matrix on the left-hand side depends only on the capacitance matrix $[C]$. If we choose a lumped capacitance matrix, the coefficient matrix is diagonal. Thus, the algorithm becomes

$$[C]\{T\}_{n+1} = \{\bar{R}\}_{n+1} \tag{8.17a}$$

$$\{\bar{R}\}_{n+1} = [[C] - \Delta t[K]]\{T\}_n + \Delta t\{R\}_n \tag{8.17b}$$

Finite element texts[7] establish the following properties for the $\theta = 0$ algorithm:

1) It is a forward difference algorithm since, for $\theta = 0$, $\{\dot{T}\}_\theta$ represents the forward difference approximation to $\{\mathrm{d}T/\mathrm{d}t\}$. The algorithm is the same as the Euler forward difference algorithm.

2) Since the coefficient matrix is diagonal, the algorithm is *explicit*. This means that we may solve Eq. (8.17a) for the temperatures T_i, $i = 1, 2, \ldots$ by direct division. We do not have to solve simultaneous equations.

3) The algorithm is first-order accurate. This means that the numerical error in the algorithm due to the difference approximation varies linearly with the time step Δt. If we reduce Δt by one-half, the error reduces by one-half.

4) The algorithm is conditionally stable. The time step must be smaller than a critical time step Δt_{cr} for numerical stability. If the time step Δt is larger than Δt_{cr}, nonphysical oscillations will develop in the solution. These oscillations will grow with time, and the solution may "blow up."

5) For a one-dimensional conduction element, the critical time step is given by

$$\Delta t_{\mathrm{cr}} = \frac{1}{2}\frac{L^2}{\kappa} \tag{8.18}$$

where L is the element length, and κ is the thermal diffusivity. This result shows the important role of the finite element mesh size in determining the critical time step. The smaller elements determine the critical time step.

$\theta = 1/2$. If we substitute $\theta = 1/2$ in Eq. (8.16) then the algorithm becomes

$$[\bar{K}]\{T\}_{n+1} = \{\bar{R}\}_{n+1} \tag{8.19a}$$

where

$$[\bar{K}] = [C] + \tfrac{1}{2}\Delta t[K] \tag{8.19b}$$

$$\{\bar{R}\}_{n+1} = \left[[C] - \tfrac{1}{2}\Delta t[K]\right]\{T\}_n + \tfrac{1}{2}\Delta t\{R\}_n + \tfrac{1}{2}\Delta t\{R\}_{n+1} \tag{8.19c}$$

The $\theta = 1/2$ algorithm has the following properties:

1) It is a central difference algorithm since, for $\theta = 1/2$, $\{\dot{T}\}_\theta$ represents the central difference approximation to $\{\mathrm{d}T/\mathrm{d}t\}$. The algorithm is the *Crank–Nicolson* scheme.

2) The coefficient matrix $[\bar{K}]$ contains off-diagonal terms; hence, the algorithm is *implicit*. Equation (8.19a) is a set of simultaneous equations that we must solve at each time step to obtain $\{T\}_{n+1}$.

3) The algorithm is second-order accurate. If we reduce the time step by one-half, the error reduces by one-fourth.

4) The algorithm is unconditionally stable. If the time step is too large, nonphysical oscillations may appear in the computed response, but the solution will never "blow up."

5) Either a consistent or lumped capacitance matrix may be used.

A consistent capacitance matrix predicts more accurate temperatures than a lumped capacitance matrix, provided that an adequate mesh is employed to represent the spatial temperature distribution. However, the consistent formulation requires a smaller time step than the lumped formulation. For complex, built-up structures, computational experience indicates an insignificant loss in accuracy by the lumped approach. Further discussion of lumped vs consistent approaches appears in Ref. 7.

For a nonlinear transient problem, we may use either an explicit Euler or an implicit Crank–Nicolson time-marching scheme. With the Euler scheme, the matrices may be updated at each time step to account for nonlinearities. Usually, a smaller time step is required for nonlinear problems. With the Crank–Nicolson scheme, Newton–Raphson iteration is performed at each step to account for nonlinearities. With this approach, a much larger time step can be used than with the Euler scheme. Of course, the computations at each step are also more expensive because simultaneous equations must be formed and solved at each iteration.

The selection of an explicit or implicit algorithm is not clear-cut since there are several tradeoffs between the algorithms. Often the choice depends on the software available or problem-solving experience at a particular job site.

8.2.3 Conduction with Radiation

Consider the structure shown in Fig. 8.1, with radiation heat transfer on surface S_4. The radiation boundary condition (8.3d) states that the surface heat transfer consists of heat loss by emitted radiation proportional to the fourth power of the surface temperature, $\sigma\varepsilon T_s^4$, and heat gain by incident radiation from external sources, αq_R. This equation assumes the radiation surface to be diffuse, gray, and opaque. A diffuse surface emits and reflects incident radiation uniformly in all directions; a gray surface emits and absorbs energy independently of wavelength. A medium with such a surface is called a *gray body* and, for a gray body, $\varepsilon = \alpha$, where ε may be temperature-dependent. An opaque medium does not transmit or scatter radiation. For an opaque medium, $\rho + \alpha = 1$, where ρ is the surface reflectivity. Note that we have previously used ρ to denote density but, in this section, for radiation exchanges,we use the same symbol for reflectivity. The usage should be clear from the context.

Radiation without surface heat exchanges. We assume that the surface S_4 emits energy to deep space according to $\sigma\varepsilon T_s^4$ and absorbs radiant energy from constant heat flux q_R. To incorporate these effects into the finite element formulation, we consider the heat load vectors $\{R_\sigma\}$ and $\{R_R\}$ given in Eqs. (8.9h) and (8.9i), respectively. Because of the T^4 term, we must perform a nonlinear solution. Adopting the Newton–Raphson algorithm, we seek the contribution of the radiation load vectors to the element unbalanced heat load vector and the element Jacobian. The contribution of Eqs. (8.9h) and (8.9i) to the heat load at the ith element node is

$$R_i = -\int_{S_4} \sigma\varepsilon T^4 N_i \, dS + \int_{S_4} \alpha q_R N_i \, dS$$

and the corresponding contribution to the unbalanced heat load vector as indicated by Eq. (8.12) is

$$F_i = \int_{S4} \sigma \varepsilon T^4 N_i \, \mathrm{d}S - \int_{S4} \alpha q_R N_i \, \mathrm{d}S \tag{8.20a}$$

The elements of the Jacobian are defined by Eq. (8.14). Thus,

$$J_{ij} = \frac{\partial F_i}{\partial T_j} = 4 \int_{S4} \sigma \varepsilon T^3 \frac{\partial T}{\partial T_j} N_i \, \mathrm{d}S + \int_{S4} \sigma \frac{\mathrm{d}\varepsilon}{\mathrm{d}T_j} T^4 N_i \, \mathrm{d}S - \int_{S4} \frac{\mathrm{d}\alpha}{\mathrm{d}T_j} q_R N_i \, \mathrm{d}S$$

From Eq. (8.4a), $\partial T / \partial T_j = N_j$; hence,

$$J_{ij} = 4 \int_{S4} \sigma \varepsilon T^3 N_i N_j \, \mathrm{d}S + \int_{S4} \sigma \frac{\mathrm{d}\varepsilon}{\mathrm{d}T_j} T^4 N_i \, \mathrm{d}S - \int_{S4} \frac{\mathrm{d}\alpha}{\mathrm{d}T_j} q_R N_i \, \mathrm{d}S \tag{8.20b}$$

If the emissivity and absorptivity are constant, then we may define the radiation Jacobian as

$$[J_R] = 4 \int_{S4} \sigma \varepsilon T^3 \{N\}[N] \, \mathrm{d}S \tag{8.21a}$$

The T^3 term in the integrand of Eq. (8.21) causes the radiation Jacobian to depend on the element's nodal temperatures. It also causes evaluation of the element Jacobian to be cumbersome and, as an alternative, a lumped Jacobian matrix may be used. The diagonal elements of the lumped Jacobian are defined by

$$J_{R_{ii}} = 4T_i^3 \int_{S4} \sigma \varepsilon N_i \, \mathrm{d}S \qquad \text{(lumped)} \tag{8.21b}$$

and the off-diagonal elements are zero. This approximation is similar to a lumped capacitance matrix, and it offers significant practical benefits; see Ref. 7.

Element radiation matrices. To illustrate the foregoing element radiation matrices, consider the one-dimensional, two-node element shown in Fig. 8.2. We may use the linear temperature interpolation for a rod element, and $\mathrm{d}S = p\,\mathrm{d}x$, where p is the perimeter of the radiating surface area. Assuming constant properties, the unbalanced load vector from Eq. (8.20a) is

$$\{F\} = \sigma \varepsilon \int_0^L (N_1 T_1 + N_2 T_2)^4 \begin{Bmatrix} N_1 \\ N_2 \end{Bmatrix} p \, \mathrm{d}x - \alpha q_R \int_0^L \begin{Bmatrix} N_1 \\ N_2 \end{Bmatrix} p \, \mathrm{d}x$$

After evaluation,

$$\{F\} = \frac{\sigma \varepsilon p L}{30} \begin{Bmatrix} 5T_1^4 + 4T_1^3 T_2 + 3T_1^2 T_2^2 + 2T_1 T_2^3 + T_2^4 \\ T_1^4 + 2T_1^3 T_2 + 3T_1^2 T_2^2 + 4T_1 T_2^3 + 5T_2^4 \end{Bmatrix} - \frac{\alpha q_R P L}{2} \begin{Bmatrix} 1 \\ 1 \end{Bmatrix} \tag{8.22}$$

The radiation Jacobian, Eq. (8.21a), is

$$[J_R] = 4\sigma \varepsilon \int_0^L (N_1 T_1 + N_2 T_2)^3 \begin{Bmatrix} N_1 \\ N_2 \end{Bmatrix} [N_1 N_2] p \, \mathrm{d}x$$

which produces

$$[J_R] = \frac{\sigma \varepsilon p L}{30} \left\{ \begin{matrix} 10T_1^3 + 6T_1^2 T_2 + 3T_1 T_2^2 + T_2^3 & 2T_1^3 + 3T_1^2 T_2 + 3T_1 T_2^2 + 2T_2^3 \\ 2T_1^3 + 3T_1^2 T_2 + 3T_1 T_2^2 + T_2^3 & T_1^3 + 3T_1^2 T_2 + 6T_1 T_2^2 + 10T_2^3 \end{matrix} \right\} \tag{8.23}$$

In contrast, the lumped Jacobian, Eqs. (8.21b), gives the much simpler approximation,

$$[J_R] = 2\sigma \varepsilon p L \begin{bmatrix} T_1^3 & 0 \\ 0 & T_2^3 \end{bmatrix} \tag{8.24}$$

Each of these last two matrices shows the strong nonlinearity of an element radiation Jacobian.

In computations, Eqs. (8.23) and (8.24) produce virtually identical nodal temperatures. The reason is shown by considering Eqs. (8.12) and (8.13). The Jacobian in Eq. (8.13a) is used in the iteration process to "drive" the unbalanced heat load vector $\{F\}$ toward zero. When $\{F\}$ goes to zero, Eq. (8.12) shows that the nonlinear equilibrium equations are satisfied regardless of the Jacobian used. Written on the blackboard one day in the graduate students' room at the University of Virginia was, "The Jacobian can be a watermelon." Well, that is a *bit* of an exaggeration, but it makes the point!

Radiation exchanges within enclosures. For many aerospace applications, radiation exchanges occur between multiple surfaces within enclosures. We assume that the thermal radiation is unaffected by the intervening medium. A concept fundamental to the computation of radiation heat exchanges between surfaces is the radiation view factor (also known as the shape factor, configuration factor, geometrical factor, or angle factor). Consider the surface i with surface area A_i shown in Fig. 8.4. Radiant energy leaving surface i, if diffusely distributed, flows

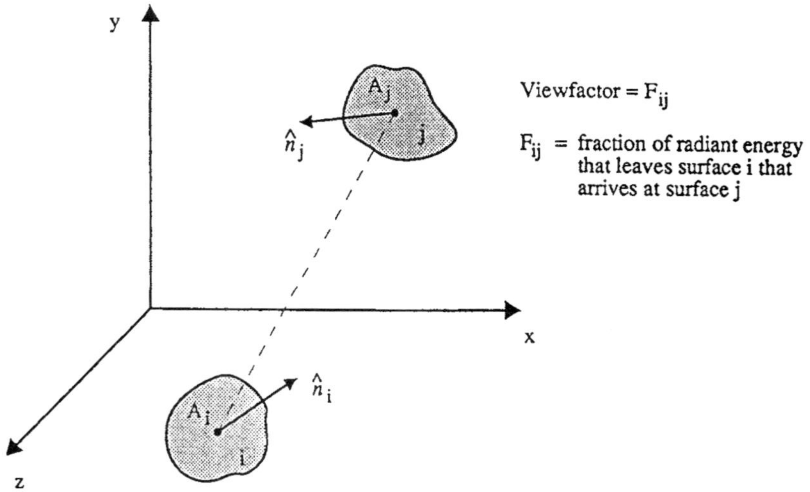

Fig. 8.4 Radiation view factor notation.

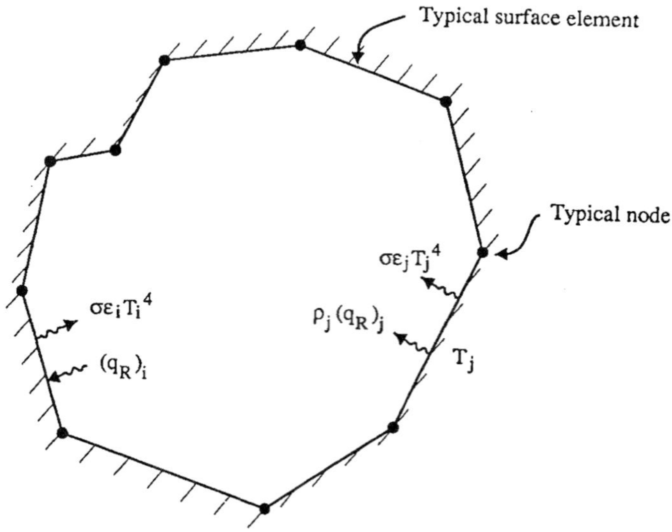

Fig. 8.5 Enclosure with *N* isothermal surfaces.

into the hemispherical space above the surface. A fraction of this energy impinges on surface j with area A_j. The *view factor* F_{ij} is the fraction of the radiation energy leaving i that arrives at j. Techniques for computing radiation view factors are discussed in Ref. 10. A useful property of view factors is the reciprocity rule,

$$F_{ij} A_i = F_{ji} A_j \tag{8.25}$$

Another useful property of view factors is based on energy conservation. For radiation within an enclosure, the radiation energy leaving a surface i in an enclosure must impinge on the surfaces of the enclosure. Thus, energy conservation requires that all fractions of energy leaving surface i total to unity:

$$\sum_{j=1}^{N} F_{ij} = 1 \tag{8.26}$$

where N denotes the number of surfaces.

An enclosure with N surfaces is shown in Fig. 8.5. We assume that 1) each surface is isothermal, 2) each surface is gray, 3) emitted and reflected radiation from each surface is diffusely distributed, 4) radiant heat fluxes are distributed uniformly over each surface, and 5) each surface is opaque so that the reflectivity $\rho = 1 - \varepsilon$. Within the enclosure, each surface emits, reflects, and absorbs radiant energy. We seek a method for computing the absorbed portion of the radiant energy q_R. On a typical surface i, the net heat flow out of the surface is

$$q_i = \sigma \varepsilon_i T_i^4 - \alpha_i (q_R)_i \tag{8.27}$$

The radiation incident on surface i consists of fractions of energy emitted and reflected from all surfaces in the enclosure. The heat flux B_j leaving a surface j consists of the energy emitted and reflected from the surface. Note that we previously used B_i to denote the derivative of an interpolation function. In this section on radiation exchanges, we use the same symbol to denote a heat flux. The usage should be clear from the context. Thus,

$$B_j = \sigma \varepsilon_j T_j^4 + \rho_j (q_R)_j \tag{8.28}$$

The energy incident on surface i is now expressed as the sum of energies leaving all surfaces by using the view factors F_{ij}. Thus, the radiation heat flow on surface i is

$$(q_R)_i A_i = \sum_{j=1}^{N} F_{ij} A_j B_j$$

Using the reciprocity rule, Eq. (8.25) enables us to express this result as

$$(q_R)_i = \sum_{j=1}^{N} F_{ij} B_j \tag{8.29}$$

To obtain the heat fluxes $(q_R)_i$, we now eliminate B_j between Eqs. (8.28) and (8.29). To do this, we first write each equation in matrix notation. From Eq. (8.28),

$$\{B\} = \{\sigma \varepsilon T^4\} + [\rho]\{q_R\}$$

where $[\rho]$ denotes a diagonal matrix of reflectivities. Next, from Eq. (8.29),

$$\{q_R\} = [F]\{B\}$$

Combining the two equations, we obtain

$$[[I] - [F][\rho]]\{q_R\} = [F]\{\sigma \varepsilon T^4\} \tag{8.30}$$

where $[I]$ denotes the identity matrix. Equation (8.30) consists of N simultaneous equations for computing N unknown values of $\{q_R\}$. The N surface temperatures are the average surface temperatures of the elements defining the enclosure.

Radiation with surface heat exchanges. To compute the temperature distribution for a structure with radiation heat exchanges, we perform either a nonlinear steady-state or transient analysis. Let us assume that we perform a Newton–Raphson iteration to handle the radiation nonlinearity. The radiation on each surface is included by using the unbalanced heat load vector, Eq. (8.20a), and the radiation Jacobian, Eq. (8.20b). For the unbalanced heat load vector, we need updated values for $\{q_R\}$ at each iteration. We obtain these heat fluxes by solving Eqs. (8.30) with the latest set of surface temperatures.

The iterative solution involves two sets of simultaneous equations: the $N \times N$ set of equations for the surface heat fluxes $\{q_R\}$ and the standard set of equations for the nodal temperatures $\{T\}$. The $N \times N$ set of equations (8.30) has a full asymmetric coefficient matrix, but the Jacobian in Eq. (8.13a) is a banded asymmetric matrix. If surface reflections are negligible, that is, if $[\rho] = 0$, then Eqs. (8.30) uncouple, and the incident heat fluxes $\{q_R\}$ on each surface are computed directly.

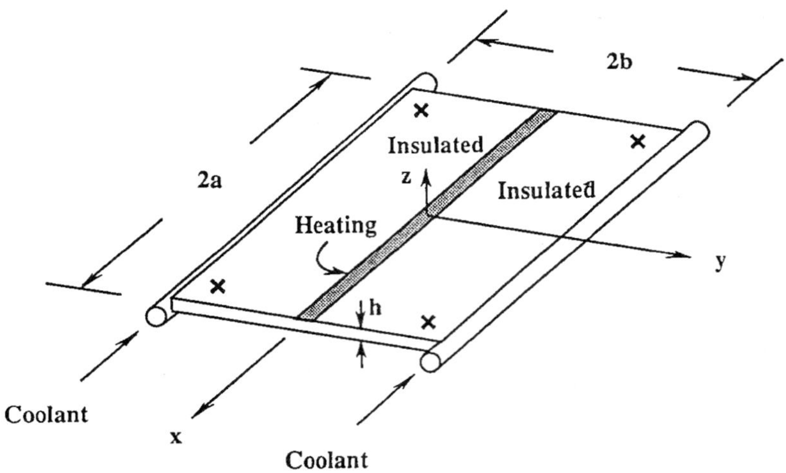

Fig. 8.6 Point-supported Hastelloy-X panel.

8.2.4 Heat-Transfer Problems

To demonstrate finite element thermal modeling and to present some typical results, analyses of three structural heat-transfer problems are presented. Each problem was analyzed using student-developed computer programs that implement the finite element theory described in the preceding sections.

Example 8.1 Transient Response of a Locally Heated Plate

A recent paper by Thornton et al.[11] describes thermal buckling tests of locally heated plate. In an M.S. thesis, Hernan[12] compares finite element temperature predictions with the experimental data. Figure 8.6 diagrams a typical test configuration. A test plate is made of Hastelloy-X of length $2a = 15$ in., width $2b = 10$ in., and thickness $h = 0.125$ in. The plate has point supports at its four corners; these are represented by x in the figure. A plate is heated along its longitudinal centerline by radiation from a quartz lamp while the long edges of the plate are maintained at a constant temperature of 65°F by chilled water flowing through coolant tubes. The plate is encased in an insulation blanket except for the narrow 0.5-in. heated region, and it is instrumented with thermocouples to measure temperatures. In a typical test, a plate is subjected to sudden heating until the center temperature reaches a selected peak value.

Figure 8.7a shows the heat-transfer model for a plate, and Figure 8.7b is a schematic representation of the incident heat flux as a function of time and position. Experimental data show that the temperature gradient through the plate thickness is negligible, and conduction in the plane of the plate is represented by the graded mesh of triangular elements shown in Fig. 8.8. The finite element model consists of 403 nodes and 720 triangular elements. In the finite element model, temperatures are prescribed for nodes along the cooled edges, and the incident flux q_R is prescribed as a function of x and y for surface elements over the heated region. Energy losses due to convection, $h(T - T_e)$, and radiation, $\sigma \varepsilon T^4$, may occur along the uninsulated heated region, and these losses are modeled with surface elements. The temperature dependence of the thermal conductivity and specific heat was included in the nonlinear transient analysis.

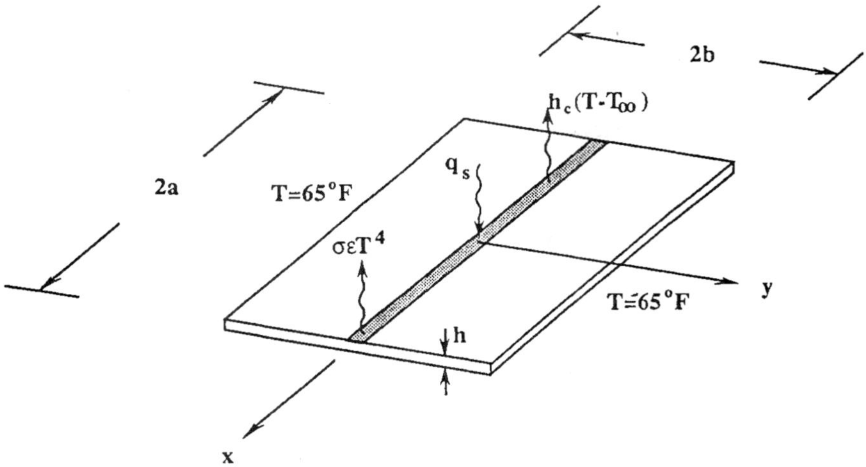

a) Mathematical heat-transfer model

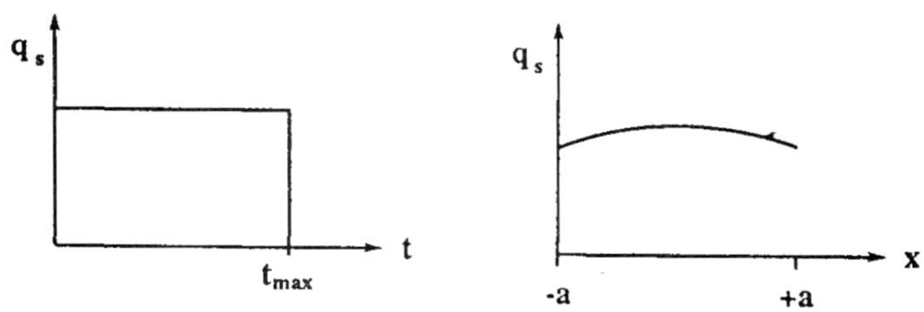

b) Schematic representation of applied heat flux as a function of time and as a function of x

Fig. 8.7 Schematics of experimental plate.

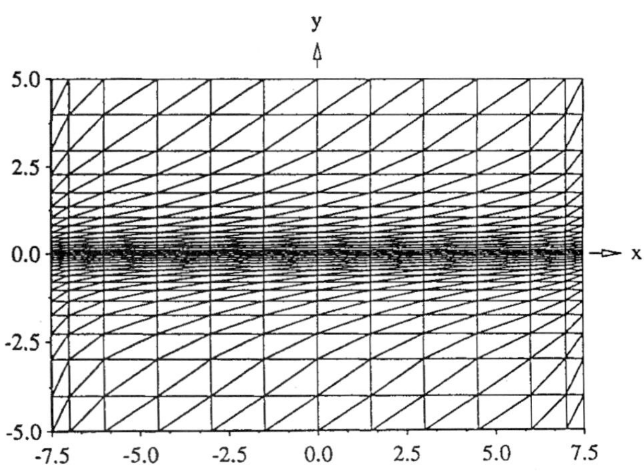

Fig. 8.8 Finite element mesh of Hastelloy-X panel (x and y in inches).

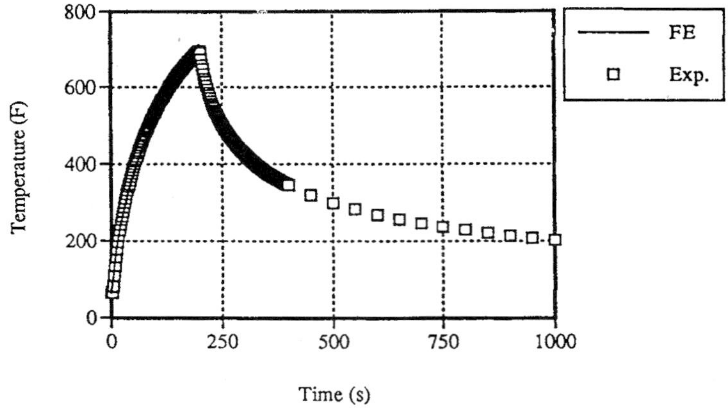

a) Test 4 temperature histories

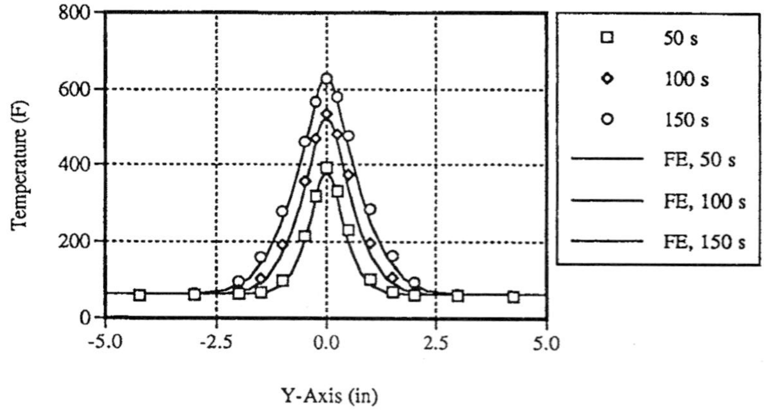

b) Test 4 temperature profiles

Fig. 8.9 Finite element and experimental results.

Two series of five tests each were analyzed by Hernan,[12] and representative results for one test are presented here. Figure 8.9a presents a temperature history at the center of the plate, and Figure 8.7b shows temperature distributions at three times in a response. The temperature distributions shown are along the y axis, but there is very little variation of the temperatures with x. The analysis predicts the response for $0 \leq t \leq 200$ s when the peak temperature of 700°F is reached. The temperature history comparison in Fig. 8.9a shows quite good agreement with the experimental data points covering the solid line representing the finite element predicted temperatures. Figure. 8.9b also shows that the temperature distributions are predicted very well for the three times in the response. The computations showed that, up to 200 s, the convection and radiation heat losses were small and could be neglected.

Example 8.2: Curved Thermal Protection System Panel

Metallic thermal protection systems (TPSs) have been under study by NASA as an alternative to the present Space Shuttle tile TPS system. A curved TPS test panel

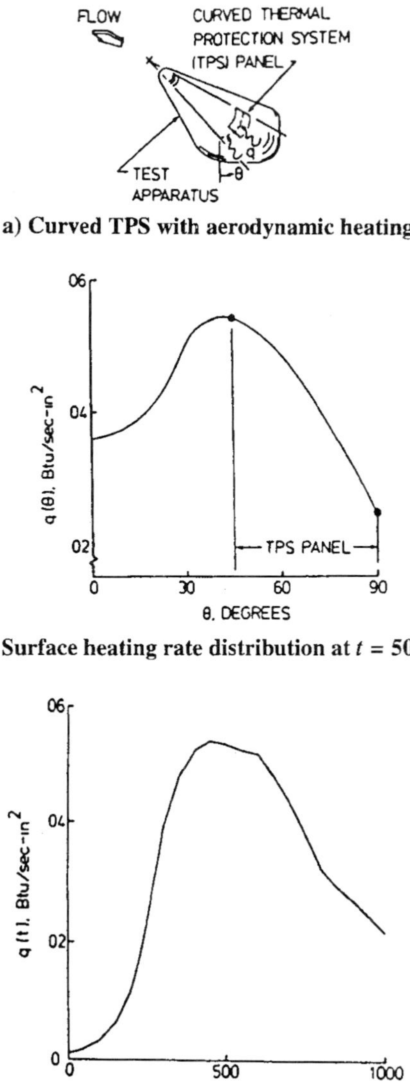

a) Curved TPS with aerodynamic heating

b) Surface heating rate distribution at t = 500 s

c) Surface heating rate history at θ = 45 deg

Fig. 8.10 Aerodynamic heating of curved thermal protection system test panel.

subject to wind-tunnel aerodynamic heating is shown in Fig. 8.10a. The test panel experiences aerodynamic heating that varies significantly over its width, Fig. 8.10b, and strongly with time, Fig. 8.10c. Some details of the sandwich construction of the test panel are shown in Fig. 8.11a. The thermal model represents three-dimensional conduction in the inner and outer face sheets and insulation. Significant energy losses from the outer skin by radiation mandate a nonlinear transient analysis. The finite element model consists of 235 quadrilateral elements to represent the face sheets and end closures. Each face sheet consists of two layers of quadrilateral

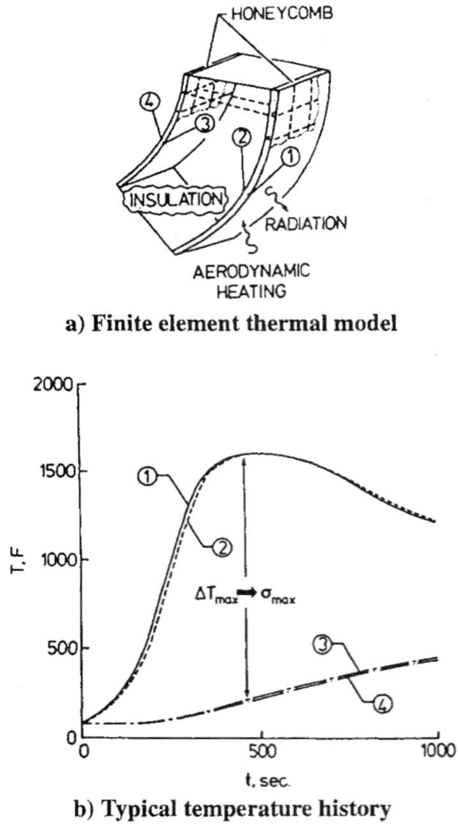

Fig. 8.11 Thermal analysis of curved thermal protection system test panel.

elements to represent the skins connected by a grid of quadrilateral elements oriented perpendicular to the face sheets to connect the inner and outer skins. These latter elements model conduction through the honeycomb. Conduction in the insulation between the face sheets is modeled with 36 one-dimensional elements.

Figure. 8.11b shows typical temperature histories at four points on the panel. The results for points 1 and 2 as well as points 3 and 4 show that there are only very small temperature drops across the honeycomb face sheets. However, there is a major temperature drop from the outer to inner face sheet. The figure shows that this temperature difference reaches a maximum at $t = 450$ s. We should perform the quasistatic structural analysis with the temperature distribution at this time to determine the maximum thermal stresses and deformations. Maximum surface temperatures on the outer skin are about 1500°F, and energy losses due to radiation are significant. These energy losses tend to smooth spatial temperature distributions.

Example 8.3: Conduction and Radiation in a Wing Box

To illustrate finite element analysis with radiation exchanges, temperatures in an idealized titanium wing box were computed. The wing box (Fig. 8.12) is aero-

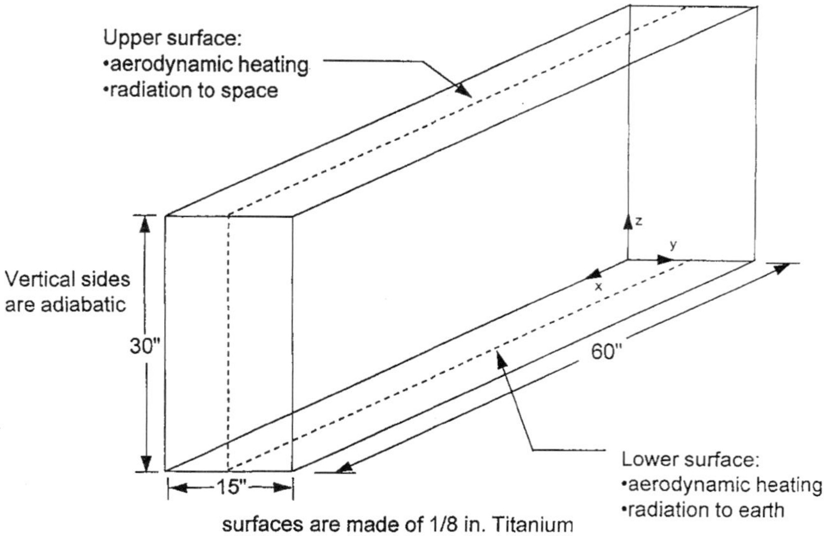

Fig. 8.12 Combined conduction and radiation in a wing box.

dynamically heated on its upper and lower surfaces, with specified heat fluxes corresponding to the wing at an angle of attack so that the lower wing surface experiences higher heat flux than the upper surface. The upper and lower aerodynamic surfaces lose energy by radiation to space. However, all four vertical external surfaces of the wing box are adiabatic and have zero heat loss. The upper and lower surfaces as well as the four sides of the box are made of 1/8-in.-thick titanium and experience two-dimensional conduction in their planes. Radiation exchanges occur between all internal surfaces.

The wing box was represented by a uniform finite element mesh with 702 nodes and 1400 triangular elements. Each triangle modeled conduction with surface radiation. Steady-state temperatures were computed using the Newton–Raphson iteration scheme represented by Eqs. (8.13). At each iteration, incident radiation heat fluxes on the internal surfaces of the elements were computed using Eq. (8.30).

Temperature profiles for the centerline of the wing box are shown in Figs. 8.13 and 8.14. Figure 8.13 shows the temperature distributions on the upper and lower surfaces of the box, and Fig. 8.14 shows a typical temperature distribution on a side surface. Because of higher heating, the lower surface exhibits higher temperatures, as expected. The effect of the internal radiation is significant. If the internal radiation is neglected, the average temperature of the lower surface is increased, and the average temperature of the upper surface is decreased. Radiation accounts for a surface-temperature change of about 10°F.

8.3 Structural Analysis

We assume that the structural temperature distribution is known from a thermal analysis. Our task now is to compute structural deformations and stresses. From the thermal analysis, we know nodal temperatures for the structural finite element model. For a steady-state thermal problem, we have a single temperature vector $\{T\}$. For a transient thermal problem, we have several temperature vectors at suc-

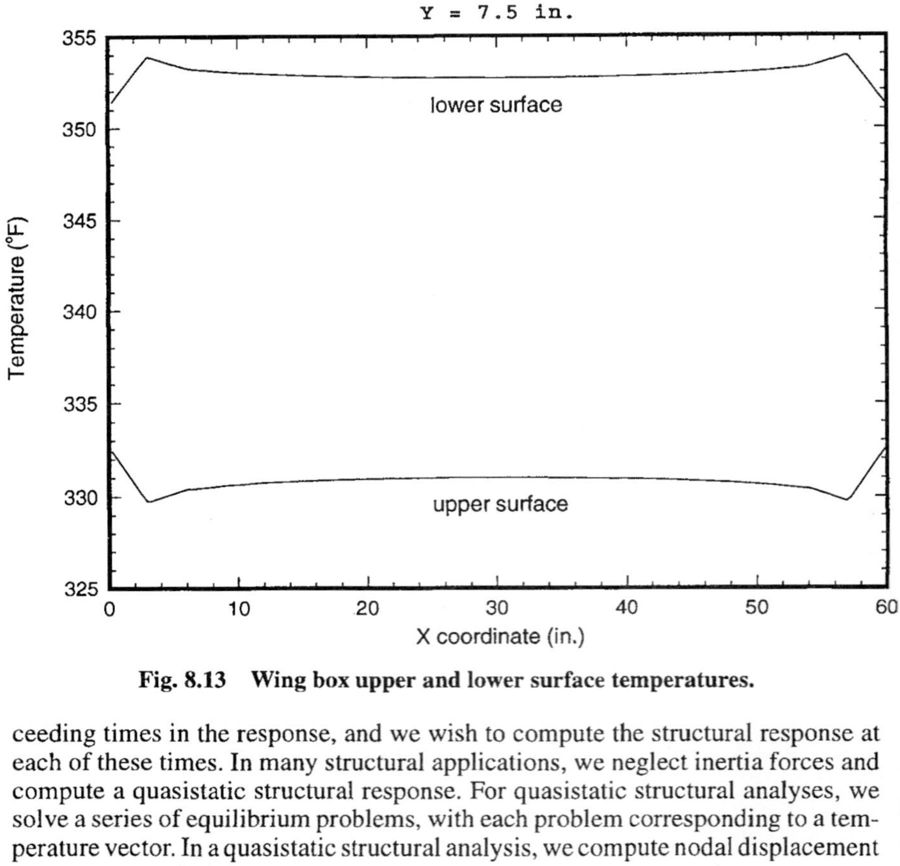

Fig. 8.13 Wing box upper and lower surface temperatures.

ceeding times in the response, and we wish to compute the structural response at each of these times. In many structural applications, we neglect inertia forces and compute a quasistatic structural response. For quasistatic structural analyses, we solve a series of equilibrium problems, with each problem corresponding to a temperature vector. In a quasistatic structural analysis, we compute nodal displacement components and element stresses.

The temperature distribution affects the structural problem through 1) temperature-dependent material properties and 2) equivalent nodal forces and moments.

Material properties such as the modulus of elasticity and the coefficient of thermal expansion may change appreciably if the structure experiences a temperature change of a few hundred degrees. Thus, we need to accommodate temperature-dependent properties in our analyses. Property data are usually available in the form of tables or simple algebraic equations from curve fits. Often in the finite element structural analysis, the temperature-dependent properties are based on the average element temperature and assumed constant within an element. But, because each element usually has a different average temperature, material properties vary throughout the structural model. Note that, in a structural analysis, temperature-dependent properties do *not* cause the analysis to be nonlinear because temperature plays the role of a specified parameter in the structural analysis. A structural analysis is nonlinear if the constitutive law is nonlinear, say, due to plasticity, or if the structure experiences large displacements.

Equivalent nodal forces and moments are integrals of the temperature distribution over an element. These integrals are similar to the integrals we encounter for thermal forces and moments in analytical approaches. A contribution of our finite element formulations is to provide the integrals for equivalent nodal loads appropriate to each element. We develop finite formulations for the basic struc-

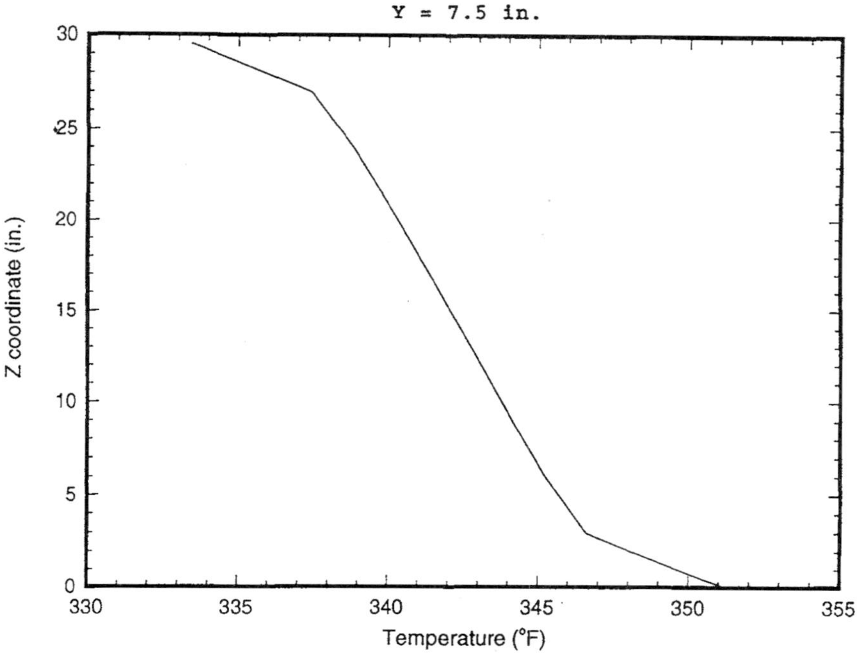

Fig. 8.14 Wing box side surface temperatures.

tural components—rods, beams, and plates—as well as a general formulation for an elastic continuum. Quasistatic behavior is assumed throughout this chapter.

8.3.1 Rods

Problem statement. We begin with the analytical formulation for the rod presented in Sec. 6.2. The fundamental equations are equilibrium, strain-displacement, and Hooke's law:

$$\frac{\partial}{\partial x}(\sigma_x A) + F_x A = 0 \tag{8.31a}$$

$$\varepsilon_x = \frac{\partial u}{\partial x} \tag{8.31b}$$

$$\sigma_x = E\varepsilon_x - E\alpha(T - T_{\text{ref}}) \tag{8.31c}$$

where the stress is σ_x, the cross-sectional area is A, and body force per unit volume is F_x.

Finite element formulation. We assume a one-dimensional element with r nodes. If $r = 2$, we have a linear element with two nodes; if $r = 3$, we have a quadratic element with three nodes. The interpolation equation for the axial displacement is

$$u(x) = [N_1 \quad N_2, \ldots, \; N_r]\{u\} \tag{8.32}$$

where $N_i(x), i = 1, 2, \ldots, r$, are the interpolation functions, and $\{u\}$ is a vector of nodal displacements. The interpolation equation for the element strain ε_x is obtained by differentiation,

$$\varepsilon_x(x) = \frac{\partial u}{\partial x} = [B_1 \quad B_2, \ldots, \quad B_r]\{u\} \tag{8.33}$$

where $B_i = \mathrm{d}N_i/\,\mathrm{d}x, i = 1, 2, \ldots, r$, are the strain interpolation functions.

To derive the general element equations, we multiply the equilibrium equation (8.31a) by a weighting function N_i and integrate over the length of the element. For convenience, we assume that the element is located in a local coordinate system with $0 \leq x \leq L$, where L is the element length. Then we may write

$$\int_0^L \frac{\partial}{\partial x}(\sigma_x A)N_i \,\mathrm{d}x + \int_0^L F_x A N_i \,\mathrm{d}x = 0 \qquad i = 1, 2, \ldots, r \tag{8.34}$$

where we have r element equations since we have r unknown nodal displacements, $u_1, u_2, \ldots, u_r$. The next step is to integrate the stress term by parts. The integration introduces the element boundary conditions and reduces the order required for the interpolation functions. After the integration, we have

$$[\sigma_x A N_i]_0^L - \int_0^L B_i \sigma_x A \,\mathrm{d}x + \int_0^L F_x A N_i \,\mathrm{d}x = 0 \qquad i = 1, 2, \ldots, r \tag{8.35}$$

where we have introduced $B_i = \mathrm{d}N_i/\mathrm{d}x$. From Eqs. (8.31b) and (8.31c), we may write the element stress as

$$\sigma_x = E\frac{\partial u}{\partial x} - E\alpha\Delta T$$

or

$$\sigma_x = E[B]\{u\} - E\alpha\Delta T \tag{8.36}$$

where $\Delta T = T(x) - T_{\text{ref}}$. Substituting the stress in Eq. (8.35) and writing the result in matrix form yield

$$\begin{Bmatrix} -\sigma_x(0)A \\ 0 \\ \vdots \\ 0 \\ \sigma_x(L)A \end{Bmatrix} - \int_0^L EA\{B\}[B] \,\mathrm{d}x\{u\} + \int_0^L E\alpha\Delta T\{B\} \,\mathrm{d}x$$

$$+ \int_0^L F_x A\{N\} \,\mathrm{d}x = 0 \tag{8.37}$$

In going from Eqs (8.35) to (8.37), we evaluated the boundary term, taking into account that the interpolation functions have the value either of 0 or 1 at the element nodes. The resulting term in braces in Eq. (8.37) represents forces at the element's two end nodes. Finally, we write Eq. (8.37) in the more compact form,

$$[K]\{u\} = \{F_c\} + \{F_T\} + \{F_B\} \tag{8.38a}$$

where

$$[K] = \int_0^L EA\{B\}[B]\,\mathrm{d}x \tag{8.38b}$$

$$\{F_c\} = \begin{Bmatrix} -F_1 \\ 0 \\ \vdots \\ 0 \\ F_r \end{Bmatrix} \tag{8.38c}$$

$$\{F_T\} = \int_0^L E\alpha\Delta T\{B\}\,\mathrm{d}x \tag{8.38d}$$

$$\{F_B\} = \int_0^L F_x A\{N\}\,\mathrm{d}x \tag{8.38e}$$

Equation (8.38a) represents element equilibrium, where $[K]$ is the element stiffness matrix. The right-hand side has the element's nodal forces. The vector $\{F_c\}$ represents concentrated nodal forces at the element's ends, the vector $\{F_T\}$ represents equivalent nodal forces due to the prescribed temperature distribution, and $\{F_B\}$ represents equivalent nodal forces due to the distributed body force. These element matrices are assembled to form the global matrices for a rod. The concentrated nodal forces cancel in pairs at interior nodes but may exist at the ends of the rod. When global boundary conditions for the rod are imposed, the end forces will either be prescribed or, if displacements are specified, be unknown reactions.

Element matrices. For a rod, elements usually are linear with two nodes or quadratic with three nodes. The most popular element is the linear element where the displacement and strain interpolation functions are given in Sec. 8.2.1. Note that, because the terms of $[B]$ are constant, the strain ε_x by Eq. (8.33) is constant within the element. Note, however, that Eq. (8.36) shows that the stress σ_x varies within the element since $\Delta T = T(x) - T_{\text{ref}}$.

The element's matrices are evaluated by direct integration. For convenience, we assume that the material properties are evaluated at the average element temperature T_a and are constant within the element. In addition, we assume a constant area A. Then the element stiffness matrix is

$$[K] = E(T_a)A\int_0^L \begin{Bmatrix} -1/L \\ 1/L \end{Bmatrix} [-1/L \quad 1/L]\,\mathrm{d}x$$

$$[K] = \frac{E(T_a)A}{L}\begin{bmatrix} 1 & -1 \\ -1 & 1 \end{bmatrix} \tag{8.39}$$

The thermal load vector is given by

$$\{F_T\} = E(T_a)\alpha(T_a)\int_0^L [T(x) - T_{\text{ref}}]\,\mathrm{d}x \begin{Bmatrix} -1/L \\ 1/L \end{Bmatrix}$$

which we rewrite as

$$\{F_T\} = F_T \begin{Bmatrix} -1 \\ 1 \end{Bmatrix} \tag{8.40a}$$

where

$$F_T = \frac{E(T_a)\alpha(T_a)}{L} \int_0^L [T(x) - T_{\text{ref}}]\, \mathrm{d}x \tag{8.40b}$$

The last result shows that the equivalent nodal forces depend on an element's temperature distribution. A popular approach is to assume a coincident two-node thermal element with a linear temperature distribution. Alternatively, we may select a higher-order temperature distribution and evaluate the equivalent nodal forces using Eq. (8.40). If we have an element body force such as weight or centrifugal force, we include its effects by the load vector given in Eq. (8.38e). If the body force is constant, then,

$$\{F_B\} = F_x A \int_0^L \begin{Bmatrix} 1 - x/L \\ x/L \end{Bmatrix} \mathrm{d}x = \frac{F_x A L}{2} \begin{Bmatrix} 1 \\ 1 \end{Bmatrix} \tag{8.41}$$

This last equation shows that the resultant force $F_x AL$ is divided equally between the element's nodes. In solving a problem, the foregoing element matrices are assembled to form the global matrices. Then the nodal boundary conditions of either specified displacements or forces are imposed. Subsequently, we solve the resulting simultaneous equations for the nodal displacements $\{u\}$. With the nodal displacements known, we may then compute the stresses within each element. From Eq. (8.36), using the linear interpolation for displacement given in Sec. 8.2.1, we may write

$$\sigma_x = E(T_a)\frac{u_2 - u_1}{L} - E(T_a)\alpha(T_a)[T(x) - T_{\text{ref}}] \tag{8.42}$$

Equation (8.42) shows, as we mentioned earlier, that the stress varies within the element because of the temperature variation. Often, the stress is calculated at the element's centroid.

Truss structures. The element matrices derived in the preceding section represent equilibrium of an element in a local coordinate system. The element has two degrees of freedom, u_1 and u_2, which are the nodal displacements along the local x axis. For a member of a general three-dimensional truss, each node has three possible displacements—u, v, w—in the global XYZ coordinate system (Fig. 8.15). Thus, in global coordinates, the element with two nodes has six degrees of freedom. To analyze a three-dimensional truss, element matrices must be transformed to the global system before they are assembled in the system matrices. The transformation equations are

$$[K]_{\text{Global}} = [R]^T [K]_{\text{Local}} [R] \tag{8.43a}$$

$$\{F\}_{\text{Global}} = [R]^T \{F\}_{\text{Local}} \tag{8.43b}$$

where $[R]$ is the transformation matrix. For a three-dimensional truss element,

$$[R] = \begin{bmatrix} l & m & n & 0 & 0 & 0 \\ 0 & 0 & 0 & l & m & n \end{bmatrix} \tag{8.44}$$

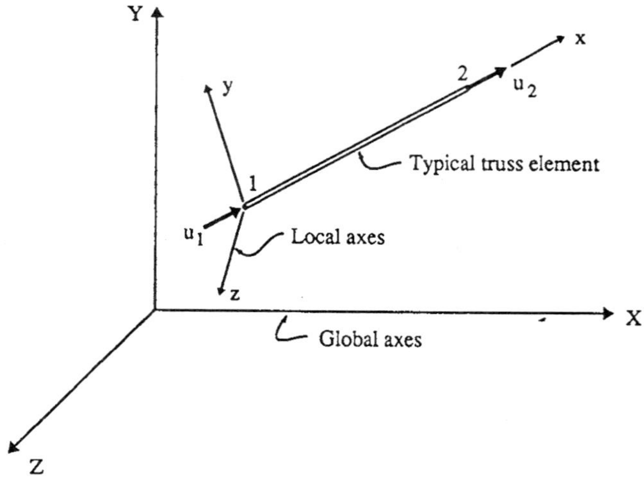

Fig. 8.15 Truss element with local and global coordinate systems.

where l, m, n are the cosines of the angles between the local x axis and the global X, Y, Z axes. After the element matrices are transformed to global coordinates, the system matrices are assembled, boundary conditions are imposed, and the system equations are solved for the nodal displacements. From the nodal displacements in global coordinates, we may compute the element's local nodal displacements from

$$\{u\}_{\text{Local}} = [R]\{u\}_{\text{Global}} \tag{8.45}$$

With the local displacements, we can then compute the element stress using Eq. (8.42). For a two-dimensional truss in the global X–Yplane, the local x and global z axes are perpendicular so that the direction cosine n is zero. The transformation matrix $[R]$ in Eq. (8.44) reduces to a 2×4 array from a 2×6. Otherwise, all steps in the analysis are identical.

In this section, we develop a general formulation for a rod element with thermal effects and give specific equations for a two-node linear element. The equations apply also to cables or tethers, provided they are in tension and tensile forces are small. Large tensile forces induce an additional stiffening effect, which must be taken into account in the formulation; see Example 8.4. Reference 13 discusses finite element formulations for isothermal cables.

8.3.2 Beams

Problem statement. We consider a Bernoulli–Euler beam for which the coordinate planes coincide with the principal axes of the cross section. In this case, bending in the x–y and x–z planes occurs independently. For simplicity, we consider bending only in the x–y plane. The generalization for a three-dimensional beam element is straightforward.

The fundamental beam equations appear in Sec. 6.3.1. From Eq. (6.18), we may write the equations as

$$M = -EI\frac{\mathrm{d}^2 v}{\mathrm{d}x^2} - M_T \tag{8.46a}$$

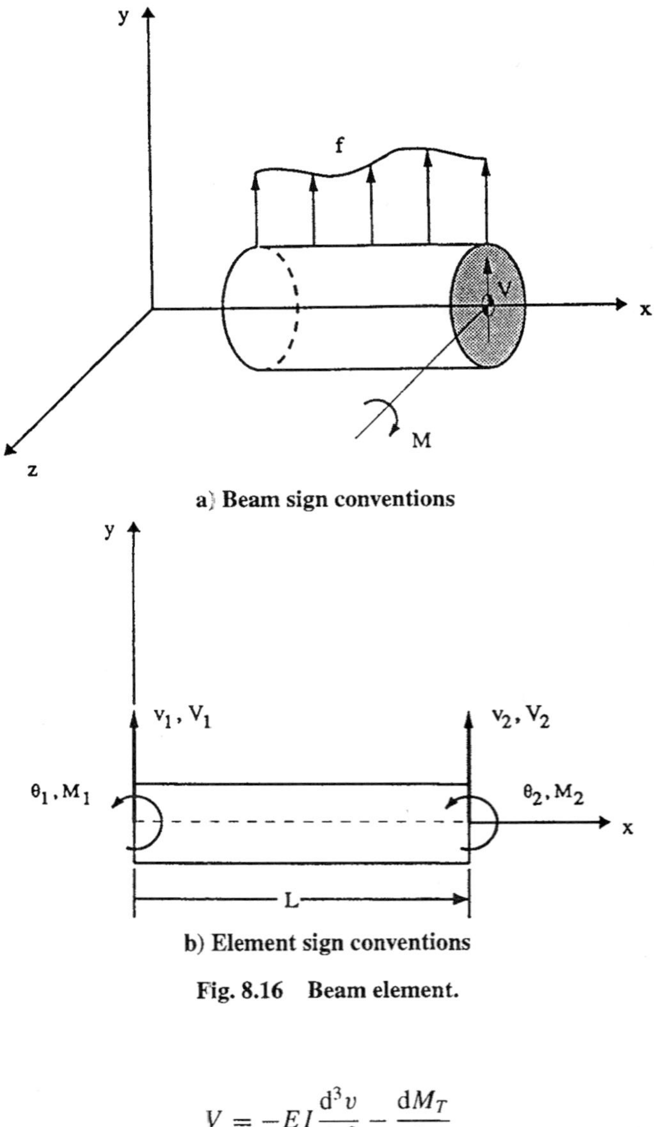

a) Beam sign conventions

b) Element sign conventions

Fig. 8.16 Beam element.

$$V = -EI\frac{\mathrm{d}^3 v}{\mathrm{d}x^3} - \frac{\mathrm{d}M_T}{\mathrm{d}x} \tag{8.46b}$$

$$EI\frac{\mathrm{d}^4 v}{\mathrm{d}x^4} + \frac{\mathrm{d}^2 M_T}{\mathrm{d}x^2} = f \tag{8.46c}$$

The sign conventions for the bending moment M, the shear force V, and the distributed force f are shown in Fig. (8.16a). The thermal moment is defined by Eq. (6.12c) as

$$M_T = \int_A E\alpha\Delta T y \,\mathrm{d}A \tag{8.47}$$

where $\Delta T = T(x, y, z) - T_{\mathrm{ref}}$, and the integration takes place over the beam cross-sectional area A. The beam's normal stress $\sigma_x$is given by Eq. (6.9b),

$$\sigma_x = -Ey\frac{\mathrm{d}^2 v}{\mathrm{d}x^2} - E\alpha\Delta T \tag{8.48}$$

The basic objective of a finite element solution is to solve Eq. (8.46c) for the transverse deflection. Once the deflection is known, we can evaluate the bending stress σ_x from Eq. (8.48).

Finite element formulation. In our formulation of the rod element, we use the displacement u as the basic unknown, and we interpolate $u(x)$ from nodal displacement values. As we assemble the rod element stiffness matrices to form the global stiffness matrix, the basic requirement is continuity of nodal displacements. In finite element terminology, the requirement of continuity of the dependent variable at nodes is called C^0 (read C zero) continuity. For second-order differential equations (as with the rod), we require C^0 continuity.

The differential equation for the beam displacement, Eq. (8.46c), is fourth-order, and we must have higher-order continuity than C^0 at the nodes. More specifically, we must have continuity of the dependent variable and its first derivative, which is called C^1 continuity. Because, for beams, the first derivative of the displacement is the slope, we take as the nodal degrees of freedom the displacement v and slope $\theta = \mathrm{d}v/\mathrm{d}x$. Figure 8.16b shows the nodal degrees of freedom for a typical beam element. We have two unknowns, the displacement and slope at each node, hence, there are a total of four degrees of freedom, denoted as $v_1, \theta_1, v_2, \theta_2$. The interpolation equation for the transverse displacement v for the two-node element shown in Fig. 8.16b is

$$v(x) = [N_1(x) \quad N_2(x) \quad N_3(x) \quad N_4(x)] \begin{Bmatrix} v_1 \\ \theta_1 \\ v_2 \\ \theta_2 \end{Bmatrix} \tag{8.49a}$$

where the interpolation functions are defined as

$$\begin{aligned} N_1 &= 1 - 3\frac{x^2}{L^2} + 2\frac{x^3}{L^3} \\ N_2 &= x - 2\frac{x^2}{L} + \frac{x^3}{L^2} \\ N_3 &= \frac{3x^2}{L^2} - \frac{2x^3}{L^3} \\ N_4 &= -\frac{x^2}{L} + \frac{x^3}{L^2} \end{aligned} \tag{8.49b}$$

The interpolation functions $N_i, i = 1, 2, 3, 4$, are in the local coordinate system, and L is the beam element length.

To derive the finite element equilibrium equations, we multiply Eq. (8.46c) by a weighting function N_i and integrate over the length of the element. Then, by the method of weighted residuals, we have

$$\int_0^L \left(EI\frac{\mathrm{d}^4 v}{\mathrm{d}x^4} + \frac{\mathrm{d}^2 M_T}{\mathrm{d}x^2} - f \right) N_i \,\mathrm{d}x = 0 \qquad i = 1, 2, 3, 4 \tag{8.50}$$

where we have four element equations since we have four degrees of freedom. The next step is to integrate by parts. Here we integrate the derivative terms in Eq. (8.50) by parts twice to introduce the element boundary conditions. After the integration by parts, we obtain

$$\int_0^L EI\frac{\mathrm{d}^2N_i}{\mathrm{d}x^2}\frac{\mathrm{d}^2v}{\mathrm{d}x^2}\,\mathrm{d}x = \left[-\left(EI\frac{\mathrm{d}^3v}{\mathrm{d}x^3}+\frac{\mathrm{d}M_T}{\mathrm{d}x}\right)N_i + \left(EI\frac{\mathrm{d}^2v}{\mathrm{d}x^2}+M_T\right)\frac{\mathrm{d}N_i}{\mathrm{d}x}\right]_0^L$$
$$-\int_0^L M_T\frac{\mathrm{d}^2N_i}{\mathrm{d}x^2}\,\mathrm{d}x + \int_0^L fN_i\,\mathrm{d}x \qquad i=1,2,3,4 \tag{8.51}$$

Finally, we introduce the interpolation for $v(x)$ from Eq. (8.49a) and write the result in matrix notation. The result is

$$[K]\begin{Bmatrix} v_1 \\ \theta_1 \\ v_2 \\ \theta_2 \end{Bmatrix} = \begin{Bmatrix} V_1 \\ M_1 \\ V_2 \\ M_2 \end{Bmatrix} + \{F_T\} + \{F_B\} \tag{8.52a}$$

where the stiffness matrix is given by

$$[K] = \int_0^L EI\left\{\frac{\mathrm{d}^2N}{\mathrm{d}x^2}\right\}\left[\frac{\mathrm{d}^2N}{\mathrm{d}x^2}\right]\mathrm{d}x \tag{8.52b}$$

and the equivalent nodal load vectors are

$$\{F_T\} = -\int_0^L M_T\left\{\frac{\mathrm{d}^2N}{\mathrm{d}x^2}\right\}\mathrm{d}x \tag{8.52c}$$

$$\{F_B\} = \int_0^L f\{N\}\,\mathrm{d}x \tag{8.52d}$$

The load vector $\{F_T\}$ represents equivalent nodal loads due to the thermal moment M_T, and the load vector $\{F_B\}$ represents equivalent nodal loads due to the distributed force f. In Eq. (8.52a), the vector on the right-hand side contains nodal shear forces V_1 and V_2 and nodal moments M_1 and M_2. This vector arises from evaluation of the boundary term in Eq. (8.51) at $x = 0$ and $x = L$ for $i = 1, 2, 3, 4$.

Element matrices. To obtain the element matrices, we evaluate the integrals given in Eq. (8.52) using the interpolation function given in Eqs. (8.49). If the elastic modulus is assumed to be evaluated at an average element temperature, the stiffness matrix is

$$[K] = \frac{E(T_a)I}{L^3}\begin{bmatrix} 12 & 6L & -12 & 6L \\ 6L & 4L^2 & -6L & 2L^2 \\ -12 & -6L & 12 & -6L \\ 6L & 2L^2 & -6L & 4L^2 \end{bmatrix} \tag{8.53}$$

To evaluate the thermal load vector in Eq. (8.52c), we must know the dependence of the thermal moment M_T on x. If we assume that the thermal moment is constant, we can evaluate the thermal load vector directly,

$$\{F_T\} = -M_T \int_0^L \left\{ \frac{\mathrm{d}^2 N}{\mathrm{d}x^2} \right\} \mathrm{d}x$$

which yields

$$\{F_T\} = M_T \begin{Bmatrix} 0 \\ 1 \\ 0 \\ -1 \end{Bmatrix} \tag{8.54}$$

Note that the equivalent thermal load vector consists of self-equilibrating thermal moments.

The equivalent nodal loads due to the distributed force f depend also on how f varies with x. If we assume f constant, then we can evaluate Eq. (8.52d) to obtain

$$\{F_B\} = \frac{fL}{2} \begin{Bmatrix} 1 \\ L/6 \\ 1 \\ -L/6 \end{Bmatrix} \tag{8.55}$$

Equation (8.55) shows that equivalent nodal forces are one-half the resultant force fL. In addition, however, there are nodal moments of magnitude $fL^2/12$ that follow from the finite element mathematics.

The element stress σ_x is evaluated by substituting the interpolation equation (8.49) into Eq. (8.48). The result may be written in the form

$$\sigma_x = -E[N''(x)] \begin{Bmatrix} v_1 \\ \theta_1 \\ v_2 \\ \theta_2 \end{Bmatrix} y - E\alpha[T(x, y, z) - T_{\mathrm{ref}}] \tag{8.56a}$$

where

$$[N''(x)] = \left[\left(-\frac{6}{L^2} + \frac{12x}{L^3} \right) \quad \left(-\frac{4}{L} + \frac{6x}{L^2} \right) \left(\frac{6}{L^2} - \frac{12x}{L^3} \right) \quad \left(-\frac{2}{L} + \frac{6x}{L^2} \right) \right] \tag{8.56b}$$

In the beam element, the stress varies along the length and through the depth so that we need to evaluate Eq. (8.56) at several points to determine the maximum tensile and compressive stresses.

The extension of the above formulation for a three-dimensional beam element is given by Przemieniecki.[14] He includes the effects of shear deformations in the stiffness matrix and provides the transformation matrices required if the beam element's local x axis is inclined to the global XYZ system. A finite element formulation for the isothermal large-deflection beam problem is given by Reddy.[6] Since the large-deflection problem is nonlinear, an iterative solution is required. We describe this type of formulation for plates in Sec. 8.3.5. Finite element analysis of isothermal thin-walled beams is discussed by Gellin and Lee.[15]

8.3.3 Thermoelastic Solids

We have presented finite element formulations for the rod and beam. We developed the formulations beginning with the specialized governing differential equations that describe rod and beam behavior. Here we develop the finite element formulation for a general three-dimensional thermoelastic solid. The formulation begins with the general equations of thermoelasticity. The approach follows the displacement formulation discussed in Chapter 5 because our basic unknowns are the displacement components u, v, w.

Problem statement. The problem is formulated in terms of the equations of motion, the strain-displacement equations, and the constitutive equations of a thermoelastic solid. For convenience, we summarize the equations here; where appropriate, we write the equations in the matrix forms appropriate for finite element equations.

Equilibrium equations. From Eq. (3.18), the three equilibrium equations are

$$\frac{\partial \sigma_x}{\partial x} + \frac{\partial \sigma_{xy}}{\partial y} + \frac{\partial \sigma_{xz}}{\partial z} + B_x = 0 \tag{8.57a}$$

$$\frac{\partial \sigma_{xy}}{\partial x} + \frac{\partial \sigma_y}{\partial y} + \frac{\partial \sigma_{yz}}{\partial z} + B_y = 0 \tag{8.57b}$$

$$\frac{\partial \sigma_{xz}}{\partial x} + \frac{\partial \sigma_{yz}}{\partial y} + \frac{\partial \sigma_z}{\partial z} + B_z = 0 \tag{8.57c}$$

In finite element equations, we write the stress components as a vector $\{\sigma\}$, where

$$\{\sigma\}^T = \begin{bmatrix} \sigma_x & \sigma_y & \sigma_z & \sigma_{xy} & \sigma_{xz} & \sigma_{yz} \end{bmatrix} \tag{8.57d}$$

Strain-displacement equations. From Eq. (3.10), the six strain-displacement equations are

$$\begin{aligned} \varepsilon_x &= \frac{\partial u}{\partial x} & \varepsilon_{xy} &= \frac{1}{2}\left(\frac{\partial u}{\partial y} + \frac{\partial v}{\partial x}\right) & \varepsilon_{xz} &= \frac{1}{2}\left(\frac{\partial u}{\partial z} + \frac{\partial w}{\partial x}\right) \\ & & \varepsilon_y &= \frac{\partial v}{\partial y} & \varepsilon_{yz} &= \frac{1}{2}\left(\frac{\partial v}{\partial z} + \frac{\partial w}{\partial x}\right) \\ & & & & \varepsilon_z &= \frac{\partial w}{\partial z} \end{aligned} \tag{8.58a}$$

In finite element equations, we write the strain components as a vector $\{\varepsilon\}$, where

$$\{\varepsilon\}^T = \begin{bmatrix} \varepsilon_x & \varepsilon_y & \varepsilon_z & \varepsilon_{xy} & \varepsilon_{xz} & \varepsilon_{yz} \end{bmatrix} \tag{8.58b}$$

Constitutive equations. From Eq. (3.23), we may write the six equations of generalized Hooke's law in the matrix form

$$\{\sigma\} = [C]\{\varepsilon\} - [C]\{\alpha\}\Delta T \tag{8.59a}$$

where the elasticity matrix expressed in terms of E and υ is

$$[C] = \frac{E}{(1+\upsilon)(1-2\upsilon)} \begin{bmatrix} 1-\upsilon & \upsilon & \upsilon & 0 & 0 & 0 \\ \upsilon & 1-\upsilon & \upsilon & 0 & 0 & 0 \\ \upsilon & \upsilon & 1-\upsilon & 0 & 0 & 0 \\ 0 & 0 & 0 & 1-2\upsilon & 0 & 0 \\ 0 & 0 & 0 & 0 & 1-2\upsilon & 0 \\ 0 & 0 & 0 & 0 & 0 & 1-2\upsilon \end{bmatrix} \tag{8.59b}$$

and the vector of thermal expansion coefficients is

$$\{\alpha\}^T = [\alpha \quad \alpha \quad \alpha \quad 0 \quad 0 \quad 0] \tag{8.59c}$$

where, of course, $\Delta T = T(x, y, z) - T_{\text{ref}}$.

Boundary conditions. We specify either the displacements on a portion of the surface S_1 or the surface tractions on a portion of the surface S_2. In the development of the finite element equations, we use the relationship between the surface tractions T_x, T_y, T_z and stress tensor given in Eq. (5.5b). In matrix notation,

$$\begin{Bmatrix} T_x \\ T_y \\ T_z \end{Bmatrix} = \begin{bmatrix} \sigma_x & \sigma_{xy} & \sigma_{xz} \\ \sigma_{xy} & \sigma_y & \sigma_{yz} \\ \sigma_{xz} & \sigma_{yz} & \sigma_z \end{bmatrix} \begin{Bmatrix} l \\ m \\ m \end{Bmatrix} \quad \text{on} S_2 \tag{8.60}$$

where l, m, n are the components of a unit vector normal to the surface S_2.

Finite element formulation. The solution domain Ω is divided into M elements of r nodes each. We express the displacements within an element in terms of the nodal displacements with the interpolation equation,

$$\begin{Bmatrix} u(x,y,z) \\ v(x,y,z) \\ w(x,y,z) \end{Bmatrix} = [N(x,y,z)]\{u\} \tag{8.61a}$$

where the interpolation matrix $[N]$ is a rectangular array of dimensions $3 \times 3r$. Alternatively, we may represent each row in the preceding equation by a summation,

$$u = \sum_{i=1}^{r} N_i(x,y,z)u_i$$
$$v = \sum_{i=1}^{r} N_i(x,y,z)v_i \tag{8.61b}$$
$$w = \sum_{i=1}^{r} N_i(x,y,z)w_i$$

which shows that there are r interpolation functions, $N_i(x, y, z)$, $i = 1, 2, \ldots, r$. The displacement vector $\{u\}$ has $3r$ terms and has the form

$$\{u\}^T = [u_1 \quad v_1 \quad w_1 \quad \cdots \quad u_r \quad v_r \quad w_r] \tag{8.61c}$$

Using the strain-displacement relations (8.58), we may express the strains within an element by

$$\{\varepsilon\} = [B(x, y, z)]\,\{u\} \tag{8.62}$$

where $[B]$ is a $6 \times 3r$ array whose elements are derivatives of the interpolation functions defined in Eq. (8.61a).

We focus our attention now on a single finite element of domain Ω_e, and we use the MWR to derive the element matrix equations. Because we have $3r$ nodal unknowns shown in Eq. (8.61c), we need $3r$ element equations. We multiply each of the equilibrium equations (8.57) by the interpolation functions N_i and integrate over the element domain Ω_e. Thus, we may write

$$\begin{aligned}
&\int_{\Omega_e}\left[\frac{\partial\sigma_x}{\partial x}+\frac{\partial\sigma_{xy}}{\partial y}+\frac{\partial\sigma_{xz}}{\partial z}+B_x\right]N_i\,\mathrm{d}\Omega=0\\
&\int_{\Omega_e}\left[\frac{\partial\sigma_{xy}}{\partial x}+\frac{\partial\sigma_y}{\partial y}+\frac{\partial\sigma_{yz}}{\partial z}+B_y\right]N_i\,\mathrm{d}\Omega=0\\
&\int_{\Omega_e}\left[\frac{\partial\sigma_{xz}}{\partial x}+\frac{\partial\sigma_{yz}}{\partial y}+\frac{\partial\sigma_z}{\partial z}+B_z\right]N_i\,\mathrm{d}\Omega=0 \qquad i=1,2,\ldots,r
\end{aligned} \tag{8.63}$$

Following our standard procedure, we integrate the derivatives of the stresses by parts using Gauss's theorem. The integration introduces a surface integral on the element boundary S_e where the boundary conditions are specified. We may write the results as

$$\begin{aligned}
&-\int_{\Omega_e}\left(\sigma_x\frac{\partial N_i}{\partial x}+\sigma_{xy}\frac{\partial N_i}{\partial y}+\sigma_{xz}\frac{\partial N_i}{\partial z}\right)\mathrm{d}\Omega+\int_{\Omega_e}N_iB_x\,\mathrm{d}\Omega\\
&\quad+\int_{S_e}N_i\left(\sigma_xl+\sigma_{xy}m+\sigma_{xz}n\right)\,\mathrm{d}S=0\\
&-\int_{\Omega_e}\left(\sigma_{xy}\frac{\partial N_i}{\partial x}+\sigma_y\frac{\partial N_i}{\partial y}+\sigma_{yz}\frac{\partial N_i}{\partial z}\right)\mathrm{d}\Omega+\int_{\Omega_e}N_iB_y\,\mathrm{d}\Omega\\
&\quad+\int_{S_e}N_i\left(\sigma_{xy}l+\sigma_ym+\sigma_{yz}n\right)\,\mathrm{d}S=0\\
&-\int_{\Omega_e}\left(\sigma_{xz}\frac{\partial N_i}{\partial x}+\sigma_{yz}\frac{\partial N_i}{\partial y}+\sigma_z\frac{\partial N_i}{\partial z}\right)\mathrm{d}\Omega+\int_{\Omega_e}N_iB_z\,\mathrm{d}\Omega\\
&\quad+\int_{S_e}N_i\left(\sigma_{xz}l+\sigma_{yz}m+\sigma_zn\right)\,\mathrm{d}S=0 \qquad i=1,2,\ldots,r
\end{aligned} \tag{8.64}$$

We notice that the integrands of the surface integrals in Eqs. (8.64) are the components of the surface tractions given in Eqs. (8.60). Introducing these definitions, we may rewrite the last result as

$$\int_{\Omega_e} \begin{bmatrix} \frac{\partial N_i}{\partial x} & \frac{\partial N_i}{\partial y} & \frac{\partial N_i}{\partial z} \end{bmatrix} \begin{Bmatrix} \sigma_x \\ \sigma_{xy} \\ \sigma_{xz} \end{Bmatrix} \mathrm{d}\Omega = \int_{\Omega_e} N_i B_x \, \mathrm{d}\Omega + \int_{S_e} N_i T_x \, \mathrm{d}S$$

$$\int_{\Omega_e} \begin{bmatrix} \frac{\partial N_i}{\partial x} & \frac{\partial N_i}{\partial y} & \frac{\partial N_i}{\partial z} \end{bmatrix} \begin{Bmatrix} \sigma_{xy} \\ \sigma_y \\ \sigma_{yz} \end{Bmatrix} \mathrm{d}\Omega = \int_{\Omega_e} N_i B_y \, \mathrm{d}\Omega + \int_{S_e} N_i T_y \, \mathrm{d}S$$

$$\int_{\Omega_e} \begin{bmatrix} \frac{\partial N_i}{\partial x} & \frac{\partial N_i}{\partial y} & \frac{\partial N_i}{\partial z} \end{bmatrix} \begin{Bmatrix} \sigma_{xz} \\ \sigma_{yz} \\ \sigma_z \end{Bmatrix} \mathrm{d}\Omega = \int_{\Omega_e} N_i B_z \, \mathrm{d}\Omega + \int_{S_e} N_i T_z \, \mathrm{d}S$$

$$i = 1, 2, \ldots, r$$

and then consolidate this result into the single matrix equation,

$$\int_{\Omega_e} [B]^T \{\sigma\} \, \mathrm{d}\Omega = \int_{\Omega_e} [N]^T \{B_F\} \, \mathrm{d}\Omega + \int_{S_e} [N]^T \{T\} \, \mathrm{d}S \tag{8.65a}$$

where the vector of body force components is

$$\{B_F\}^T = \begin{bmatrix} B_x & B_y & B_z \end{bmatrix} \tag{8.65b}$$

and the vector of surface-traction components is

$$\{T\}^T = \begin{bmatrix} T_x & T_y & T_z \end{bmatrix} \tag{8.65c}$$

The last step in the development of the finite element equations is to substitute for the stress vector using Eq. (8.59a). In the process, we introduce the strain interpolation from Eq. (8.62). The final result may be written in the standard form

$$[K]\{u\} = \{F_u\} + \{F_c\} + \{F_B\} + \{F_T\} \tag{8.66a}$$

where the element matrices are defined by the integrals:

$$[K] = \int_{\Omega_e} [B]^T [C][B] \, \mathrm{d}\Omega \tag{8.66b}$$

$$[F_u] = \int_{S_1} [N]^T \{T\} \, \mathrm{d}S \tag{8.66c}$$

$$[F_c] = \int_{S_2} [N]^T \{T\} \, \mathrm{d}S \tag{8.66d}$$

$$[F_B] = \int_{\Omega_e} [N]^T \{B_F\} \, \mathrm{d}\Omega \tag{8.66e}$$

$$[F_T] = \int_{\Omega_e} [B]^T [C]\{\alpha\} \Delta T \, \mathrm{d}\Omega \tag{8.66f}$$

The coefficient matrix $[K]$ is the element stiffness matrix. The integration of the surface tractions over the surface of the element has been broken into an integration over the surface S_1, where displacements are specified, and an integration over the surface S_2, where the tractions are specified. The forces $\{F_u\}$ are concentrated nodal forces corresponding to specified nodal displacements. These forces are, in fact, the nodal reactions, and they may be determined as part of the solution to the system's set of algebraic equations. Thus, the integral given in Eq. (8.66c) is never evaluated. However, if we have known surface tractions on the element surface S_2, then we determine the equivalent nodal forces $\{F_c\}$ by evaluating Eq. (8.66d). The vector $\{F_B\}$ represents the equivalent nodal forces from distributed body forces, and the vector $\{F_T\}$ represents the equivalent nodal forces from the temperature distribution.

Equations (8.66) provide a general finite element representation for the equilibrium of a thermoelastic solid. If we look back to our results for the rod and beam in Eqs. (8.38) and (8.52), respectively, we see that those results are special cases of Eqs. (8.66). Each finite equilibrium equation contains a stiffness matrix and load vectors. The structure's temperature distribution affects an element's stiffness matrix $[K]$ implicitly through the temperature dependence of the elastic material properties. The structure's temperature dependence affects an element's nodal loads explicitly by introducing a load vector $\{F_T\}$ that is defined in terms of the integration of the temperature over an element's volume. The vector $\{F_T\}$ also includes temperature-dependent elastic properties and, most important, the coefficient of thermal expansion.

Once we have evaluated element matrices using the integral definitions presented earlier, we assemble these contributions to form the global equilibrium equations. Then the specified nodal displacement boundary conditions are imposed. Subsequently, the system equilibrium equations are solved for the nodal displacements. After nodal displacements are determined, we may evaluate each element's stresses from Eq. (8.59a),

$$\{\sigma\} = [C][B]\{u\} - [C]\{\alpha\}\Delta T \tag{8.67}$$

Element matrices. For three-dimensional solids, the evaluation of an element's stiffness matrix and load vectors becomes a much more difficult task than we encountered for the one-dimensional rod and beam elements. An element's interpolation functions need have only C^0 continuity, but the more complex geometry makes the evaluation of the integrals more challenging. Only for the simplest elements, such as a tetrahedron, can the integrations be performed analytically to yield closed-form results. For popular three-dimensional elements, such as hexahedrons or triangular prisms, the element matrices are evaluated using numerical integration. The most popular approach is to map a general curved-side element into a simpler shape, such as a cube. Then element matrices are evaluated by numerical integration over the simpler geometry. A popular approach involves *isoparametric* elements,where the mapping uses the element's interpolation functions in the shape transformation equations. The integrals defining the element matrices are normally evaluated using Gauss integration. Finite element texts[4–7] describe these procedures in detail.

8.3.4 Large Deflections of Plates

Problem statement. The finite element formulation is based on the von Kármán plate theory described in Sec. 7.3. The theory rests on classical isotropic, elastic plate bending assumptions, including neglecting shear deformations. The theory is developed from the equilibrium equations, Hooke's law, and the nonlinear strain-displacement relations assumed by von Kármán.

Equilibrium equations. From Eq. (7.11), the membrane equilibrium equations are

$$\frac{\partial N_x}{\partial x} + \frac{\partial N_{xy}}{\partial y} = 0 \tag{8.68a}$$

$$\frac{\partial N_{xy}}{\partial x} + \frac{\partial N_y}{\partial y} = 0 \tag{8.68b}$$

and, from Eqs. (7.24) and (7.12), we may write the bending equilibrium equation as

$$\frac{\partial^2 M_x}{\partial x^2} + 2\frac{\partial^2 M_{xy}}{\partial x \partial y} + \frac{\partial^2 M_y}{\partial y^2} + \frac{\partial}{\partial x}\left(N_x \frac{\partial w}{\partial x}\right) + \frac{\partial}{\partial y}\left(N_{xy}\frac{\partial w}{\partial x}\right)$$
$$+ \frac{\partial}{\partial x}\left(N_{xy}\frac{\partial w}{\partial y}\right) + \frac{\partial}{\partial y}\left(N_y \frac{\partial w}{\partial y}\right) + q = 0 \tag{8.68c}$$

where N_x, N_{xy}, N_y are the membrane forces per unit length; M_x, M_{xy}, M_y are moments per unit length; $w(x, y)$ is the transverse deflection; and q denotes the transverse pressure loading.

Membrane stress resultants. The membrane stress resultants given in Eq. (7.23) may be expressed in matrix notation as

$$\{N\} = \{N_m\} + \{N_{nl}\} - \{N_T\} \tag{8.69a}$$

where

$$\{N\} = \begin{bmatrix} N_x & N_y & N_{xy} \end{bmatrix}^T \tag{8.69b}$$

$$\{N_m\} = [A]\{\varepsilon_m\} \tag{8.69c}$$

$$\{N_{nl}\} = [A]\{\varepsilon_{nl}\} \tag{8.69d}$$

$$\{N_T\} = \frac{E\alpha}{1-\upsilon}\int_{-(h/2)}^{h/2} \begin{Bmatrix} \Delta T(x, y, z) \\ \Delta T(x, y, z) \\ 0 \end{Bmatrix} \mathrm{d}z \tag{8.69e}$$

where $\{N_m\}$ denotes membrane forces due to the linear membrane strain components $\{\varepsilon_m\}$; $\{N_{nl}\}$ denotes membrane forces due to the nonlinear strain components $\{\varepsilon_{nl}\}$; and $\{N_T\}$ denotes membrane forces induced by a temperature change $\Delta T(x, y, z)$. The elastic matrix $[A]$ is defined by

$$[A] = \frac{Eh}{1-\upsilon^2}\begin{bmatrix} 1 & \upsilon & 0 \\ \upsilon & 1 & 0 \\ 0 & 0 & \dfrac{1-\upsilon}{2} \end{bmatrix} \tag{8.70}$$

In the preceding equations, h denotes the constant plate thickness. The material properties may be temperature-dependent.

Moment stress resultant. The moment stress resultants given in Eq. (7.6d–f) may be written in matrix form as

$$\{M\} = [D]\{\kappa\} - \{M_T\} \tag{8.71a}$$

where

$$\{M\} = \begin{bmatrix} M_x & M_y & M_{xy} \end{bmatrix}^T \tag{8.71b}$$

$$\{M_T\} = \frac{E\alpha}{1-\upsilon} \int_{-(h/2)}^{h/2} \begin{Bmatrix} \Delta T(x, y, z) \\ \Delta T(x, y, z) \\ 0 \end{Bmatrix} z\,\mathrm{d}z \tag{8.71c}$$

where $\{\kappa\}$ denotes the plate's curvatures, and $\{M_T\}$ denotes the thermal moments. The plate curvatures in Eq. (8.7la) are defined in Eq. (7.3). The elastic flexural rigidity is defined by

$$[D] = \frac{Eh^3}{12(1-\upsilon^2)} \begin{bmatrix} 1 & \upsilon & 0 \\ \upsilon & 1 & 0 \\ 0 & 0 & \dfrac{1-\upsilon}{2} \end{bmatrix} \tag{8.72}$$

Strain-displacement relations. The linear and nonlinear strain components given in Eqs. (8.69c) and (8.69d) are defined by

$$\{\varepsilon_m\} = \begin{Bmatrix} \dfrac{\partial u_0}{\partial x} \\ \dfrac{\partial v_0}{\partial y} \\ \dfrac{\partial u_0}{\partial y} + \dfrac{\partial v_0}{\partial x} \end{Bmatrix} \tag{8.73a}$$

$$\{\varepsilon_{nl}\} = \begin{Bmatrix} \dfrac{1}{2}\left(\dfrac{\partial w}{\partial x}\right)^2 \\ \dfrac{1}{2}\left(\dfrac{\partial w}{\partial y}\right)^2 \\ \dfrac{\partial w}{\partial x}\dfrac{\partial w}{\partial y} \end{Bmatrix} \tag{8.73b}$$

where u_0, v_0 are the in-plane displacements of points on the middle surface.

In the classical von Kármán plate theory that we describe in Sec. 7.3, the preceding equations are reduced to a pair of partial differential equations where the stress function F is introduced to satisfy the in-plane equilibrium equations. The form of the von Kármán plate theory just presented facilitates development of the finite element formulation.

Finite element formulation. Finite element formulations for the nonlinear von Kármán equations have been given by a number of authors, particularly for isothermal plates subjected to mechanical loads. A standard approach adopted by structural engineers is to use the principle of virtual work. The approach we present develops the formulation using the method of weighted residuals. The approach consists first of multiplying the equilibrium equations by a test function (we use an interpolation function as the test function) and integrating each equation over the plate surface. Then, each equation is integrated by parts. The first-order equations (8.68a) and (8.68b) are integrated by parts once, and the second-order equation (8.68c) is integrated by parts twice. The integration by parts introduces appropriate boundary conditions on the plate edges. The integral formulation of the partial differential equations and boundary conditions is sometimes called the *weak* form of the problem statement. Further details of the weak form for the von Kármán plate theory are given in the Appendix.

For development of the finite element equations, the in-plane displacements and the linear strains are interpolated from nodal values by

$$\begin{Bmatrix} u_0 \\ v_0 \end{Bmatrix} = [N^m]\{\delta_m\} \tag{8.74a}$$

$$\{\varepsilon_m\} = [B^m]\{\delta_m\} \tag{8.74b}$$

where $[N^m]$ denotes the interpolation matrix of the membrane displacements, $[B^m]$ the strain-displacement interpolation matrix, and the $\{\delta_m\}$ the in-plane nodal displacements at the middle surface. The plate transverse displacement, slopes, and curvatures are interpolated from nodal values by

$$w = [N^b]\{\delta_b\} \tag{8.74c}$$

$$\begin{Bmatrix} \dfrac{\partial w}{\partial x} \\ \dfrac{\partial w}{\partial y} \end{Bmatrix} = [G]\{\delta_b\} \tag{8.74d}$$

$$\{\kappa\} = -[B^b]\{\delta_b\} \tag{8.74e}$$

where $[N^b]$ denotes the bending displacement interpolation matrix, $[G]$ the slopes' interpolation matrix, $[B^b]$ the curvature interpolation matrix, and $\{\delta_b\}$ the nodal transverse displacements and slopes.

With the weak form from Eqs. (A.4), in the Appendix and these definitions, the finite element equilibrium equations may be developed. The equations may be written in the form

$$\left[\begin{array}{c|c} [K_m] & \frac{1}{2}[K_{mb}] \\ \hline \frac{1}{2}[K_{mb}]^T & [K_b] + \frac{1}{2}[K_{bl}] + \frac{1}{3}[K_{bn\ell}] - [K_{bT}] \end{array}\right] \left\{\begin{array}{c} \delta_m \\ \hline \delta_b \end{array}\right\} = \left\{\begin{array}{c} F_T \\ \hline F_q + F_{\mathrm{TM}} \end{array}\right\} \tag{8.75}$$

where the components of the stiffness matrix are defined by

$$[K_m] = \int_\Omega [B^m]^T [A][B^m]\,\mathrm{d}\Omega \tag{8.76a}$$

$$[K_{mb}] = \int_\Omega [B^m]^T [A][C][G]\,\mathrm{d}\Omega \tag{8.76b}$$

$$[K_b] = \int_\Omega [B^b]^T [D][B^b]\,\mathrm{d}\Omega \tag{8.76c}$$

$$[K_{b\ell}] = \int_\Omega [G]^T [N_m][G]\,\mathrm{d}\Omega \tag{8.76d}$$

$$[K_{bn\ell}] = \int_\Omega [G]^T [N_w][G]\,\mathrm{d}\Omega \tag{8.76e}$$

$$[K_{bT}] = \int_\Omega [G]^T [N_T][G]\,\mathrm{d}\Omega \tag{8.76f}$$

$$[C] = \begin{bmatrix} \dfrac{\partial w}{\partial x} & 0 \\ 0 & \dfrac{\partial w}{\partial y} \\ \dfrac{\partial w}{\partial y} & \dfrac{\partial w}{\partial x} \end{bmatrix} \tag{8.76g}$$

and

$$[N_m] = \begin{bmatrix} N_x^m & N_{xy}^m \\ N_{xy}^m & N_y^m \end{bmatrix} \tag{8.76h}$$

$$[N_w] = [C]^T[A][C] \tag{8.76i}$$

$$[N_T] = \begin{bmatrix} N_x^T & N_{xy}^T \\ N_{xy}^T & N_y^T \end{bmatrix} \tag{8.76j}$$

The components of the load vectors are defined by

$$\{F_T\} = \int_\Omega [B^m]^T \{N_T\}\, \mathrm{d}\Omega \tag{8.77a}$$

$$\{F_q\} = \int_\Omega q[N^b]\, \mathrm{d}\Omega \tag{8.77b}$$

$$\{F_{\mathrm{TM}}\} = -\int_\Omega [B^b]^T \{M_T\}\, \mathrm{d}\Omega \tag{8.77c}$$

The matrix $[K_m]$ is a constant matrix representing the in-plane membrane stiffness, and the matrix $[K_b]$ is a constant matrix representing the plate bending stiffness. The matrix $[K_{mb}]$ depends on the unknown plate slopes and represents coupling between membrane and bending displacements. The matrix $[K_{b\ell}]$ depends on the unknown midplane displacements and represents bending stiffness induced by the linear midplane strains. The matrix $[K_{bn\ell}]$ depends on the unknown plate slopes and represents bending stiffness induced by the nonlinear midplane strains. The matrix $[K_{bT}]$ depends on temperature and, because of the minus sign, represents a loss of bending stiffness induced for an increase in temperature by in-plane thermal forces.

The load vector $\{F_T\}$ represents equivalent in-plane nodal forces due to the thermal forces. The load vector $\{F_q\}$ represents equivalent transverse nodal forces and moments due to the surface pressure q. Finally, $\{F_{\mathrm{TM}}\}$ represents equivalent nodal forces and moments due to thermal moments. If there is no temperature gradient through the plate thickness, the thermal moment is zero; therefore, the load vector $\{F_{\mathrm{TM}}\}$ is not required. The form of the finite element equations given in Eqs. (8.75–8.77) was given in a paper by Mei and Locke,[16] who studied the acoustic response of thermally buckled plates.

Elements. In the large-deflection plate formulation, we have membrane and bending behavior. Thus, we need interpolation functions for the membrane displacements and strains as well as the bending displacement, slopes, and curvatures.

The membrane problem is the same as a plane stress problem where the displacements must satisfy C^0 continuity. For the bending problem, the displacement interpolation functions must satisfy C^1 continuity.

To model plates of arbitrary shapes, we probably will choose either a triangular or quadrilateral element. If we choose a triangular element, the membrane problem can be represented by the interpolation functions for the classical constant strain triangle (CST), which is described in all finite element texts. The choice of a bending triangle is more difficult because the development of triangular elements with C^1 continuity is challenging, and researchers have developed several alternative formulations. A survey[17] of plate bending elements lists over 150 references and identifies nearly 90 different plate bending elements. In the next paragraph, we describe briefly a triangular element that will be used in Example 8.6.

In local coordinates, the triangular element has five degrees of freedom at each node: two midplane displacements u_0 and v_0 and three bending degrees of freedom, the transverse displacement w, and rotations θ_x and θ_y. The in-plane stiffness is represented by the classical constant strain triangle. The bending stiffness is represented by the discrete Kirchhoff theory (DKT) bending element. The DKT element is formulated by beginning with Mindlin theory and then imposing the Kirchhoff assumption at specific, or "discrete," points of the element. The element was introduced first by Stricklin et al.[18] and has since been tested extensively.[19] In this latter reference, Batoz provides an expression for the $[B^b]$ matrix of Eq. (8.74e). Details of the interpolation functions and transformation equations are given in Ref. 19. Explicit formulations for the bending stiffness matrix of the DKT element have also been published.[20,21]

Solution algorithm. The most popular solution procedure solves the nonlinear equilibrium equations in Eq. (8.75) by Newton–Raphson iteration. The procedure involves equations of the form

$$[K_T]\{\Delta\delta\} = \{\Delta F\} \tag{8.78}$$

In this equation, $[K_T]$ is the Jacobian; in structures literature, it is called the tangent stiffness matrix, and its form is given by Yang[22] as

$$[K_T] = \left[\begin{array}{c|c} [K_m] & [K_{mb}] \\ \hline [K_{mb}]^T & [K_b] + [K_{b\ell}] + 3/2[K_{bn\ell}] - [K_{bT}] \end{array}\right] \tag{8.79}$$

The only differences between $[K_T]$ of Eq. (8.79) and $[K]$ of Eq. (8.75) are the coefficients of $[K_{bl}]$ and $[K_{bnl}]$. In Eq. (8.78), $\{\Delta\delta\}$ is the increment of the global displacement vector, and $\{\Delta F\}$ is the load imbalance defined as

$$\{\Delta F\} = \{F\} - [K(\delta)]\{\delta\} \tag{8.80}$$

where $[K]$ is given by Eq. (8.75). The solution proceeds by forming $\{\Delta F\}$ with the stiffness matrix $[K]$ based on current values of the global displacement vector $\{\delta\}$. Next, Eq. (8.78) is solved to find the incremental global displacement vector $\{\Delta\delta\}$. A new value for the global displacement vector is then found by adding $\{\Delta\delta\}$ to the previous displacement vector,

$$\{\delta\}_{\text{new}} = \{\delta\}_{\text{old}} + \{\Delta\delta\} \tag{8.81}$$

The process is terminated when convergence is achieved.

As the iterations continue, the value of $[K(\delta)]\{\delta\}$ in Eq. (8.80) approaches that of the load vector $\{F\}$. The Euclidean norm of either $\{\Delta\delta\}$ or $\{\Delta F\}$ could be used as a measure of convergence; however, the units of the individual terms of these vectors are not consistent. A more rigorous scalar error measure suggested by Cook et al.[21] is

$$e = \frac{\{\delta\}^T\{\Delta F\}}{\{\delta\}^T\{F\}} \tag{8.82}$$

This definition has physical significance as the ratio of work done by the load imbalance to the work done by the loads applied to the structure.

8.3.5 Thermal-Structural Problems

To demonstrate some unique aspects of thermal-structural problems, we present analyses of three problems. The first and last problems present numerical results from student-developed programs that implement finite element theory described in this chapter.

Example 8.4: Cable-Stiffened Structure

To increase structural stability and to provide additional stiffness, prestressed cables and membranes have been considered for some designs of orbiting space structures. In Ref. 23, Thornton et al. describe a finite element thermal-structural analysis of a cable-stiffened hoop-column antenna. Figure 8.17 diagrams the complete analysis procedure. The series of analyses include 1) a prestress analysis, 2) a heating rate analysis, 3) a thermal analysis, and 4) a structural analysis. To provide details of the nonlinear structural analysis of prestressed, heated cables, we highlight the finite element formulation, describe the solution procedure, and present illustrative results for an orbiting structure.

Finite element formulation. We wish to develop the finite element equations to model a prestressed cable with a known temperature distribution. A cable element

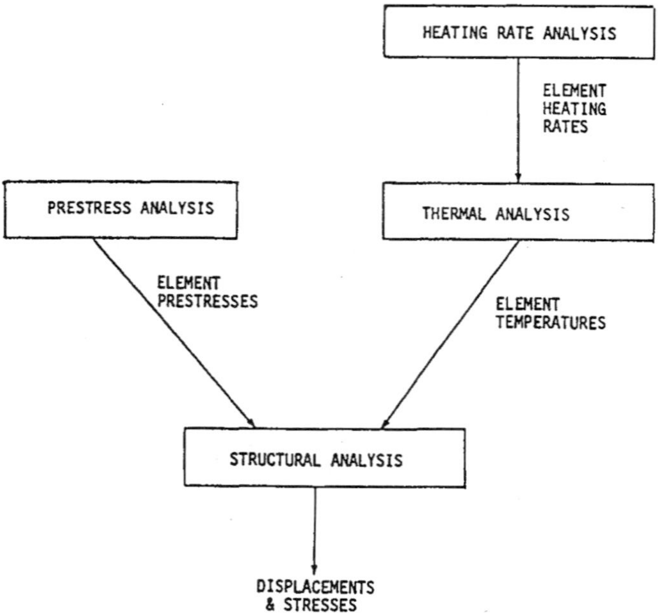

Fig. 8.17 Analysis procedure for cable-stiffened orbiting structure.

differs from the standard small-deflection rod element because 1) the element has an initial prestress, 2) the element cannot support compressive stresses, and 3) as the structure deforms, a cable element may experience large rotations. As in our previous rod analysis, we consider the cable element in a local coordinate system. The rod equilibrium equation (8.31a) applies; but the strain-displacement equation (8.31b) and Hooke's law given in Eq. (8.31c) no longer apply. Instead, we use the nonlinear strain-displacement equation (3.9),

$$\varepsilon_x = \frac{\partial u}{\partial x} + \frac{1}{2}\left[\left(\frac{\partial u}{\partial x}\right)^2 + \left(\frac{\partial v}{\partial x}\right)^2 + \left(\frac{\partial w}{\partial x}\right)^2\right]$$

and, in the spirit of von Kármán (see Sec. 7.3), we neglect $(\partial u/\partial x)^2$ to obtain

$$\varepsilon_x = \frac{\partial u}{\partial x} + \frac{1}{2}\left(\frac{\partial v}{\partial x}\right)^2 + \left(\frac{\partial w}{\partial x}\right)^2$$

If we include a known prestress σ_0, Hooke's law becomes

$$\sigma_x = \sigma_0 + E\varepsilon_x - E\alpha\Delta T$$

where $\Delta T = T(x) - T_{\text{ref}}$.

We assume an element with two nodes but, because the element can have large rotations, each node can have three nodal displacements in the local coordinate system. Thus, the element has six degrees of freedom. We interpolate the displacements linearly from the nodal displacements, that is,

$$u = [N_1 \quad N_2]\begin{Bmatrix} u_1 \\ u_2 \end{Bmatrix} \qquad v = [N_1 \quad N_2]\begin{Bmatrix} v_1 \\ v_2 \end{Bmatrix} \qquad w = [N_1 \quad N_2]\begin{Bmatrix} w_1 \\ w_2 \end{Bmatrix}$$

where $N_1 = 1 - x/L$ and $N_2 = x/L$. Substituting these interpolations in the nonlinear strain equation, we may write the results in the standard form

$$\varepsilon_x = [B]\{u\}$$

where

$$[B] = \begin{bmatrix} -\dfrac{1}{L} & -\dfrac{\theta}{2L} & -\dfrac{\psi}{2L} & \dfrac{1}{L} & \dfrac{\theta}{2L} & \dfrac{\psi}{2L} \end{bmatrix}$$

$$\{u\}^T = [u_1 \quad v_1 \quad w_1 \quad u_2 \quad v_2 \quad w_2]$$

$$\theta = \frac{v_2 - v_1}{L} \qquad \psi = \frac{w_2 - w_1}{L}$$

In these last four equations, we see that the strain depends nonlinearly on the nodal displacements since $[B]$ depends on v_1, v_2, w_1, and w_2. The variables θ and ψ we recognize as the cable rotations due to the transverse displacements.

We may develop the finite element equilibrium equations by returning to Eq. (8.35). This equation was derived by the method of weighted residuals followed by an integration by parts. Let us now write Eq. (8.35) in the form

$$\int_0^L B_i \sigma_x A \, \mathrm{d}x = F_{ci} + F_{Bi} \qquad \mathrm{i} = 1, 2, \ldots, 6$$

where F_{ci} represents concentrated nodal forces from the boundary terms, and F_{Bi} equivalent nodal forces due to the axial body force F_x. If we now combine Hooke's law with the strain-displacement equation and substitute the result into the preceding equation, we obtain

$$\int_0^L B_i \sigma_0 A \, \mathrm{d}x + \int_0^L EAB_i[B] \, \mathrm{d}x\{u\} - \int_0^L B_i EA\alpha\Delta T \, \mathrm{d}x = F_{ci} + F_{Bi} \qquad i = 1, 2, \ldots, 6$$

which we write in matrix notation as

$$[K]\{u\} + \{F_\sigma\} + \{F_T\} = \{F_C\} + \{F_B\}$$

where

$$[K] = \int_0^L EA\{B\}[B] \, \mathrm{d}x$$

$$\{F_\sigma\} = \int_0^L \sigma_0 A\{B\} \, \mathrm{d}x$$

$$\{F_T\} = \int_0^L EA\Delta T\{B\} \, \mathrm{d}x$$

$$\{F_B\} = \int_0^L F_x A\{N\} \, \mathrm{d}x$$

The matrix equation represents element equilibrium. The integrals that define the stiffness matrix $[K]$, as well as the vectors $\{F_\sigma\}$ and $\{F_T\}$, depend on the $[B]$ matrix. The terms of $[B]$ depend on the transverse nodal displacements, so that we have a nonlinear set of equations.

Element matrices. Whereas the $[B]$ matrix depends on the nodal displacements, it does not depend on x, and so the integrals are evaluated easily. In this case, some of the terms of $\{F_\sigma\}$ and $\{F_T\}$ also depend on the nodal displacements and contribute to the element stiffness matrix. Terms independent of nodal displacements are transferred to the right-hand side and become element load vectors. In the evaluation of the thermal load vector, we assume that the temperature change ΔT is constant. The final equation is

$$[K(u)]\begin{Bmatrix} u_1 \\ v_1 \\ w_1 \\ u_2 \\ v_2 \\ w_3 \end{Bmatrix} = EA\alpha\Delta T \begin{Bmatrix} -1 \\ 0 \\ 0 \\ 1 \\ 0 \\ 0 \end{Bmatrix} + A\sigma_0 \begin{Bmatrix} 1 \\ 0 \\ 0 \\ -1 \\ 0 \\ 0 \end{Bmatrix} + \begin{Bmatrix} F_1 \\ 0 \\ 0 \\ F_2 \\ 0 \\ 0 \end{Bmatrix}$$

where we omit the body forces. The terms of the upper triangle of the symmetric stiffness matrix are

$$K_{11} = -K_{14} = K_{44} = \frac{EA}{L}$$

$$K_{12} = -K_{15} = -K_{24} = K_{45} = \frac{EA}{L}\frac{\theta}{2}$$

$$K_{13} = -K_{16} = K_{46} = \frac{EA}{L}\frac{\psi}{2}$$

$$K_{22} = -K_{25} = K_{55} = \frac{EA}{L}\frac{1}{2}\left[\frac{\sigma_0}{E} - \alpha\Delta T + \frac{\theta^2}{2}\right]$$

$$K_{23} = -K_{26} = K_{56} = \frac{EA}{L}\frac{\theta\psi}{4}$$

$$K_{33} = -K_{36} = K_{66} = \frac{EA}{L}\frac{1}{2}\left[\frac{\sigma_0}{E} - \alpha\Delta T + \frac{\psi^2}{2}\right]$$

$$K_{34} = -\frac{EA}{L}\frac{\psi}{2}$$

$$K_{35} = -\frac{EA}{L}\frac{\theta\psi}{4}$$

Note that the stiffness matrix depends on the prestress σ_0 and temperature change ΔT. This is a consequence of the nonlinear formulation. The prestress σ_0 contributes positively to the diagonal elements in the stiffness matrix. This agrees with intuition since we expect the prestress to tighten the cable. Some authors call this effect "stress-stiffening." In contrast, for a positive change in ΔT, the diagonal terms of the stiffness matrix decrease. Again, this result agrees with intuition because, for a thermal expansion, we expect the cable to slacken. Recall that the element stiffness matrix and load vectors are defined in a local element coordinate system.

Solution method. An accurate solution method for solving the nonlinear system equations is the Newton–Raphson iteration. We use the same approach as that presented in Eqs. (8.76–8.79) for the nonlinear large-deflection plate problem. Further details of the element Jacobian and residual load vector appear in a problem at the end of the chapter. Before the element Jacobian and residual load vector are assembled into the system arrays, they must be transformed into the global *XYZ* system using the transformation equations given in Eq. (8.43).

NODE	1	2	3	4	5
X	3.37	10.1	23.8	38.0	52.7
Y	-9.6	-9.4	-8.9	-6.8	-3.7

(DIMENSIONS ARE IN METERS)

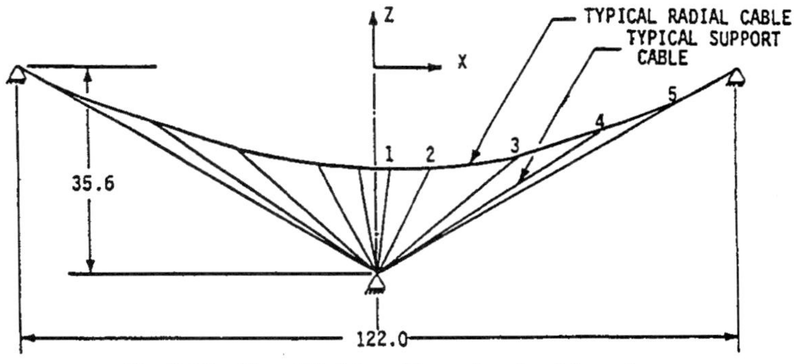

Fig. 8.18 Simplified two-dimensional cable system.

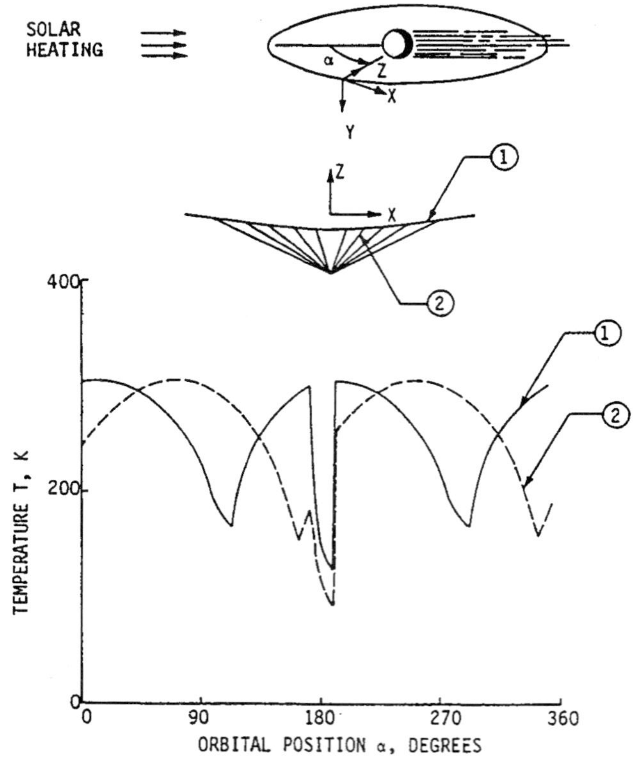

Fig. 8.19 Temperature histories for typical members of simplified cable system.

Numerical example. In Ref. 23, the finite element approach is used to solve for deformations and stresses in an orbiting hoop-column antenna. For illustrative purposes, we present results from a similar analysis on a simpler two-dimensional structure. A simplified pretensioned cable system is shown in Fig. 8.18. The structure is assumed to be in an Earth-facing geosynchronous orbit. Isothermal elements are used because members are made of graphite–epoxy, which has low thermal conductivity. Figure 8.19 shows temperature histories for two typical members. For the data used, the member temperatures are close to radiation equilibrium temperatures throughout the orbit. When the structure enters the Earth's shadow, member temperatures drop suddenly and approach much lower radiation equilibrium temperatures. When the structure leaves the Earth's shadow, member temperatures rise abruptly in response to the abrupt changes in heat loads. The range of member temperatures during an orbit is 320 to 90 K.

Figure 8.20 shows the displacement histories of two typical nodes. The displacement histories tend to follow the temperature histories of the connected element. Stress histories (not shown) of the members indicate that thermal stresses are small compared to member prestresses.

Example 8.5: Simply Supported Beam

As an illustration of a very simple beam problem that we can solve by hand, consider the beam shown in Fig. 8.21. The beam is simply supported, and the

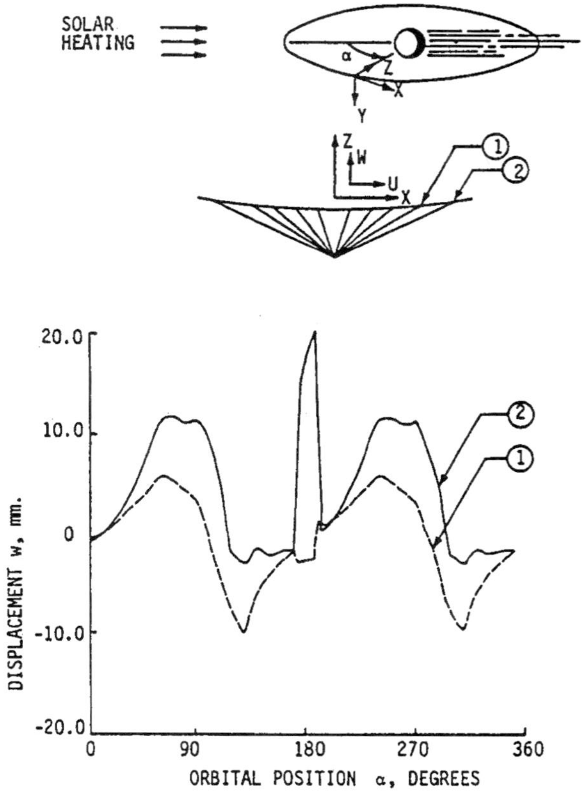

Fig. 8.20 Displacement histories of nodes of simplified cable system.

thermal moment is constant. We wish to determine 1) beam slopes at $x = 0$ and $x = L$, 2) the equation for the beam deflection, 3) the beam reactions, and 4) the equation for the normal stress σ_x.

Because the beam has just two supports and no concentrated transverse forces, we may model it with a single finite element. From Eqs. (8.52a), (8.53), and (8.54), we may write the element's equilibrium equations as

$$\frac{EI}{L^3}\begin{bmatrix} 12 & 6L & -12 & 6L \\ 6L & 4L^2 & -6L & 2L^2 \\ -12 & -6L & 12 & -6L \\ 6L & 2L^2 & -6L & 4L^2 \end{bmatrix}\begin{Bmatrix} v_1 \\ \theta_1 \\ v_2 \\ \theta_2 \end{Bmatrix} = \begin{Bmatrix} V_1 \\ M_1 \\ V_2 \\ M_2 \end{Bmatrix} + \begin{Bmatrix} 0 \\ M_T \\ 0 \\ -M_T \end{Bmatrix}$$

The preceding equation describes the element's equilibrium before the boundary conditions are imposed. In this problem, the simple supports require that the transverse displacement and bending moment at each support are zero. Thus, we have $v_1 = M_1 = v_2 = M_2 = 0$. For these conditions, we may partition the preceding equation into two sets of equations:

$$\frac{EI}{L^3}\begin{bmatrix} 4L^2 & 2L^2 \\ 2L^2 & 4L^2 \end{bmatrix}\begin{Bmatrix} \theta_1 \\ \theta_2 \end{Bmatrix} = \begin{Bmatrix} M_T \\ -M_T \end{Bmatrix}$$

$$\frac{EI}{L^3}\begin{bmatrix} 6L & 6L \\ -6L & -6L \end{bmatrix}\begin{Bmatrix} \theta_1 \\ \theta_2 \end{Bmatrix} = \begin{Bmatrix} V_1 \\ V_2 \end{Bmatrix}$$

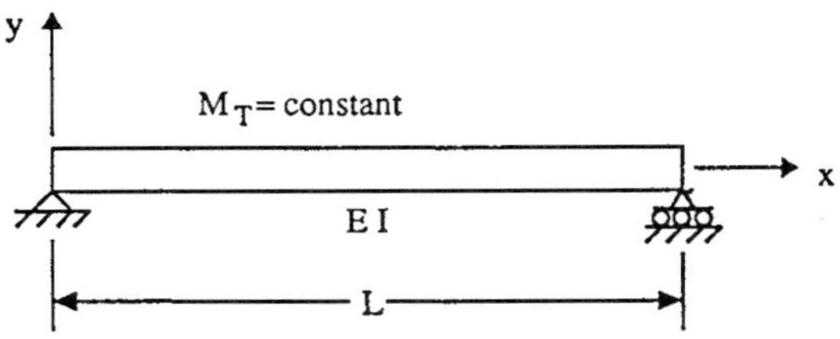

Fig. 8.21 Simply supported beam with constant thermal moment.

Because the thermal moment is known, we may solve the first of these sets for θ_1 and θ_2. Thus, we find

$$\theta_1 = -\theta_2 = \frac{1}{2}\frac{M_T L}{EI}$$

Then, with θ_1 and θ_2 known, we solve the second set for the unknown reactions V_1 and V_2. This step shows

$$V_1 = V_2 = 0$$

This result makes sense because there are no external applied loads; hence, there should be no reactions. The beam deflection is given by Eq. (8.49a) with $v_1 = v_2 = 0$. Thus,

$$v(x) = N_2(x)\theta_1 + N_4(x)\theta_2$$

and, substituting for N_2 and N_4 from Eq. (8.49b) and introducing θ_1 and θ_2 from the equation at the top of the page,

$$v(x) = \frac{M_T}{2EI}\left(xL - x^2\right)$$

The beam's normal stress is given by Eq. (8.48). Substituting θ_1 and θ_2, we obtain

$$\sigma_x = \frac{M_T}{I}y - E\alpha(T - T_{\text{ref}})$$

In this example, the single-element model provides an exact analytical solution. Why is this true?

Example 8.6 Large Deflections of a Simply Supported Plate

In Chapter 10, we apply the large-deflection finite element formulation developed in Sec. 8.3.4 to the study of thermal buckling of plates. Since there are very few analytical solutions for thermally induced large-deflection plate problems, we present here a finite element analysis for a simply supported plate subjected to uniform pressure. For this problem, Levy developed an analytical solution that Kolenski[25] used to validate the finite element analysis.

The problem studied is a 16-in.-square steel plate, 0.1 in. thick. The edges of the plate are simply supported and immovable. The immovable boundary condition restrains in-plane motion so that membrane forces develop from the onset of deflection.

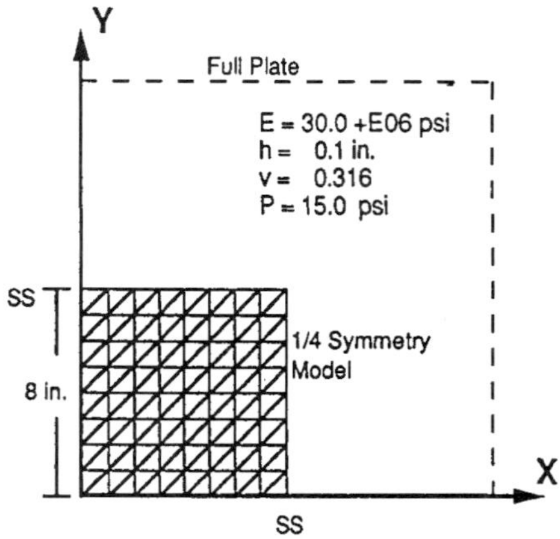

Fig. 8.22 Finite element model for simply supported plate with immovable edges.

The finite element model consists of 128 triangular elements and 81 nodes as shown in Fig. 8.22. Symmetry permits the model to include only one quarter of the plate. The pressure load of 15 psi was applied in increments of 0.25 psi with modified Newton–Raphson iterations at each load step. In the modified Newton–Raphson approach, the Jacobian is held constant throughout each iteration cycle. This approach offers computational savings because the cost of computing and assembling the global Jacobian is minimized, but a few more iterations may be required.

The finite element solution for the deflection at the plate center is compared with Levy's solution in Fig. 8.23. The agreement between the two solutions is excellent through the load range up to a maximum deflection of two plate thicknesses. The stiffening of the plate with increased deflection is apparent. The figure also shows the discrepancy that occurs with a linear solution that neglects the membrane forces.

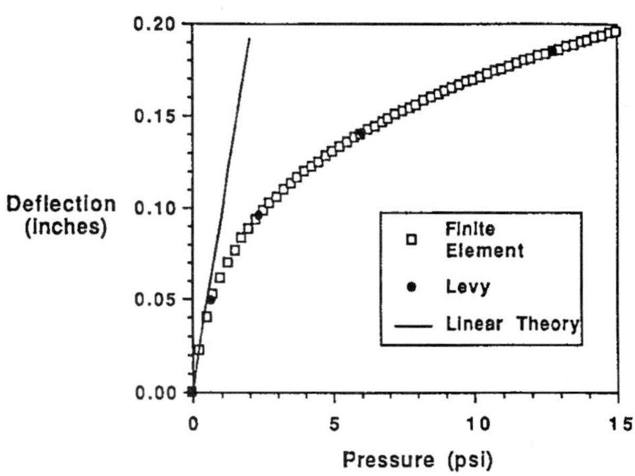

Fig. 8.23 Load-deflection solutions for a simply supported plate with immovable edges.

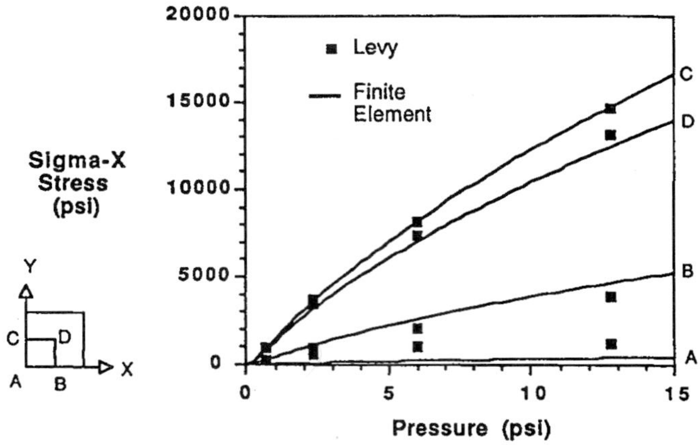

Fig. 8.24 Load-stress solutions for a simply supported plate with immovable edges.

Figure 8.24 compares the σ_x membrane stress at four points on the plate. The finite element results agree very well with the Levy values, with a maximum error of 1% at point C and 4% at point D. Levy's stress values were calculated exactly at the corners of the quarter-plate, whereas the finite element stresses were evaluated at the element centroids nearest the corners. The figure shows that appreciable membrane forces develop because of the restraint of the in-plane displacements at the immovable supports. Other computations not shown demonstrate that, if in-plane displacements are unrestrained, considerably larger transverse displacements occur but membrane stresses are greatly reduced.

8.4 Closing Remarks

In this chapter, we describe finite element approaches for determination of temperatures in structures and thermal-stress analysis. The description assumes some finite element background and places emphasis on the practical aspects of formulating and analyzing problems for real structures. The treatment is from a researcher's viewpoint, where theory is implemented in special computer programs to investigate the methodology or some aspect of thermally induced behavior. Undoubtedly, most engineers do not have the luxury of developing their own computer programs; the pressures of deadlines and costs mandate the use of commercially available software. Yet an understanding of the theoretical foundations on which the software is based is extremely important for making informed engineering decisions. Moreover, an understanding of finite element methodology and solution algorithms has an important role in effective use of engineering software. The host of a television program on woodworking once said, "Knowledge is the ultimate power tool." Certainly, this remark applies as well to modern engineering software tools.

In the remaining chapters, we use finite element analysis as we investigate thermally induced vibrations, thermal buckling, and thermoviscoplasticity. Many of the problems we discuss are time-dependent or nonlinear. These analyses were not tractable in precomputer days, and much progress in understanding the phenomena is possible because of finite element analysis.

References

[1]Turner, M. J., Clough, R. W., Martin, H. C., and Topp, L. C., "Stiffness and Deflection Analysis of Complex Structures," *Journal of the Aeronautical Sciences*, Vol. 23, No. 9, 1956, pp. 805–823, 854.

[2]Clough, R. W., "The Finite Element Method after Twenty-Five Years: A Personal View," *Computers and Structures*, Vol. 12, No. 4, 1980, pp. 361–370.

[3]Noor, A. K., "Bibliography of Books and Monographs on Finite Element Technology," *Applied Mechanics Reviews*, Vol. 44, June 1991, pp. 307–317.

[4]Zienkiewicz, O. C., and Taylor, R. L., *The Finite Element Method*, 4th ed. Vol. 1—*Basic Formulation and Linear Problems*, McGraw–Hill, New York, 1989.

[5]Zienkiewicz, O. C., and Taylor, R. L., *The Finite Element Method*, 4th ed., Vol. 2—*Solid and Fluid Mechanics, Dynamics and Nonlinearity*, McGraw–Hill, New York, 1991.

[6]Reddy, J. N., *An Introduction to the Finite Element Method*, 2nd ed. McGraw–Hill, New York, 1993.

[7]Huebner, K. H., Thornton, E. A., and Byrom, T. G., *The Finite Element Method for Engineers*, 3rd ed. Wiley, New York, 1995.

[8]Hsu, T. R., *The Finite Element Method in Thermomechanics*, Allen and Unwin, Boston, 1986.

[9]Thornton, E. A., and Wieting, A. R., "A Finite Element Thermal Analysis Procedure for Several Temperature-Dependent Parameters," *Journal of Heat Transfer*, Vol. 100, No. 3, 1978, pp. 551–553.

[10]Emery, A. F., Johnson, O., Lobo, M., and Abrous, A., "A Comparative Study of Methods for Computing the Diffuse Radiation Viewfactors for Complex Structures," *Journal of Heat Transfer*, Vol. 113, May 1991, pp. 413–422.

[11]Thornton, E. A., Coyle, M. F., and McLeod, R. N., "Experimental Study of Plate Buckling Induced by Spatial Temperature Gradients," *Journal of Thermal Stresses*, Vol. 17, No. 2, 1994, pp. 191–212.

[12]Hernan, P. R., "Finite Element Analysis of Plate Thermal Buckling Tests—Comparison of Analyses and Experiments," M.S. Thesis, Univ. of Virginia, Charlottesville, May 1994.

[13]Leonard, J. W., *Tension Structures: Behavior and Analysis*, McGraw–Hill, New York, 1988.

[14]Przemieniecki, J. S., *Theory of Matrix Structural Analysis*, McGraw–Hill, New York, 1968.

[15]Gellin, S., and Lee, G. C., "Finite Elements Available for the Analysis of Non-curved Thin-Walled Structures," *Finite Element Analysis of Thin-Walled Structures*, edited by J. W. Bull, Elsevier Applied Science, London, 1988, Chap. 1, pp. 1–45.

[16]Mei, C., and Locke, J. E., "Finite Element Large Deflection Response of Thermally Buckled Plates," *Journal of Sound and Vibration*, Vol. 158, Nov. 1992.

[17]Hrabok, M. M., and Hrudey, T. M., "A Review and Catalog of Plate Bending Finite Elements," *Computers and Structures*, Vol. 19, No. 3, 1989, pp. 479–495.

[18]Stricklin, J. A., Haisler, W., Tisdale, P., and Gunderson R., "A Rapidly Converging Triangular Plate Element," *AIAA Journal*, Vol. 7, No. 1, 1969, pp. 180–181.

[19]Batoz, J. L., Bathe, K. J., and Ho, L. W., "A Study of Three-Node Triangular Plate Bending Elements," *International Journal for Numerical Methods in Engineering*, Vol. 15, No. 12, 1980, pp. 1771–1812.

[20]Jeyachandrabose, C., Kirkhope, J., and Babu, C. R., "An Alternative Explicit Formulation for the DKT Plate-Bending Element," *International Journal for Numerical Methods in Engineering*, Vol. 21, No. 7, 1985, pp. 1289–1293.

[21]Cook, R. D., Malkus, D. S., and Plesha, M. E., *Concepts and Applications of Finite Element Analysis*, 3rd ed., Wiley, New York, 1989.

[22]Yang, T. Y., and Han, A. D., "Buckled Plate Vibrations and Large Amplitude Vibrations Using High-Order Triangular Elements," *AIAA Journal*, Vol. 21, No. 5, May 1983, pp. 758–766.

[23]Thornton, E. A., Dechaumphai, P., and Pandey, A. K., "Finite Element Thermal-Structural Analyses of a Cable-Stiffened Orbiting Antenna," *Journal of Spacecraft and Rockets*, Vol. 23, No. 6, 1986, pp. 620–624.

[24]Levy, S., "Bending of Rectangular Plates with Large Deflections," NACA TN-846, Washington, DC, 1942.

[25]Kolenski, J. D., "Large Deflection Thermal Buckling Analysis of Plates by the Finite Element Method," M. S. Thesis, Univ. of Virginia, Charlottesville, 1992.

[26]Stricklin, J. A., Haisler, W. E., and Riesemann, W. A., "Geometrically Nonlinear Structural Analysis by Direct Stiffness Method," *Journal of Structural Division, Proceedings of the ASCE*, Vol. 97, No. 1, 1971, pp. 2299–2314.

Problems

8.1. Consider steady-state heat transfer in the three-member truss shown in Fig. P8.1. In a crude finite element model, each truss member is represented by a two-node rod element. Member 2 experiences surface heating, and member 3

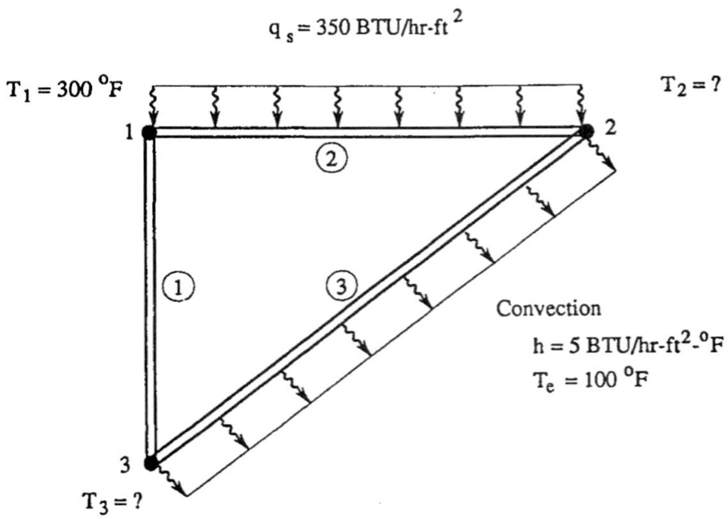

Element	k (BTU/hr-ft-$^\circ$F)	A (ft^2)	p (ft)	L (ft)
1	60	0.02	0.56	4
2	60	0.02	0.14	5
3	80	0.02	0.56	6.4

Fig. P8.1

experiences convective heat transfer to the environment. Compute a) temperatures at nodes 2 and 3, b) the nodal heat flow at node 1, and c) heat fluxes for all elements.

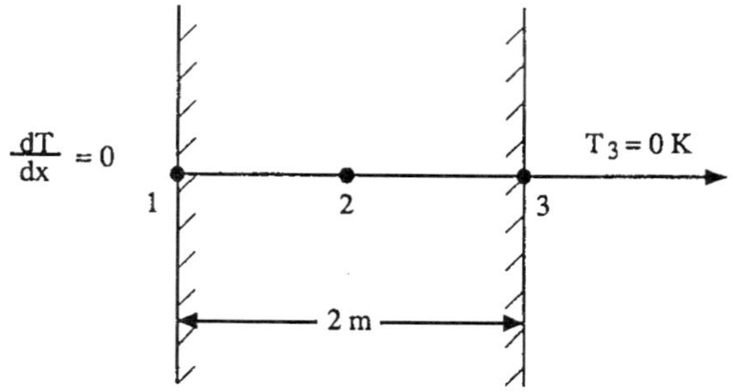

Fig. P8.2

8.2. Consider nonlinear steady-state heat conduction in the slab shown. The problem is described by the heat-conduction equation,

$$\frac{\mathrm{d}}{\mathrm{d}x}\left[k(T)\frac{\mathrm{d}T}{\mathrm{d}x}\right] + Q = 0$$

where $k(T) = k_0(1 + \beta T)$. The boundary conditions are $\mathrm{d}T/\mathrm{d}x(0) = 0$ and $T(L) = 0$.

a) Develop a finite element formulation for the problem, and derive the element Jacobian and residual load vector.

b) Use Newton–Raphson iteration to compute the temperatures at nodes 1 and2.

c) Derive the analytical solution for the problem, and assess the accuracy of the finite element solution.

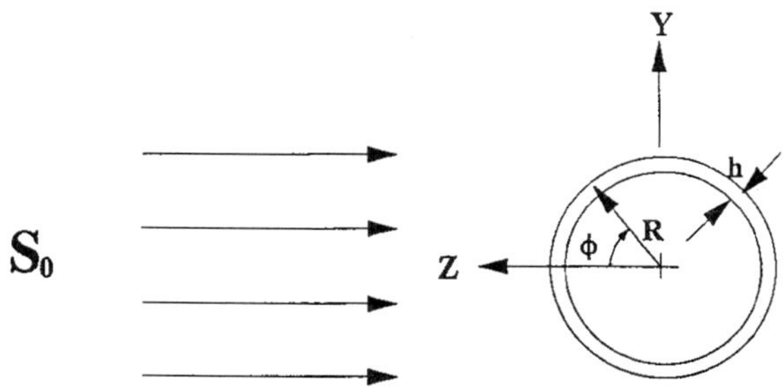

Fig. P8.3

8.3. A thin-wall tube initially at a uniform temperature acts as a boom in a solar array for a spacecraft in Earth orbit. At time equal to zero, the boom is exposed suddenly to solar heating. Conduction along the tube length is negligible so that the temperature varies only around the tube parameter, i.e., $T = T(\theta, t)$. The problem consists of solving the one-dimensional energy equation,

$$\rho c h\frac{\partial T}{\partial t} - \frac{kh}{R^2}\frac{\partial^2 T}{\partial \theta^2} + \sigma\varepsilon T^4 = \alpha q_s$$

where h is the tube-wall thickness, and R is the average tube radius. To determine the transient temperature distribution:

a) Develop a finite element formulation for the problem.

b) Derive the element matrices for an element with two nodes.

c) Describe a solution algorithm that combines Newton–Raphson iteration with an implicit Crank–Nicolson time-marching scheme.

8.4. Two rods with the properties shown are attached at their ends to rigid supports. The nodal temperatures shown are specified, and temperatures may be assumed to vary linearly between nodes. Compute a) the displacement of node 2, b) the reactions at nodes 1 and 3, c) the rod stresses, and d) the rod forces.

Node	T_i, °F
1	150
2	375
3	240

$T_{ref} = 70°F$
$A_1 = 2\,\text{in.}^2$
$A_2 = 1.25\,\text{in.}^2$
$E = (11.4-0.0045T)\times 10^6$ psi
$\alpha = 7\times(12.4+0.0022T)\times 10^6\ 1/°F$

where T is in degrees Fahrenheit.

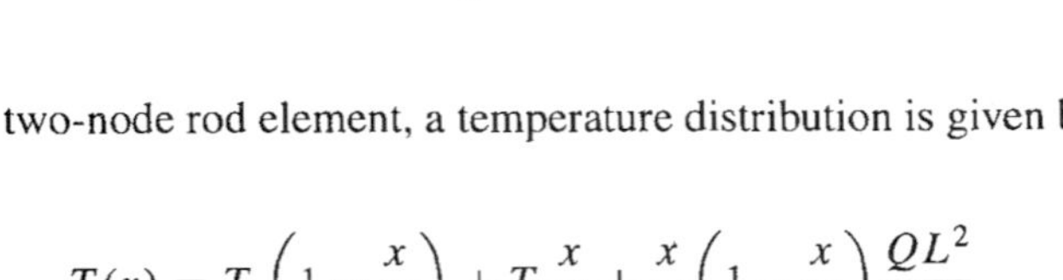

Fig. P8.4

8.5. For a two-node rod element, a temperature distribution is given by

$$T(x) = T_1\left(1 - \frac{x}{L}\right) + T_2\frac{x}{L} + \frac{x}{L}\left(1 - \frac{x}{L}\right)\frac{QL^2}{2k}$$

where T_1 and T_2 are nodal temperatures. The internal heat-generation rate Q and the thermal conductivity k are constant. Evaluate the rod element equivalent thermal load vector $\{F_T\}$.

8.6. A two-member truss has the properties shown. Nodal temperatures are specified as shown, and temperatures may be assumed to vary linearly over each truss member. Compute a) the displacement components at node 1, and b) the stresses in the truss members.

Node	T_i, °F
1	900
2	300
3	240

$T_{\text{ref}} = 70°\text{F}$
$A_1 = 6\,\text{in.}^2$
$A_2 = 4\,\text{in.}^2$
$E = 30\times10^6\,\text{psi}$
$\alpha = 7\times10^6\,1/°\text{F}$

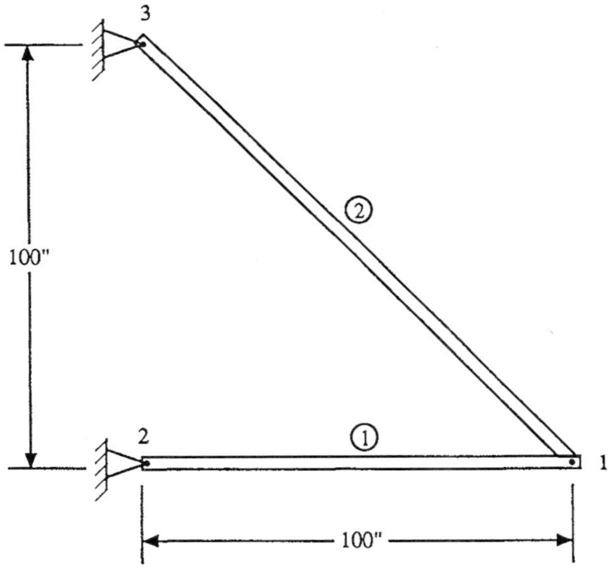

Fig. P8.6

8.7. The three-member truss shown in Fig. P8.7 has identical members of length L and the same material with constant E and α. Member 1 experiences a change in temperature of ΔT, but the other member temperatures are unchanged. Determine a) the displacement of node 2, and b) the maximum normal stress σ_x in the truss.

8.8. The simply supported beam shown in Fig. P8.8 has a steady-state linear temperature distribution through the beam depth, i.e., $T = a + by$, where a and b are constants. Use a single finite element to show that the beam normal stress σ_x is zero.

8.9. The cantilever beam shown in Fig. P8.9 is subjected to a constant thermal moment. Use a single finite element to determine a) the deflection and slope at $x = L$, and b) the equation for the deflected shape of the beam.

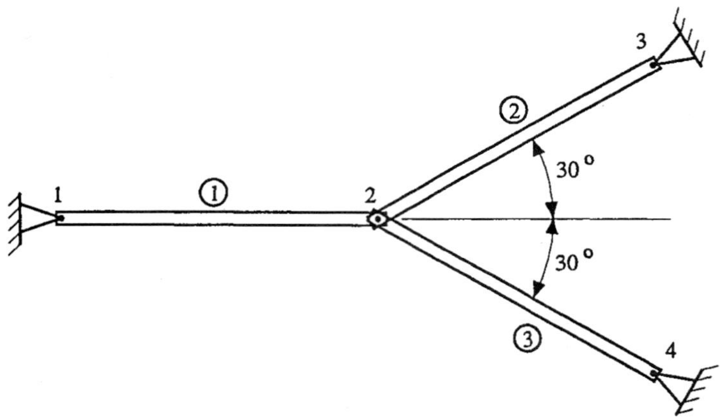

Fig. P8.7

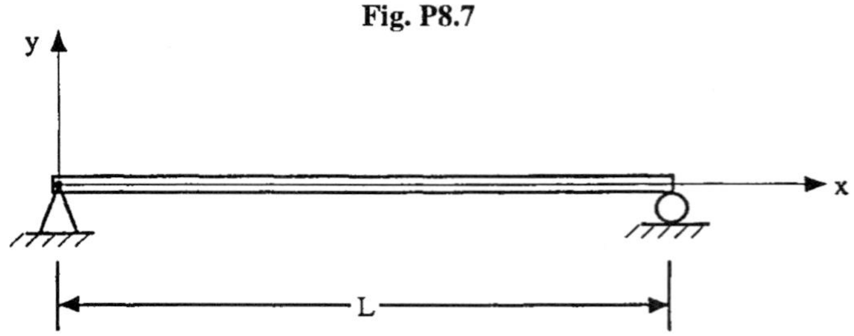

Fig. P8.8

8.10. Show that the finite element solution to the preceding cantilever beam problem is the "exact" solution to the problem.

8.11. Use a two-element model to determine a) the slope θ_2 at the center support, and b) the reactions at the supports. The beam is subjected to constant thermal moments given by

$0 < x < L$	M_T
$L < x < 2L$	$2M_T$

Sketch the beam's deflected shape.

8.12. Consider a two-dimensional triangular element either in a state of plane stress or plane strain. The element thickness is denoted by h. The element inter-

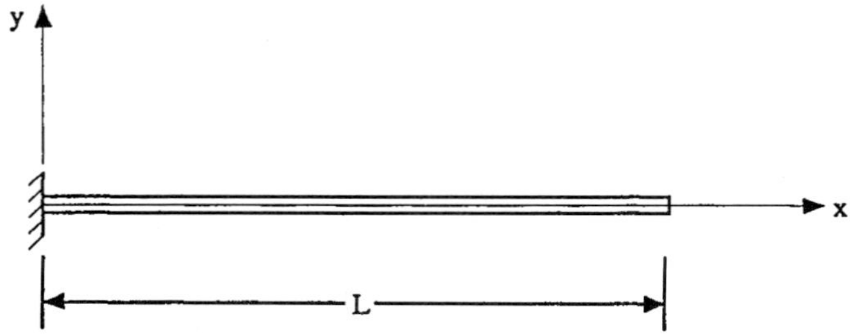

Fig. P8.9

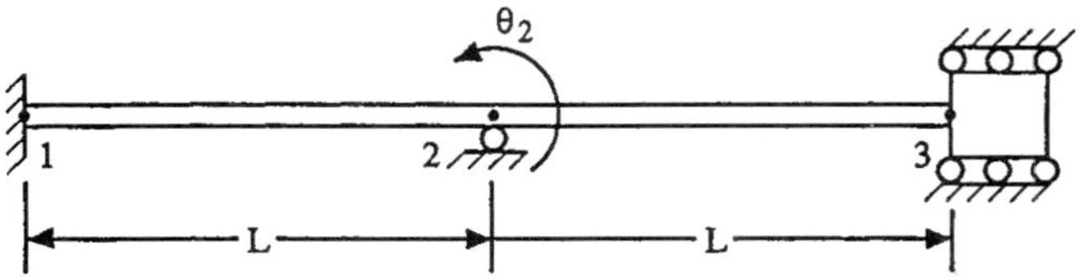

Fig. P8.11

polation functions are given by

$$N_i = (1/2A)(a_i + b_i x + c_i y) \qquad i = 1, 2, 3$$

where A is area of the element, and

$$a_i = x_j y_k - x_k y_j$$

$$b_i = y_j - y_k$$

$$c_i = x_k - x_j \qquad i, j, k = 1, 2, 3$$

where $x_i, y_i, i = 1, 2, 3$ are the element's nodal coordinates in the global coordinate system. Develop a) the strain interpolation matrix $[B]$, b) the element stiffness matrix $[K]$ in terms of $[B]$ and the elasticity matrix $[C(T_a)]$, and c) the equivalent thermal load vector $\{F_T\}$. if the element temperature distribution is $T(x, y) = [N(x, y)]\{T\}$, where $T_i, i = 1, 2, 3$ are the nodal temperatures.

8.13. Answer the questions in the preceding problem for a three-dimensional tetrahedral element. The interpolation functions for a four-node tetrahedral element are

$$N_i = (1/6V)(a_i + b_i x + c_i y + d_i z) \qquad i = 1, 2, 3, 4$$

where V is the element volume, and the coefficients a_i, b_i, c_i, d_i are computed from the element's nodal coordinates. The element's temperature is $T(x, y, z) = [N(x, y, z)]\{T\}$, where $T_i, i = 1, 2, 3, 4$ are the nodal temperatures.

8.14. For the von Kármán plate equation, use the weak form given in Eqs. (A.4) in the Appendix to derive the matrix form of the equilibrium equations given in Eq. (8.75).

8.15. Example 8.4 gives the nonlinear equilibrium equations for a cable element. Show that the terms in the element's Jacobian are given by

$$J_{11} = -J_{14} = J_{44} = \frac{AE}{L}$$

$$J_{12} = -J_{15} = J_{45} = -J_{24} = \frac{AE\theta}{L}$$

$$J_{13} = -J_{16} = J_{46} = -J_{34} = \frac{AE\psi}{L}$$

$$J_{22} = -J_{25} = J_{55} = \frac{AE}{L}\left[e - \alpha\Delta T + \frac{\sigma_0}{E} + \frac{3}{2}\theta^2 + \frac{\psi^2}{2}\right]$$

$$J_{23} = -J_{26} = J_{56} = \frac{AE\theta\psi}{L}$$

$$J_{33} = J_{66} = \frac{AE}{L}\left[e - \alpha\Delta T + \frac{\sigma_0}{E} + \frac{3}{2}\psi^2 + \frac{\theta^2}{2}\right]$$

where $e = (u_2 - u_1)/L$.

8.16. Figure P8.16 shows a nonlinear spring-rod system. When the applied load is zero, the rod has an original length L and makes an angle θ with the horizontal. For an applied vertical load P, show that the downward vertical displacement u at node 2 can be computed from the exact equation,

$$P = ku + \frac{EA}{L}(L\sin\theta - u)\left[\frac{L}{(L\cos\theta)^2 + (L\sin\theta - u)^2} - 1\right]$$

where k is the spring constant, and EA/L is the rod stiffness.

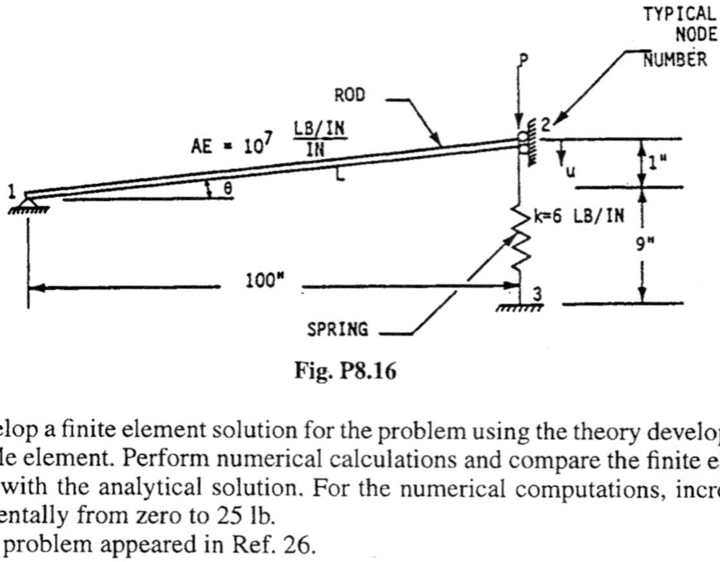

Fig. P8.16

Develop a finite element solution for the problem using the theory developed for the cable element. Perform numerical calculations and compare the finite element results with the analytical solution. For the numerical computations, increase P incrementally from zero to 25 lb.

This problem appeared in Ref. 26.

9
Thermally Induced Vibrations

9.1 Introduction

On April 24, 1990, the Hubble Space Telescope (HST) was launched into low Earth orbit aboard the Space Shuttle *Discovery*. The next day, the HST was deployed successfully from the *Discovery* payload bay and began its long-awaited operations. Within a short time, initial imaging data were acquired, and the now famous problem of the spherical aberration of the spacecraft mirror was discovered.

During this initial checkout period, another spacecraft problem[1–4] came to light: a pointing "jitter" induced by thermally driven bending of the solar arrays. The vibrations were induced when the solar-array structure reacted to temperature changes occurring as the spacecraft passed from shadow to sunlight. The solar arrays (Fig. 9.1), developed by the European Space Agency, are based on a design purchased from Hughes Aircraft Company. The arrays consist of thin solar panels that were furled initially within a drum and later unfurled by 20-ft. Bi-STEMs. A Bi-STEM consists of two concentric thin-walled tubes stored on a drum in a flat configuration (see Fig. 9.2). Early analyses of the jitter problem indicated that the solar array oscillations were related to thermally induced bending of the Bi-STEMs. On the Space Shuttle *Endeavour's* mission in December 1993, new solar arrays designed to eliminate the persistent pointing jitter problem were installed.

Thermally induced vibrations of space structures are not a new phenomena. Such oscillations became known during the flight of the OGO-IV spacecraft in the 1960s.[5] On that flight, a 60-ft experiment boom sustained a solar-induced large-amplitude oscillation that severely compromised spacecraft performance. A coupled thermal-structural analysis predicted torsional–flexural oscillations consistent with the observed satellite phenomena. Table 9.1 lists 22 spacecraft in which thermally induced vibrations are known to have compromised spacecraft performance. These data were taken from the open literature; there are probably unpublished accounts of additional spacecraft that have experienced similar problems. Military satellites and Soviet spacecraft are prime examples.

The investigation of thermally induced vibration phenomena is an interesting story. Today, virtually all thermally induced vibration problems are associated with spacecraft. Yet the study of thermally induced vibrations began with basic applied mechanics research initiated in the mid-1950s before *Sputnik* became the first artificial Earth satellite on October 4, 1957. It was not until anomalous spacecraft behavior began to occur in the 1960s that the basic theory found practical application. The actual behavior of spacecraft appendages was much more complex than anticipated in the early theoretical development. In fact, new unexpected nonlinear dynamic behavior was encountered. The problem of understanding the unusual thermally induced phenomena is exacerbated because actual motions of spacecraft appendages have rarely been observed. Typically, conclusions about the phenomena are inferred from the spacecraft's sensors and the control system's

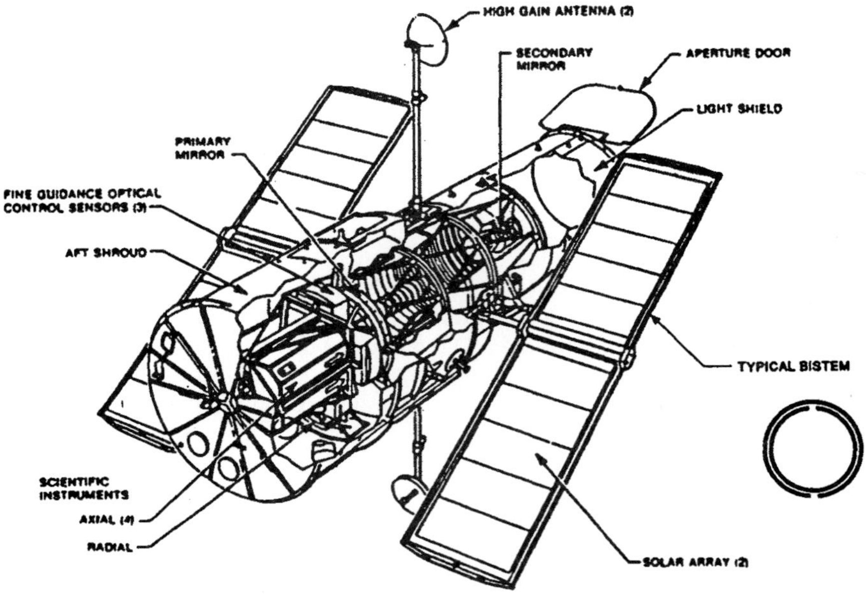

Fig. 9.1 Hubble space telescope.

response. The study of thermally induced vibrations is intrinsically difficult because of the multidisciplinary nature of the problem, where the thermal problem may be coupled nonlinearly to the structure's dynamic behavior. Thus, the occurrence of unexpected nonlinear dynamic behavior poses a significant engineering challenge to designers, analysts, and experimentalists. In spite of the practical importance of thermally induced vibrations to the successful operation of spacecraft, recurrent vibration problems from the 1960s to the present day suggest that the phenomena is either underemphasized or not well understood in spacecraft design.

In this chapter, we tell the story of thermally induced vibrations from the inception of the basic research to the present. We begin with a historical review of

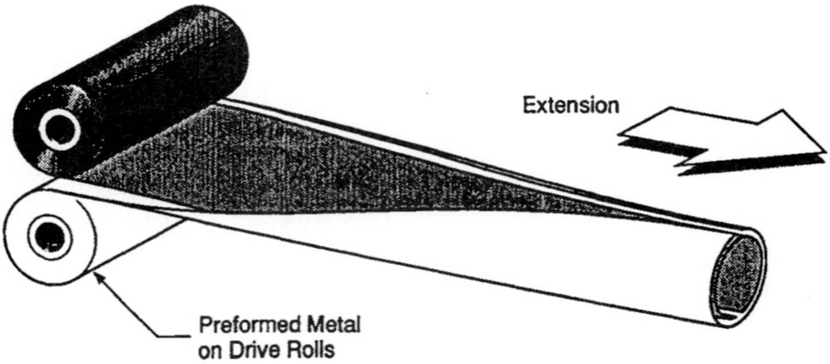

Fig. 9.2 Bi-STEM extendable boom.

Table 9.1 Spacecraft with thermally induced disturbances

Spacecraft	Date	Description
Ulysses	1990–1991	Spin-stabilized solar probe experiences unexpected nutation due to themally induced vibrations of beryllium–copper axial boom
Upper Atmosphere Research Satellite	1991	"Thermal snap" at orbital night/day transition attributed to rapid heating of large single solar array
Hubble Space Telescope	1990	"Jitter" phenomena attributed to thermally induced vibrations of FRUSA-type solar arrays
LANDSAT-4/5	1980s	Thermal elastic shock experienced during orbital night/day transition due to disturbance of large single solar array.
Communications Technology Satellite	1978	Three-axis-stabilized satellite experienced "thermal elastic shock" during orbital night/day transitions
Voyager	1977	Low-frequency oscillations of PRA booms during LEO operations attributed to "thermal flutter"
Apollo 15	1971	Thermally induced vibrations of twin Bi-STEM booms filmed during 64th lunar orbit by astronauts
Explorer 45	1971	Unexpected nutation of spin-stabilized satellite attributed to thermal bending of four radial booms
NRL 161, 163, and 164	1969	Thermally induced vibrations of STEM booms used for gravity-gradient stabilization leads to large satellite motions
GGSE III–VI	1968–1969	NRL gravity-gradient-stabilized satellite STEM booms experience thermal flutter
OGO-IV	1967	NASA Orbiting Geophysical Observatory Satellite STEM booms experience strong thermally induced vibrations during orbital night/day transition
OV-10	1966	U.S. Air Force gravity-gradient satellite using STEM booms equipped with tip masses experiences thermal flutter and complete inversion in orientation
Explorer XX	1964	NASA spin-stabilized satellite experiences spin decay due to interaction between STEM boom thermal bending and solar-radiation pressure
1964-83D and 1963-22A	1964/1962	Dynamic thermal bending of STEM booms on APL gravity-gradient satellites
Aloutte I	1962	Canadian spin-stabilized satellite experiences spin-rate decay due to thermal-structural response of STEM booms similar to *Explorer XX*.

thermally induced vibration research from the 1950s to the 1990s. Then we describe an analysis of thermally induced, uncoupled flexural vibrations of beams. Subsequently, we simulate thermally induced, coupled flexural vibrations of a boom in a laboratory experiment and demonstrate analytically the phenomenon of thermal flutter. Finally, we present analyses of thermally induced oscillations of solar arrays for the Hubble Space Telescope.

9.2 Historical Review

The historical review is divided into six sections, in which we present pertinent literature chronologically.

9.2.1 Thermally Induced Vibrations of Beams, Plates, and Shells

In earlier chapters, we presented conventional quasistatic thermal-stress analysis of aerospace structures in which we neglected inertia forces in the structure's equations of motion. When the associated thermal problem is unsteady, the structural analysis proceeds through a succession of equilibrium analyses, using as input spatial temperature distributions at succeeding times. Such quasistatic analyses assume that dynamic magnification effects due to the structure's inertia forces are negligible.

In 1956, Boley provided the first analytical investigation of thermally induced response of beams including inertia forces. In his initial paper, subsequent papers,[6,7,14] and a thermal-stress analysis text,[8] he established fundamental concepts of thermally induced vibration behavior. A typical problem consists of a thin beam or plate subjected to uniform step heating on one surface. The transient temperature distribution through the thickness causes a time-dependent thermal moment that acts as the dynamic forcing function. The equation of motion is solved analytically by separation of variables in terms of natural frequencies and normal modes.

Boley identified a key parameter for the occurrence of thermally induced oscillations: the ratio of the thermal response time of the structure to the structural response time. Using the notation of his 1972 paper,[14] the Boley nondimensional parameter is

$$B = \frac{t_T}{t_S} \tag{9.1}$$

where t_T is the characteristic thermal response time (e.g., h^2/κ_T, where h is the beam or plate thickness, and κ_T is the thermal diffusivity); t_S is the characteristic structural response time (the lowest natural period). Boley presents an approximate formula for the deflection magnification factor as

$$\frac{v_{\text{dyn}}}{v_{\text{st}}} = 1 + \frac{1}{\sqrt{1 + B^2}} \tag{9.2}$$

where v_{dyn} is the maximum dynamic deflection considering inertia forces, and v_{st} is the maximum quasistatic deflection. His exact calculations, as well as the approximate result given by Eq. (9.2), indicate a maximum dynamic displacement amplification factor of 2.

For most Earthbound structures, $B \gg 1$, given that the thermal response time is typically much larger than the characteristic structural response time. This fact justifies using quasistatic thermal-structural analysis. Structures designed for the Earth environment must be stiff enough to withstand gravitational loads and thus have relatively high frequencies and short natural periods. For a space environment with very low gravitational loads, however, structures are much more flexible, with low natural frequencies often much less than 1 Hz and long natural periods. In such circumstances, the Boley parameter B may be less than 1, and thermally induced oscillations may occur in response to rapidly applied heating.

After Boley's original research, several subsequent papers presented analytical studies of linear, thermally induced vibrations of beams, plates, and shells. Jones[9] investigated thermoelastic vibrations of beams, including shear effects, rotatory inertia, and axial forces. Kraus[10] analyzed thermally induced vibrations of thin, nonshallow spherical shells. In 1966, Lyons[11] noted that there had been no experimental evidence to support the analytical work and argued incorrectly that instantaneous surface heating could not cause thermally induced transverse oscillations. Lyons was correct, however, in his observation concerning the lack of fundamental experiments. The first documented experimental demonstration of thermally induced vibrations was for a spacecraft boom and did not occur until June 1968 (see Sec. 9.2.4 for a description of this experiment). The first application[12] of finite element techniques to thermally induced vibrations occurred in 1968. Mason used time-dependent nodal loads for beam and rectangular plate elements to solve beam and plate problems subjected to surface heating; he verified the approach by comparison with analytical solutions.

Analytical and computational studies[13–19] of thermally induced oscillations of beams and plates have continued to the present day. Stroud and Mayers[13] investigated the influence of temperature-dependent properties on the dynamic response of plates and showed that neglecting temperature dependence is an unconservative assumption. Thermoelastic coupling for vibrations of Bernoulli–Euler beams was investigated by Seibert and Rice,[15] who found the coupling effect insignificant for such beams. They also investigated thermally induced oscillations for Timoshenko beams and concluded that a step heat input may cause significant vibrations in short as well as long beams. Bargmann[16] surveyed developments for thermally induced waves and vibrations and included an extensive bibliography.

Jadeja and Loo[17] investigated the vibrations of a rectangular plate subjected to sinusoidal surface heating and mentioned the role of such vibrations in terms of potential fatigue failures. Biswas[18] analyzed thermally induced vibrations of a triangular plate on an elastic foundation using the Laplace transform method. Manolis and Beskos[19] developed a numerical method for determining the dynamic response of planar beam structures to rapidly applied thermal loads. They used the Laplace transform and inverted the transformed response numerically. Numerical results were presented for beams and a gridwork of beam elements.

Takeuti and Furukawa[20] considered thermal-shock problems for a plate for 1) thermally induced vibrations due to the inertia force and 2) a coupled thermoelastic formulation. Actually, they did not consider plate bending but, instead, analyzed a finite thickness slab as a one-dimensional thermoelasticity problem. Thus, their results did not apply to the low-frequency vibrations characteristic of plate bending.

Das[21] described a complex variables approach for determining the vibration response of polygonal plates due to thermal shock. He presented numerical results for triangular, square, and circular plates.

In the first of two papers, Nakajo and Hayashi[22] studied thermally induced bending vibrations for circular plates analytically, using separation of variables.

In the second paper,[24] they continued their study of circular plates by 1) the analytical approach, 2) the finite element method, including large deflections, and 3) experimental study. Their experimental study is one of the very few published studies that attempts to compare theoretical results with experimental data. The experimental study is described briefly, and few details are given. No comparisons of displacement histories are given. For a simply supported plate, the experimental data for the peak dynamic deflection agree better with analysis than the data for a clamped plate do. The experimental amplitude of the peak dynamic response of the clamped plate is 20 times the analytical result and 60 times the finite element predictions.

In a survey chapter, Tauchert[23] described thermally induced plate vibration solutions for isotropic and composite materials. In later papers, Tauchert[25,26] and Chang et al.[28] studied thermally induced vibrations of metallic and composite plates. Tauchert presented analytical solutions, and Chang and his coauthors used the finite element method. Numerical calculations demonstrated vibration histories for aluminum and graphite–epoxy plates. In the computations for an aluminum plate with dimensions $10 \times 10 \times 0.06$ in., an incident heat flux of 10 Btu/in.^{2}-s was used by both authors to excite the vibration response. Note that 10 Btu/in.^{2}-s $= 1.6 \times 10^6$ W/m^2. The solar flux in Earth orbit is about 1350 W/m^2; hence, for the plate studied, thermally induced vibrations would probably be negligible for the much lower heat-flux values encountered in spacecraft applications.

Namburu and Tamma[27] used a finite element approach with time-marching to analyze flexural configurations and assess the role of nonlinear thermal effects.

The investigations of thermally induced vibrations of beams, plates, and shells that we review in this section are based on the assumption that the applied heating and temperature are independent of the structure's deformation. The dynamic response of an elastic structure is the sum of two parts: 1) the quasistatic deformation and 2) the thermally induced vibration. When damping is included, the vibrations decay with time. For sustained heating, after the vibrations decay, the structure returns to the quasistatic deformation shape. For suddenly applied heating, the peak dynamic displacement may be, at most, two times the quasistatic displacement.

Based on the assumption that the structure's heating and temperature are independent of the structural deformation, thermally induced vibrations are always stable. Later sections describe applications in which the applied heating and temperature depend on the structure's deformation. In this circumstance, thermally induced vibrations may be unstable. When heating and/or temperature are coupled with deformations yielding unstable vibrations, we call the phenomenon *thermal flutter*.

9.2.2 Storable Tubular Extendable Member (STEM)

A storable tubular extendable member (STEM) is a flat strip of thin metal that assumes a tubular shape when extended from a storage drum. The principle employed in a STEM is similar to that of a carpenter's steel tape. STEMs are manufactured from flat tape that is formed into a tube and heat-treated to relieve stresses. The tape is flattened elastically as it is wound on the drum so that, when extended, it curls into the tubular shape. Two concentric STEMs constitute a Bi-STEM (Fig. 9.2). In a 1965 article,[29] Rimrott described the types and design considerations for STEMs.

The STEM was developed specifically for use as an extendable boom for spacecraft requiring mechanical systems that can be deployed and retracted in orbit.

Open-section booms such as STEMs or Bi-STEMs have been used as communication antennas, gravity-gradient stabilizing booms, solar-panel actuators, and instrument carriers. A brochure[30] prepared by the Astro Aerospace Corporation presents STEM design information. Currently available STEM tube diameters range from 0.28 to 3.12 in. Materials used include beryllium–copper and stainless steel. Lengths of such extendable spacecraft booms range from a few feet to 600 ft.

Because STEMs are structural members of open cross section, torsional stiffness is relatively low. Cross sections warp during rotation. The static torsional behavior of such slit, overlapped thin tubes was investigated by Rimrott and Elliott[31] in 1966. The following year, Frisch published a quasistatic analysis[32] of a gravity-gradient boom in Earth orbit. The analysis considered a thin-walled, overlapped thin tube subjected to solar heating. A thermal model was developed, and tube temperature distributions were computed. A combined torsion-bending structural analysis was presented, and boom equilibrium shapes were predicted. Shortly thereafter, Florio and Hobbs[33] described an investigation of the temperature distributions in gravity-gradient rods. They considered an overlapped thin-walled tube and developed an analytical representation for the tube's circumferential temperature distribution. Predicted temperature distributions were compared with experimental data obtained from tests conducted in the NASA Goddard Solar Simulation Facility. Two years later, Graham[34] presented an equation for the radiation heat transfer within the interior of a long cylinder. He noted that, for conventional STEMs, internal radiation heat transfer may often be neglected but, for some STEMs of the overlapping type, internal radiation may have to be considered. The thermal analysis of Frisch and the analysis of Florio and Hobbs assume that 1) incident heat fluxes are independent of tube deformations and 2) temperatures are steady.

An early indication that thermally induced deformations of STEMs could cause unexpected spacecraft behavior is described in the 1967 paper by Etkin and Hughes.[35] Two satellites, *Alouette I* and *Explorer XX*, exhibited anomalous spin behavior. Each satellite demonstrated a progressive loss of spin from its initial value of 1.5 rpm that could not be explained on the basis of structural, mechanical, aerodynamic, or electrodynamic actions. The common features of the satellites were STEM antennas. The satellites typically had four booms, ranging in length from 37.5 to 75 ft. The article shows that the spin decay was caused by a torque induced by solar-radiation pressure on the thermally deformed booms. The unexpected STEM behavior in this case was caused by the combined action of solar-radiation pressure and solar-heating–induced bending. The asymmetric, thermally induced bending led to the undesirable torque that resulted in the loss of angular velocity.

9.2.3 Thermally Induced Vibrations and Stability of Spacecraft Booms

From September 1964 to July 1972, the United States operated six Orbiting Geophysical Observatory (OGO) spacecraft that returned significant data on various geophysical phenomena. On the OGO satellites, instruments sensitive to magnetic fields generated by electrical equipment on the satellite itself were mounted on booms (STEMs) extending from the spacecraft. The OGO spacecraft were the first scientific satellites to be three-axis-stabilized, with all three axes held relatively stable in space. OGO I–III (1964–1965) each experienced attitude-control problems within a few days and resorted to a backup spin-stabilized control mode. OGO IV (1967) maintained nominal altitude control for 18 months.

By spring 1968, structural engineers had learned that, at various orbital times, the OGO satellites were experiencing oscillations of increasing amplitudes. The oscillations began when a satellite moved from the Earth's shadow into sunlight and subsequently decayed when the satellite re-entered the Earth's shadow. Engineers began to conjecture that the unstable oscillations were the result of coupling solar-thermal radiation and elastic deformation of the booms.

In June 1968, Beam[36] of NASA Ames demonstrated unstable thermally induced oscillations in the laboratory. The experiment was performed using a 3-ft-long boom, fixed at one end and with an attached mass at the other. The boom was a slit, overlapped tube of beryllium–copper with a diameter of 0.5 in. and a wall thickness of 0.002 in. The boom was heated along its length by several photo flood lamps. The boom-mass system was stable for small-amplitude motions but, for large amplitudes, the motions were unstable, with an eventual limit-cycle oscillation in a coupled bending-torsion mode. The limit-cycle oscillation amplitudes consisted of tip-mass deflections of several inches normal to the plane of the heat lamps and tip-mass rotations of ±90 deg. Beam also demonstrated thermal flutter for a 19-ft-long boom with and without tip masses. Thus, for the first time, the phenomenon of thermally induced unstable oscillations was demonstrated in the laboratory. Before 1968, Boley and other researchers had performed computations demonstrating thermally induced stable oscillations, but the new phenomenon, thermal flutter, was more serious and had important practical significance to satellite operations.

Before 1968 analysts had not yet simulated thermal flutter. Interestingly, however, just two months after Beam's thermal-flutter experiment, Augusti[37] presented a paper on the stability of heated structures. He studied radiantly heated rigid links with elastic springs and demonstrated that dynamic instability was possible. In the next few months, several researchers, including Beam, began analytical studies of the spacecraft boom problem.

In December 1968, the Air Force sponsored a symposium on gravity-gradient attitude stabilization. Several papers, among them Refs. 38 and 39, described analyses of thermal flutter. Donohue and Frisch[38] described a simulation of thermally induced bending-torsion oscillations of a 60-ft OGO IV open-section boom and concluded that the abnormal behavior of the OGO IV spacecraft was explained by the thermal oscillation theory. Koval et al.[39] developed a mathematical model and presented numerical results for a case representative of an OGO IV boom. Bending-torsion oscillations with sustained tip motion of ±8 to 9 ft were predicted normal to the plane of solar radiation. The oscillations occur at the fundamental bending frequency of the boom. Both papers concluded that the key feature of the booms that permitted buildup of thermally induced oscillations was their low torsional stiffness.

In August 1969, Yu[40] published a paper on thermally induced bending of a flexible boom. Yu modeled a boom as a cantilever beam with tip mass. The boom was assumed to bend in a plane without twisting. The analysis demonstrated for the first time that thermal flutter could occur in bending without torsion, provided that the solar-heating vector is aligned at an appropriate angle with respect to the beam axis. Actually, Yu's conclusion concerning the required orientation of the sun was not correct because of an error in his equations. The required sun orientation is actually the reverse of his original prediction. The correct orientation was pointed out in a paper by Graham[41] in November of 1970 and in comments by Augusti[42] and by Jordan[43] in early 1971.

A boom subjected to solar radiation may experience unstable bending oscillations if the boom is pointed away from the sun. Graham[41] derived a stability

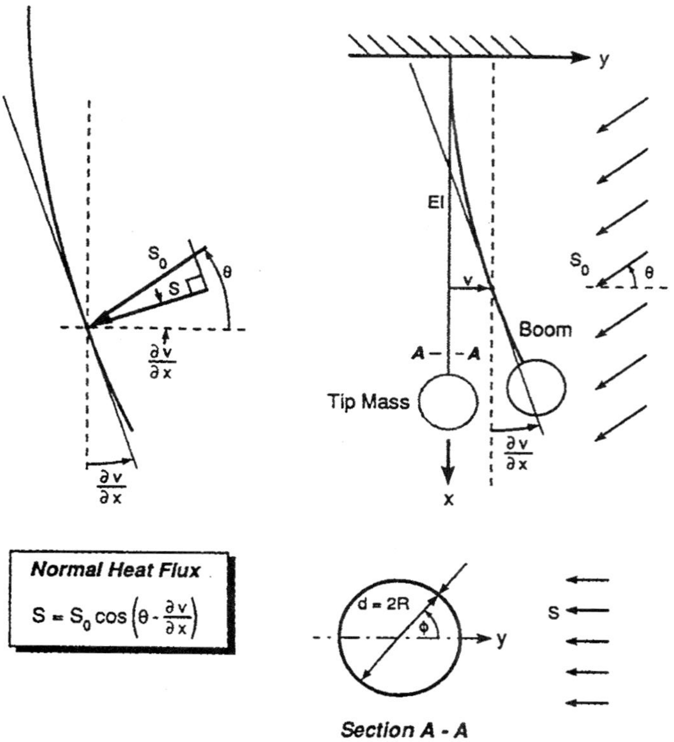

Fig. 9.3 Model of spacecraft boom with solar heat flux.

criterion showing that, in addition to the sun orientation angle, other parameters such as damping and surface absorptivity affect stability when the boom is pointed away from the sun. Figure 9.3 plots the geometry of a cantilever beam with a tip mass; the solar-induced motion may be unstable if the direction of radiation is inclined toward the fixed end, that is, for $\theta > 0$, but the motion is stable if the radiation is normal or inclined toward the tip of the beam, that is, for $\theta \leq 0$.

In 1970, Frisch presented a detailed study[44,45] of thermally induced transverse bending plus torsional oscillations of spacecraft booms (STEMs) with reference to flight data. He showed that increasing the torsional rigidity can eliminate the problem.

Vigneron[46] investigated the thermal time constants for heat transfer in booms and discussed an experimental study of the thermal response. He suggested that the internal radiation heat transfer can be represented conveniently by using an average wall temperature. Ahmed[47] discussed the quasistatic problem of thermally induced twist of thin-walled, open-section beams and noted that the twist behavior depends heavily on the mechanical end conditions of the boom.

Until 1971, the aforementioned studies had predicted thermal flutter of spacecraft booms and had inferred that anomalous behavior of several spacecraft was consistent with thermally induced boom oscillations. Actual boom oscillations due to solar heating in space had not been observed.

The *Apollo 15* mission was launched on July 26, 1971. In lunar orbit, two silver-plated stainless steel, 2-in.-diam Bi-STEMs were extended 25 ft away from the *Apollo 15* spacecraft. The booms supported gamma-ray and mass spectrometers.

The tip of the mass-spectrometer boom was photographed during the 64th lunar orbital revolution of the spacecraft. Kumar and Ahmed[48] described the experiment, measurements taken from the film, and comparisons with predictions. Thermally induced bending oscillations were observed to start at the near-side terminator position and to last about 3.5 min. The observed period of vibration was about 5.5 s, corresponding to a frequency of 0.18 Hz. The measured amplitude of vibration was 0.67 in. Because the spacecraft remained in sunlight for about 74 min, the relatively short duration of the oscillations suggests stable behavior. Kumar and Ahmed also discussed quasistatic bending and torsion of the boom and presented comparisons with data determined from the film. The *Apollo 15* mission is apparently the only space mission for which solar-induced oscillations have been observed directly.

After the late 1960s and early 1970s, during which time thermal flutter of spacecraft booms was intensely investigated, research slowed considerably. In 1973, however, Beam and Yagoda[49] presented a detailed study of the torsional static stability and response of open-section tubes subjected to thermal radiation heating. And Farrell,[50] in 1977, presented a formulation for thermal curvature of satellite booms.

9.2.4 Thermal Flutter in the Laboratory

As previously discussed, in 1968, Beam[36] succeeded in producing torsional thermal flutter in the laboratory; almost a decade later, in 1977, Rimrott and Abdel-Sayed[51] successfully produced flexural thermal flutter in the laboratory. Thermally induced oscillations are further discussed by Rimrott in Refs. 52 and 53.

A very important result from the Rimrott investigations is the distinction between the heat-flux environment in space and the heat-flux environment in the laboratory. In space, the radiation heat flux is constant on a spacecraft because the heat source, the sun, is essentially located at infinity. In the laboratory, the radiation heat flux might decrease rapidly with distance if the heat source is a heat lamp.

For flexural thermal flutter, the temperature gradient through the beam depth is determined by the heat flux normal to the beam surface. With reference to Fig. 9.3 for the space environment, the heat flux S normal to the beam surface is given by

$$S = S_0 \cos\left(\theta - \frac{\partial v}{\partial x}\right) \qquad \text{(space)} \tag{9.3}$$

where S_0 is the solar flux, θ the angle of orientation of the solar vector, and $\partial v/\partial x$ the slope of the deflected beam. The dependence of the incident heat flux, and consequently beam temperatures, on the beam slope couples the thermal and structural problems and introduces the possibility of thermal flutter. In contrast, for a beam in the laboratory, Rimrott[52] argues that the variation of the normal heat flux occurs with displacement rather than slope. Rimrott (Fig. 9.4) approximates the laboratory flux by a linear function,

$$S = S_0(1 - v/\beta^*) \qquad \text{(laboratory)} \tag{9.4}$$

where v is beam displacement, and β^* is a parameter determined from a lamp heat flux vs distance test.

Because of differences between the normal heat fluxes for space and the laboratory, the flexural thermal phenomenon of space cannot be simulated on Earth using heat lamps. Yet Rimrott showed in experiments and by analysis that flexural thermal flutter can be demonstrated in the laboratory. The laboratory flutter,

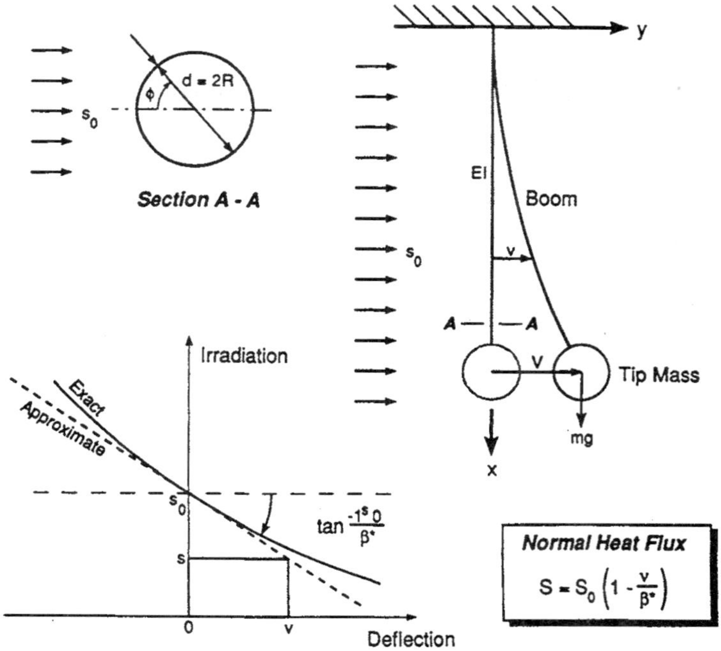

Fig. 9.4 Laboratory model of spacecraft boom.

however, occurs by a different mechanism than in space, and the stability criterion for laboratory heating is different than for solar heating. For example, in a laboratory environment, thermal flutter does not depend strongly on the incident flux angle. Unstable oscillations can be induced by a heat flux normal to the beam axis in the laboratory, which does not occur in space.

9.2.5 Japanese Research

Since 1985, Japanese engineers have conducted an intensive research program focused on thermally induced oscillations of spacecraft booms. Analytical and experimental studies[54–60] of both torsional and flexural oscillations have been conducted. With one exception, Ref. 57, the available reports are in Japanese with English abstracts. The research has focused on establishing stability conditions for thermal flutter using analytical solutions and validating the stability criteria experimentally. Torsional and flexural thermal-flutter experiments have been performed in air and in vacuum chambers. Experimental results are a particularly valuable contribution since, with the exception of Beam[36] and Rimrott,[51] there have been few experiments. Data for validation of computational methods are scarce.

9.2.6 Thermally Induced Motions of Spacecraft Appendages

In the 1960s, engineers began designing spacecraft with large, flexible appendages such as antennas or solar arrays. Increasing attention was focused on the dynamics of spacecraft, including the interaction of the flexible appendages with spacecraft motions. In 1965 at a symposium held at NASA Ames Research Center, two papers describing solar effects on stabilization booms were presented. Zimmerman[61] reports solar effects on four extendable booms (STEMs) on the Applications Technology Satellite. Mobley and Fischell[62] describe vibrations experienced on two gravity-gradient–stabilized satellites. The vibrations are attributed to a 100-ft beryllium-copper boom. The paper discusses silver-plating the booms as a means to reduce temperature gradients and vibration amplitudes. In 1967, Etkin and Hughes[35] investigated the de-spin mechanism of spacecraft with flexible antennas.

Further research on the influence of thermal distortions on gravity-gradient satellites was conducted by Kanning in 1967 and Goldman in 1975. Kanning[63] discusses the effects of quasistatic distortions on a satellite's attitude. Goldman[66,67] addresses a similar problem and describes the development of a computer program to simulate the behavior. Although the vibration behavior of gravity-gradient–stabilized satellites is attributed to the thermal distortions of the booms, the behavior is distinct from the thermal-flutter phenomena experienced by other satellites.

A 1973 paper[64] by Cherchas and Hughes deals with a study of the attitude stability of a dual-spin satellite with a large, flexible solar array. The interesting feature of the paper is that the design of the solar array is similar to the one used on the Hubble Space Telescope. The array studied is the Hughes Aircraft FRUSA (flexible rolled-up solar array) design. The supporting beams are STEM or Bi-STEM booms. The vibration modes of the solar arrays were analyzed by treating the booms as cantilever beams, and the solar-panel sheets were represented as pretensioned membranes. The dimensions of the array were 18×6 ft. Natural vibration modes consisting of bending in two orthogonal planes and torsion are identified. The vibration frequencies of the panel are in the range of 1–2 Hz and are found to be a strong function of the membrane tension. The effect of the vibration modes on the spacecraft stability and nutation damping was investigated, but thermal effects were not included.

Tsuchiya[65,68] studied the effects of thermally induced vibrations of flexible appendages on the motion of spacecraft. An appendage is assumed to be a thin beam cantilevered from the spacecraft. Tsuchiya investigated the interaction between the thermally induced vibration and the nutational frequency for a spinning spacecraft[65] and a spacecraft containing a rotor.[68]

A 1980 paper[69] by Ayer and Soosaar presents an overview of the major sources of environmental disturbances that affect structural distortions of space systems. Thermal effects are discussed, and it was noted that, for large, extremely flexible platforms, the transition from eclipse to sudden solar exposure may excite thermally induced vibrations. The work of these authors reflects recognition that thermally induced dynamic effects for structures other than booms may be significant. Frisch[70] describes a computational approach for thermally induced response of flexible structures. This approach is based on using widely accessible general purpose computer programs such as SINDA, TRASYS, DISCOS, and NASTRAN. Further investigations of large space structures were conducted by Krishna and Bainum[71] as well as Hamsath et al.,[73] who studied the effect of solar-radiation pressure on the dynamics and control of orbiting beams and plates that undergo thermal deflections due to solar heating.

The book on spacecraft attitude dynamics by Hughes[72] describes the effects of thermally induced motions for several satellites.

Papers by Dennehy et al.[74] and Zimbelman et al.[75] discuss the prediction of the impulsive torque exerted on a satellite by rapid heating or cooling of a large solar array. The thermal "snap" phenomenon had been observed during normal operations of the LANDSAT-4 satellite and the Communications Technology Satellite.

9.2.7 Assessment

In this section, we have described the history of thermally induced vibrations. Although the initial analytical investigations began before the advent of satellites, spacecraft appendages represent an important application of the theory. The majority of theoretical studies assume that the structure's heating and temperature distributions are independent of structural deformations. Vibration behavior for this condition is always stable. Classical analytical techniques or modern numerical methods work well for these problems.

The more challenging problem occurs when the structure's heating and temperature distribution depend on structural deformations. This leads to thermal-flutter phenomena in which the vibrations can become unstable. As we have seen, thermally induced vibrations and thermal flutter have contributed to the anomalous behavior of satellites and spacecraft for over 30 years. One basic problem is that it is difficult to evaluate structural dynamics behavior by ground testing. This is particularly true with regard to thermal flutter and other nonlinear behavior induced by coupling between heating and structural deformations.

In the next two sections, we describe analyses of uncoupled and coupled thermally induced flexural vibrations in more detail.

9.3 Uncoupled, Thermally Induced Flexural Vibrations

In Sec. 9.2.1, we describe several papers that present analytical solutions to thermally induced vibration problems. The basic approach is the modal expansion approach introduced by Boley in his original 1956 paper. In this section, we illustrate the modal expansion approach for thermally induced vibrations of a cantilever beam. Similar analytical solutions for more complicated structural components may be found in the references.

9.3.1 Problem Statement

We consider a Bernoulli–Euler beam where the coordinate planes coincide with the cross section's principal axes. In this case, bending of the beam can occur only in the x–y plane. From Sec. 6.3.1, Eq. (6.18c), we may write the beam's equilibrium equation as

$$EI\frac{\partial^2 v}{\partial x^4} + \frac{\partial^2 M_T}{\partial x^2} = f_y$$

where f_y is the distributed force per unit length. By D'Alembert's principle, we may let f_y be the inertia force,

$$f_y = -\rho A\frac{\partial^2 v}{\partial t^2}$$

Hence, the equation of motion is

$$EI\frac{\partial^4 v}{\partial x^4} + \rho A\frac{\partial^2 v}{\partial t^2} = -\frac{\partial^2 M_T}{\partial x^2} \tag{9.5}$$

From Chapter 6, Eqs. (6.18a–b) the bending moment and shear force are given by

$$M = -EI\frac{\partial^2 v}{\partial x^2} - M_T \tag{9.6a}$$

$$V = -EI\frac{\partial^3 v}{\partial x^3} - \frac{\partial M_T}{\partial x} \tag{9.6b}$$

where we use partial derivatives because all quantities are a function of x and t. The thermal moment is defined as

$$M_T = \int_A E\alpha(T - T_{\text{ref}})y\,\mathrm{d}A \tag{9.7}$$

where the integration takes place over the beam cross-sectional area A.

If we assume that the beam's motion does not affect the temperature distribution, then the energy equation and the equation of motion (9.5) are uncoupled and may be solved independently. Thus, we may solve the energy equation (4.1) first to obtain the beam's temperature distribution as a function of position and time. Then, for the prescribed temperature, we solve the foregoing structural equations for the beam's dynamic response . For simplicity, let us assume that the beam's temperature does not vary along the beam; hence, we take $T = T(y, z, t)$. After the integration over the beam cross section, the thermal moment depends only on time; $M_T = M_T(t)$. Then the equation of motion (9.5) reduces to the homogeneous equation

$$EI\frac{\partial^4 v}{\partial x^4} + \rho A\frac{\partial^2 v}{\partial t^2} = 0 \tag{9.8}$$

The boundary conditions for a cantilever beam of length L are

$$v(0, t) = 0 \tag{9.9a}$$

$$\frac{\partial v}{\partial x}(0, t) = 0 \tag{9.9b}$$

$$M(L, t) = -EI\frac{\partial^2 v}{\partial x^2}(L, t) - M_T(t) = 0 \tag{9.9c}$$

$$V(L, t) = -EI\frac{\partial^3 v}{\partial x^3}(L, t) = 0 \tag{9.9d}$$

We see that the temperature enters the problem through the moment boundary condition at $x = L$ and note that this is a nonhomogeneous boundary condition. If the motion starts with the beam at rest, the initial conditions are

$$v(x, 0) = 0 \qquad \frac{\partial v}{\partial t}(x, 0) = 0 \tag{9.10}$$

The mathematical objective is to solve for $v = v(x, t)$. Once we know the displacement as a function of x and t, we may evaluate the bending moment and shear force from Eqs.(9.6). With the stress resultants known, we may compute the beam stresses from the standard stress equations.

Thermally induced deformation occurs for applied heating that causes a temperature gradient through the beam depth. If the heating is applied at a low rate, the temperature gradient develops slowly and the beam deforms slowly to a quasistatic equilibrium position. But, if the heating is applied at a high rate, the temperature gradient develops rapidly. The associated thermal moment accelerates the beam from rest. As the beam acquires momentum, it moves past the quasistatic equilibrium until elastic restoring forces decelerate it to bring its velocity to zero at the peak displacement. Thereafter, as the motion becomes a balance between elastic and inertia forces, a vibration ensues. In the equation of motion (9.8), there are no dissipative forces, and so the response is undamped. The equations predict that vibrations continue indefinitely with undiminished amplitudes. In practice, energy dissipation occurs, and the amplitude of the vibration decreases with time. For sustained heating, the vibrations about the quasistatic equilibrium position diminish until the beam comes to rest at the quasistatic deformation shape.

Note that, in the preceding formulation, we assume that material properties are constant, which is a reasonable assumption for small temperature changes. For larger temperature changes, the dynamic response is affected by material properties changing with temperature. Thermal prestresses may also have a significant influence on the vibration characteristics of structures. These situations are beyond the scope of this chapter as we center our attention on vibrations that result from temperature changes. The bibliography at the end of this chapter lists sources that describe effects of temperature on the vibration characteristics of structures.

9.3.2 *Quasistatic Response*

We obtain the quasistatic response by neglecting the inertia force in Eq.(9.8) and solving the resulting fourth-order equation subject to the four boundary conditions. The solution is straightforward because we are able to integrate the differential equations four times while imposing a boundary condition at each step. The result is

$$v_{\text{QS}}(x,t) = -\frac{1}{2}\frac{M_T(t)x^2}{EI} \tag{9.11}$$

This solution has an important role in the mathematical solution in that it satisfies all the boundary conditions, including the nonhomogeneous boundary condition in Eq.(9.9c). This solution also has an important role in the physical response because the vibrations occur about the quasistatic response. For a positive thermal moment with the top of the beam hotter than the bottom, the top fibers of the beam expand more than the bottom fibers, causing the beam to deflect down, as indicated by the minus sign.

9.3.3 *Free Vibration Response*

For free vibrations, we assume an isothermal beam and a harmonic response,

$$v(x,t) = \phi(x)e^{i\omega t}$$

where $\phi(x)$ is the eigenfunction, $i = \sqrt{-1}$, and ω is the natural frequency. Substituting the assumed solution in Eq. (9.8) shows that the eigenfunction satisfies

$$\frac{\mathrm{d}^4\phi}{\mathrm{d}x^4} - \beta^4\phi = 0 \tag{9.12a}$$

where

$$\beta^4 = \frac{\rho A\omega^2}{EI} \tag{9.12b}$$

The boundary conditions are

$$\phi(0) = \frac{d\phi}{dx}(0) = 0 \tag{9.12c}$$

$$\frac{d^2\phi}{dx^2}(L) = \frac{d^3\phi}{dx^3}(0) = 0 \tag{9.12d}$$

The eigenvalues β and the eigenfunctions ϕ are given by Young and Felgar.[76] The eigenvalues form an infinite set, $\beta_n, n = 1, 2, 3, \ldots$, where β_n are the roots of

$$\cos\beta_n L \cosh\beta_n L + 1 = 0 \tag{9.13}$$

and the corresponding eigenfunctions are

$$\phi_n(x) = \cosh\beta_n x - \cos\beta_n x - \alpha_n(\sinh\beta_n x - \sin\beta_n x) \tag{9.14}$$

where α_n is a coefficient that depends on $\beta_n L$. An important property of the eigenfunctions is the orthogonality condition

$$\int_0^L \phi_m\phi_n\, dx = 0 \qquad m \neq n \tag{9.15}$$

This relationship means that we may represent an arbitrary function $f(x)$ by the series expansion

$$f(x) = \sum_{n=1}^{\infty} a_n\phi_n(x) \tag{9.16a}$$

where

$$a_n = \frac{\int_0^L f(x)\phi_n(x)\, dx}{\int_0^L \phi_n^2\, dx} \tag{9.16b}$$

We will obtain the dynamic response by means of a series, or modal, expansion of the eigenfunctions.

9.3.4 Dynamic Response

The key step is to represent the dynamic response as the sum of the quasistatic solution and an infinite series of the eigenfunctions. Thus, we take

$$v(x,t) = v_{QS}(x,t) + \sum_{n=1}^{\infty}\phi_n(x)T_n(t) \tag{9.17}$$

where $T_n(t)$ is a function of time to be determined. The solution in this form satisfies all the boundary conditions. To determine $T_n(t)$, we substitute Eq. (9.17) into the equation of motion (9.8). The result is

$$\rho A\frac{\partial^2 v_{QS}}{\partial t^2} + \sum_{n=1}^{\infty}\left(EI\frac{d^4\phi_n}{dx^4}T_n + \rho A\phi_n\frac{d^2T_n}{dt^2}\right) = 0 \tag{9.18}$$

since $\partial^4 v_{QS}/\partial x^4 = 0$. To perform the separation of variables in this last equation, we represent the quasistatic solution as an infinite series of the eigenfunctions using Eq. (9.16). Thus, we write

$$v_{QS} = M_T(t) \sum_{n=1}^{\infty} a_n \phi_n(x) \tag{9.19}$$

where we evaluate the coefficients of a_n using Eq. (9.16b) with $f(x) = -x^2/2EI$. Substituting Eq. (9.19) into Eq. (9.18) and using Eq. (9.12a), we find that $T_n(t)$ is a solution of

$$\frac{d^2 T_n}{dt^2} + \omega_n^2 T_n = -a_n \frac{d^2 M_T}{dt^2} \tag{9.20}$$

This last equation has a solution in the form

$$T_n(t) = A_n \sin \omega_n t + B_n \cos \omega_n t + T_{p_n}(t)$$

where A_n and B_n are constants, and $T_{p_n}(t)$ is a particular solution to Eq. (9.20). If we have the thermal moment $M_T(t)$ as an explicit function of time, we may solve Eq. (9.20) for the particular solution. The final solution for the response is obtained by substituting this last result in Eq. (9.17). Thus, in general,

$$v(x,t) = v_{QS}(x,t) + \sum_{n=1}^{\infty} \phi_n(x)[A_n \sin \omega_n t + B_n \cos \omega_n t + T_{p_n}(t)] \tag{9.21}$$

where, for a given $M_T(t)$, we will know all terms in the solution except the two coefficients A_n and B_n. We determine these coefficients by imposing the two initial conditions given in Eq. (9.10).

We can see from Eq. (9.21) that the solution for uncoupled thermally induced vibrations has the general form of the sum of a quasistatic deflection and a dynamic response. We represent the dynamic response as a modal expansion in terms of the beam's vibration modes and natural frequencies. The specific form of the solution depends on the transient temperature distribution.

In his original paper,[6] Boley obtained a solution in the form of Eq. (9.21) for a simply supported beam of rectangular cross section. The top surface of the beam experiences a suddenly applied heat flux, and the bottom surface of the beam is insulated. We present Boley's solution for $T(x,t)$ in Eq. (4.21). Based on this temperature distribution, Boley develops the form of Eq. (9.21) that applies for a simply supported beam. From this solution written in nondimensional form, he establishes the important role of the parameter given in Eq. (9.1). In a later paper,[14] Boley develops the approximation given in Eq. (9.2).

Solutions to uncoupled problems of this type are important because they identify key parameters for the problem and establish conditions for which thermally induced vibrations occur. The Boley parameter B establishes the important role of the ratio of the thermal response time to the structural response time. Thermally induced vibrations will be significant only if B is of the order of one or less. As we note in Sec. 9.1, this usually means that vibrations occur for structures with low natural frequencies, say, <1 Hz. We present numerical results for thermally induced vibrations in the sections that follow.

9.4 Coupled, Thermally Induced Flexural Vibrations

As described in Sec. 9.2.3, the engineering analysis of spacecraft booms has evolved over two decades with contributions from several researchers. We base the following discussion of coupled, thermally induced flexural vibrations on research reported by Sumi et al.[56] We consider thermally induced flexural oscillations for laboratory heating. Assumptions customarily used in the analysis of spacecraft booms are given, and engineering models for the thermal and structural problems are formulated. The criterion for the development of thermal flutter is presented. Numerical results demonstrate the behavior. Comparisons with Sumi's experimental data demonstrate the validity of the computational model.

9.4.1 Problem Statement

In a laboratory experiment to simulate the behavior of spacecraft booms, a thin-walled tube of circular cross section is rigidly attached at its top and supports a concentrated mass at its bottom (Fig. 9.5). The tube and mass are initially at rest. Then a heat lamp is turned on that applies a heat flux S_0 along the length of the tube. The instantaneous application of the heat flux causes the incident surface to experience an increase in temperature while the unheated (back) side remains at the initial temperature. The temperature gradient causes the beam to bend rapidly away from the lamp, and a vibration ensues.

According to Eq. (9.4), the incident heat flux experienced by the tube surface depends on the distance of the beam from the lamp. As the beam moves away from the lamp, the heat flux decreases; as the beam moves toward the lamp, the heat flux increases. Thus, beam temperatures depend on beam deflections, and vice versa. A consequence of this thermal-structural coupling is that, under certain conditions, the vibration amplitude increases with time; that is, thermal flutter occurs.

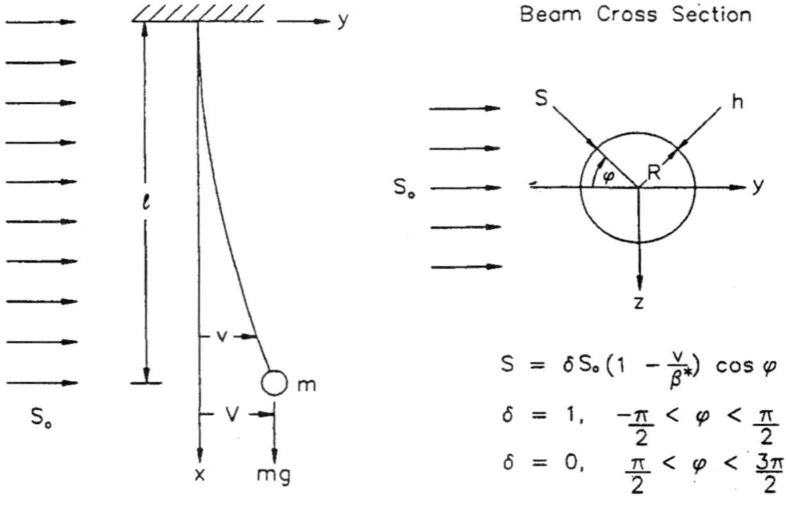

Fig. 9.5 Mathematical model of cantilever beam with tip mass subjected to laboratory heating.

9.4.2 *Thermal Analysis*

Uncoupled thermal response. In Sec. 4.8, we determine the temperature distribution in a thin-walled tube with radiation. For convenience, we summarize the key results here. With reference to Fig. 9.5, we consider a tube of radius R with wall thickness h. The tube is subjected to an incident heat flux S that is applied instantaneously at time equal to zero. The tube wall temperature is represented as

$$T(\phi, t) = \bar{T}(t) + T_m(t)\cos\phi \tag{9.22}$$

where $\bar{T}$ is the average temperature, and $T_m(t)$ is a perturbation temperature. The solution for the transient response for $\bar{T}(t)$ is given in Eqs. (4.65) and (4.66) where the steady-state value of $\bar{T}$ is

$$\bar{T}_{SS} = \left[\frac{1}{\pi}\frac{\alpha_s S}{\sigma\varepsilon}\right]^{\frac{1}{4}} \tag{9.23}$$

where α_s is the surface absorptivity, σ the Stefan–Boltzmann constant, and ε the surface emissivity. The solution for the transient response for $T_m(t)$ is given in Eq. (4.67), where the steady-state value of T_m is given in Eq. (4.69),

$$T^* = \frac{1}{2}\frac{\alpha_s S}{\rho c h}\tau \tag{9.24}$$

where ρ is the density, and c is the specific heat. The parameter τ is defined by Eq. (4.68),

$$\frac{1}{\tau} = \frac{k}{\rho c R^2} + \frac{4\sigma\varepsilon}{\rho c h}\left(\frac{1}{\pi}\frac{\alpha_s S}{\sigma\varepsilon}\right)^{\frac{3}{4}} \tag{9.25}$$

where k is the thermal conductivity.

Coupled thermal response. In the determination of the structural response, the perturbation temperature $T_m(t)$ is important because it contributes to the temperature variation around the tube as shown by Eq. (9.22). This temperature variation induces the thermal moment. When the incident heat flux varies with the beam displacement according to Eq. (9.4), we need to develop a new solution for $T_m(t)$. From Eq. (4.63), the perturbation temperature satisfies

$$\frac{\mathrm{d}T_m}{\mathrm{d}t} + \frac{1}{\tau}T_m = \frac{1}{2}\frac{\alpha_s}{\rho c h}S$$

Substituting for the incident normal heat flux S from Eq. (9.4) and introducing Eq. (9.24), we may write the differential equation as

$$\frac{\partial T_m}{\partial t} + \frac{1}{\tau}T_m = \frac{T^*}{\tau}\left(1 - \frac{v}{\beta^*}\right) \tag{9.26}$$

where we use a partial derivative since we now have $T_m = T_m(x, t)$ because of the displacement $v(x, t)$ on the right-hand side. The general solution to Eq. (9.26) is

$$T_m(x, t) = \frac{T^*}{\tau}\int_0^t \exp\left(\frac{p - t}{\tau}\right)\left[1 - \frac{v(x, t)}{\beta^*}\right]\mathrm{d}p \tag{9.27}$$

Equation (9.27) couples the temperature distribution with the unknown beam deflection $v(x, t)$.

9.4.3 Structural Analysis

The structural model is based on the following assumptions:

1) Bending vibrations occur without torsion in a single plane.
2) Material properties are constant.
3) Linear, elastic Bernoulli–Euler beam theory applies.
4) Rotatory inertia of the tip mass may be neglected.

We note from Fig. 9.5 that the beam is oriented vertically, and the bending moment depends on the restoring force of the tip mass. Using Eq. (9.6a) we may write the second order equation of motion,

$$EI\frac{\partial^2 v}{\partial x^2} + m\frac{\mathrm{d}^2 v}{\mathrm{d}t^2}(\ell - x) + mg(V - v) + M_T(x, t) = 0 \tag{9.28}$$

where $I = \pi R^3 h$, m is the tip mass, and g is the acceleration of gravity. The displacement of the tip mass is $V(t) = v(\ell, t)$. The thermal moment is given by

$$M_T(x, t) = \int_A E\alpha\Delta T(x, \phi, t) y \,\mathrm{d}A \tag{9.29}$$

where α is the coefficient of thermal expansion, $\Delta T(x, t)$ is the temperature change, and the integration is carried out over the area of the tube. Note that, because of coupling, the temperature varies along the beam. Equation (9.28) is solved subject to boundary conditions

$$v(0, t) = \frac{\partial v}{\partial x}(0, t) = 0 \tag{9.30a}$$

and initial conditions

$$v(x, 0) = \frac{\partial v}{\partial t}(x, 0) = 0 \tag{9.30b}$$

A useful approach for studying the dynamic response is to convert the equation of motion into a single-degree-of-freedom system. The beam deflection is represented approximately by using the shape function for a cantilever with a concentrated end force. Thus,

$$v(x, t) = \left(\frac{3}{2}\frac{x^2}{\ell^2} - \frac{1}{2}\frac{x^3}{\ell^3}\right) V(t) \tag{9.31}$$

where l is the beam length. With this approximation, Eq. (9.28) is integrated twice to produce

$$\ddot{V} + \left(\omega_0^2 + \frac{6g}{5\ell}\right) V = \frac{F(t)}{m} \tag{9.32}$$

where a dot denotes differentiation with respect to time; the frequency ω_0 is given by

$$\omega_0 = \sqrt{\frac{3EI}{m\ell^3}} \tag{9.33}$$

and the forcing function is

$$F(t) = -\frac{3}{\ell^3}\int_0^{\ell}\int_0^{x} M_T(u,t)\,du\,dx \tag{9.34}$$

To approximate energy dissipation, we introduce equivalent viscous damping to yield

$$\ddot{V} + 2\zeta\omega_0\dot{V} + \left(\omega_0^2 + \frac{6g}{5\ell}\right)V = \frac{F(t)}{m} \tag{9.35}$$

where ζ is the damping ratio. The forcing function $F(t)$ depends on the temperature distribution around the circumference of the tube. To evaluate it, the thermal moment is obtained by substituting $\Delta T = T_m(x,t)\cos\phi$ in Eq. (9.29) and performing the integration over the cross section. Then $T_m(x,t)$ is expressed in terms of $v(x,t)$ through Eq. (9.27). Next, the forcing function is evaluated by performing the two integrations indicated in Eq. (9.34). Finally, the damped equation of motion (9.35) becomes

$$\ddot{V} + 2\zeta\omega_0\dot{V} + \left(\omega_0^2 + \frac{6g}{5\ell}\right)V$$

$$-\frac{3EI\alpha}{m\ell d}\frac{T^*}{\tau}\int_0^t \exp\left(\frac{p-t}{\tau}\right)\left(1 - \frac{V}{5\beta^*}\right)dp = 0 \tag{9.36}$$

where d is the tube diameter. Equation (9.36) describes the thermally induced oscillations of the tip mass, including the dependence of the heat flux on the beam motion. Solutions to Eq. (9.36) demonstrate thermal flutter.

9.4.4 Thermal Flutter Condition

The stability condition for flexural thermal flutter is derived by nondimensionalizing Eq. (9.36). The new nondimensional variables are $\bar{V} = V/\ell$ and $\bar{t} = \omega_0 t$. In addition, four new nondimensional parameters are introduced:

$$\begin{aligned} \kappa &= 1/\omega_0\tau \qquad & \eta &= \alpha T^*\ell/d \\ \beta &= \beta^*/\ell \qquad & \gamma &= mg\ell^2/EI \end{aligned} \tag{9.37}$$

With these new terms, the equation of motion is written in nondimensional form. Next, it is differentiated with respect to time to yield the third-order equation:

$$\dddot{\bar{V}} + (2\zeta + \kappa)\ddot{\bar{V}} + \left(1 + \frac{2}{5}\gamma + 2\kappa\zeta\right)\dot{\bar{V}} + \kappa\left(1 + \frac{2}{5}\gamma\frac{\eta}{5\beta}\right)\bar{V} = \kappa\eta \tag{9.38}$$

We wish to determine the condition for which solutions of Eq. (9.38) diminish with time. Such solutions represent stable vibrations. The condition may be established by studying the Laplace transform of the differential equation. Using the Routh–Hurwitz stability criterion,[77] we examine the signs of a series of determinants of coefficients of the polynomial produced by taking the Laplace transform of the left-hand side of the differential equation. For a third-order equation, the polynomial has the form

$$a_0 s^3 + a_1 s^2 + a_2 s + a_3 s = 0$$

and the condition for the differential equation to have stable solutions is that

$$a_1 a_2 - a_0 a_3 > 0$$

The Routh–Hurwitz stability criterion applied to Eq. (9.38) shows that oscillatory solutions with increasing amplitudes will occur if

$$2\zeta\left(1 + \frac{2}{5}\gamma + 2\zeta\kappa + \kappa^2\right) - \frac{\kappa\eta}{5\beta} < 0 \tag{9.39}$$

Equation (9.39) is the condition for the occurrence of flexural thermal flutter. Key parameters are 1) the critical damping ratio ζ; 2) the parameter κ, which is proportional to the reciprocal of the Boley parameter given in Eq. (9.1); 3) the parameter η, which depends on the heat flux; 4) the nondimensional lamp parameter β^*; and 5) the nondimensional parameter γ, which depends on the gravity-restoring force. The stability criterion is valuable for understanding the conditions for thermal flutter in planning experiments and computational studies. We illustrate the role of the stability criterion in subsequent numerical computations.

9.4.5 Computational Approach

The derivation of the flexural thermal-flutter condition, Eq. (9.39), is based on an approximate temperature distribution. To assess the effects of the approximate temperature distribution on the dynamic response and to demonstrate the thermal-flutter phenomena, the dynamic response was determined computationally. The computational approach[78] uses a finite element analysis to determine the temperature distribution, and the equivalent single-degree-of-freedom model to compute displacement histories of the tip mass.

Thermal analysis. Although conduction heat transfer along the tube length is negligible, temperatures vary along the length because the incident heat flux varies with the beam deflection. However, temperatures at each beam cross section may be computed independently. A typical finite element discretization at a cross section is shown in Fig. 9.6, and a representative distribution of cross sections is plotted in Fig. 9.7. At each cross section, the temperature distribution is computed from a finite element analysis of Eq. (4.59). Linear elements are used, and the transient response is computed with an explicit Euler time-marching algorithm. From a typical time t_n, temperatures are advanced using the algorithm

$$[C]\{T\}_{n+1} = \{R\}_{n+1} \tag{9.40}$$

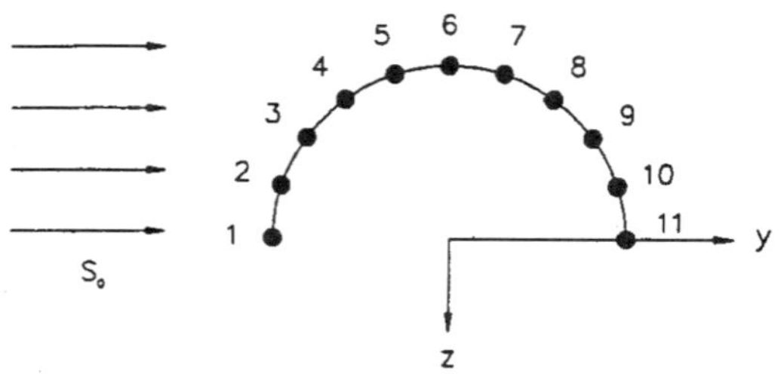

Fig. 9.6 Finite element discretization of typical beam cross section.

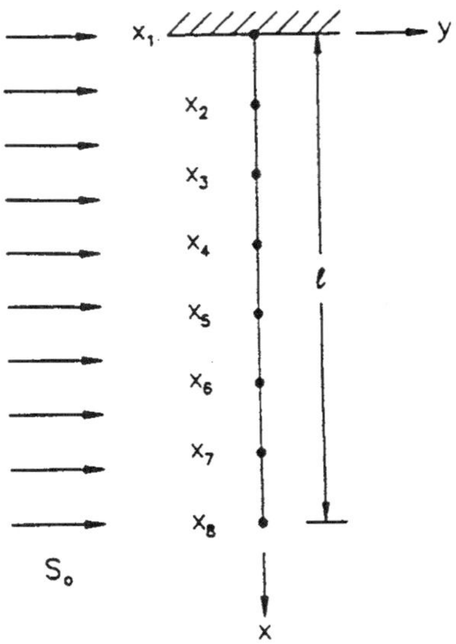

Fig. 9.7 Distribution of cross sections along beam length.

where $[C]$ is a diagonal capacitance matrix, $\{T\}_{n+1}$ are the unknown nodal temperatures, and $\{R\}_{n+1}$ is a thermal load vector. The thermal load vector contains contributions from conduction, radiation, the surface heat flux, and temperatures at t_n. The algorithm is described in Sec 8.2.2.

Structural analysis. Solutions for the tip-mass displacement history were computed from the equivalent single-degree-of-freedom model given in Eq. (9.35). At typical times t_n and t_{n+1} in the thermal response, the structural analysis proceeds through the following steps:

1) By using nodal temperatures computed from Eq. (9.40), thermal moments at the cross sections are computed by evaluating Eq. (9.29), assuming a linear temperature variation between nodes.

2) The forcing function $F(t)$ is evaluated using Eq. (9.34), assuming a linear variation of thermal moment between cross sections.

3) An analytical solution to Eq. (9.35) is used to advance $V(t_n)$ to $V(t_{n+1})$, assuming a linear variation of $F(t)$ between times t_n and t_{n+1}.

9.4.6 Numerical Results

Sumi et al.[60] experimentally studied thermally induced bending vibrations of a cantilever beam with tip mass. By varying the levels of incident heat flux and using three different tip masses, they produced cases of thermal flutter consistent with the predictions of the stability condition of Eq. (9.39). They did not, however, predict responses or correlate these predictions with measured strain histories; rather, their emphasis was on validating the stability criterion.

The data for the computations presented here are taken from the experimental program of Ref. 60. Stable and unstable responses are demonstrated. The thermal-structural data for the beam with tip mass are listed in Table 9.2. By exposing the beam to two different levels of heat flux, stable and unstable oscillations are predicted. The heat-flux levels and nondimensional parameters for the two cases are shown in Table 9.3. Computational results are presented in Figs. 9.8–9.11.

Table 9.2 Thermal-structural data (from Ref. 60) for beam with tip mass

Parameter	Symbol	Value
Beam length	ℓ	690 mm
Tube diameter	d	3.42 mm
Wall thickness	h	80 μm
Modulus of elasticity	E	193 GN/m^2
Coefficient of thermal expansion	α	17.3×10^{-6} 1/K
Mass density	ρ	7930 kg/m^3
Specific heat	c	502.4 J/kg-K
Thermal conductivity	k	16.3 W/m-K
Tip mass	m	150 g
Heat lamp parameter	β^*	89.7 mm
Absorptivity	α_s	1.0 (assumed)
Emissivity	ϵ	1.0 (assumed)

Figure 9.8 presents a plot of the stability criterion and data points corresponding to the beam-mass system subjected to different heat-flux levels. The stability criterion is presented as a plot of the nondimensional load parameter η vs the rise time parameter κ. For this beam-mass system with a low value of κ, stability of the system is determined primarily by the heat-flux level S_0. For values of S_0 less than 2000 W/m^2, the system is stable; for higher values, the system is unstable.

The thermal response (Fig. 9.9) shows tube temperature distributions at $t = 20$ s in a response. The temperature distributions at various cross sections demonstrate how the heating varies along the beam length. The undeflected end has higher temperatures, and the beam end with tip mass experiences lower temperatures. This behavior occurs at early response times because the structural response consists of a quasistatic beam deformation with a superimposed oscillation. Thus, the deformed end of the beam moves away from the heat lamp and experiences lower heat fluxes. The temperature distribution at a cross section is consistent with trends assumed

Table 9.3 Data for computations

	Symbol	Case 1	Case 2
Heat flux	S_0	1000 W/m^2	7400 W/m^2
Nondimensional parameters:			
Damping ratio	ζ	0.001	0.001
Response time	κ	0.368	0.381
Load level	η	0.00387	0.276
Lamp behavior	β	0.13	0.13
Gravity effect	γ	2.89	2.89
Stability criterion		Eq. (9.39) < 0, stable	Eq. (9.39) > 0, unstable

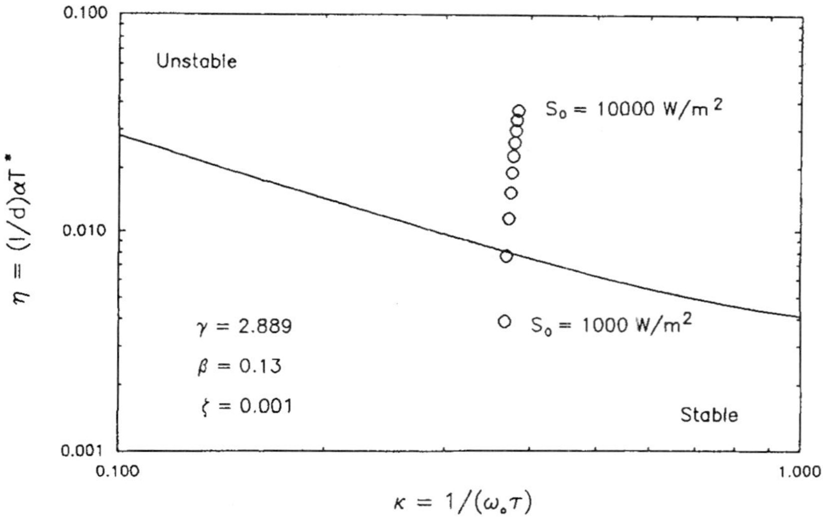

Fig. 9.8 Stability of thermally induced oscillations of cantilever beam with tip mass.

in the thermal analysis. There is a small temperature gradient from the heated to unheated sides of the beam superimposed on an average temperature level. The average tube temperature rises with time until radiation equilibrium is reached.

Although not discernible in Fig. 9.9, the temperature difference across the beam depth oscillates with time as a result of the heat-flux/displacement coupling. Oscillatory thermal moments are thus induced. If heat-flux levels are low, temperature differences are small, and motion will be damped out. If heat-flux levels are high, larger temperature differences are experienced, introducing significant oscillatory thermal moments. For low values of κ, the oscillatory thermal moments are in phase with the displacement response, and a resonance-type dynamic amplification of the

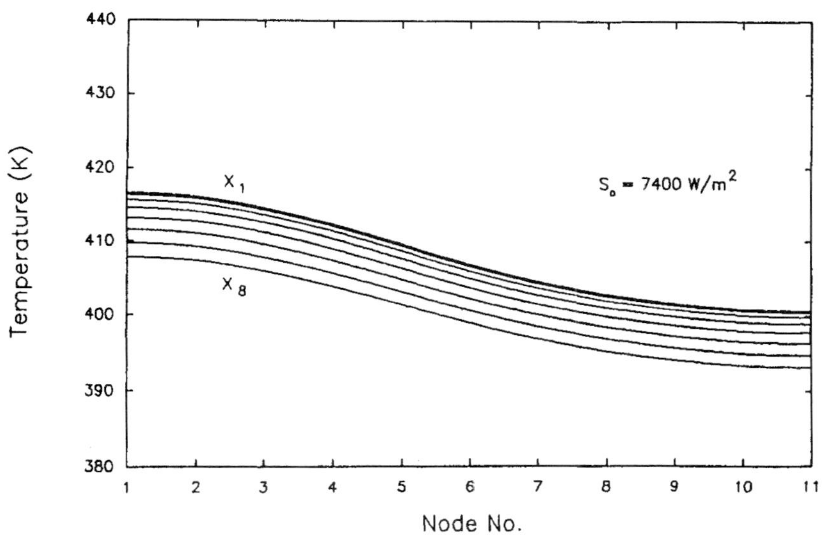

Fig. 9.9 Temperature distributions at the beam cross sections at t = 20 s.

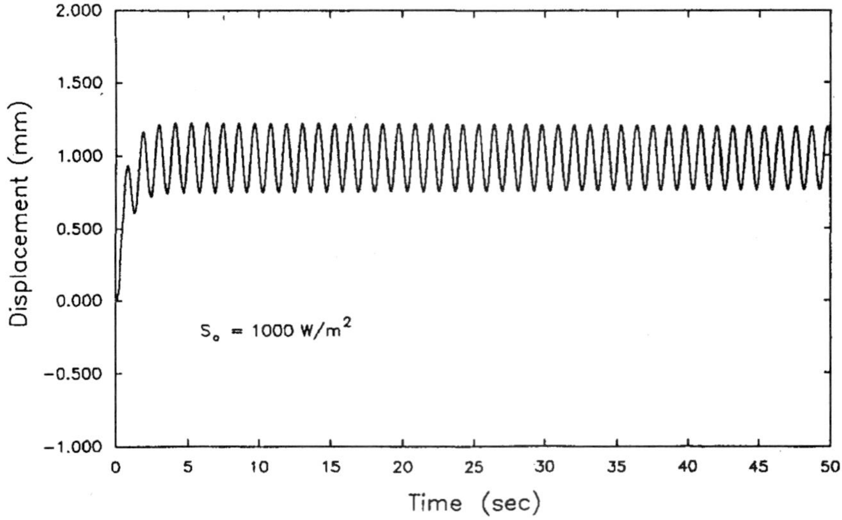

Fig. 9.10 Tip-mass displacement history for thermally induced stable oscillations.

displacement occurs. Since heat-flux levels increase with increasing displacement, the oscillation subsequently grows with time.

Figures 9.10 and 9.11 display displacement histories for stable and unstable oscillations. Early in the response in each case, the beam moves away from equilibrium to a quasistatic deformed position. Oscillations occur about this deformed position. In the stable case shown in Fig. 9.10, the oscillations damp out, and the beam will remain in the deformed position until the heating is terminated. In the unstable case shown in Fig. 9.11, the oscillations grow continuously with time. If

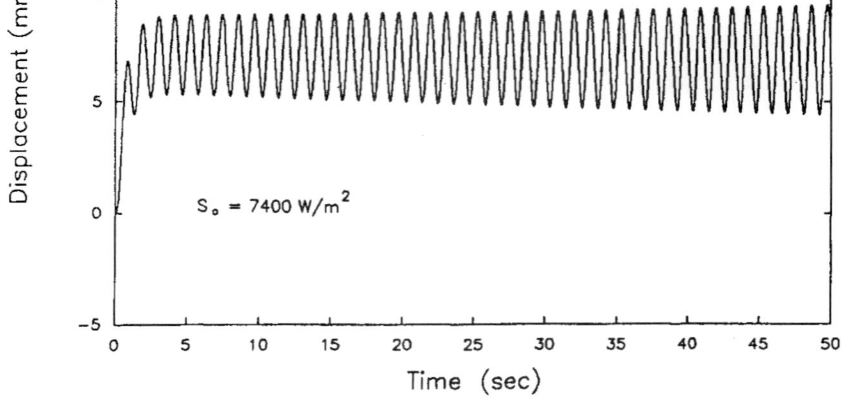

Fig. 9.11 Tip-mass displacement history for thermally induced unstable oscillations (thermal flutter).

a geometric or material nonlinearity were present in the mathematical model, a displacement limit cycle would be reached.

The computations of the dynamic response produced results consistent with the stability criterion. The results support the approximations that were made in the analytical thermal analysis. Although the finite element thermal analysis was based on fewer assumptions, the stability criterion accurately predicts the heat-flux level that induced thermal flutter in the example considered.

9.5 Hubble Space Telescope Solar Arrays

As described in Sec. 9.1, the Hubble Space Telescope's thermally induced vibration problem led to replacement of the solar arrays on the December 1993 flight of the Space Shuttle *Endeavour*. When the astronauts began to replace the solar arrays, they discovered that one of the Bi-STEM booms had buckled so that the solar array could not be retracted. This problem led to the dramatic television coverage of astronaut Kathryn Thornton releasing the damaged solar array to Earth orbit. Eventually, it re-entered the Earth's atmosphere and was destroyed by aerodynamic heating.

There were clearly some major problems with the thermal-structural design of the HST solar arrays that led to the persistent vibrations as well as to the buckled solar-array boom. Not many details are available to the public. The paper by Foster et al.[3] cited earlier describes some aspects of the vibration behavior. A recent book by Chaisson[80] describes the sequence of events that led to the identification of solar arrays as the culprit in the HST pointing jitter problem. The book also describes a classified meeting in which the author learns that similar vibration problems of solar arrays had been experienced on surveillance satellites.

We present here an analysis of an analytical model of a single solar array. The analysis is based on the study described in Ref. 79. The analysis does not explain all the phenomena experienced by the HST solar arrays because of the assumptions made in the analytical model. However, the analysis describes the basic phenomena of thermally induced vibrations and identifies engineering parameters for determining thermal flutter. Quantitative results predicted from the analytical model are consistent with published data for the HST solar arrays.

9.5.1 Problem Statement

The HST solar-array in-orbit configuration (Fig. 9.1) consists of two identical wings. Each wing has two flexible solar blankets that are deployed from a drum mounted on a shaft cantilevered from the spacecraft. Before deployment, the blankets and booms are furled in the drum. Each solar blanket is deployed by a rotating actuator mechanism that pushes the two rolled-up Bi-STEM booms from the drum. The Bi-STEM booms are connected at their deployed end by a spreader bar to which the solar blanket is attached. A Bi-STEM (two-element storable tubular extendable member) is made from thin stainless steel tapes formed into circular cross sections that are flattened and stored on spools within the drum mechanism. The stored tapes resemble a carpenter's steel tape except that, when the two tapes are deployed, they nest to form a circular cross section as shown in Fig. 9.2. The storage drum houses a torque mechanism that maintains blanket tension. The spreader bar houses a mechanism that compensates for slight differences in boom lengths. Thus, during orbital operations, the solar blanket experiences tension, and each Bi-STEM boom experiences compression.

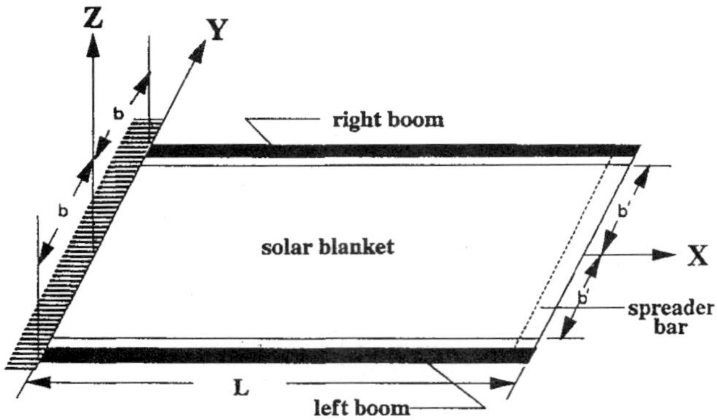

Fig. 9.12 Model of solar array.

The mathematical model and coordinate system employed in subsequent analyses are shown in Fig. 9.12. The model assumes the following:

1) Symmetric solar heating with the solar vector parallel to the x–z plane; only flexural motions without torsion may occur.

2) The solar blanket is a membrane subjected to uniform tension in the x direction; the membrane tensile force F_x (per unit length) is assumed constant; the membrane may experience transverse deflection that varies in the x direction, but the deformation does not vary in the y direction.

3) The right and left Bi-STEM booms are identical cantilevered beams, each subjected to an axial compressive force P.

4) The spreader bar is a rigid member of length $2b$ and supports the membrane tensile forces F_x over a length $2b'$.

5) The solar blanket is inextensible; solar-blanket thermal expansions or contractions are neglected.

6) The Bi-STEM booms are one-piece, thin-walled circular tubes for determination of their temperature distributions.

Equilibrium of forces on the spreader bar in the x direction requires that $P = b' F_x$. Based on these basic assumptions, analytical models are developed for thermal and structural analyses.

9.5.2 Uncoupled Thermal-Structural Analyses

In uncoupled thermal-structural analyses, temperatures are computed assuming that structural motions do not affect incident, absorbed heat fluxes; that is, the temperature distribution in a boom is assumed independent of structural deformations.

Thermal analysis. Although a Bi-STEM consists of two concentric tubes, for the thermal analysis here, a Bi-STEM is idealized as a single thin-walled tube. We describe the thermal analysis of a thin-walled tube in Sec. 4.8, and we summarize key results in this chapter in Sec. 9.4.2.

Structural analysis. The structural analysis is performed using equations of motion for the booms, the solar blanket, and the spreader bar. The equations are solved subject to boundary conditions at the cantilevered support and to interface conditions at the spreader bar.

As a result of symmetry, the motions of the two booms are synchronous, and only one boom needs to be considered.

The equation of motion for a boom is

$$EI\frac{\partial^4 w_b}{\partial x^4} + P\frac{\partial^2 w_b}{\partial x^2} + \frac{\partial^2 M_T}{\partial x^2} + \rho A\frac{\partial^2 w_b}{\partial t^2} = 0 \tag{9.41a}$$

where w_b is the boom deflection, E the modulus of elasticity, I the cross-section moment of inertia, A the cross-section area, and M_T the thermal moment defined by

$$M_T = \int_A E\alpha_T \Delta T z \,\mathrm{d}A \tag{9.41b}$$

where α_T is the coefficient of thermal expansion, and $\Delta T(\phi, t)$ denotes a beam's cross-sectional temperature gradient. The equation of motion of the solar blanket is

$$F_x\frac{\partial^2 w_{\mathrm{sb}}}{\partial x^2} = \sigma_{\mathrm{sb}}\frac{\partial^2 w_{\mathrm{sb}}}{\partial t^2} \tag{9.42}$$

where $w_{\mathrm{sb}}(x, t)$ is the solar-blanket deflection, and σ_{sb} is the mass per unit area of the solar blanket. The equation of motion of the spreader bar is

$$2V_Z(L, t) + \int_{-b'}^{b'} F_x\frac{\partial w_{\mathrm{sb}}}{\partial x}(L, t)\,\mathrm{d}y + M_s\frac{\mathrm{d}^2 w_s}{\mathrm{d}t^2} = 0 \tag{9.43a}$$

where V_Z is the boom shear force defined by

$$V_Z = -EI\frac{\partial^3 w_b}{\partial x^3} - P\frac{\partial w_b}{\partial x} - \frac{\partial M_T}{\partial x} \tag{9.43b}$$

In Eq. (9.43a), $w_s(t)$ is the spreader bar deflection, and M_s is the spreader bar mass.

The boom boundary conditions are

$$w_b(0, t) = 0 \qquad \frac{\partial w_b}{\partial x}(0, t) = 0 \qquad M_y(L, t) = 0 \tag{9.44a}$$

where M_y is the boom bending moment defined by

$$M_y = -EI\frac{\partial^2 w_b}{\partial x^2} - M_T \tag{9.44b}$$

The solar-blanket boundary conditions are

$$w_{\mathrm{sb}}(0, t) = 0 \qquad w_{\mathrm{sb}}(L, t) = w_b(L, t) = w_s(t) \tag{9.44c}$$

The initial conditions are that the displacements and velocities at $t = 0$ are zero. Equations (9.41–9.44) formulate the thermal-structural boundary/initial-value problem. The solution is described in terms of 1) the buckling modes, 2) the free vibration frequencies and modes, and 3) a modal expansion for the dynamic response.

Boom buckling. Because the booms are subjected to compressive forces P, buckling may occur. Critical buckling modes are determined by solving Eq. (9.41a) subject to the boom boundary conditions, assuming static behavior and neglecting thermal effects. The result is

$$P_{\mathrm{cr}} = n^2 \frac{\pi^2 EI}{L^2} \qquad n = 1, 2, 3, \ldots \tag{9.45}$$

where P_{cr} denotes Euler buckling loads. The result is identical to that of a pinned-end beam column. This agreement occurs because the line of action of the compressive force P passes through the end supports of the boom as in the case of a simply supported beam.

Vibration modes. Vibration frequencies and mode shapes are determined by solving Eqs. (9.41–9.43) subject to the boundary conditions, assuming harmonic motion and neglecting thermal effects. Assuming

$$w_b = \phi(x) \sin \omega t$$

$$w_{\mathrm{sb}} = \psi(x) \sin \omega t$$

where $\phi(x)$ and $\psi(x)$ are the eigenfunctions, and ω is the natural frequency, the analysis yields

$$\phi(x) = \frac{D_1}{D_2} \sin \delta x - \cos \delta x - \frac{D_1}{D_2}\frac{\delta}{\varepsilon} \sinh \varepsilon x + \cosh \varepsilon x \tag{9.46a}$$

$$\psi(x) = \phi(L) \sin\left(\sqrt{\frac{\sigma_{\mathrm{sb}}}{F_x}}\omega x\right) \Big/ \sin\left(\sqrt{\frac{\sigma_{\mathrm{sb}}}{F_x}}\omega L\right) \tag{9.46b}$$

where

$$\begin{aligned} D_1 &= \delta^2 \cos \delta L + \varepsilon^2 \cosh \varepsilon L \\ D_2 &= \delta^2 \sin \delta L + \delta\varepsilon \sinh \varepsilon L \end{aligned} \tag{9.46c}$$

and

$$\begin{aligned} \delta &= \left\{\frac{P}{2EI}\left[\left(1 + \frac{4\rho A\omega^2 EI}{P^2}\right)^{\frac{1}{2}} + 1\right]\right\}^{\frac{1}{2}} \\ \varepsilon &= \left\{\frac{P}{2EI}\left[\left(1 + \frac{4\rho A\omega^2 EI}{P^2}\right)^{\frac{1}{2}} - 1\right]\right\}^{\frac{1}{2}} \end{aligned} \tag{9.46d}$$

For free vibrations, the frequency equation is derived by substituting Eqs. (9.46a–b) into Eqs. (9.43),

$$EI\phi'''(L) + P\phi'(L) - P\psi'(L) + \tfrac{1}{2}M_s\omega^2\phi(L) = 0 \tag{9.47}$$

where a prime superscript denotes differentiation with respect to x. For a given column force P, Eq. (9.47) is solved numerically to obtain the natural frequencies.

Dynamic response. A modal analysis procedure for determining thermally induced vibrations of a beam was developed originally in Ref. 7. The same basic approach is followed here for the solar array, but details differ, and a more complex orthogonality conditions is required. The thermally induced dynamic response can be viewed as the sum of a quasistatic deflection of the array and vibrations about the quasistatic deflection. The response for a boom or solar blanket can be written as

$$w(x,t) = w_{Qs}(x,t) + \tilde{w}(x,t) \tag{9.48}$$

where w_{Qs} is the quasistatic response, and $\tilde{w}$ is the vibration response. The quasistatic response for a boom and the solar blanket are represented as

$$\begin{aligned} w_{bQs}(x,t) &= M_T(t) f(x) \\ w_{sbQs}(x,t) &= M_T(t) g(x) \end{aligned} \tag{9.49}$$

where $M_T(t)$ is the thermal moment, and $f(x)$ and $g(x)$ are the quasistatic deflected shapes. From Eq. (9.41b), the thermal moment may be evaluated using the temperature gradient from Eq. (9.22). With these substitutions, the integral is evaluated over the tube cross section shown in Fig. 9.13b. Thus,

$$M_T(t) = \frac{EI\alpha_T T_m(t)}{R} \tag{9.50}$$

where $T_m(t)$ is given by Eq. (4.67). Note that, in response to sudden boom heating, the thermal-bending moment monotonically increases, approaching a steady-state value for large values of time.

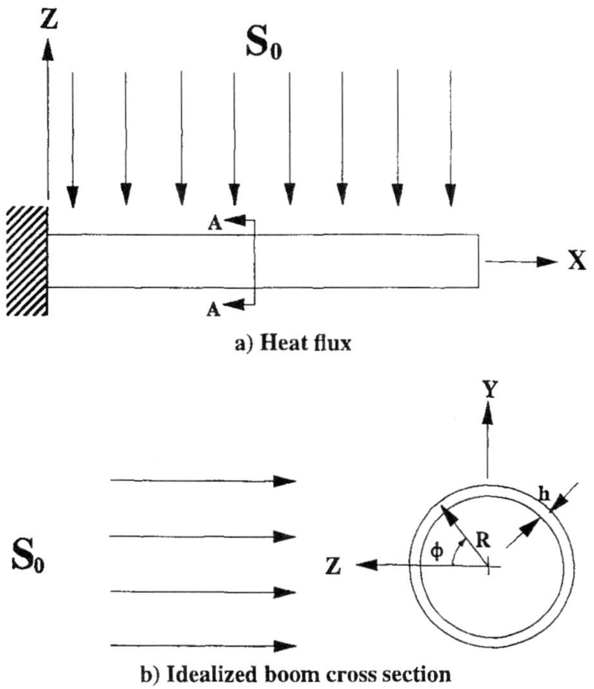

Fig. 9.13 Uncoupled thermal model of solar-array boom.

For the quasistatic response, the inertia forces in the equations of motion (9.41–9.43) are neglected. Solving these equations with the boundary conditions yields

$$w_{\mathrm{bQs}}(x,t) = \frac{M_T(t)}{P}\left[\cos\lambda x - 1 + \frac{(\cos\lambda L - 1)}{\sin\lambda L}(\lambda x - \sin\lambda x)\right] \tag{9.51a}$$

where $\lambda^2 = P/EI$ and

$$w_{\mathrm{sbQs}}(x,t) = \frac{M_T(t)}{P}\left[\frac{\lambda(\cos\lambda L - 1)}{\sin\lambda L}x\right] \tag{9.51b}$$

Because the thermal moment monotonically increases to a steady-state value, the quasistatic deflections do likewise. The steady-state deflected positions of the booms and solar blanket are maintained as long as the solar flux is applied.

Modal expansions of the quasistatic response are used to obtain the dynamic response. The functions of x defined by Eqs. (9.49) are represented by

$$f(x) = \sum_{n=1}^{\infty} a_n \phi_n(x) \tag{9.52a}$$

$$g(x) = \sum_{n=1}^{\infty} a_n \psi_n(x) \tag{9.52b}$$

where $\phi_n(x)$ and $\psi_n(x)$ are the eigenfunctions given in Eqs. (9.46). The coefficients are evaluated using the orthogonality condition

$$\int_0^L r_b(x)\phi_n(x)\phi_m(x)\,\mathrm{d}x + \int_0^L r_{\mathrm{sb}}(x)\psi_m(x)\psi_n(x)\,\mathrm{d}x = 0 \qquad m \neq n \tag{9.53}$$

where the weighting functions $r_b(x)$ and $r_{\mathrm{sb}}(x)$ are the mass distributions for the booms and solar blankets, respectively,

$$r_b(x) = \rho A + \tfrac{1}{2}M_s\delta(x - L)$$

$$r_s(x) = \sigma_{\mathrm{sb}}b'$$

where $\delta(x - L) = 0$ for $0 < x < L$, and $\delta(x - L) = 1$ for $x = L$. With the orthogonality condition, the coefficients of the expansions a_n can be evaluated in the usual way:

$$a_n = \frac{\rho A \int_0^L \phi_n(x) f(x)\,\mathrm{d}x + b'\sigma_{\mathrm{sb}} \int_0^L \psi_n(x) g(x)\,\mathrm{d}x + (M_s/2)\phi_n(L) f(L)}{\rho A \int_0^L \phi_n^2(x)\,\mathrm{d}x + b'\sigma_{\mathrm{sb}} \int_0^L \psi_n^2(x)\,\mathrm{d}x + (M_s/2)\phi_n^2(L)}$$

The expression for a_n shows that all elements of the solar array, that is, the boom, the solar blanket, and the spreader bar, affect the model expansion of the quasistatic shapes. The dynamic response of a boom is then represented, based on Eqs. (9.48) and (9.49), as

$$w_b(x,t) = M_T(t)\sum_{n=1}^{\infty} a_n\phi_n(x) + \sum_{n=1}^{\infty}\phi_n(x)T_n(t)$$

and the time functions $T_n(t)$ are evaluated using standard separation of variable techniques. The solar-blanket deflection is represented similarly. The final solutions for the dynamic response are

$$w_b(x,t) = M_T(t)f(x) + \frac{\alpha_T T^* EI}{R}\sum_{n=1}^{\infty}\frac{a_n\phi_n(x)}{1+\omega_n^2\tau^2}\bar{T}_n(t) \tag{9.54a}$$

$$w_{\text{sb}}(x,t) = M_T(t)g(x) + \frac{\alpha_T T^* EI}{R}\sum_{n=1}^{\infty}\frac{a_n\psi_n(x)}{1+\omega_n^2\tau^2}\bar{T}_n(t) \tag{9.54b}$$

where a new time function has been introduced, defined by

$$\bar{T}_n(t) = e^{-t/\tau} - \omega_n\tau\,\sin\omega_n t - \cos\omega_n t \tag{9.55}$$

As mentioned previously, the dynamic response consists of the nonoscillatory quasistatic response represented by the first terms on the right-hand sides of Eqs. (9.54a–b) plus the vibratory response represented by the second terms. In this analysis, damping has been neglected so that the vibratory terms persist for large time. If damping were included, the vibratory terms would diminish with increasing time so that the solar-array deflected shape would approach the quasistatic shape. A key parameter that determines the relative magnitude of thermally induced vibrations is the product $\omega_n\tau$. From Eqs. (4.67) and (4.68), τ can be seen as a measure of the thermal response time. Writing ω_n in terms of a modal natural period shows that $\omega_n\tau$ represents the ratio of the thermal response time to the structural response time. When $\omega_n\tau$ is of the order of unity, thermally induced oscillations become large.

9.5.3 Coupled Thermal-Structural Analyses

In the preceding section, we present uncoupled thermal-structural analyses based on the assumption that the absorbed heat flux is not affected by the boom's deformations. In this section, the effect of the boom's deformation is taken into account, and the thermal-structural analyses are coupled.

Thermal analysis. A deformed boom with the incident heat flux vector S_0 is shown in Fig. 9.14. To provide generality, the heat-flux vector is inclined to the vertical by the angle θ. When the energy conservation equation is written, the heat flux absorbed by the tube is the component normal to the surface. As discussed in Sec. 9.24, because of bending, a normal to the beam surface has rotated through a small angle equal to the beam slope. Thus, the incident normal heat flux to the

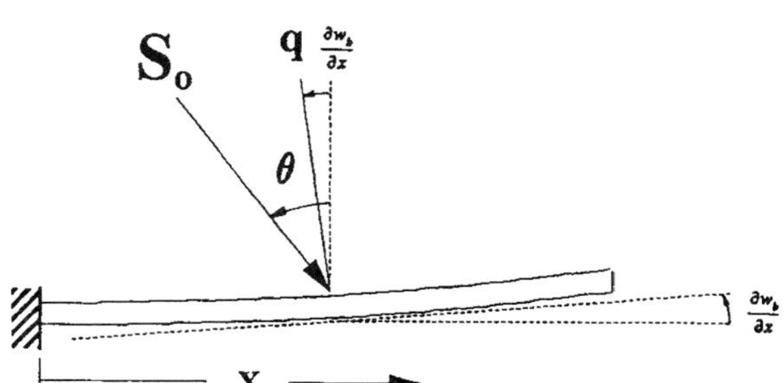

Fig. 9.14 Heat flux for coupled thermal-structural analysis.

surface varies with x along the beam and can be expressed by

$$S = S_0 \cos\left(\theta - \frac{\partial w_b}{\partial x}\right) \tag{9.56}$$

Equation (9.56) is identical to Eq. (9.3), introduced in Sec. 9.2.4, except for the difference in notation for the beam displacement. The conservation of energy equation (4.59) then becomes

$$\frac{\partial T}{\partial t} - \frac{k}{\rho c R^2}\frac{\partial^2 T}{\partial \phi^2} + \frac{\sigma\varepsilon}{\rho c h}T^4 = \frac{\alpha S_0}{\rho c h}\delta \cos\phi \cos\left(\theta - \frac{\partial w_b}{\partial x}\right) \tag{9.57}$$

Because the absorbed energy depends on the beam slope, the temperature varies along the beam length. The approximate solution for the temperature, Eq. (9.22), now takes the form

$$T(x,\phi,t) = \bar{T}(x,t) + T_m(x,t)\cos\phi \tag{9.58}$$

where now all temperatures depend on x. Proceeding with the same approximations as before produces two similar differential equations,

$$\frac{\partial \bar{T}}{\partial t} + \frac{\sigma\varepsilon}{\rho c h}\bar{T}^4 = \frac{1}{\pi}\frac{\alpha S_0}{\rho c h}\cos\left(\theta - \frac{\partial w_b}{\partial x}\right) \tag{9.59a}$$

$$\frac{\partial T_m}{\partial t} + \frac{1}{\tau}T_m = \frac{1}{2}\frac{\alpha S_0}{\rho c h}\cos\left(\theta - \frac{\partial w_b}{\partial x}\right) \tag{9.59b}$$

At steady state, the incident heat flux approaches a constant value that is approximated by neglecting the boom slope. The steady-state average temperature is then found from Eq. (9.59a) as

$$\bar{T}_{ss} = \left(\frac{1}{\pi}\frac{\alpha S_0 \cos\theta}{\sigma\varepsilon}\right)^{\frac{1}{4}} \tag{9.60}$$

An approximate perturbation temperature function was obtained previously by assuming τ to be constant; see Eq. (9.25). With this assumption, Eq. (9.59b) can be solved to yield

$$T_m(x,t) = \frac{e^{-(t/\tau)}T^*}{\tau}\int_0^t e^{\frac{p}{\tau}}\cos\left(\theta - \frac{\partial w_b}{\partial x}\right)\mathrm{d}p \tag{9.61}$$

The thermal moment can then be obtained from Eq. (9.41b),

$$M_T(x,t) = \frac{EI\alpha_T T^*}{R\tau}e^{-(t/\tau)}\int_0^t e^{\frac{p}{\tau}}\cos\left(\theta - \frac{\partial w_b}{\partial x}\right)\mathrm{d}p \tag{9.62}$$

Structural analysis. The thermal-structural boundary/initial-value problem is formulated in Eqs. (9.41–9.44). Because of the dependence of the thermal moment on the beam slope that appears in the integrand of Eq. (9.62), a modal representation of the response could not be obtained. Instead an approximate solution based on the method of weighted residuals (MWR) was developed.

Approximate solution. The solutions for the boom deflection and the solar blanket are taken in the form

$$\begin{aligned} w_b(x,t) &= N_b(x)U(t) \\ w_{sb}(x,t) &= N_{sb}(x)U(t) \end{aligned} \tag{9.63}$$

where the approximating functions N_b and N_{sb} are assumed to satisfy the conditions

$$N_b(0) = 0 \qquad N_b'(0) = 0$$
$$N_{\text{sb}}(0) = 0 \qquad N_{\text{sb}}(L) = N_b(L)$$

These functions are nondimensional to allow the time function $U(t)$ to represent the deflection of the array end as a function of time. The method of weighted residuals is based on

$$\int_0^L R(x,t)W(x)\,\mathrm{d}x = 0 \tag{9.64}$$

where $R(x,t)$ is the residual from substituting an approximate solution such as Eqs. (9.63) in a differential equation, and $W(x)$ is a weighting function. Using Galerkin's form of MWR, the weighting functions are selected as the approximating functions. Thus, for the boom, Eq. (9.41a) can be written as

$$\int_0^L \left[EI\frac{\partial^4 w_b}{\partial x^4} + P\frac{\partial^2 w_b}{\partial x^2} + \frac{\partial^2 M_T}{\partial x^2} + \rho A\frac{\partial^2 w_b}{\partial t^2}\right] N_b(x)\,\mathrm{d}x = 0 \tag{9.65a}$$

and, for the solar blanket, Eq. (9.42) becomes

$$\int_0^L \left[F_x\frac{\partial^2 w_{\text{sb}}}{\partial x^2} - \sigma_{\text{sb}}\frac{\partial^2 w_{\text{sb}}}{\partial t^2}\right] N_{\text{sb}}(x)\,\mathrm{d}x \tag{9.65b}$$

The balance of the MWR solution consists of 1) integrating the spatial derivatives by parts, 2) introducing the boundary conditions, and 3) introducing the approximations for w_b and w_{sb} from Eqs. (9.63). The result is an ordinary differential equation for the unknown function $U(t)$,

$$\ddot{U} + \omega_0^2 U = \frac{F(t)}{M} \tag{9.66a}$$

where ω_0 is an approximation to the first-mode natural frequency

$$\omega_0 = \sqrt{K/M} \tag{9.66b}$$

and

$$K = EI\int_0^L (N_b'')^2\,\mathrm{d}x - P\int_0^L (N_b')^2\,\mathrm{d}x + P\int_0^L (N_{\text{sb}}')^2\,\mathrm{d}x \tag{9.66c}$$

$$M = \rho A\int_0^L N_b^2\,\mathrm{d}x - \sigma_{\text{sb}} b'\int_0^L N_{\text{sb}}^2\,\mathrm{d}x + \frac{1}{2}M_s N_b^2(L) \tag{9.66d}$$

$$F(t) = -\int_0^L N_b'' M_T(x,t)\,\mathrm{d}x \tag{9.66e}$$

where K, M, and F are the stiffness, mass, and force, respectively, for the equivalent single-degree-of-freedom (SDOF) system defined by Eq. (9.66a). The equivalent SDOF remains difficult to solve analytically because the thermal moment appearing in Eq. (9.66e) contains the unknown function $U(t)$ inside an integral as shown in Eq. (9.62). However, with Laplace transforms, a solution may be obtained. First, the stability of the solution will be investigated and a stability criterion established.

Stability criterion. The stability of the dynamic response is determined by obtaining the Laplace transform of the differential equation of motion and applying the Routh–Hurwitz criterion. With damping and substitution for the thermal moment using Eq. (9.62), the equation of motion becomes

$$\ddot{U} + 2\zeta\omega_0\dot{U} + \omega_0^2 U = \frac{F(t)}{M} \tag{9.67a}$$

$$F(t) = \frac{-EI\alpha_T T^*}{R\tau} e^{-(t/\tau)} \int_0^L \int_0^t N_b''(x) e^{p/\tau} \cos\left(\theta - \frac{\partial w_b}{\partial x}\right) \mathrm{d}p \tag{9.67b}$$

where ζ is the damping ratio. Because of the coupling between the structural and the thermal response, which is represented by the cosine term in the expression for the thermal moment in Eq. (9.62), Eq. (9.67a) is nonlinear and difficult to solve analytically. However, Eq. (9.67a) can be linearized by approximating the cosine term based on the assumption that the boom's slope is small,

$$\cos\left(\theta - \frac{\partial w_b}{\partial x}\right) = \cos\theta + \frac{\partial w_b}{\partial x}\sin\theta$$

Thus, the applied load is approximated by

$$F(t) = \frac{-EI\alpha_T T^*}{R\tau} e^{-(t/\tau)} \int_0^L \int_0^t N_b''(x) e^{p/\tau} [\cos\theta + N_b'(x) U(p) \sin\theta] \, \mathrm{d}p$$

After the Laplace transform, the transfer function is defined as

$$\frac{U(s)}{G(s)} = \frac{1/M}{q(s)} \tag{9.68a}$$

where the input $G(s)$ is

$$G(s) = \frac{-EI\alpha_T T^* \cos\theta \int_0^L N_b''(x)\,\mathrm{d}x}{R\tau} \frac{1}{s} \tag{9.68b}$$

The characteristic equation $q(s)$, which determines the stability of the solution, is

$$q(s) = s^3 + \left(2\zeta\omega_0 + \frac{1}{\tau}\right)s^2 + \left(\omega_0^2 + \frac{2\zeta\omega_0}{\tau}\right)s + \frac{\omega_0^2}{\tau} + \frac{EI\alpha_T T^* \sin\theta}{\tau M R} \int_0^L N_b''(x) N_b'(x)\,\mathrm{d}x \tag{9.69}$$

The transfer function can be nondimensionalized to obtain a dimensionless characteristic equation to permit an effective parameter study. A dimensionless $q(s)$ can be written as

$$\bar{q}(s) = \bar{s}^3 + (2\zeta + \kappa)\bar{s}^2 + (1 + 2\zeta\kappa)\bar{s} + (\kappa + \kappa\eta) \tag{9.70a}$$

where

$$\bar{s} = s/\omega_0 \tag{9.70b}$$

$$\kappa = 1/\omega_0 \tau \tag{9.70c}$$

$$\eta = \alpha_T T^* (B/R) \sin\theta \tag{9.70d}$$

$$B = \frac{EI \int_0^L N_b''(x) N_b'(x)\, \mathrm{d}x}{K} \tag{9.70e}$$

In the last equation, K is the stiffness, Eq. (9.66c). For a stable response, the Routh–Hurwitz criterion requires that

$$\eta < \frac{2\zeta\kappa^2 + 4\zeta^2\kappa + 2\zeta}{\kappa} \tag{9.71}$$

Equation (9.71) indicates that the stability of the response becomes critical when η is equal to the right-hand side of the equation,

$$\eta_{\mathrm{cr}} = \frac{2\zeta\kappa^2 + 4\zeta^2\kappa + 2\zeta}{\kappa}$$

These critical values represent the boundary between stable and unstable responses. Thus, for stability, η must be less than the critical value. Because ζ and κ are positive quantities, η_{cr} is always positive. However, negative and zero incident angles yield $\eta \leq 0$ and, thus, η is always less than η_{cr} for these solar-heating orientations. Thus, the stability criterion shows that negative and zero incident angles always produce stable responses. However, for $\theta > 0$, the coupled thermal-structural vibrations may be unstable; that is, thermal flutter may occur.

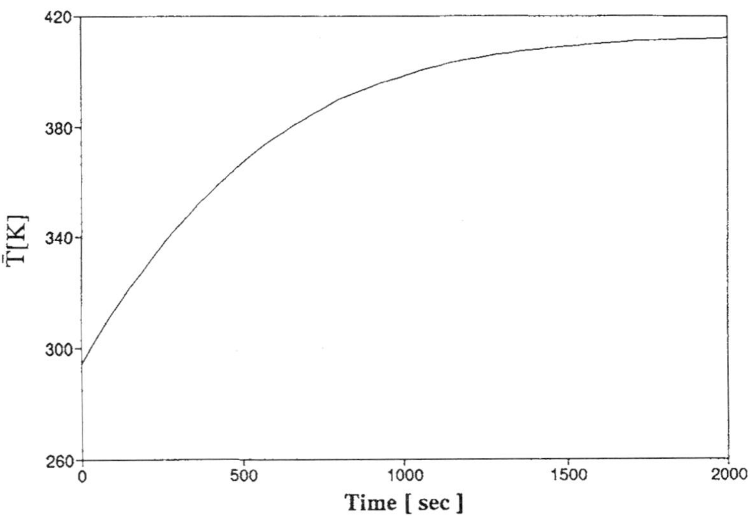

Fig. 9.15 Boom average temperature response.

Table 9.4 Hubble Space Telescope solar-array data (from Ref. 81)

Parameter	Value	Units
α	0.5	——
α_T	$1.692E-5$	K^{-1}
S_0	$1.35E+3$	W/m^2
σ	$5.670E-8$	W/m^2-K^4
ε	0.13	——
c	$5.020E+2$	J/Kg-K
ρ	$7.010E+3$	kg/m^3
h	$2.35E-4$	m
k	$1.661E+1$	W/m-k
R	$1.092E-2$	m
E	$1.930E+11$	N/m^2
EI	$1.711E+2$	N-m^2
L	5.91	m
σ_{sb}	1.589	kg/m^2
M_s	1.734	kg
b	1.428	m
b'	1.194	m
A	$1.613E-5$	m^2
P	14.75	N

Dynamic response. The dynamic response is obtained by inversion of the Laplace transform of $U(s)$ given in Eq. (9.68) to obtain $U(t)$. Then approximate solutions for $w_b(x, t)$ and $w_{sb}(x, t)$ are given by Eqs. (9.63). Finally, the perturbation temperature $T_m(x, t)$ can be determined by Eq. (9.61). These steps were performed numerically and are described in the numerical results section.

9.5.4 Numerical Results

The numerical calculations use data (Table 9.4) for the Hubble Space Telescope solar array that were obtained from Ref. 81. The tube material is stainless steel.

Uncoupled thermal analysis. Numerical results for predicted temperatures due to the suddenly applied heating are presented to assess the assumptions made in the analytical development and to provide insight into the thermal response. Figures 9.15 and 9.16 present the predicted average temperature response and the perturbation temperature response, respectively. These figures show that $T_m/\bar{T} = 0.03$, which supports the assumption made in the derivation that $T_m/\bar{T} < 1$. The perturbation temperature T_m was computed for two different values of $\bar{T}$ in the equation for τ, Eq. (4.67). Although $\bar{T}$ is assumed to be $\bar{T}_{ss}$ in Eq. (4.67), Fig. 9.16 shows that assuming $\bar{T}$ to be T_0 is also reasonable. The two curves are similar in shape, and the predicted values for $T_m(t)$ are in good agreement.

The average temperature $\bar{T}$ (Fig. 9.15) approaches steady state in approximately 1500 s. In contrast, the perturbation temperature (Fig. 9.16) nears steady state within 120 s. The perturbation temperature rises at a much faster rate than the

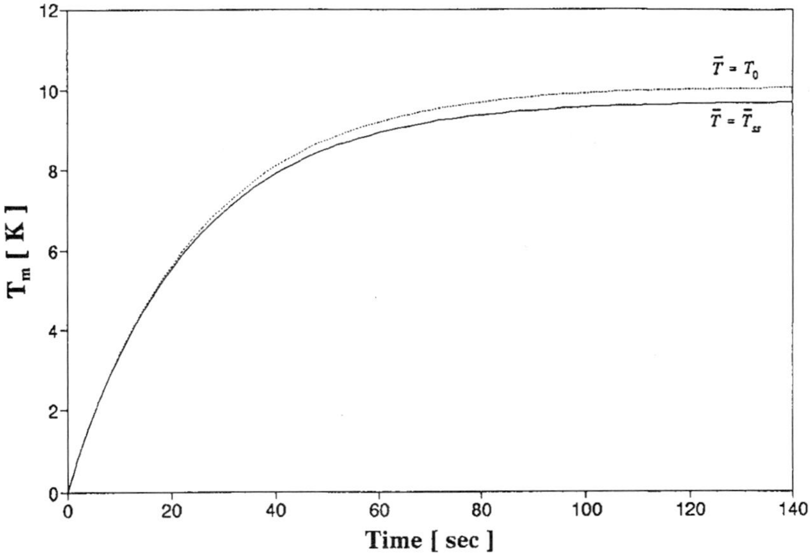

Fig. 9.16 **Boom perturbation temperature response.**

average temperature; hence, a temperature gradient develops in the boom in a relatively short time after the boom is exposed to solar radiation.

Temperature profiles for the upper half of the boom cross section at various times are shown in Fig. 9.17. These also demonstrate the development of the boom thermal gradients. To provide further validation for the thermal analytical solution, a finite element analysis of Eq. (4.59) was also performed. Figure 9.18 compares perturbation temperature profiles from the analytical and finite element solutions at $t = 60$ s. The two results are in good agreement, providing further

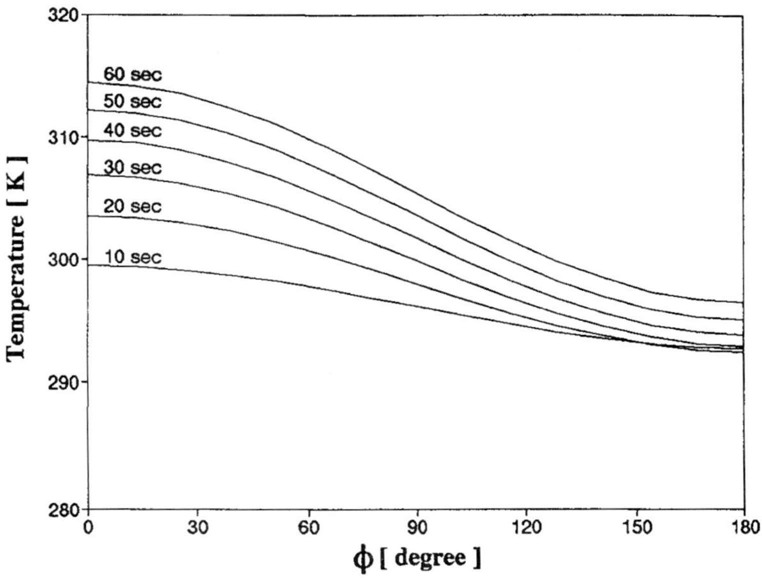

Fig. 9.17 Boom temperature distributions.

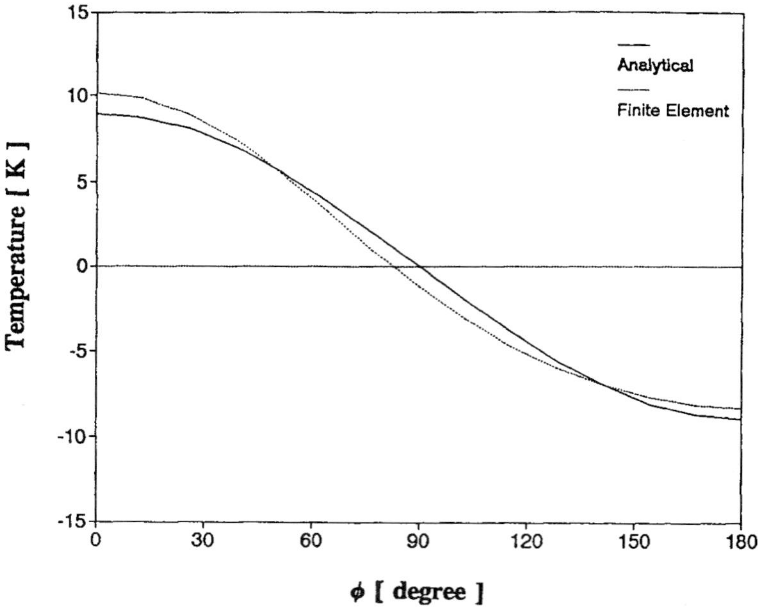

Fig. 9.18 Comparison of analytical and finite element predicted perturbation temperature distributions at t = 60 s.

validation for the approximations made in developing the analytical solution. This result is significant because the temperature gradient induces boom bending and, thus, the vibration of the solar array. The temperature gradient ΔT across a boom is $2T_m$, or about 20 K (36°F).

Uncoupled structural analysis. *Boom buckling.* The minimum critical buckling load for a boom was computed using Eq. (9.45) and the data presented in Table 9.4. The critical buckling load is $P_{\text{cr}} = 48.3$ N. From Table 9.4, the nominal boom force due to the solar-blanket tension is $P = 14.8$ N, or 31% of P_{cr}.

Vibration modes. The vibration frequencies were computed from Eq. (9.47) for various values of P/P_{cr}. The variation of the frequencies of the first three modes with P/P_{cr} is presented in Fig. 9.19. The frequencies are not shown for small values of P/P_{cr} because the frequency equation applies only if there is membrane tension. The figure shows that the first-mode frequency decreases with increasing axial force and approaches zero as expected when $P/P_{\text{cr}} = 1$. The mode shapes and frequencies for the HST solar array are displayed in Fig. 9.20. Note that the frequency of the first mode is relatively low, $f_1 = 0.0967$ Hz.

Quasistatic response. The quasistatic response of the HST solar array was computed using Eqs. (9.51), and the shape of the boom is shown at various times in the response in Fig. 9.21. The solar-panel deflection varies linearly with x and is not shown. The quasistatic response develops smoothly with time and approaches a maximum deflection of $w_b(L) = -35R$ in about 120 s as the temperature gradient approaches steady state. The minus sign results because the top surfaces of the booms are heated and the booms deflect in the negative direction in Fig. 9.21. Thus, because of a temperature gradient $\Delta T = 20$ K, the maximum solar-array deflection is predicted to be $35R = 0.38$ m (15 in.). Reference 3 predicts a temperature gradient $\Delta T = 30$ K. The present analysis underestimates

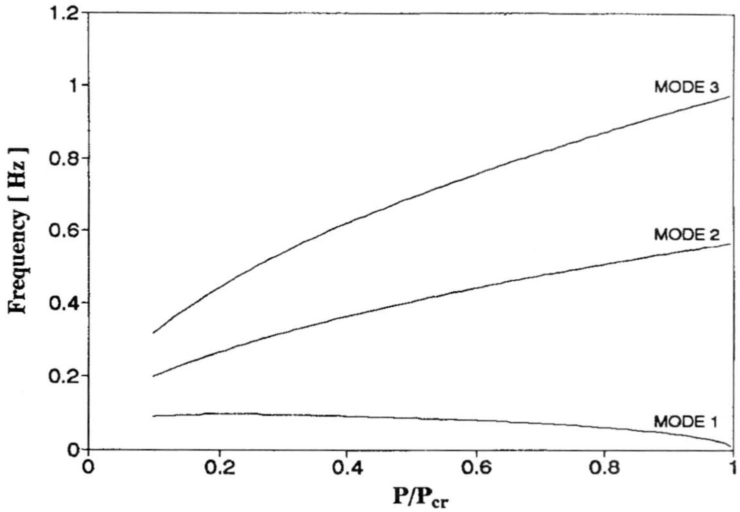

Fig. 9.19 Variation of modal frequencies with axial force ratio P/P_{cr}.

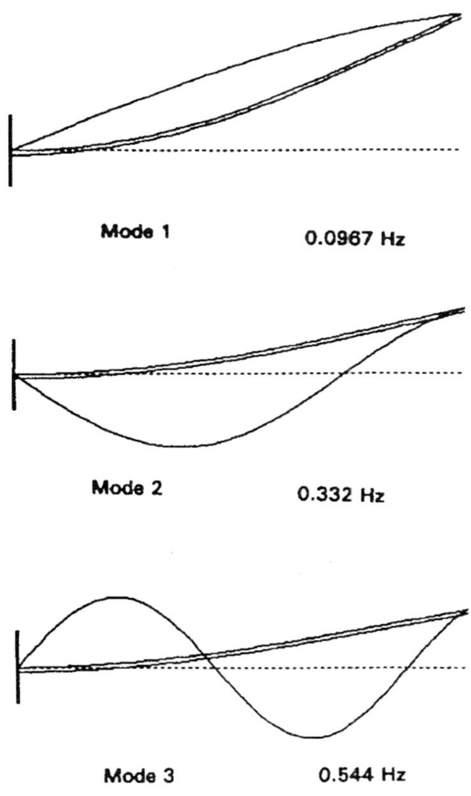

Fig. 9.20 First three vibration modes of HST solar array.

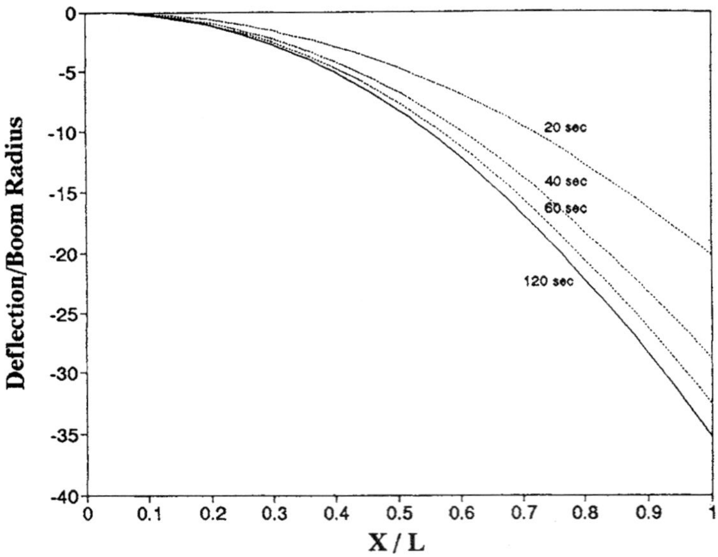

Fig. 9.21 Quasistatic deflection response of HST solar array.

the temperature gradient because a Bi-STEM is assumed to be a solid thin-walled tube. In Ref. 3, a finite difference thermal model was employed, and a Bi-STEM was modeled thermally as separate slit tubes with heat transfer between the tubes represented by a contact resistance.

Dynamic response. The dynamic response of the HST solar array is shown in Fig. 9.22. The figure shows time histories of the boom-tip deflection for three ratios of P/P_{cr}. The responses consist of a thermally induced undamped oscillation

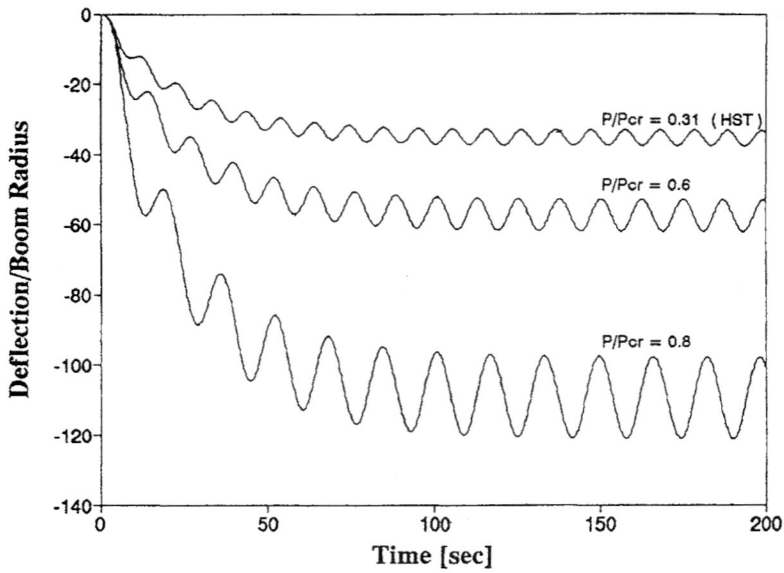

Fig. 9.22 Uncoupled deflection histories for HST solar array.

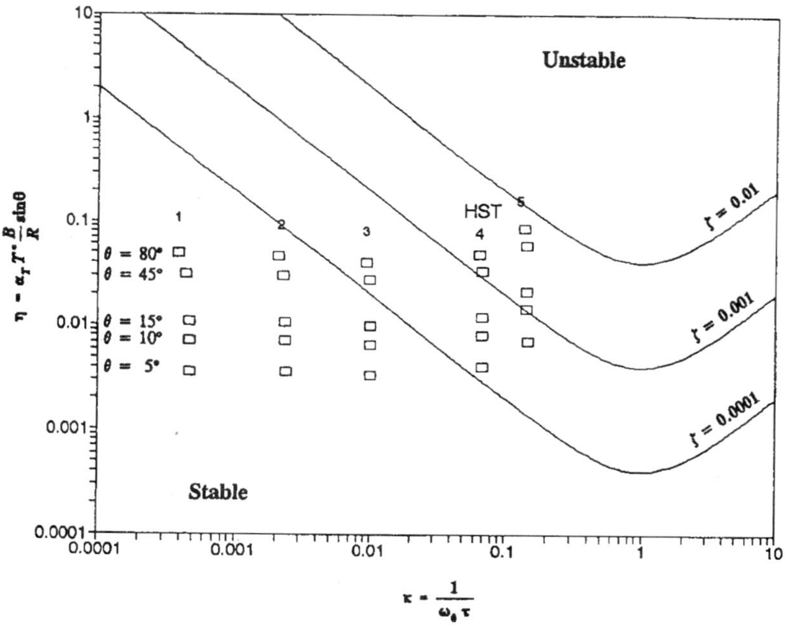

Fig. 9.23 Stability boundaries for HST solar array.

superimposed on the quasistatic deflection. The responses shown use the first three modes in the modal analysis. For the ratio of $P/P_{cr} = 0.31$, the value used in the HST design, the amplitude of the oscillation is about $2.5R$, or 0.027 m (1.1 in.). Thus, the amplitude of the thermally induced oscillations predicted by the uncoupled analysis is relatively small in comparison to the quasistatic deflections. The higher ratios of P/P_{cr} lower the bending stiffness and the first-mode frequency of the boom and lead to larger quasistatic deflections and increased vibration amplitudes. Thus, the P/P_{cr} ratio is a critical factor in the thermally induced dynamic response of the solar arrays.

Coupled thermal-structural analysis. The coupled thermal-structural analysis depends on selection of the approximating functions $N_b(x)$ and $N_{sb}(x)$ used in Eqs. (9.63). With these functions selected, the various integrals that define parameters such as K, Eq. (9.66c), M, Eq. (9.66d), and B, Eq. (9.70e), may be evaluated. Several approximating functions were considered, including the quasistatic deflection shapes and the eigenfunctions from the uncoupled analysis. The eigenfunctions for the first mode were selected because these functions yielded the best approximation for the first-mode frequency ω_0, Eq. (9.66b). For $0.1 < P/P_{cr} < 0.75$, the maximum error in the approximate frequency is less than 1%.

Stability criterion. Using the stability criterion given in Eq. (9.71), the stability boundaries for the coupled thermal-structural response of the solar arrays were computed, and results are shown in Fig. 9.23. On the vertical axis of the figure is the nondimensional parameter η, which is a measure of the thermally induced response as it is proportional to the coefficient of expansion α_T and the temperature T^* [see Eq. (9.70d)], and the sine of the angle θ, which measures the inclination of

Table 9.5 Stainless steel booms

Boom	R, m	h, m
1	$3.962E-2$	$6.323E-4$
2	$2.540E-2$	$4.528E-4$
3	$1.702E-2$	$3.379E-4$
4 (HST)	$1.092E-2$	$2.351E-4$
5	$9.525E-3$	$1.811E-4$

the solar vector from the vertical (see Fig. 9.14). Recall that, for $\theta \leq 0$, the response is unconditionally stable. On the horizontal axis is the nondimensional parameter κ, which is a ratio of the structural response time to the thermal response time. The three curves shown for the damping ratios $\zeta = 0.01, 0.001, 0.0001$ are stability boundaries. For a given damping ratio, designs below a stability boundary are stable and designs above a stability boundary are unstable. Figure 9.23 demonstrates that damping is a critical factor; increasing damping corresponds to a larger region of stable vibrations. Another critical parameter is the response time ratio κ. When $\kappa = 1$, the stability boundaries are at a minimum. Thus, a thermally induced vibration of a solar array is more prone to be unstable when its thermal response time equals its structural response time.

To investigate the effects of boom parameters on the stability of the response, five booms that vary in radius and thickness were analyzed. These were selected from stainless steel Bi-STEMs manufactured by the Astro Aerospace Corporation.[30] Table 9.5 lists the dimensions of five boom designs. These designs were selected because their critical buckling loads were larger than the HST solar-blanket tension. Boom 4 is the HST design. The square symbols shown in Fig. 9.23 correspond to these designs for various values of the solar incidence angle θ.

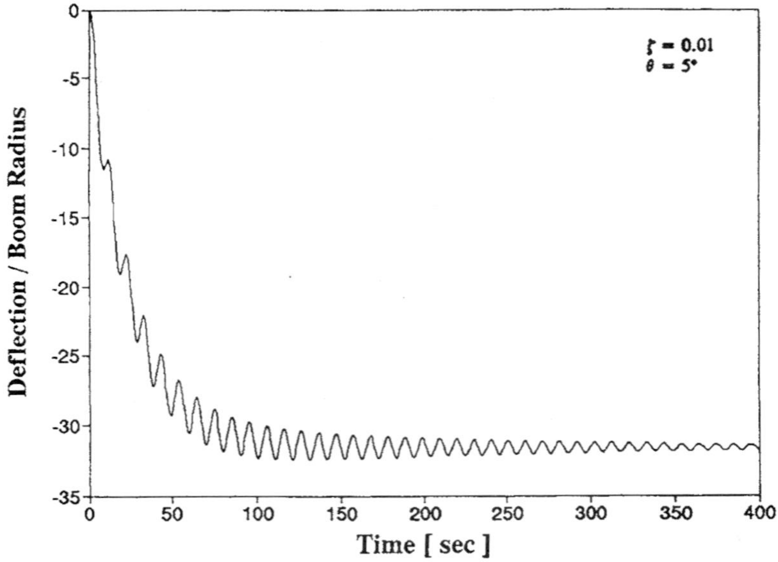

Fig. 9.24 Stable boom deflection history for HST solar array from coupled thermal-structural analysis.

The stability of the five designs depends on the solar vector angle of incidence and the damping ratio. Because the solar arrays will nominally be facing the sun, the angle θ would typically be small, say, $\theta < 10$ deg. Then the stability of the response is determined by the damping ratio. Under such conditions, all five designs, including the HST, are predicted to be stable for $\zeta \geq 0.1\%$ damping. However, for very small damping, say, $\zeta = 0.01\%$, the HST design and design 5 are predicted to be unstable.

The analysis thus raises questions about the stability of thermally induced oscillations of the HST solar arrays. The analysis suggests the possibility of unstable oscillations, but a definitive conclusion cannot be reached without further knowledge of the in-orbit solar vector incident angle and the damping ratio.

Dynamic response. The coupled dynamic response can be determined from the inverse Laplace transform of $U(s)$ in Eq. (9.68a). To demonstrate the coupled thermal-structural responses, computations were performed for stable and unstable cases. Figure 9.24 presents a stable response for the HST solar array for a solar incident angle $\theta = 5$ deg and relatively high damping of $\zeta = 1\%$. The oscillation about the quasistatic deflection decays rapidly, and oscillations would be negligible after about 6 min. In contrast, Fig. 9.25 presents an unstable response for the HST solar array for an incident angle $\theta = 80$ deg and relatively small damping of $\zeta = 0.01\%$. The stability analysis (see Fig. 9.23) predicts this case to be unstable and, indeed, this is evident in the growing oscillations demonstrated in the dynamic response. Note that, because of the large incident angle, only a small component of the solar flux is absorbed by the boom, which results in the relatively small magnitude of the total response shown in Fig. 9.25.

The mechanism that causes the unstable oscillation is the time-dependent character or the incident normal heat flux. The time-varying angle between the incident

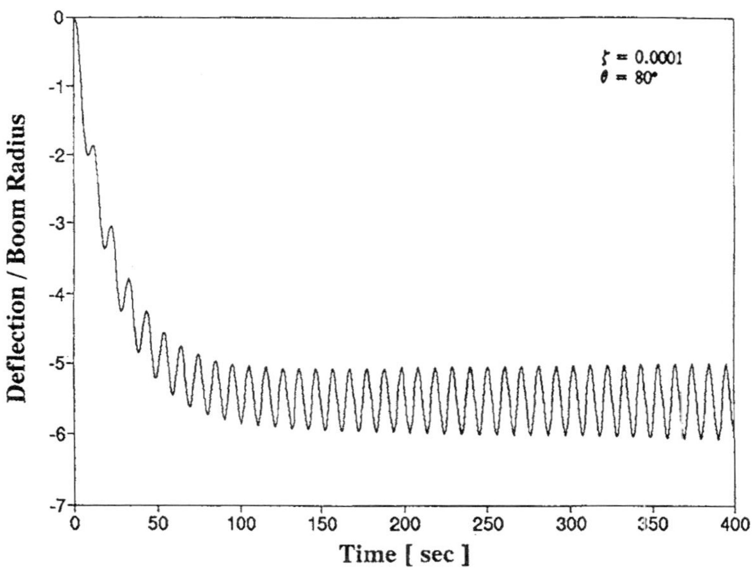

Fig. 9.25 Unstable boom deflection history for HST solar array from coupled thermal-structural analysis.

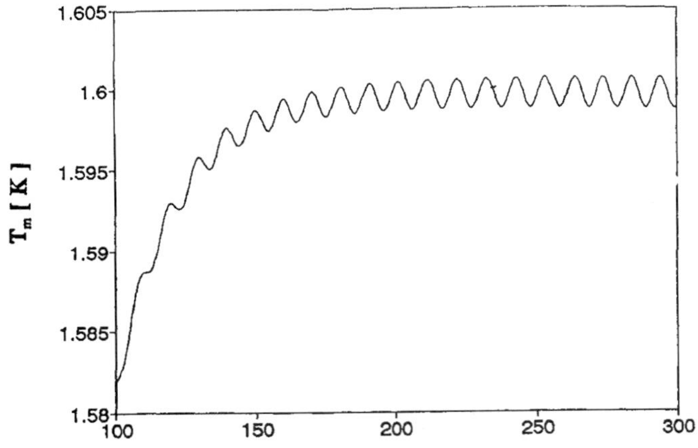

Fig. 9.26 Unstable perturbation temperature history for HST solar array from coupled thermal-structural analysis.

solar vector and a normal to the beam surface determines the amount of flux absorbed by the beam. For $\theta > 0$, Fig. 9.14, when the beam deforms toward the solar vector, the incident angle is reduced and more flux is absorbed. When the beam deforms away from the solar vector, the incident angle is increased and less flux is absorbed. This time-varying flux induces an oscillating temperature gradient, which serves as a dynamic forcing function. The time-varying temperature gradient for the unstable oscillation is shown in Fig. 9.26. Although the amplitude of the temperature oscillation is small, it is increasing with time and coupled to the unstable displacement response of the boom.

The stability analysis showed that the HST solar array will experience unstable oscillation for $\theta = 5$ deg and $\zeta = 0.0001$ damping. The dynamic response was predicted for these conditions and is shown in Fig. 9.27. Although the analysis predicts the response to be unstable, amplitude growth after 400 s is not discernible. In fact, the response was computed for greater than 4000 s (not shown), and the increase in vibration amplitudes was insignificant. The result for this solar inclination shows that unstable oscillations for the HST solar array are not of practical importance.

9.5.5 Design Parameters

The analytical methods provide insight into the thermally induced behavior of a flexible, rolled-up solar array and identify key parameters that are important for understanding the static and dynamic response. Such parameters include the thermal and surface properties of the boom material that determine the magnitude to the boom temperature gradient and the thermal response time. Significant structural parameters include the boom stiffness and thermal expansion coefficient that relate to the quasistatic deflection. Important also in the design is the boom compressive force, which lowers the first-mode natural frequency and affects the structural response time. Key parameters in determining the possibility of thermal flutter include the ratio of the thermal and structural response times, the solar inclination angle, and system damping.

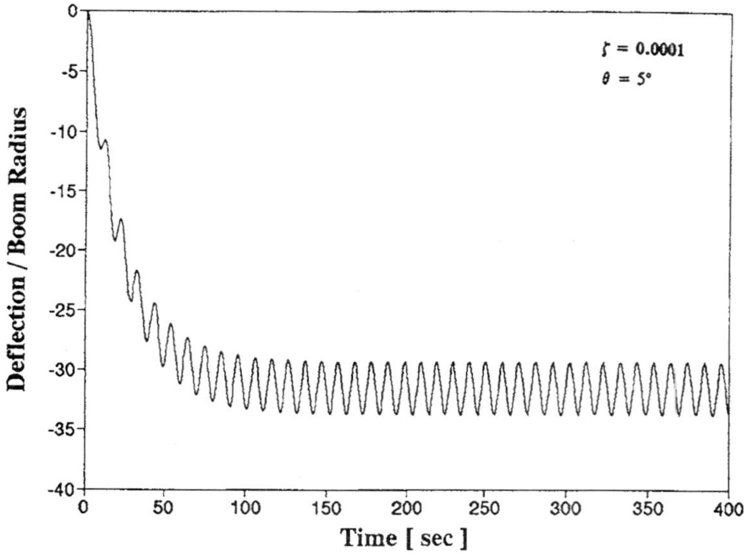

Fig. 9.27 Boom deflection history for HST solar array from coupled thermal-structural analysis.

9.6 Closing Remarks

In this chapter, we describe thermally induced vibrations. Such vibrations are caused by rapid heating of flexible structures, and the most frequently encountered problems have been experienced by spacecraft appendages. Thermally induced oscillations typically occur on very low frequency booms or solar arrays as a result of sudden heating changes at night/day transitions in the orbit. Since the spacecraft experiences no external forces due to temperature changes, linear and angular momenta are conserved. Therefore, because of appendage deformation, the entire spacecraft responds dynamically with motions about the mass center. These motions may degrade system performance either by undesirable displacements or unexpected acceleration. These motions often cause unexpected responses of the control system. Spacecraft designers should give further attention to these potential problems because the consequences of underestimating or overlooking these effects can seriously compromise the spacecraft's performance.

This chapter describes thermally induced motions primarily for structural components in Earth orbits. The scenario involves sudden solar heating as well as local member temperature gradients due to combined conduction and radiation heat transfer. Other scenarios also exist. Table 9.1 lists the thermally induced vibration problem experienced by the *Ulysses* spacecraft, which is on a mission to study the sun's polar region. Gulick[82] presents a study of thermally induced vibrations of an axial boom on a spin-stabilized spacecraft motivated by the *Ulysses* spacecraft. As another scenario, instead of local member temperature gradients, a spacecraft appendage may experience "front-to-back" temperature gradients due to self-shadowing. Thornton et al.[83] describe thermally induced vibrations of a self-shadowed split-blanket solar array. Computations are made for a solar array representative of the Space Station *Alpha* design.

References

[1]Asker, J. R., "Hubble Beset by Radiation Effects, Solar Array Vibrations, but Controllers Expect Early Resolution of Both Problems," *Aviation Week and Space Technology*, Vol. 132, No. 25, June 18, 1990, p. 26.

[2]Sawyer, K., "Hubble Space Telescope's Flutter Will Require Solar-Panel Fix," *The Washington Post*, Nov. 11, 1990, p. A6.

[3]Foster, C. L., Tinker, M. L., Nurre, G. S., and Till, W. A., "The Solar Array-Induced Disturbance of the Hubble Space Telescope Pointing System," *Proceedings, 61st Shock and Vibration Symposium* (Pasadena, CA), Vol. 4, Shock and Vibration Information Analysis Center, Arlington, VA, Oct. 1990, pp. 19–37.

[4]Nurre, G. S., Sharkey, J. P., and Waites, H. P., "Initial Performance Improvements Due to Design Modifications for the Pointing Control System on the Hubble Space Telescope," *Guidance and Control 1991*, Vol. 74, Advances in the Astronautical Sciences, edited by R. D. Culp and J. P. McQuerry, Univelt, San Diego, 1991, pp. 493–511.

[5]Thornton, E. A., and Paul, D. B., "Thermal-Structural Analysis of Large Space Structures: An Assessment of Recent Advances," *Journal of Spacecraft and Rockets*, Vol. 22, No. 4, 1985, pp. 385–393.

[6]Boley, B. A., "Thermally Induced Vibrations of Beams," *Journal of the Aeronautical Sciences*, Vol. 23, No. 2, 1956, pp. 179–181.

[7]Boley, B. A., and Barker, A. D., "Dynamic Response of Beams and Plates to Rapid Heating," *Journal of Applied Mechanics*, Vol. 24, No. 3, 1957, pp. 413–416; see Discussion, *Journal of Applied Mechanics*, Vol. 25, No. 2, 1958, pp. 309, 310.

[8]Boley, B. A., and Weiner, J. H., *Theory of Thermal Stresses*, Wiley, New York, 1960.

[9]Jones, J. P., "Thermoelastic Vibrations of a Beam," *Journal of the Acoustical Society of America*, Vol. 39, No. 3, 1966, pp. 542–548.

[10]Kraus, H., "Thermally Induced Vibrations of Thin Nonshallow Spherical Shells," *AIAA Journal*, Vol. 4, No. 3, 1966, pp. 500–505.

[11]Lyons, W. C., "Comments on Heat-Induced Vibrations of Elastic Beams, Plates and Shells," *AIAA Journal*, Vol. 4, No. 8, 1966, pp. 1502, 1503.

[12]Mason, J. B., "Analysis of Thermally Induced Structural Vibrations by Finite Element Techniques," NASA TM X-63488, Aug. 1968.

[13]Stroud, R. C., and Mayers, J., "Dynamic Response of Rapidly Heated Plate Elements," *AIAA Journal*, Vol. 9, No. 1, 1971, pp. 76–83.

[14]Boley, B. A., "Approximate Analyses of Thermally Induced Vibrations of Beams and Plates," *Journal of Applied Mechanics, Transactions of the ASME, Series E*, Vol. 39, No. 1, 1972, pp. 212–216.

[15]Seibert, A. G., and Rice, J. S., "Coupled Thermally Induced Vibrations of Beams," *AIAA Journal*, Vol. 7, No. 7, 1973, pp. 1033–1035.

[16]Bargmann, H., "Recent Developments in the Field of Thermally Induced Waves and Vibrations," *Nuclear Engineering and Design*, Vol. 27, No. 3, 1974, pp. 372–385.

[17]Jadeja, N. D., and Loo, T. C., "Heat Induced Vibration of a Rectangular Plate," *Journal of Engineering for Industry*, Vol. 96, No. 3, 1974, pp. 1015–1021.

[18]Biswas, P., "Thermally Induced Vibrations of a Rectangular Slab Resting on Elastic Foundation," *Journal of Sound and Vibration*, Vol. 59, No. 2, 1978, pp. 304–306.

[19]Manolis, G. D., and Beskos, D. E., "Thermally Induced Vibrations of Beam Structures," *Computer Methods in Applied Mechanics and Engineering*, Vol. 21, No. 3, 1980, pp. 337–355.

[20]Takeuti, Y., and Furukawa, T., "Some Considerations on Thermal Shock Problems in A Plate," *Journal of Applied Mechanics*, Vol. 48, March 1981, pp. 113–118.

[21]Das, S., "Vibrations of Polygonal Plates Due to Thermal Shock," *Journal of Sound and Vibration*, Vol. 89, No. 4, 1983, pp. 471–476.

[22]Nakajo, Y., and Hayashi, K., "Response of Circular Plates to Thermal Impact," *Journal of Sound and Vibration*, Vol. 95, No. 2, 1984, pp. 213–222.

[23]Tauchert, T. R., "Thermal Stresses in Plates—Dynamical Problems," *Thermal Stresses II*, edited by R. N. Hetnarski, North-Holland, Amsterdam, The Netherlands, 1987, Chap. 1.

[24]Nakajo, Y., and Hayashi, K., "Response of Simply Supported and Clamped Circular Plates to Thermal Impact," *Journal of Sound and Vibration*, Vol. 122, No. 2, 1988, pp. 347–356.

[25]Tauchert, T. R., "Thermal Shock of Orthotropic Rectangular Plates," *Journal of Thermal Stresses*, Vol. 12, 1989, pp. 241–258.

[26]Tauchert, T. R., "Thermally Induced Vibration of Cross-Ply Laminates," *Thermal Effects on Structures and Materials*, edited by V. Birman and D. Hui, American Society of Mechanical Engineers, New York, AMD-Vol. 110, 1990, pp. 15–20.

[27]Namburu, R. R., and Tamma, K. K., "Thermally Induced Structural Dynamic Response of Flexural Configurations Influenced by Linear/Nonlinear Thermal Effects," *Proceedings of AIAA/ASME/ASCE/AMS/ASC 32nd Structures, Structural Dynamics and Materials Conference* (Baltimore, MD), April 1991, AIAA, Washington, DC, pp. 2667–2676; AIAA 91-1175 CP.

[28]Chang, J.-S., Wang, J.-H., and Tsai, T.-Z., "Thermally Induced Vibration of Thin Laminated Plates by Finite Element Method," *Computers and Structures*, Vol. 42, No. 1, 1992, pp. 117–128.

[29]Rimrott, F. P. J., "Storable Tubular Extendable Member," *Machine Design*, Vol. 37, No. 12, 1965, pp. 156–165.

[30]"STEM Design Characteristics and Parameters," Astro Aerospace Corp., Carpenteris, CA, AAC-B-006, Sept. 1985.

[31]Rimrott, F. P. J., and Elliott, T., "Torsion of Slit, Overlapped, Thin-Walled Tubes," *Journal of Spacecraft and Rockets*, Vol. 3, No. 6, 1966, pp. 873–876.

[32]Frisch, H. P., "Thermal Bending Plus Twist of a Thin Walled Cylinder of Open Section with Application to Gravity Gradient Booms," NASA TN D-4069, Aug. 1967.

[33]Florio, F. A., and Hobbs, R. B., Jr., "An Analytical Representation of Temperature Distributions in Gravity Gradient Rods," *AIAA Journal*, Vol. 6, No. 1, 1968, pp. 99–102.

[34]Graham, J. D., "Radiation Heat Transfer Around the Interior of a Long Cylinder," *Journal of Spacecraft and Rockets*, Vol. 7, No. 3, 1970, pp. 372–374.

[35]Etkin, B., and Hughes, P. C., "Explanation of the Anomalous Spin Behavior of Satellites with Long-Flexible Antenna," *Journal of Spacecraft and Rockets*, Vol. 4, No. 9, 1967, pp. 1139–1145.

[36]Beam, R. M., "On the Phenomenon of Thermoelastic Instability (Thermal Flutter) of Booms with Open Cross Section," NASA TN D-5222, June 1969.

[37]Augusti, G., "Instability of Struts Subject to Radiant Heat," *Meccanica*, Vol. 3, No. 2, 1968, pp. 167–176.

[38]Donohue, J. H., and Frisch, H. P., "Thermoelastic Instability of Open Section Booms," Symposium on Gravity Gradient Attitude Control, sponsored by the U.S. Air Force and Aerospace Corporation (Los Angeles, CA), Dec. 1968; NASA TN D-5310, Dec. 1969.

[39]Koval, L. R., Mueller, M. R., and Paroczai, A. J., "Solar Flutter of a Thin-Walled Open-Section Boom," Symposium on Gravity Gradient Altitude Control, sponsored by the U.S. Air Force and Aerospace Corporation (Los Angeles, CA), Dec. 1968.

[40]Yu, Y. Y., "Thermally Induced Vibration and Flutter of a Flexible Boom," *Journal of Spacecraft and Rockets*, Vol. 6, No. 8, 1969, pp. 902–910.

[41]Graham, J. D., "Solar Induced Bending Vibrations of a Flexible Member," *AIAA Journal*, Vol. 8, No. 11, 1970, pp. 2031–2036.

[42]Augusti, G., "Comment on Thermally Induced Vibration and Flutter of Flexible Booms," *Journal of Spacecraft and Rockets*, Vol. 8, No. 2, 1971, pp. 202–204.

[43]Jordan, P. F., "Comment on Thermally Induced Vibration and Flutter of a Flexible Boom," *Journal of Spacecraft and Rockets*, Vol. 8, No. 2, 1971, pp. 204–208.

[44]Frisch, H., "Coupled Thermally-Induced Transverse Plus Torsional Vibrations of a Thin-Walled Cylinder of Open Cross Section," NASA TR R-333, March 1970.

[45]Frisch, H., "Thermally Induced Vibrations of Long Thin-Walled Cylinders of Open Section," *Journal of Spacecraft and Rockets*, Vol. 7, No. 8, 1970, pp. 897–905.

[46]Vigneron, F. R., "Thermal Curvature Time Constants of Thin-Walled Tubular Spacecraft Booms," *Journal of Spacecraft and Rockets*, Vol. 7, No. 10, 1970, pp. 1256–1259.

[47]Ahmed, S., "Observations on the Thermally Induced Twist of Thin Walled Open Section Booms," *Journal of Spacecraft and Rockets*, Vol. 8, No. 5, 1971, pp. 534–536.

[48]Kumar, R., and Ahmed, S., "Bending and Flexing of the Apollo 15 Mass Spectrometer Boom," *Journal of Spacecraft and Rockets*, Vol. 9, No. 12, 1972, pp. 940–942.

[49]Beam, R. M., and Yagoda, H. P., "On the Torsional Static Stability and Response of Open Section Tubes Subjected to Thermal Radiation Loading," *International Journal of Solids and Structures*, Vol. 9, No. 1, 1973, pp. 151–175.

[50]Farrell, J. L., "Thermal Curvature of Satellite Booms," *AIAA Journal*, Vol. 15, No. 9, 1977, pp. 1331–1333.

[51]Rimrott, F. P. J., and Abdel-Sayed, R., "Flexural Thermal Flutter under Laboratory Conditions," *Transactions of Canadian Society of Mechanical Engineers*, Vol. 4, No. 4, 1976–1977, pp. 189–196.

[52]Rimrott, F. P. J., "Thermal Flutter Experiments," *Zeitschrift für Angewandte Mathematik and Mechanik*, Vol. 57, No. 5, 1977, pp. T87–T89.

[53]Rimrott, F. P. J., "The Frequency Criterion for Thermally Induced Vibrations in Elastic Beams," *Ingenieur-Archiv*, Vol. 50, 1981, pp. 281–287.

[54]Murozono, M., Hashimoto, Y., and Sumi, S., "Thermally-Induced Vibration and Stability of Booms with Open Cross Section Caused by Unidirectional Radient Heating," *Journal of the Japan Society for Aeronautical and Space Sciences*, Vol. 33, No. 383, 1985, pp. 719–727 (in Japanese with English abstract).

[55]Murozono, M., and Sumi, S., "Experimental Verification of Thermally-Induced Torsional Vibration of Thin-Walled Booms with Open Cross Section," *Journal of the Japan Society for Aeronautical and Space Sciences*, Vol. 34, No. 393, 1986, pp. 545–554 (in Japanese with English abstract).

[56]Sumi, S., Murozono, M., and Imoto, T., "Thermally-Induced Bending Vibration of Thin-Walled Boom with Tip Mass Caused by Radiant Heating," *Technology Reports of Kyushu University*, Vol. 61, No. 4, Fukuoka, Japan, 1988, pp. 449–455 (in Japanese with English abstract).

[57]Murozono, M., and Sumi, S., "Thermally Induced Bending Vibration of Thin Walled Boom with Closed Section Caused by Radiant Heating," *Memoirs of the Faculty of Engineering*, Kyushu Univ., Vol. 49, No. 4, Fukuoka, Japan, 1989, pp. 273–290 (in English).

[58]Sumi, S., Murozono, M., Imoto, T., and Nakazato, S., "Thermally-Induced Bending Vibration of Thin-Walled Boom with Tip Mass-Analysis for the Experiments under Laboratory Conditions," *Technology Reports of Kyushu University*, Vol. 63, No. 1, Fukuoka, Japan, 1990, pp. 63–70 (in Japanese with English abstract).

[59]Sumi, S., Murozono, M., and Imoto, T., "Thermally Induced Bending Vibration of Thin-Walled Boom Caused by Radiant Heating," *Transactions of Japan Society of Mechanical Engineers*, Pt. C, Vol. 56, Nihon Kikaigakkai, Tokyo, 1990, pp. 300–307 (in Japanese with English abstract).

[60]Sumi, S., Murozono, M., Nakazato, S., and Tatsuyana, Y., "Thermally-Induced Bending Vibration of Thin-Walled Boom with Tip Mass-Verification by Experiments in the Atmosphere," *Technology Reports of Kyushu University*, Vol. 63, No. 2, Fukuoka, Japan, 1990, pp. 167–175 (in Japanese with English abstract).

[61]Zimmerman, B. G., "Study of Gravity-Gradient Experiment of Applications Technology Satellite," *Symposium on Passive Gravity-Gradient Stabilization*, NASA Ames Research Center, Hampton, VA, May 1965, NASA SP-107, pp. 105–111.

[62]Mobley, F. F., and Fischell, R. E., "Orbital Results from Gravity-Gradient-Stabilized Satellites," *Symposium on Passive Gravity-Gradient Stabilization*, NASA Ames Research Center, Hampton, VA, May 1965, NASA SP-107, pp. 237–251.

[63]Kanning, G., "The Influence of Thermal Distortion on the Performance on Gravity-Stabilized Satellites," NASA TN D-5435, Nov. 1969.

[64]Cherchas, D. B., and Hughes, P. C., "Attitude Stability of a Dual-Spin Satellite with a Large Flexible Solar Array," *Journal of Spacecraft and Rockets*, Vol. 10, No. 2, 1973, pp. 126–132.

[65]Tsuchiya, K., "Thermally Induced Nutational Body Motion of a Spinning Spacecraft with Flexible Appendages," *AIAA Journal*, Vol. 13, No. 4, 1975, pp. 448–453.

[66]Goldman, R. L., "Instability of a Gravity Gradient Satellite due to Thermal Distortion," NASA CR-2516, March 1975.

[67]Goldman, R. L., "Influence of Thermal Distortion on Gravity Gradient Stabilization," *Journal of Spacecraft and Rockets*, Vol. 12, No. 7, 1975, pp. 406–413.

[68]Tsuchiya, K., "Thermally Induced Vibrations of a Flexible Appendage Attached to a Spacecraft," *AIAA Journal*, Vol. 15, No. 4, 1977, pp. 505–510.

[69]Ayer, F., and Soosaar, K., "Structural Distortions of Space Systems Due to Environmental Disturbances," *AIAA International Meeting and Technology Display*, "Global Technology 2000" (Baltimore, MD), May, 1980; AIAA Paper 80-0854.

[70]Frisch, H. P., "Thermally Induced Response of Flexible Structures," *Journal of Guidance and Control*, Vol. 3, No. 1, 1980, pp. 92–94.

[71]Krishna, R., and Bainum, P. M., "Dynamics and Control of Orbiting Flexible Structures Exposed to Solar Radiation," *Journal of Guidance, Control, and Dynamics*, Vol. 8, No. 5, 1985, pp. 591–596.

[72]Hughes, P. C. , *Spacecraft Attitude Dynamics*, Wiley, New York, 1986, pp. 336–413.

[73]Hamsath, N., Bainum, P. M., and Krishna, R., "The Development of an Environmental Disturbance Model for Large Space Structures after the Onset of Thermal Shock," *Proceedings of the Astrodynamics Conference* (Williamsburg, VA), AIAA, Washington, DC, Aug. 1986, pp. 243–249; AIAA Paper 86-2123.

[74]Dennehy, C., Zimbelman, D., and Welch, R., "Sunrise/Sunset Thermal Shock Disturbance Analysis and Simulation for the TOPEX Satellite," AIAA 28th Aerospace Sciences Meeting (Reno NV), Jan. 1990; AIAA Paper 90-0470.

[75]Zimbelman, D. F., Welch R. V., and Born, G. H., "Optimal Temperature Estimation for Modeling Thermal Elastic Shock Disturbance Torque," *Journal of Spacecraft and Rockets*, Vol. 28, No. 4, 1991, pp. 448–456.

[76]Young, D., and Felgar, R. P., Jr., "Tables of Characteristic Functions Representing Normal Modes of Vibrations of a Beam," Univ. of Texas Pub. 4913, July 1949.

[77]Wylie, C. R., Jr., *Advanced Engineering Mathematics*, 2nd ed., McGraw–Hill, New York, 1960, pp. 609–614.

[78]Thornton, E. A., and Foster, R. S., "Dynamic Response of Rapidly Heated Space Structures," *Computational Nonlinear Mechanics in Aerospace Engineering*, edited by S. N. Atluri, Progress in Astronautics and Aeronautics, Vol. 146, AIAA, Washington, DC, 1992, Chap. 12, pp. 451–477.

[79]Thornton, E. A., and Kim, Y. A., "Thermally Induced Bending Vibrations of a Flexible Rolled-Up Solar Array," *Journal of Spacecraft and Rockets*, Vol. 30, No. 4, 1993, pp. 438–448.

[80]Chaisson, E. J., *The Hubble Wars*, HarperCollins, New York, 1994, pp. 92–97.

[81]Reynolds, J., "The Analysis of the Deployed Space Telescope Solar Array," ESA document TN-SA-B142, British Aerospace, Bristol, England, UK, Jan. 1983.

[82]Gulick, D. W., and Thornton, E. A., "Thermally Induced Vibrations of a Spinning Spacecraft Boom," *Acta Astronautica,* Vol. 36, No. 3, 1995, pp. 163–176.

[83]Thornton, E. A., Chini, G. P., and Gulick, D. W., "Thermally Induced Vibrations of a Self-Shadowed Split-Blanket Solar Array," *Journal of Spacecraft and Rockets,* Vol. 32, No. 2, 1995, pp. 302–311.

Problems

9.1. A spacecraft orbiting the Earth has a long, slender boom that can be modeled as a cantilever beam. While passing through the Earth's shadow, the beam cools until its temperature may be approximated as uniform with a value T_0. On emergence from the Earth's shadow, the boom is heated uniformly along its length by a solar flux S_0. The solar heating is applied instantaneously at time zero and maintained constant thereafter. The boom cross section may be represented as a thin-walled circular tube of radius R and wall thickness h. The material is stainless steel.
Using the data shown, perform a modal analysis, and determine the thermal-structural response. Your results should include:

(a) Frequencies (Hz) and mode shapes for the first five modes.

(b) Graphs of the cross-sectional temperature distributions for several times in the dynamic response.

(c) A graph of the thermal-moment history.

(d) A graph of the tip-deflection history.

Data:

$k = 16.6$ W/m-K	$R = 0.0109$ m
$c = 502$ J/kg-K	$h = 2.35 \times 10^{-4}$ m
$\rho = 7010$ kg/m^3	$L = 50$ m
$\varepsilon = 0.13$	$E = 28 \times 10^6$ lb/in.^{2}
$\alpha = 0.5$	$\alpha = 9.4 \times 10^{-6}$ 1/°F
$S_0 = 1390$ W/m^2	$\upsilon = 0.3$
$T_0 = 210$ K	$\sigma_y = 225$ ksi

9.2. The purpose of this problem is to design an experiment that will permit the study of thermally induced vibrations. The experiment consists of a simply supported beam subjected to suddenly applied surface heating. The beam has a rectangular cross section, and the unheated beam surfaces are to be insulated. The surface heating will be supplied by heat lamps that can provide a maximum heat flux $q = 50{,}000$ W/m^2. The beam will be tested in a vacuum chamber that restricts the beam length to less than 4 ft. The beam behavior should be linear and elastic. Your design of the experiment should include the following:

(a) Recommended beam dimensions (in.)

(b) Recommended heat fluxes (W/m^2)

(c) Expected temperatures (°F)

(d) Expected vibration frequencies (Hz)

(e) Expected quasistatic beam deflection (in.)

(f) Expected vibration amplitude (in.)

9.3. Develop a modal analysis formulation for flexure of a cantilever beam with a tip mass. The formulation should be similar to the theory presented in Sec. 9.3. Develop equations for the following:

(a) Determination of the beam's natural frequency. This equation should be written in terms of the ratio of the tip mass to the beam's mass, $M/\rho AL$

(b) The beam's eigenfunctions

(c) The orthogonality relation

(d) The beam's quasistatic deflection

(e) The beam's dynamic response

9.4. A spinning spacecraft in Earth orbit supports a long cantilever boom oriented normal to the spacecraft's spin axis. The length of the boom is much larger than the radius of the spacecraft. The angular velocity of the spacecraft is Ω (rad/s). The boom is a tube of closed cross section. Investigate potential thermally induced vibration problems that the rotating boom may experience.

(a) Develop equations for the incident heat flux.

(b) Develop a theory for determining the temperature distribution in the boom.

(c) Develop a theory for determining the beam's structural response.

(d) Discuss the possibility of the boom experiencing thermal flutter.

9.5. Several early spacecraft had booms with thin-walled tubes of open cross section. Many of these booms experienced thermally induced vibrations with combined bending and torsional deformations. For such a boom, develop a theoretical formulation for determining the thermal-structural response due to suddenly applied solar heating.

9.6. On the Space Shuttle *Endeavour's* mission in December 1993, the astronauts replaced the Hubble Space Telescope's solar arrays. The new solar arrays were designed to eliminate the persistent thermally induced vibrations. Discuss and illustrate the new solar-array design. Explain how the new design eliminates or minimizes the thermally induced vibration problem.

9.7. When the astronauts from the *Endeavour* recovered the Hubble Space Telescope in their December 1993 mission, they found that one solar array would not retract because of a buckled Bi-STEM. Discuss some possible causes for buckling of the Bi-STEM.

Selected Bibliography: Effects of Temperature on Structural Vibration Characteristics

Vosteen, L. F., and Fuller, K. E., "Behavior of a Cantilever Plate Under Rapid-Heating Conditions," NACA RM-L55E20c, July 1955.

Bisplinghoff, R. L., and Pian, T. H. H., "On the Vibrations of Thermally Buckled Bars and Plates," *Proceedings of IX International Congress of Applied Mechanics*, Vol. 7, Univ. of Brussels, Belgium, 1957, pp. 307–318.

McWithey, R. R., and Vosteen, L. F., "Effects of Transient Heating on the Vibration Frequencies of a Prototype of the X-15 Wing," NASA TN D-362, May 1960.

Fauconneau, G., and Marangoni, R. D., "Effect of a Thermal Gradient on the Natural

Frequencies of a Rectangular Plate," *International Journal of Mechanical Sciences*, Vol. 12, No. 2, 1970, pp. 113–122.

Bailey, C. D., "Vibration of Thermally Stressed Plates with Various Boundary Conditions," *AIAA Journal*, Vol. 11, No. 1, 1973, pp. 14–19.

Dhotarad, M. S., and Ganesan, N., "Vibration Analysis of a Rectangular Plate Subjected to a Thermal Gradient," *Journal of Sound and Vibration*, Vol. 60, No. 4, 1978, pp. 481–497. See comments by P. A. A. Laura and R. H. Gutierrez, *Journal of Sound and Vibration*, Vol. 72, No. 2, 1980, pp. 263, 264.

Buckens, F., "Vibrations in a Thermally Stressed Thin Plate," *Journal of Thermal Stresses*, Vol. 2, No. 3/4, 1979, pp. 367–385.

Mansfield, E. H., "On the Large-Deflexion Vibrations of Heated Plates," *Proceedings of the Royal Society*, London, A379, 1982, pp. 15–39.

Bailey, C. D., "Vibration and Local Instability of Thermally Stressed Plates," *Computer Methods in Applied Mechanics and Engineering*, Vol. 25, No. 3, 1981, pp. 263–278.

Kehoe, M. W., and Snyder, H. T., "Thermoelastic Vibration Test Techniques," NASA TM-101742, April 1991.

Snyder, H. T., and Kehoe, M. W., "Determination of the Effects of Heating on Modal Characteristics of an Aluminum Plate with Application to Hypersonic Vehicles," NASA TM-4274, April 1991.

10
Thermal Buckling

10.1 Introduction

In the fourth AIAA von Kármán lecture, Nicholas J. Hoff[1] described the evolution of thin-walled aircraft structures from the early days of aviation to the halcyon days of the Apollo program. A recurrent theme throughout his history of aerospace structures is the engineer grappling with buckling problems of thin-walled structures. An early paper by Professor Hoff[2] traces aerospace buckling problems back to the wood-steel-fabric biplane era in which the Euler buckling formula and beam-column theory were used to design struts of biplane wings and wing spars subjected to compressive loads. In the 1920s, solutions to buckling problems played an important role in the advancement of all-metal semimonocoque construction. For several decades thereafter, engineers struggled with buckling problems of metallic thin-walled plates and shells under mechanical loads. The results of their work led to safe and reliable aircraft that made commercial aviation a worldwide economic success. Indeed, the basic construction ideas of riveted skin-stringer metallic structures made possible by the buckling analyses of the 1920 and 1930s prevail today in modern commercial transports.

Thermal effects on aircraft took on importance after World War II with supersonic flight. Aerodynamic heating from supersonic flight induces elevated temperatures that affect material selection and alter structural design practices. For supersonic aircraft, thermal gradients and restrained thermal expansion introduced thermal buckling to structural designers. A recent paper[3] describes development of thermal structures from the early days of supersonic flight to the more recent challenges presented by hypersonic flight. The development of the National Aerospace Plane (NASP) presented engineers with their greatest thermal-structural challenges. New advanced metal-matrix composites were developed for severely elevated temperature applications. Complex, convectively cooled thin-walled structures subjected to intense local heating provided unprecedented challenges in terms of materials and structural modeling. An example of such an extreme instance of a thin-walled structure subjected to intense local heating is the problem of shock/shock interactions.

As hypersonic vehicles accelerate at high speeds in the atmosphere, shocks sweep across the vehicle, interacting with local shocks and boundary layers. These interactions expose structural surfaces to severe local pressures and heat fluxes. The first known instance of shock/shock interactions in high-speed flight occurred on the X-15 in 1967. In that flight, the intense local heating induced severe local damage on a supporting pylon for a dummy ramjet engine.[3] On the NASP, these interactions occurred on leading edges of integrated engine structures that experience intense, highly localized aerothermal loads. Poleski et al.[4] describe a three-dimensional thermal-structural analysis of a swept cowl leading edge subjected to skewed shock/shock interference heating. The analysis shows that, because of the intense localized heating, the thin elastic leading edge experiences very large local compressive stresses. The high level of these compressive stresses suggests the possibility of localized inelastic behavior or local buckling or both. Clearly, these new, complex problems present structural analysts with unprecedented thermal-buckling challenges.

One of the earliest discussions of thermal buckling of aircraft structures is the work of van der Neut.[5] Thermal buckling is also dealt with in thermal-stress books by Boley and Weiner,[6] Johns,[7] Parkus,[8] and Nowinski.[9]

We begin this chapter with analytical solutions for thermal buckling of beam columns. Next, we describe thermal buckling of plates and present several analytical solutions for plate critical buckling temperatures. This is followed by a discussion of plate problems analyzed by the finite element method. The chapter concludes with a description of an analytical solution for thermal buckling of a cylindrical shell.

10.2 Thermal Buckling of Beam Columns

Here we discuss the thermal-buckling behavior of beam columns due to temperature changes. By *beam column* we mean a beam that experiences an axial compressive force. For many thermal-buckling problems, the axial compressive force is introduced by restrained thermal expansion.

10.2.1 Problem Statement

We consider a long, slender beam of compact cross section described by the Bernoulli–Euler beam theory developed in Chapter 6. For simplicity, we assume that deformations occur in the x–y coordinate plane and that this plane coincides with a principal axis of the cross section. Because we wish to consider buckling, we use the large-deflection beam equations derived in Sec. 6.3.2. The equations are based on the equilibrium equations written for the deformed beam configuration and use a nonlinear strain-displacement relation. We neglect body forces and moments and assume thermal loads from a specified temperature distribution. The basic equations describe the in-plane and bending behavior of the beam. For the in-plane behavior, we have Eq. (6.25a),

$$\frac{\partial N_x}{\partial x} = 0 \qquad \text{(6.25 repeated)}$$

and Eq. (6.26),

$$N_x = EA\left[\frac{\partial u_0}{\partial x} + \frac{1}{2}\left(\frac{\partial v}{\partial x}\right)^2\right] - N_T \qquad \text{(6.26 repeated)}$$

where the thermal force is

$$N_T = E\alpha \int_A \Delta T \mathrm{d}A \qquad \text{(6.12a repeated)}$$

where u_0 is the in-plane displacement of points on the x centroidal axis, and N_x is the axial force positive in tension. The equilibrium equation (6.25a) tells us that the axial force N_x does not vary along the beam. For the bending behavior we have

Eqs. (6.28),

$$M = -EI\frac{\partial^2 v}{\partial x^2} - M_T \qquad \text{(6.28a repeated)}$$

$$V = -EI\frac{\partial^3 v}{\partial x^3} - \frac{\partial M_T}{\partial x} \qquad \text{(6.28b repeated)}$$

$$EI\frac{\partial^4 v}{\partial x^4} - N_x\frac{\partial^2 v}{\partial x^2} + \frac{\partial^2 M_T}{\partial x^2} = 0 \qquad \text{(6.28c repeated)}$$

where the thermal moment is

$$M_T = E\alpha \int_A \Delta T y \, dA \qquad \text{(6.12c repeated)}$$

These equations are solved subject to appropriate beam boundary conditions.

10.2.2 *Axially Restrained Beam with Uniform Temperature Change*

Let us consider first a simply supported beam of length L with restrained axial displacements (Fig. 10.1a). The boundary conditions are

$$u_0(0, t) = v(0, t) = M(0, t) = 0 \qquad (10.1a)$$

$$u_0(L, t) = v(L, t) = M(L, t) = 0 \qquad (10.1b)$$

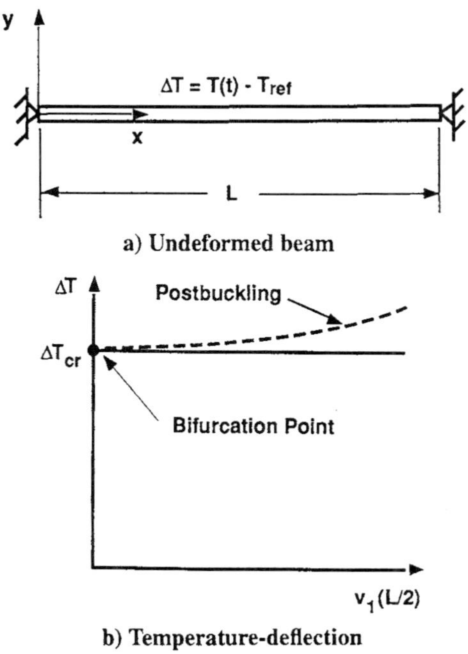

Fig. 10.1 Buckling of a restrained beam with uniform temperature change.

For a uniform temperature change, we assume that the specified temperature varies only with time, that is, $\Delta T = \Delta T(t)$. From Eq. (6.12a) repeated in the preceding section, the thermal force is given by

$$N_T(t) = E\alpha A \Delta T(t) \tag{10.2a}$$

and, from Eq. (6.12c), the thermal moment is identically zero,

$$M_T(t) \equiv 0 \tag{10.2b}$$

If the temperature change is negative, intuition tells us that the beam attempts to contract, and that the force N_x is tensile. In this situation, the beam remains straight without bending so that $v(x) \equiv 0$. On the other hand, if the temperature change is positive, the beam attempts to expand, and the force N_x is compressive. In this situation, we know that, for a critical temperature change ΔT_{cr}, the beam buckles. In the balance of this discussion, we prescribe ΔT as increasing from zero, and we determine the axial force N_x and the transverse displacement $v(x)$ as a function of ΔT. For convenience, we work with the compressive axial force as positive. Thus, we define P as the negative of N_x, that is, $P = -N_x$. At the critical temperature change ΔT_{cr}, we have $P = P_{\text{cr}}$. Let us consider now the three cases in which 1) $0 < \Delta T < \Delta T_{\text{cr}}$, 2) $\Delta T = \Delta T_{\text{cr}}$, and 3) $\Delta T > \Delta T_{\text{cr}}$. We investigate how the beam responds in each case.

$0 < \Delta T < \Delta T_{cr}$. If the temperature change ΔT is less than the critical value, the beam remains straight with $v(x) \equiv 0$. From Eq. (6.26) with $P = -N_x$, we may then write

$$\frac{\partial u_0}{\partial x} = \frac{N_T - P}{EA}$$

Integrating this last equation and imposing the boundary conditions on the axial displacements from Eq. (10.1) show us that $P(t) = N_T$. From Eq. (10.2a),

$$P(t) = E\alpha A \Delta T(t) \tag{10.3}$$

The bending equations (6.28) are satisfied identically since $v(x)$ is zero. Thus, for $\Delta T < \Delta T_{\text{cr}}$, the axial force P increases linearly with ΔT, and the beam does not bend.

$\Delta T = \Delta T_{cr}$. At the critical temperature, we have bifurcation buckling. This means that there are two possible equilibrium positions. We obtain these solutions by solving the bending problem. We solve Eq. (6.28c) in the form

$$\frac{\partial^4 v}{\partial x^4} + \lambda^2 \frac{\partial^2 v}{\partial x^2} = 0 \tag{10.4a}$$

where

$$P = \lambda^2 EI \tag{10.4b}$$

Because the thermal moment is zero, the boundary conditions for v are

$$v(0) = \frac{\partial^2 v}{\partial x^2}(0) = 0 \qquad v(L) = \frac{\partial^2 v}{\partial x^2}(L) = 0$$

The general solution to Eq. (10.4a) is

$$v = C_1 + C_2 x + C_3 \sin \lambda x + C_4 \cos \lambda x$$

The boundary conditions at $x = 0$ require that $C_1 = C_4 = 0$, and the boundary conditions at $x = L$ require that

$$C_2 L + C_3 \sin \lambda L = 0 \qquad C_3 \sin \lambda L = 0$$

The two possible solutions are

$$C_2 = C_3 = 0$$

or

$$C_2 = 0 \qquad \lambda_n L = n\pi$$

where $n = 1, 2, 3, \ldots$. The first solution means that $v(x) \equiv 0$, so that the beam does not bend but remains perfectly straight. The second solution yields the critical buckling force P_{cr}. From Eq. (10.4b),

$$P_{\text{cr}} = n^2 \pi^2 (EI/L^2) \tag{10.5}$$

which is the Euler buckling for a simply supported beam. From Eq. (10.3), the corresponding critical value of temperature is

$$\Delta T_{\text{cr}} = \frac{P_{\text{cr}}}{E\alpha A} = n^2 \pi^2 \frac{I}{\alpha A L^2} \tag{10.6}$$

From this last result, we observe that the critical buckling temperature ΔT_{cr} is independent of the modulus of elasticity. Corresponding to the critical buckling load, the beam deflects in the shape

$$v_n(x) = C_n \sin(n\pi x/L) \tag{10.7}$$

where C_n is the amplitude of the deflection. At the bifurcation point, this amplitude is indeterminant, although we may determine it in the postbuckling regime. The postbuckling regime occurs when $\Delta T > \Delta T_{\text{cr}}$.

$\Delta T > \Delta T_{cr}$. If we raise the temperature above the critical value and allow the beam to deflect in a buckling mode according to Eq. (10.7), we may determine the relationship between the temperature change ΔT and the amplitude C_n. This relationship may be found by solving Eq. (6.26). Using $N_x = -P$, we write this equation as

$$\frac{\partial u_0}{\partial x} = \frac{N_T - P}{EA} - \frac{1}{2}\left(\frac{\partial v}{\partial x}\right)^2$$

and integrate it over the beam length. Then, by imposing the conditions $u_0(0) = u_0(L) = 0$, we obtain

$$\frac{(N_T - P)L}{EA} - \frac{1}{2}\int_0^L \left(\frac{\partial v}{\partial x}\right)^2 \mathrm{d}x = 0 \tag{10.8}$$

For temperatures $\Delta T \geq \Delta T_{cr}$, the buckled force P is constant at P_{cr}, and the buckled shape is given by Eq. (10.7). Substituting for $v = v_n(x)$ in Eq. (10.8), we find, after evaluating the integral, that

$$C_n = \frac{2}{\pi}\frac{L}{n}\sqrt{\frac{N_T - P_{cr}}{EA}} \tag{10.9}$$

and

$$v_n(x) = \frac{2}{\pi}\frac{L}{n}\sqrt{\frac{N_T - P_{cr}}{EA}}\sin\frac{n\pi x}{L} \tag{10.10}$$

We may write this last result in terms of the temperature changes by using Eqs. (10.2a) and (10.6). Thus,

$$v_n(x) = \frac{2}{\pi}\frac{L}{n}\sqrt{\alpha(\Delta T - \Delta T_{cr})}\sin\frac{n\pi x}{L} \tag{10.11}$$

where $\Delta T \geq \Delta T_{cr}$. This result shows that, though the beam force reaches a maximum value of $P = P_{cr}$, the deflection continues to increase with increasing ΔT. For the lowest buckled mode with $n = 1$, the relationship between ΔT and the center beam deflection $v_1(L/2)$ is

$$\Delta T - \Delta T_{cr} = \frac{1}{\alpha}\left[\frac{\pi}{2}\frac{v_1(L/2)}{L}\right]^2 \tag{10.12}$$

The postbuckling temperature–beam deflection relationship is sketched in Fig. (10.1b).

The temperature–beam deflection behavior shown in Fig. (10.1b) summarizes the buckling behavior. We should keep in mind that this behavior represents an idealized beam rather than a real beam. The important assumptions are that 1) the beam is perfectly straight initially and 2) the temperature is uniform. Because of these assumptions, the beam remains perfectly straight until the temperature change reaches the critical value ΔT_{cr}. For temperature changes greater than ΔT_{cr}, the idealized beam may be in equilibrium in either of two configurations: 1) perfectly straight with $v(x) \equiv 0$, or 2) buckled with $v(x)$ given by a sine function. This latter configuration we denote as the postbuckling regime in Fig. 10.1b. In the postbuckling regime, the analysis shows that the beam column's axial force is constant at P_{cr} but that the beam deflection increases with increasing ΔT. In the postbuckled regime, the nonlinear ΔT vs $v_1(L/2)$ curve shows that the beam stiffens with increasing ΔT.

If the beam is not perfectly straight or if the beam has a temperature gradient through its thickness, then we no longer observe bifurcation buckling. We analyze this more realistic behavior in the next section. The bifurcation buckling case we presented here is useful, however, because it provides a simple limiting case for the complicated temperature-deflection behavior of the more realistic buckling problem.

10.2.3 Axially Restrained Beam with Thermal Bending

We consider now the same simply supported beam but with a linear temperature variation through the beam depth. The boundary conditions are given by

Eqs. (10.1), but the temperature change $\Delta T(y,t) = T(y,t) - T_{\text{ref}}$ is given by

$$\Delta T(y,t) = T_0(t) + [\Delta T_h(t)](y/h) \tag{10.13}$$

where $T_0(t)$ is the average temperature of the cross section, and $\Delta T_h(t)$ is the temperature variation through the depth h of the beam. From Eq. (6.12a), the thermal force is given by

$$N_T(t) = E\alpha A T_0(t) \tag{10.14}$$

and, from Eq. (6.12c), the thermal moment is given by

$$M_T(t) = E\alpha I \frac{\Delta T_h(t)}{h} \tag{10.15}$$

Unlike in the preceding problem, we now have a thermal moment due to the temperature gradient through the beam depth.

The presence of the thermal moment changes the character of the problem from a bifurcation buckling problem to a bending-buckling problem. In the preceding example, the beam remains straight until the axial compressive force reaches the Euler critical buckling force. Here, because of ΔT_h, the thermal moment causes the beam to bend as soon as the temperature changes. Thus, from the beginning of the temperature change, the beam deforms with a transverse deflection v. If the temperature change ΔT is positive, an increasing axial compressive force P accompanies the bending deflection v. We may expect, based on our previous experience, that, as P approaches P_{cr} the deflection v becomes increasingly large. Our objective is to find the nonlinear relationship between P and v in terms of the prescribed temperature parameters T_0 and ΔT_h. The approach is the same as in the preceding case, except that here we obtain a single relation applicable over the entire temperature range (assuming elastic behavior, of course).

The in-plane force N_x in Eq. (6.26) does not vary with x because $\partial N_x/\partial x = 0$ and, as before, we take $N_x = -P$. Then we may write Eq. (6.26) as

$$\frac{\partial u_0}{\partial x} = \frac{N_T - P}{EA} - \frac{1}{2}\left(\frac{\partial v}{\partial x}\right)^2$$

Proceeding as we did for postbuckling, we integrate this last equation over the length of the beam and introduce the boundary conditions $u_0(0) = u_0(L) = 0$. The result is the same equation we had in Eq. (10.8),

$$\frac{(N_T - P)L}{EA} - \frac{1}{2}\int_0^L \left(\frac{\partial v}{\partial x}\right)^2 \mathrm{d}x = 0 \tag{10.16}$$

The difference is that, in the preceding problem, Eq. (10.8) applies only in the postbuckling regime, but now Eq. (10.16) applies over the complete loading range. In each problem, the relation acts as a constraint equation expressing the condition that the overall change in displacements between the supports must be zero because the supports are immovable.

We turn now to the solution for $v(x,t)$. We need to solve Eq. (6.28c) and, since the thermal moment $M_T(t)$ does not depend on x, it reduces, as before, to Eq. (10.4a). The boundary conditions are slightly different. As before, we have

$$v(0) = v(L) = 0 \tag{10.17a}$$

but now the moment boundary conditions are, from Eq. (6.28a),

$$
\begin{aligned}
M(0) &= -EI\frac{\partial^2 v}{\partial x^2}(0) - M_T = 0 \\
M(L) &= -EI\frac{\partial^2 v}{\partial x^2}(L) - M_T = 0
\end{aligned}
\tag{10.17b}
$$

Using the same general solution to Eq. (10.4a), we evaluate the constants of integration using the boundary conditions expressed by Eqs. (10.17). The result is

$$
v(x) = \frac{M_T}{P}\left[\left(\frac{1-\cos\lambda L}{\sin\lambda L}\right)\sin\lambda x + \cos\lambda x - 1\right] \tag{10.18}
$$

where we have introduced P from Eq. (10.4b).

The parameter λ is unknown in Eq. (10.18), but we have the constraint relation (10.16) available for its determination. Substituting Eq. (10.18) into Eq. (10.16) and performing the indicated integration provides the transcendental equation

$$
\begin{aligned}
&\frac{(N_T - P)L}{EA} - \frac{1}{2}\frac{M_T^2\lambda^2}{P^2}\left[\frac{(1-\cos\lambda L)^2}{\sin^2\lambda L}\left(\frac{L}{2} + \frac{\sin 2\lambda L}{4\lambda}\right)\right. \\
&\left. - \frac{(1-\cos\lambda L)\sin\lambda L}{\lambda} + \left(\frac{L}{2} - \frac{\sin 2\lambda L}{4\lambda}\right)\right] = 0
\end{aligned}
\tag{10.19}
$$

In Eq. (10.19), if we prescribe the thermal force N_T and thermal moment M_T, we have the single unknown λ since $P = \lambda^2 EI$. Thus, at each time in the response, we may solve Eq. (10.19) numerically to obtain the parameter λ, which determines P. Knowing these two values, we can evaluate the beam deflection using Eq. (10.18).

We notice that, because of the transcendental character of the solution, the deflection–axial force–temperature relationship is nonlinear. In other words, the beam deflection v and the axial force P do not change proportionally to a temperature change ΔT. Notice also that the solution is not given directly in terms of the buckling force P_{cr}. It is valid over the entire range of the axial force P. We will see from numerical calculations the relationship of the solution to the critical buckling force P_{cr}.

For numerical calculations, consider the stainless steel plate shown in Fig. 10.2. The plate is heated so that it deforms with a cylindrical shape where the transverse deflection v at any given time in the response is a function only of x. We may analyze it as a beam column using the preceding equations, provided we replace EI by the plate flexural rigidity D, where $D = Eh^3/12(1-\nu^2)$. This substitution accounts for the fact that the lateral edges of a unit width of the plate, unlike those of a beam, are restrained from sideways rotation. For the stainless steel: $E = 28.4 \times 10^6$ lb/in.^{2}; $\alpha = 6.52 \times 10^{-6}$ 1/in., and $\nu = 0.3$. For a unit width of the plate, the axial buckling load is given by Eq. (10.5) as

$$
P_{cr} = \pi^2\frac{D}{L^2} = 1605 \text{ lb}
$$

and the corresponding critical temperature change is given by Eq. (10.6) as

$$
\Delta T_{cr} = \frac{P_{cr}}{E\alpha A} = 86.6^\circ\text{F}
$$

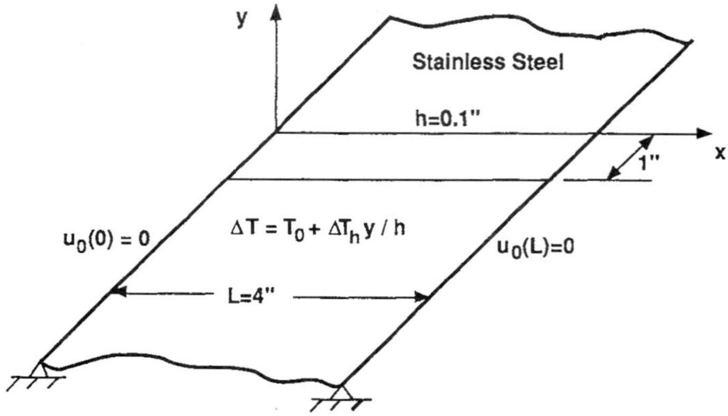

Fig. 10.2 Restrained plate with thermal bending.

We wish to determine the response of the plate for increasing ΔT as well as the effect of the thermal bending moment on the response. In the numerical calculations, the average temperature T_0 was increased linearly with time, and the temperature gradient ΔT_h was held constant. Figure 10.3 presents a plot of the axial load P vs the center deflection $v(L/2)$ for various values of ΔT_h. In each case, the *initial* center deflection of the plate was calculated from a linear quasistatic solution for simply supported ends with constant thermal moment and zero axial force.

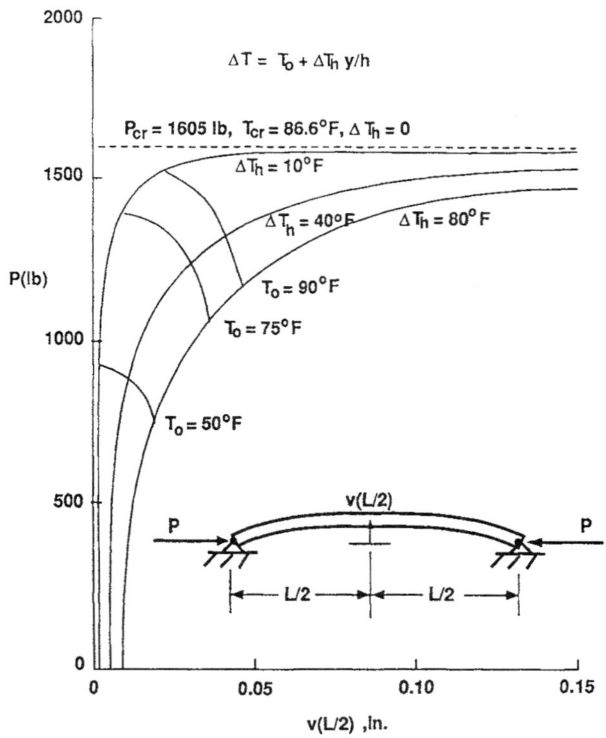

Fig. 10.3 Axial force vs center deflection for a restrained plate with thermal bending.

The figure shows that, for small ΔT_h, the axial force P increases initially with T_0, with only a small bending deflection. But, as P approaches P_{cr}, the transverse deflection then increases very rapidly, with P becoming asymptotic to P_{cr}. For larger values of ΔT_h, the load-deflection curve "bends over" much earlier in the response, and the buckling response is less pronounced. Eventually, in all cases, the axial load approaches P_{cr} from below as the deflection becomes large.

The example illustrates how a real beam column behaves and the effect of thermal bending. It also shows the important role of the bifurcation formulation for establishing the limits on the actual behavior. We may expect similar bending–buckling behavior if we have an initial beam deflection rather than a thermal-bending moment.

10.2.4 Buckling of the Hubble Space Telescope Solar Arrays

The Hubble Space Telescope (HST) was deployed in orbit from Space Shuttle *Discovery* on Flight STS-31 on April 25, 1990. Within several orbits, engineers noticed unexpected vibrational disturbances during eclipse transitions. Soon, it became apparent that the vibrations were thermally induced; the problem arose when a sudden temperature gradient formed across the boom's cross section, causing the array to deflect and oscillate about a quasistatic deflection. Section 9.5 describes an analysis of thermally induced flexural vibrations of the HST solar arrays.

Fig. 10.4 Hubble Space Telescope showing buckled Bi-STEM on the solar array.

In the first week of December 1993, seven astronauts aboard the Space Shuttle *Endeavour* on flight STS-61 repaired the HST in orbit 590 km above Earth. Of the many problems that were addressed, the replacement of the solar arrays was among the most important tasks. Early in their mission, the astronauts rendezvoused with the HST and attached it to the Space Shuttle. Early examination of the HST showed that one boom of a solar array was damaged, with a kink about midway along its length. Figure 10.4 is a photograph of the HST shortly after it was recovered, and the damaged boom is indicated clearly on the upper right-hand solar array. The mission plan called for the astronauts to retract the solar arrays into their storage canisters before replacing them, but they were unable to retract the damaged array. Television viewers may recall the dramatic scene of astronaut Kathryn Thornton unbolting the damaged array and holding it over her head as the spacecraft traveled from darkness into daylight. She then jettisoned the damaged array, and the television camera showed the orbiting solar array reacting to thrusts from maneuvering jets as the Shuttle slowly backed away.

A fundamental engineering question remains: What caused the solar array boom to fail? Figure 10.5, another view of the solar arrays, indicates pronounced twisting of each array. Is the failure of the solar-array boom related to the twisting behavior? What is the role of thermal effects? To address these questions, we focus our attention on a torsional-buckling analysis of an HST solar array.

Solar Arrays. The HST solar arrays are two identical winglike structures. Each wing consists of two solar blankets deployed from a drum mounted on a shaft cantilevered from the spacecraft. One end of the solar blanket connects to the drum while the opposite end attaches to a spreader bar. During deployment, the blanket is unfurled by a rotating actuator mechanism that deploys two rolled-up Bi-STEM booms that are connected to, and push, both ends of the spreader bar. A Bi-STEM boom is a beam whose cross section consists of two concentric tubes. Each tube is an incomplete circle and is therefore open in cross section. In their stored configuration, each STEM (storable tubular extendable member) is flattened and rolled up on a spool, resembling a carpenter's tape (see Fig. 9.2). During deployment, the stored elastic energy in the flattened metal assists the unfurling mechanisms as each STEM extends and curls back to its original shape, with seams diametrically opposed, forming a Bi-STEM. The spreader bar houses tapes, springs, and a linear potentiometer that detects and adjusts for differences in boom lengths. The drum rotates to accommodate boom thermal expansion, keeping the blanket tension constant. Thus, unlike the problems in Secs. 10.2.2 and 10.2.3, the booms are unrestrained, and thermal expansion of each boom occurs without development of a restraint force. However, the blanket tension on the spreader bar exerts a compressive force on each boom.

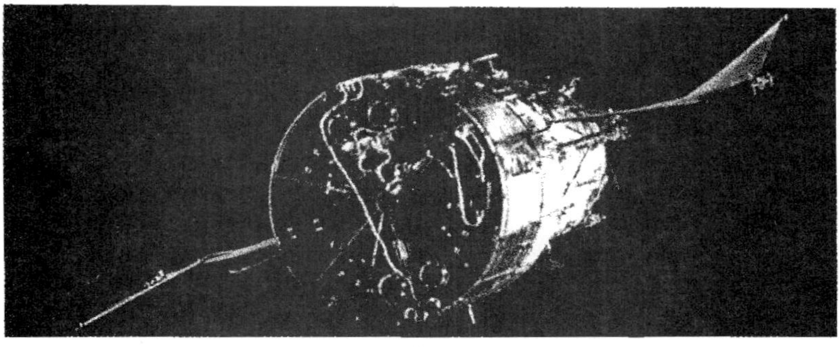

Fig. 10.5 Hubble Space Telescope showing the twist of the solar array.

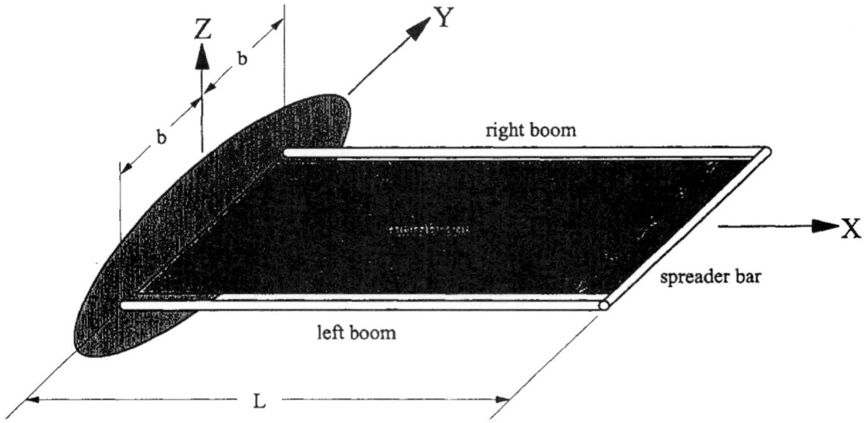

Fig. 10.6 Solar-array mathematical model.

The mathematical model and coordinate system used in the subsequent analysis are shown in Fig. 10.6. The model assumes that:

1) There is symmetric twisting about the x axis.

2) The solar blanket is an inextensible membrane whose thermal expansions and contractions are neglected.

3) The solar blanket is subjected to uniform tension in the x direction and the membrane tensile force F_x (per unit length) is constant.

4) The right and left Bi-STEM booms are identical cantilevered beams, each subjected to a constant axial compressive force P.

5) Torsional rotations are sufficiently small to restrict a boom's bending displacement to the x–z plane.

6) Thermal expansions of the booms are neglected.

7) The rigid spreader bar is of length $2b$ and supports the membrane tensile force F_x over a length $2b'$. Because P and F_x are constant, equilibrium in the x direction for the spreader bar requires that $P = b'F_x$.

Boom equations. In Sec. 6.3.3, we develop the governing differential equations for the coupled bending and torsion of a thin-walled beam with an axial force and an arbitrary temperature distribution. The set of three simultaneous fourth-order equations are given in Eqs. (6.70). The unknowns are the transverse displacements v and w and the angle of twist θ_x. Coupling between bending and twisting occurs in the general case because the beam's shear center is offset from the beam's centroid by the distances e_y and e_z.

For a Bi-STEM, each slit tube has an open cross section, but the shear center for the two concentric tubes coincides with the centroid of the boom cross section; hence, $e_y = e_z = 0$. Thus, flexure and torsion of a boom are uncoupled. Since we assume that boom flexure occurs only in the x–z plane, we have the bending displacement $v = 0$.

We consider now the role of temperature in Eqs. (6.70). In our previous discussions of boom temperature distributions, we have argued (see Secs. 4.8 and 9.5) that it is a good approximation to assume that the boom temperature is independent of x. For a hollow, thin-walled tube, we have represented the temperature distribution by

$$T(\phi, t) = \bar{T}(t) + T_m(t)\cos\phi$$

where $\bar{T}(t)$ is the average temperature of the boom, and $T_m(t)$ is a perturbation temperature representing the temperature gradient through the boom depth. For this temperature distribution, the thermal moments of a thin-walled tube are given in Sec. 9.5 as

$$M_{zT} = 0 \qquad M_{yT}(t) = \frac{EI\alpha T_m(t)}{R}$$

Because the bending displacement v and the thermal moment M_{zT} are zero, the first bending equation (6.70a) is satisfied identically. In the remaining bending equation, Eq. (6.70b), the derivative of the thermal moment $\partial^2 M_{yT}/\partial x^2 = 0$. (Recall, however, that the thermal moment still appears in the moment boundary condition.) In the torsion equation, Eq. (6.70c), temperature appears in the integral through σ_{xT} and the thermal torque term. The latter term, $\partial^2 M_{xT}/\partial x^2$, is equal to zero because the temperature is independent of x. The former term can be shown to be zero because σ_{xT}, given by Eq. (6.63c), is zero. This result can be shown by direct evaluation or by reasoning that, because the temperature is linear through the beam depth, the thermal stress is zero. Because of these simplifications, the torsion equation is independent of temperature.

Neglecting the body force and body moment and taking $N_x = -P$, Eqs. (6.70b) for a boom reduce to

$$EI\frac{\partial^4 w}{\partial x^4} + P\frac{\partial^2 w}{\partial x^2} = 0 \qquad \text{(bending)}$$

$$E\Gamma\frac{\partial^4 \theta_x}{\partial x^4} - \left(GJ - \frac{PI_E}{A}\right)\frac{\partial^2 \theta_x}{\partial x^2} = 0 \qquad \text{(torsion)}$$

Thus, if we neglect the effect of torsional deformations on temperature and assume that material properties are constant, the only way temperature affects the problem is through the thermal-bending moment. Because the solar arrays normally face the sun, we may assume that the thermal moments in the two booms are equal. Thus, the thermal moments cause symmetric flexure of the solar array and are not a factor in torsional buckling. Hence, we conclude that thermal effects have no major role in torsional buckling of the solar arrays. In the subsequent analysis, we neglect thermal bending and analyze torsional buckling as an isothermal problem.

Torsional buckling. In the torsional-buckling analysis, static equilibrium equations for the booms and the solar blanket are solved by applying appropriate boundary conditions and continuity conditions at the spreader bar. Then, equilibrium of the spreader bar gives an equation for the critical buckling loads. As a result of symmetry about the x axis, we can analyze the behavior of only one boom.

The torsional-buckling analysis begins with the boom bending equation given above, written in the form

$$\frac{d^4 w}{dx^4} + \lambda^2\frac{d^2 w}{dx} = 0 \tag{10.20a}$$

where

$$\lambda^2 = P/EI \tag{10.20b}$$

The boom's flexural boundary conditions are

$$w(0) = 0 \qquad \frac{\mathrm{d}w}{\mathrm{d}x}(0) = 0 \qquad M(L) = 0 \tag{10.20c}$$

where M is the boom bending moment defined by

$$M = -EI\frac{\mathrm{d}^2 w}{\mathrm{d}x^2} \tag{10.20d}$$

After the boundary conditions are imposed, the solution is given by

$$w(x) = \alpha_1(\sin\lambda x - \tan\lambda L\cos\lambda x + \tan\lambda L - \lambda x) \tag{10.21}$$

where α_1 is a constant that remains to be determined.

A boom composed of the two concentric open cross sections has low resistance to an applied torque. The torsional response is described by the boom torsion equation given above, written in the form

$$\frac{\mathrm{d}^4\theta_x}{\mathrm{d}x^4} - \beta^2\frac{\mathrm{d}^2\theta_x}{\mathrm{d}x^2} = 0 \tag{10.22a}$$

where

$$\beta^2 = \frac{GJ}{E\Gamma} - \frac{PI_E}{EA\Gamma} \tag{10.22b}$$

The boom's torsional boundary conditions are

$$\theta_x(0) = 0 \qquad \frac{\mathrm{d}\theta_x}{\mathrm{d}x}(0) = 0 \qquad \frac{\mathrm{d}\theta_x}{\mathrm{d}x}(L) = 0 \tag{10.22c}$$

where the latter two conditions mean that the cross section is restrained from warping at each end. There are two possible solutions to Eq. (10.22a), depending on the sign of β^2 in Eq. (10.22b). If $\beta^2 > 0$, the solution after the boundary conditions are imposed is

$$\theta_x(x) = \gamma_1\left[\sinh\beta x + \frac{(1-\cosh\beta L)}{\sinh\beta L}\cosh\beta x - \beta x - \frac{(1-\cosh\beta L)}{\sinh\beta L}\right] \tag{10.23}$$

where γ_1 is a constant to be determined. If $\beta^2 < 0$, the hyperbolic functions in Eq. (10.23) are replaced by the trigonometric sine and cosine. The constants α_1 and γ_1 are determined by considering equilibrium equations for the solar blanket and the spreader bar. These equations, together with continuity equations between the displacements and rotations of the booms, solar blanket, and spreader bar, provide the remainder of the solution.

We model the solar blanket as a membrane with constant tension F_x, but we neglect the tension F_y since the membrane has a high aspect ratio, $L/b' \gg 1$, and the transverse edges of the membrane are free. The equilibrium equation of the membrane is

$$F_x \frac{\partial^2 w_m}{\partial x^2} = 0 \tag{10.24a}$$

where w_m is the membrane deflection. The membrane boundary conditions are

$$w_m(0, y) = 0 \qquad w_m(L, y) = w_s(y) \tag{10.24b}$$

where w_s is the deflection of the spreader bar. Since we assume that the spreader bar is rigid, we may write

$$w_s(y) = y\theta_x(L) \tag{10.24c}$$

where $\theta_x(L)$ denotes the rotation of the spreader bar. By introducing θ_x in the last equation, we assume that the angle of twist of the booms equals the rotation of the spreader bar. For small rotations, we may also impose the condition that the deflection at the end of the spreader bar must equal the boom's tip deflection. Thus,

$$\theta_x(L) = [w(L)]/b \tag{10.25}$$

Solving Eq. (10.24a) and using the last three equations permits us to write the membrane deflection as

$$w_m(x, y) = \alpha_1(\tan\lambda L - \lambda L)\frac{x}{L}\frac{y}{b} \tag{10.26}$$

The final equation comes from the moment equilibrium of the spreader bar. We refer to the free-body diagram shown in Fig. 10.7 and obtain

$$M_x(L) + bV(L) - P\frac{\partial w}{\partial x}(L) + \frac{F_x}{2}\int_{-b'}^{b'} \frac{\partial w_m}{\partial x}(L, y)y\,\mathrm{d}y = 0 \tag{10.27}$$

where the boom shear force is defined by

$$V(L) = -EI\frac{\mathrm{d}^3 w}{\mathrm{d}x^3}(L) \tag{10.28a}$$

and the boom torque is defined by

$$M_x(L) = \left(GJ - \frac{PI_E}{A}\right)\frac{\mathrm{d}\theta_x}{\mathrm{d}x}(L) - E\Gamma\frac{\mathrm{d}^3\theta_x(L)}{\mathrm{d}x^3} \tag{10.28b}$$

We may now combine the foregoing equations to obtain a transcendental equation for the critical buckling load. We first substitute Eqs. (10.21) and (10.23) into Eq. (10.25) to obtain the relationship between α_1 and γ_1. This step yields the relation

$$\gamma_1 = \frac{\alpha_1}{b}\frac{\tan\lambda L - \lambda L}{\left[\sinh\beta L - \dfrac{(1-\cosh\beta L)^2}{\sinh\beta L} - \beta L\right]} \tag{10.29}$$

Then we substitute Eq. (10.21) into Eq. (10.28a) to obtain $V(L)$ and substitute Eq. (10.23) into Eq. (10.28b) to obtain the torque $M_x(L)$. Finally, we substitute these

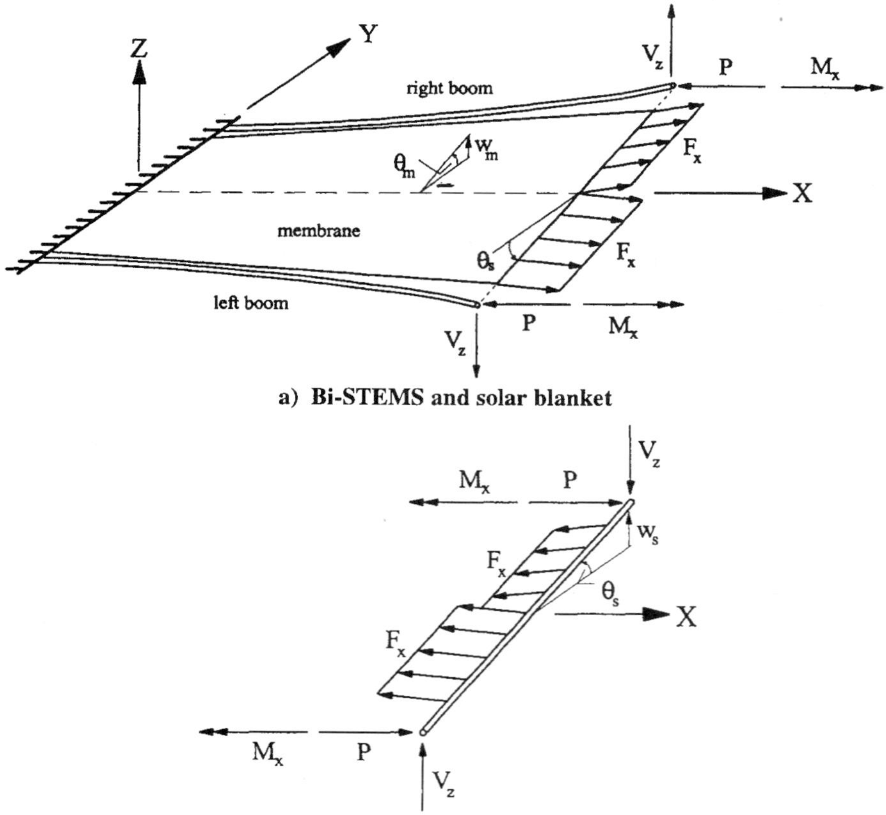

a) Bi-STEMS and solar blanket

b) Spreader bar

Fig. 10.7 Free-body diagrams of solar arrays.

results and Eq. (10.26) into the spreader bar moment equilibrium equation. This last step yields the transcendental equation,

$$\frac{\dfrac{\beta}{b}\left(\dfrac{PI_E}{A} - GJ\right)(\tan\lambda L - \lambda L)}{\left[\sinh\beta L - \dfrac{(1-\cosh\beta L)^2}{\sinh\beta L} - \beta L\right]} + \frac{F_x(b')^3}{3bL}(\tan\lambda L - \lambda L) + bP\lambda = 0 \tag{10.30}$$

where we recall $P = b'F_x$. Recalling the definitions of λ from Eq. (10.20b) and β from Eq. (10.22b), we see that Eq. (10.30) is a function only of P and physical parameters of the solar array. We may solve it numerically for the critical torsional-buckling forces.

Numerical results. The numerical results use data (Table 10.1) for the HST solar arrays that were obtained from Refs. 10 and 11. Table 10.2 tabulates the first three critical buckling values of P obtained from Eq. (10.30), and Fig. 10.8 shows the corresponding mode shapes. Based on data from Ref. 10, the boom force P at deployment was 14.75 N. Table 10.2 shows that the critical value of

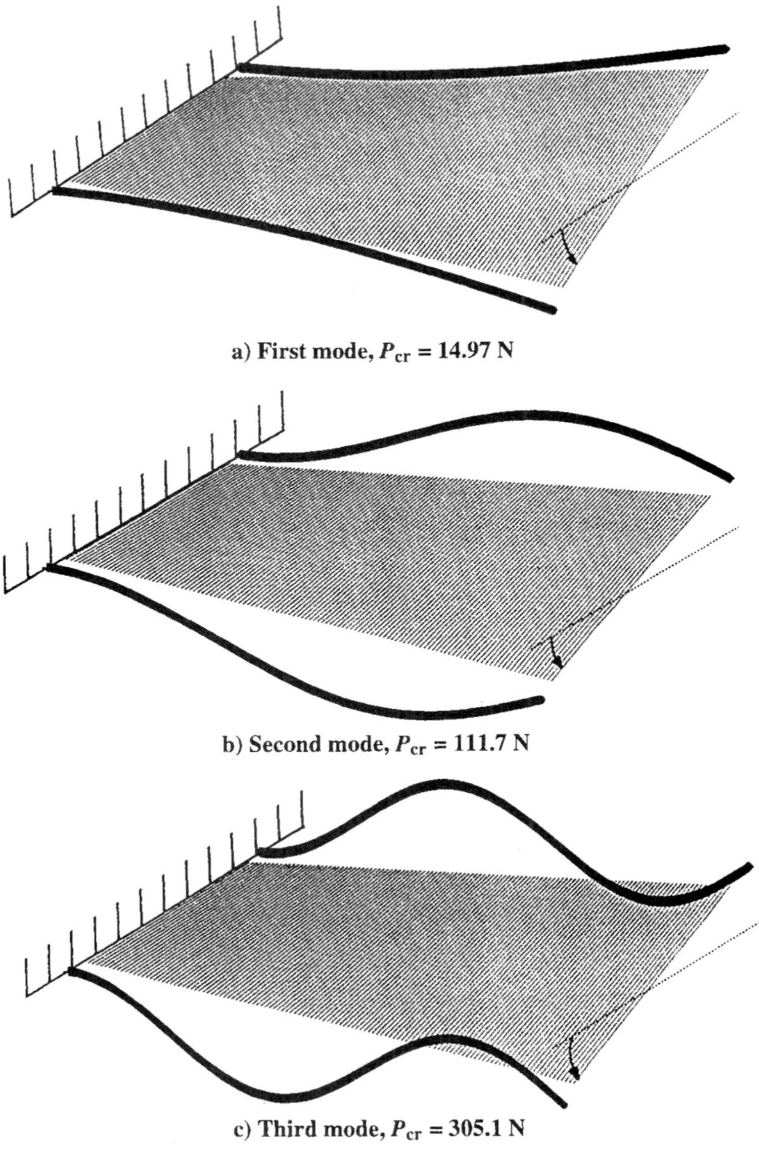

a) First mode, P_{cr} = 14.97 N

b) Second mode, P_{cr} = 111.7 N

c) Third mode, P_{cr} = 305.1 N

Fig. 10.8 Torsional-buckling mode shapes.

the boom force for the first torsional-buckling mode is 14.97 N. Thus, the solar array was deployed with a compressive boom force almost equal to the critical torsional-buckling load. Because the boom force was near its critical buckling load, any asymmetries in the solar array would cause a large initial angle of twist as shown in Fig. 10.5. Although full details of the HST solar array's behavior are not available, we hypothesize that the large initial deformation due to torsional buckling, combined with large, periodic quasistatic deformations plus thermally induced dynamic deformations, are likely to have contributed to the damaged boom shown in Fig. 10.4.

Table 10.1 Solar array properties (Refs. 10 and 11)

Γ	$= 2.586 \times 10^{-12}\,\mathrm{m}^6$
A	$= 1.613 \times 10^{-5}\,\mathrm{m}^2$
b'	$= 1.194\,\mathrm{m}$
b	$= 1.428\,\mathrm{m}$
E	$= 193.0 \times 10^{10}\,\mathrm{N/m}^2$
EI	$= 171.1\,\mathrm{N/m}^2$
G	$= 7.5 \times 10^{10}\,\mathrm{N/m}^2$
I_E	$= 1.948 \times 10^{-9}\,\mathrm{m}^4$
J	$= 8.67 \times 10^{-14}\,\mathrm{m}^4$
L	$= 5.91\,\mathrm{m}$
P	$= 14.75\,\mathrm{N}$

10.3 Thermal Buckling of Plates

A 1993 article[12] by the author provides a comprehensive review of past research on thermal buckling of plates and shells. Here we summarize some conclusions reached from this review.

10.3.1 Historical Review

Discussions of thermal plate buckling may be divided into three parts: 1) the plane stress problem, 2) bifurcation buckling, and 3) postbuckling. Some of the earliest research was conducted at NACA by Heldenfels and Roberts[13] and Gossard et al.[14] We describe this work in Sec. 7.4 as examples of plate problems. Since these original investigations, numerous researchers have studied thermal buckling and postbuckling of plates. (We present some of these solutions in Sec. 10.3.3.) Until the 1980s, most studies considered isotropic plates but, in the 1970s, research was initiated for anisotropic plates. Almost all the work has assumed perfectly flat plates and elastic behavior. There is a need for further study of buckling of plates with initial imperfections and inelastic behavior. Aerospace designers also need rapid postbuckling analysis methods for both mechanical and thermal loads. With a few exceptions, the investigations have been analytical or computational; there is a scarcity of experimental data available for validation of these investigations. Further experimental studies of thermally induced buckling are needed for both isotropic materials and laminated composites. Applications to aircraft structures suggest that experimental programs be conducted for stiffened plates with spatial temperature variations.

Table 10.2 Solar-array torsional-buckling loads

Mode	Buckling load, P_{cr}, N
1	14.97
2	111.7
3	305.1

10.3.2 Problem Statement

We consider thin, isotropic rectangular plates described by the plate theory developed in Chapter 7. Because we wish to analyze buckling problems, we use the large-deflection plate equations derived in Sec. 7.3. The equations are based on equilibrium equations written for the deformed plate and use nonlinear strain-displacement equations in the spirit of von Kármán. We neglect the transverse distributed loading and permit thermal loads from a specified temperature distribution. The basic equations describe the in-plane and bending behavior of the plate. From Chapter 7, for the in-plane behavior, we specify the membrane forces in terms of a stress function $F(x, y)$, Eq. (7.13),

$$N_x = \frac{\partial^2 F}{\partial y^2} \qquad N_{xy} = -\frac{\partial^2 F}{\partial x \partial y} \qquad N_y = \frac{\partial^2 F}{\partial x^2}$$

With these definitions, the in-plane equilibrium equations are satisfied identically. The strain-compatibility equation yields the first of the von Kármán equations for the membrane behavior, Eq. (7.29),

$$\nabla^4 F + (1-\nu)\nabla^2 N_T + Eh\left[\frac{\partial^2 w}{\partial x^2}\frac{\partial^2 w}{\partial y^2} - \left(\frac{\partial^2 w}{\partial x \partial y}\right)^2\right] = 0$$

where the thermal force for the plate is defined as

$$N_T = \frac{E\alpha}{1-\nu}\int_{(-h/2)}^{h/2} \Delta T \,\mathrm{d}z$$

where $\Delta T = T(x,y,z,t) - T_{\mathrm{ref}}$. We assume quasistatic behavior, and so time is a parameter. The bending behavior is described by the second of the von Kármán equations, Eq. (7.31),

$$D\nabla^4 w = \frac{\partial^2 F}{\partial x^2}\frac{\partial^2 w}{\partial x^2} - 2\frac{\partial^2 F}{\partial x \partial y}\frac{\partial^2 w}{\partial x \partial y} + \frac{\partial^2 F}{\partial x^2}\frac{\partial^2 w}{\partial y^2} - \nabla^2 M_T$$

which is a statement of vertical equilibrium written for the deformed plate configuration. The plate thermal bending moment is defined by

$$M_T = \frac{E\alpha}{1-\nu}\int_{(-h/2)}^{h/2} \Delta T z \,\mathrm{d}z$$

Expressions for the stresses and stress resultants in terms of the in-plane displacements u_0, v_0 and the transverse displacement w appear in Eqs. (7.22) and (7.23). Boundary conditions for the von Kármán equations are discussed in Chapter 7. With the exception of a different equation for the Kirchhoff shear force, the boundary conditions are the same as for small-deflection plate theory.

10.3.3 Analytical Solutions

If a flat isotropic plate is heated or cooled *uniformly* and is *free* to expand or contract in its plane, buckling cannot occur because membrane forces do not develop. However, if the plate is restrained in its plane or if a nonuniform spatial

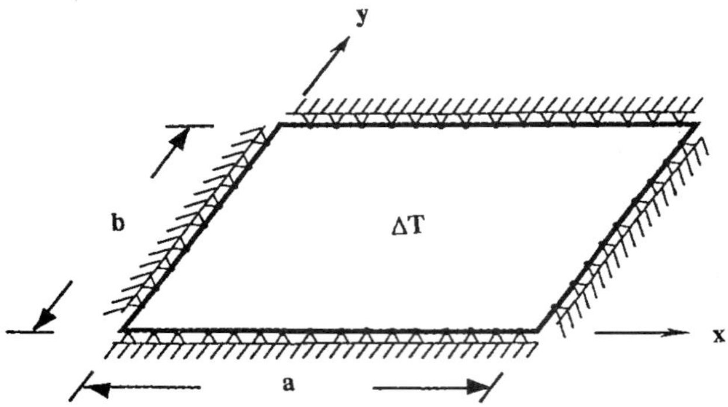

Fig. 10.9 Uniformly heated plate with in-plane edge restraints.

temperature distribution occurs, then membrane forces develop. If compressive forces occur, then there is a possibility of thermal buckling.

In this section, we describe analytical solutions for bifurcation-buckling temperatures of flat rectangular plates. The problems considered include restrained and unrestrained in-plane displacements as well as uniform, parabolic, tentlike, and sinusoidal temperature distributions.

Fully restrained plate with uniform temperature. One of the most fundamental thermal-buckling problems is a simply supported rectangular plate with fully restrained in-plane displacements (Fig. 10.9). If the plate temperature change is uniform, that is, if ΔT is independent of x, y, z, we may obtain a relatively simple solution for the critical temperature for bifurcation buckling. If the temperature change is positive, the immovable edges restrain the thermal expansion completely, and we may evaluate the membrane forces directly. From Eqs. (7.6a–c), we have

$$N_x = \frac{Eh}{1-\nu^2}\left(\frac{\partial u_0}{\partial x} + \nu\frac{\partial v_0}{\partial y}\right) - N_T$$

$$N_y = \frac{Eh}{1-\nu^2}\left(\frac{\partial v_0}{\partial y} + \nu\frac{\partial u_0}{\partial x}\right) - N_T$$

$$N_{xy} = \frac{Eh}{2(1+\nu)}\left(\frac{\partial u_0}{\partial x} + \nu\frac{\partial v_0}{\partial y}\right)$$

where u_0 and v_0 are the plate midplane displacements. Since plate in-plane displacements are fully restrained, we know that $u_0 = v_0 \equiv 0$. In addition, because the temperature is uniform, we may evaluate the integral for the thermal force, Eq. (7.8) (repeated in the preceding section), to obtain

$$N_T = -\frac{E\alpha h}{1-\nu}\Delta T$$

With these results, we find that the membrane forces are given by

$$N_x = N_y = -\frac{E\alpha h}{1-\nu}\Delta T \tag{10.31a}$$

$$N_{xy} = 0 \tag{10.31b}$$

Thus, the membrane forces N_x and N_y are equal, compressive, and uniform throughout the plate. The shear force N_{xy} is zero as we should expect for a uniform temperature field and a fully restrained plate.

To determine the bifurcation temperature, we must consider plate bending. We use Eq. (7.31) and express it directly in terms of the membrane forces by using their definitions in terms of the stress functions given by Eq. (7.13). Thus, we may rewrite Eq. (7.31) as

$$D\nabla^4 w = N_x \frac{\partial^2 w}{\partial x^2} + 2N_{xy}\frac{\partial^2 w}{\partial x \partial y} + N_y \frac{\partial w}{\partial y^2} - \nabla^2 M_T \tag{10.32}$$

Since the temperature is uniform, the thermal moment is zero. Substituting Eq. (10.31), the plate bending equation reduces to

$$D\nabla^4 w = -\frac{E\alpha h}{1-\nu}\Delta T\left(\frac{\partial^2 w}{\partial x^2} + \frac{\partial^2 w}{\partial y^2}\right) \tag{10.33}$$

Since, for bending, the plate edges are simply supported, the boundary conditions require the transverse deflection and bending moment to be zero on each edge. The moment boundary conditions are simplified because the thermal moment is zero for uniform temperature. The boundary conditions are

$$w(0, y) = w(a, y) = \frac{\partial^2 w}{\partial x^2}(0, y) = \frac{\partial^2 w}{\partial x^2}(a, y) = 0 \tag{10.34a}$$

$$w(x, 0) = w(x, b) = \frac{\partial^2 w}{\partial y^2}(x, 0) = \frac{\partial^2 w}{\partial y^2}(x, b) = 0 \tag{10.34b}$$

Because of the form of Eqs. (10.34), a solution of the form

$$w = w_m \sin\frac{m\pi x}{a}\sin\frac{n\pi y}{b} \tag{10.35}$$

is taken since it satisfies all the boundary conditions. Substituting Eq. (10.35) into Eq. (10.33), we find that either

$$w_m = 0 \tag{10.36a}$$

or

$$D\left[\left(\frac{m\pi}{a}\right)^2 + \left(\frac{n\pi}{b}\right)^2\right] = \frac{E\alpha h\Delta T}{1-\nu} \tag{10.36b}$$

These equations express the bifurcation character of the solution. The plate either remains perfectly flat with $w \equiv 0$, or it may be in equilibrium at a critical temperature ΔT_{cr} with a deflection shape given by Eq. (10.35). From Eq. (10.36b), the critical buckling temperature is found to be

$$\Delta T_{\text{cr}} = \frac{\pi^2(1-\nu)D}{E\alpha h}\left(\frac{m^2}{a^2} + \frac{n^2}{b^2}\right) \tag{10.37}$$

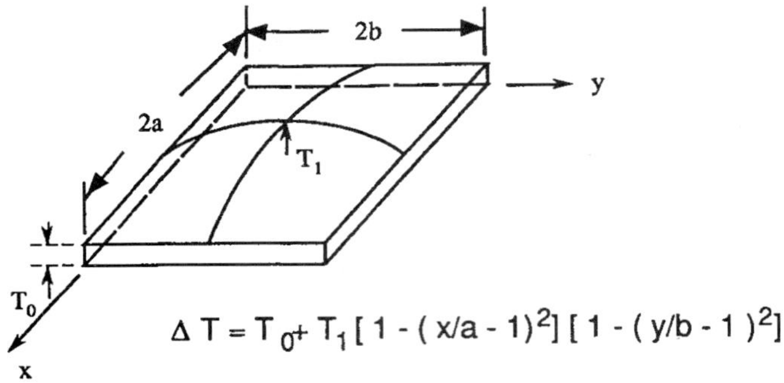

Fig. 10.10 Simply supported plate with parabolic temperature distribution.

Substituting $D = Eh^3/12(1 - \nu^2)$ and noting that the minimum value of ΔT_{cr} occurs when $m = n = 1$, we obtain,

$$\Delta T_{cr} = \frac{\pi^2 h^2}{12(1+\nu)\alpha b^2}\left(1 + \frac{b^2}{a^2}\right) \tag{10.38}$$

Note that the critical buckling temperature is independent of the material's elastic modulus. For thin plates that are fully restrained, Eq. (10.38) predicts quite small

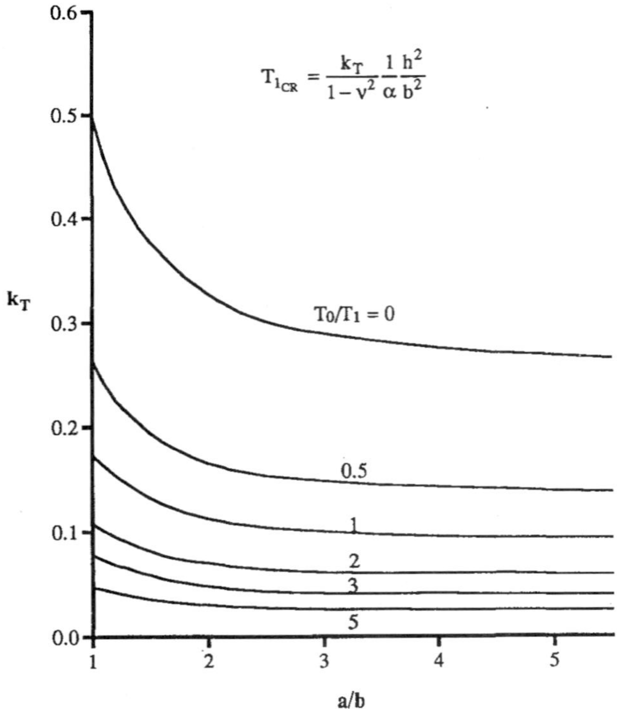

Fig. 10.11 Buckling coefficient for a simply supported fully restrained plate with parabolic temperature distribution (from Ref. 15).

critical buckling temperatures. For example, consider a Hastelloy-X plate with $a = 15$ in., $b = 10$ in., $h = 0.125$ in., $\alpha = 7.9 \times 10^{-6}$ 1/°F, and $\nu = 0.316$. Equation (10.38) predicts a critical buckling temperature $\Delta T_{cr} = 17.8$°F! This temperature, of course, is impractically low because it is extremely difficult to restrain a plate completely against thermal expansion.

Restrained plate expansion with parabolic temperature distribution. In a related analysis, Klosner and Forray[15] studied buckling of simply supported plates under an arbitrary symmetrical temperature distribution. In-plane displacements of the plate are assumed to be restrained elastically by supporting edge members. The critical buckling temperature was determined by a Rayleigh–Ritz procedure. Miura[16] used a similar procedure to obtain the critical buckling temperature when in-plane displacements were restrained completely. Critical buckling temperatures were computed in these papers for a restrained, simply supported plate (Fig. 10.10) of dimensions $(2a \times 2b)$ and the parabolic temperature distribution

$$T = T_0 + T_1 \left[1 - \left(1 - \frac{x}{a}\right)^2\right] \left[1 - \left(1 - \frac{y}{a}\right)^2\right] \tag{10.39}$$

The results of these two papers show that the critical buckling temperature is given by

$$(\Delta T_1)_{cr} = \frac{k_T}{1 - \nu^2} \frac{1}{\alpha} \frac{h^2}{b^2} \tag{10.40}$$

where the buckling coefficient k_T depends on the plate aspect ratio a/b and the temperature ratio T_0/T_1. The variation of k_T with these parameters is shown in Fig. 10.11.

Elastically restrained plate with uniform temperature. Wilcox and Clemmer[17] considered the problem of a flat rectangular plate elastically restrained in its plane. As in the first example, the plate experiences a uniform temperature change ΔT. The plate (Fig. 10.12) is restrained elastically by uniform springs with stiffnesses k_x at $x = 0, a$ and k_y at $y = 0, b$. Otherwise, for bending, the edges are simply supported. Wilcox and Clemmer considered both bifurcation buckling and postbuckling.

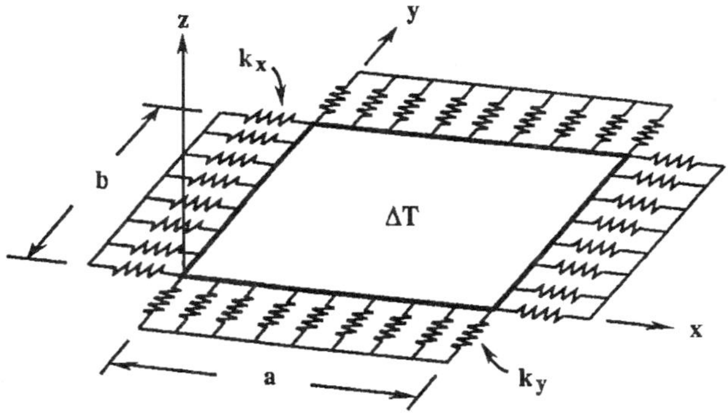

Fig. 10.12 Uniformly heated plate with elastically restrained edges.

The bifurcation-buckling analysis showed that

$$\alpha \Delta T_{cr} = \frac{\dfrac{\pi^2}{12(1-\nu^2)}\left(\dfrac{h}{a}\right)^2 \left[1+\left(\dfrac{a}{b}\right)^2\right]^2 \left[\left(1+\dfrac{2}{a}\dfrac{E}{k_x}\right)\left(1+\dfrac{2}{b}\dfrac{E}{k_y}\right)-\nu^2\right]}{(1+\nu)\left[1+\left(\dfrac{a}{b}\right)^2\right]+\dfrac{2}{b}\dfrac{E}{k_y}+\left(\dfrac{a}{b}\right)^2\dfrac{2}{a}\dfrac{E}{k_x}} \tag{10.41}$$

Note that if $k_x = k_y = \infty$, the solution reduces to Eq. (10.38) for the fully restrained plate.

To determine the postbuckling behavior of the plate, Wilcox and Clemmer solved the von Kármán equations (7.29) and (7.31) using a Fourier series expansion. For a uniform temperature distribution, the plate center deflection w_0 is given by

$$\beta_1\left(\frac{w_0}{h}\right)^3 + \beta_2\frac{w_0}{h} = 0 \tag{10.42a}$$

where

$$\beta_1 = \frac{\pi^4}{16}\left[1+\left(\frac{a}{b}\right)^4\right] + \frac{\pi^4}{8}\left[\frac{1+\dfrac{2}{b}\dfrac{E}{k_y}+\left(\dfrac{a}{b}\right)^4\left(1+\dfrac{2}{a}\dfrac{E}{k_x}\right)+2\nu\left(\dfrac{a}{b}\right)^2}{\left(1+\dfrac{2}{a}\dfrac{E}{k_x}\right)\left(1+\dfrac{2}{b}\dfrac{E}{k_y}\right)-\nu^2}\right] \tag{10.42b}$$

$$\beta_2 = \frac{\pi^4}{12(1-\nu^2)}\left[1+\left(\frac{a}{b}\right)^2\right]^2\left(1-\frac{\Delta T}{\Delta T_{cr}}\right) \tag{10.42c}$$

Figure 10.13 shows the nondimensional center displacement w_0/h for a square

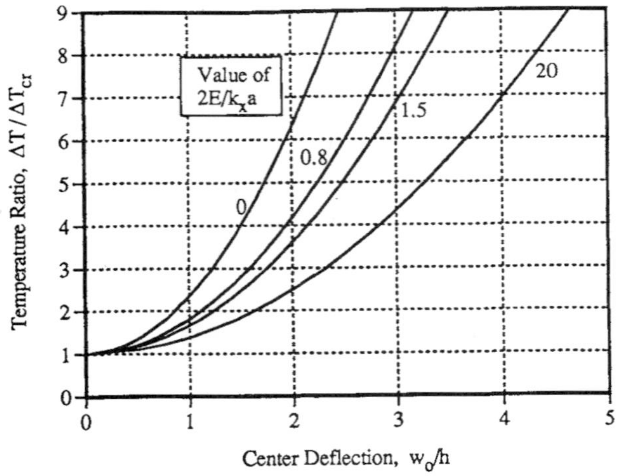

Fig. 10.13 Postbuckling response of an elastically restrained square plate with uniform temperature.

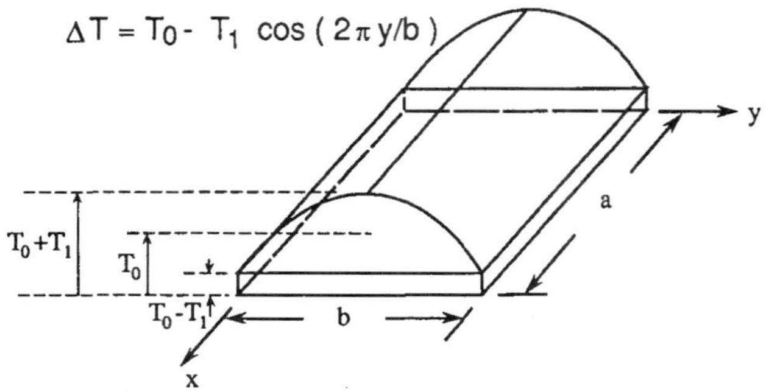

Fig. 10.14 Simply supported plate with sinusoidal temperature distribution.

plate plotted as a function of the temperature ratio $\Delta T/\Delta T_{\text{cr}}$ for various values of the edge restraint stiffness.

The figure shows the postbuckling deflection beginning at the bifurcation point where $\Delta T/\Delta T_{\text{cr}} = 1$. Above a temperature ratio of 1, the curves demonstrate the characteristic postbuckling behavior of plates as the slopes of the curves increase with increasing deflection. This means that the plate is "stiffening" because of stretching of the middle plane as plate deflections become large. The curves are shown for various values of $2E/k_x a$. As the spring support stiffness increases, $2E/k_x a$ decreases. Thus, the curve on the left in Fig. 10.13 corresponds to $k_x = \infty$; it shows the greatest stiffening effect. The other curves tend to have less stiffening effects as the in-plane stiffness decreases since the membrane forces become progressively smaller.

Free expansion of a plate with "tentlike" temperature distribution. In the three preceding problems, thermal buckling occurs because of restrained thermal expansion. In the Heldenfels[13] problem, the plate is free to expand, but the nonuniform temperature distribution causes the plate to buckle. In Sec. 7.4, Example 7.4, we describe the membrane stress distribution induced by the "tentlike" temperature distribution. The nonuniform temperature distribution produces biaxial stresses, and the free-edge boundary condition induces a combined tensile-compressive stress state because the plate has to be self-equilibrating. The analysis shows that there is a large central region of the plate with compressive stresses. Gossard et al.[14] perform analyses of bifurcation buckling and the postbuckling response. For a plate aspect ratio of $a/b = 1.57$, an approximate critical buckling temperature was found to be given by

$$(\Delta T_1)_{\text{cr}} = \frac{5.39\pi^2 h^2}{12(1-\nu^2)\alpha b^2} \tag{10.43}$$

We describe Gossard's postbuckling solution in Sec. 7.4, Example 7.5.

Free expansion of a plate with sinusoidal temperature distribution. Boley and Weiner[6] investigate buckling of a simply supported plate with a sinusoidal temperature distribution in the y direction (Fig. 10.14). Edges are assumed to remain straight, but displacements may occur normal to an edge. The in-plane boundary conditions are

$$u(0, y) = 0 \qquad v(x, 0) = 0$$
$$u(a, y) = u_0 \qquad v(x, b) = v_0$$

An exact solution for the critical buckling temperature is developed by first solving for the stress function F that satisfies the compatibility equation (7.29) with w equal to zero. From the stress function, the membrane forces are obtained. The plate equilibrium equation (10.32) is then solved for bifurcation buckling. Boley and Weiner show that the solution leads to an infinite determinant for computing the critical buckling temperature. The solution has the form

$$(\Delta T_1)_{cr} = k_1 \frac{\pi^2 h^2}{6(1 - \nu^2)\alpha b^2} \tag{10.44}$$

where k_1 is a function of m, the number of half-waves in the x direction, and the aspect ratio a/b. The text[6] gives a graph of k_1 vs a/b for varying values of m and determines that the minimum value of $k_1 = 3.848$ for an aspect ratio greater than or equal to 1.

Boley and Weiner also solve the postbuckling problem by an iterative procedure that consists of determining a set of successive approximations for F and w. Their text[6] gives a graph of the solution showing the center deflection as a function of temperature.

10.3.4 Finite Element Solutions

In Sec. 8.3.4, we present a finite element formulation for solving the von Kármán equations. In Example 8.6, the finite element approach was validated for mechanical loads by comparison with an analytical solution for the large deflection of a simply supported plate with and without restraints on the in-plane displacement. Here we present finite element solutions for thermal buckling of rectangular plates.

Gossard problem. For further evaluation of the finite element approach but with thermal loads, we analyze the postbuckling behavior of the rectangular plate with the "tentlike" temperature distribution. Gossard et al.[14] developed approximate analytical solutions (see Sec. 7.4, Example 7.5) and provided experimental data for comparison.

Figure 7.8 is a schematic view of the temperature distribution. The experimental boundary conditions are stated to be simple supports although no practical details are given.[14] For the finite element analysis, the symmetry of the boundary conditions and temperature distribution are exploited by using a one-quarter plate model. Figure 10.15 shows the boundary conditions, plate properties, and dimensions for the finite element model. The mesh consisted of 121 nodes and 200 triangular elements. Gossard's original plate had an initial center deflection $w_{ic} = 0.045$ in.

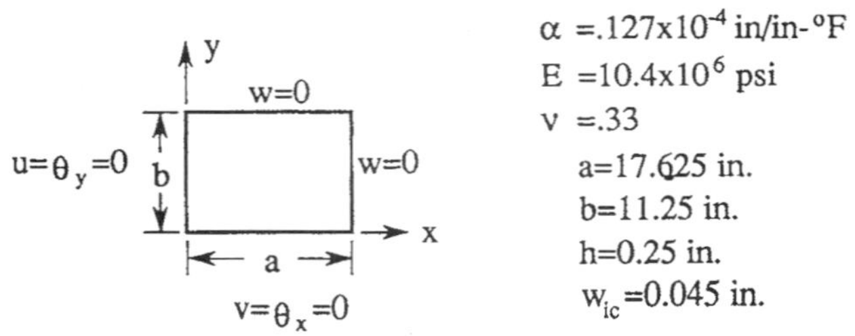

Fig. 10.15 Boundary conditions and data for finite element analysis of a Gossard square plate.

relative to the edges. For the finite element analysis, an initial shape given by

$$w_{ic}(x, y) = w_{ic} \cos \frac{\pi x}{2a} \cos \frac{\pi y}{2b} \tag{10.45}$$

was used. The z coordinates of the nodes were calculated from this equation, and the elements were located on this initial surface.

The results of the finite element analysis are shown in Fig. 10.16, along with Gossard's approximate analytical solutions for $w_{ic} = 0$ and $w_{ic} = 0.045$ in. The graph shows the plate center temperature change plotted vs the center displacement; the center deflection is measured relative to the initial plate deflection. The critical buckling temperature $\Delta T_{cr} = 197°$F was calculated using Eq. (10.43). The dotted line represents the postbuckling deflection path derived by Gossard for a perfectly flat plate ($w_{ic} = 0$), and the dashed line represents an approximate deflection path derived by Gossard for a plate with an initial deflection for $w_{ic} \neq 0$. The finite element results are denoted by the solid line and are in excellent agreement with Gossard's approximate solution for an imperfect plate. As we have discussed, the postbuckling solution provides an upper bound on the actual buckling behavior. The smaller the initial imperfection, the more closely the temperature-deflection curve approaches the critical buckling temperature for small deflections.

Transient buckling test of Hastelloy-X plate. In Sec. 8.2.4, Example 8.1, we describe finite element computations for transient temperatures in a rectangular Hastelloy-X plate. The temperatures were computed for a laboratory test as part of an experimental study of plate buckling induced by spatial temperature gradients.[18] In a recent thesis, Hernan[19] compares finite element predicted plate buckling displacement histories with the experimental data.

Let us briefly review the plate test that is described in Sec. 8.2.4. Figure 8.6 shows schematically a typical test configuration. A Hastelloy-X plate has dimensions of 15 × 10 in. and a thickness of 0.125 in. The plate has point supports near its four corners. It is heated transiently along its longitudinal centerline by a focused quartz lamp while the long edges of the plate are maintained at constant temperature. In a typical test, a plate is subjected to sudden heating until the centerline temperature reaches a peak value. As the Gossard problem showed, the plate's initial deflected shape has an important role in the buckling response. Figure 10.17 shows the initial

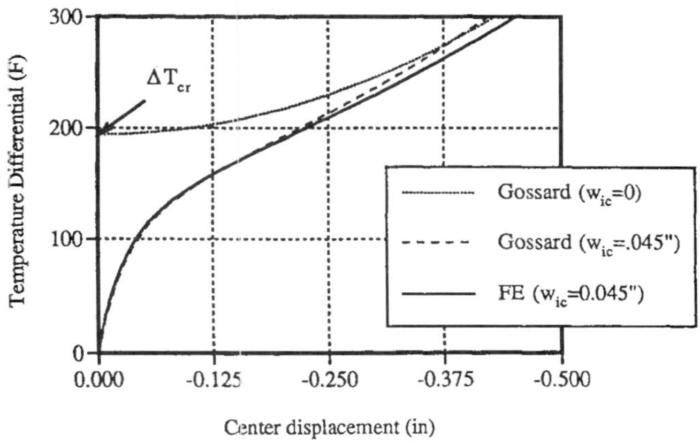

Fig. 10.16 Buckling response of a Gossard plate.

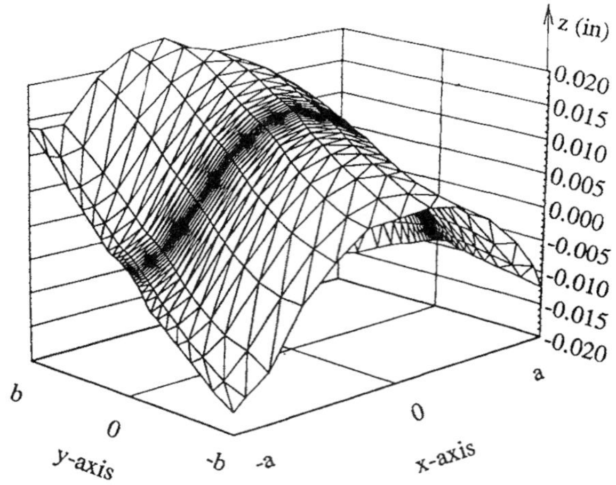

Fig. 10.17 Initial deformation of a test plate.

configuration of a plate in the test fixture. The heat lamp is mounted above the plate. Because of the initial upward "bow" of the plate, the plate deforms upward during the test.

The finite element mesh is also shown in Fig. 10.17. Although the plate is heated symmetrically and temperatures are symmetrical about the plate's longitudinal centerline, the lack of symmetry in the initial deflections requires a finite element model of the complete plate. The mesh consists of 403 nodes and 720 triangular elements.

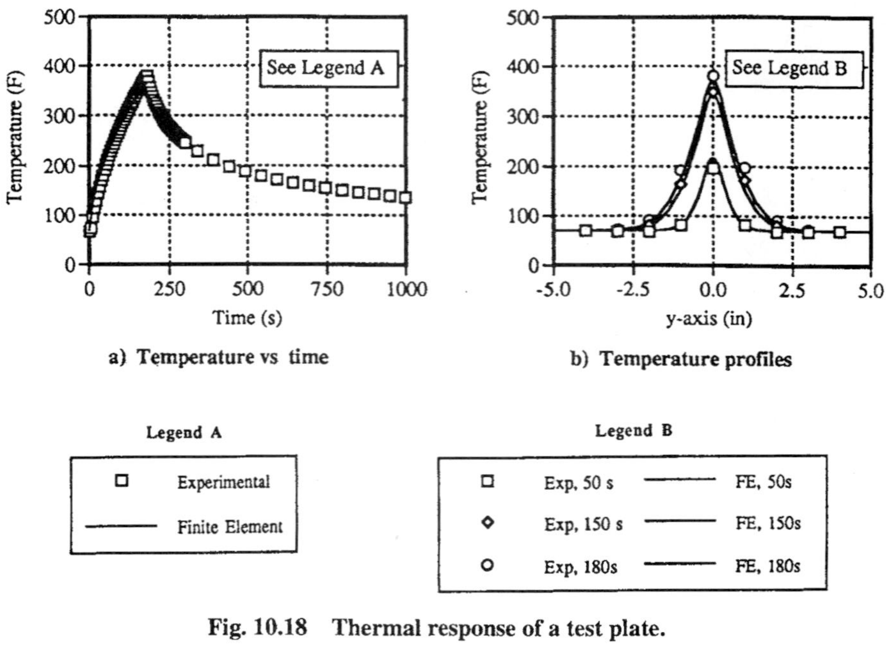

Fig. 10.18 Thermal response of a test plate.

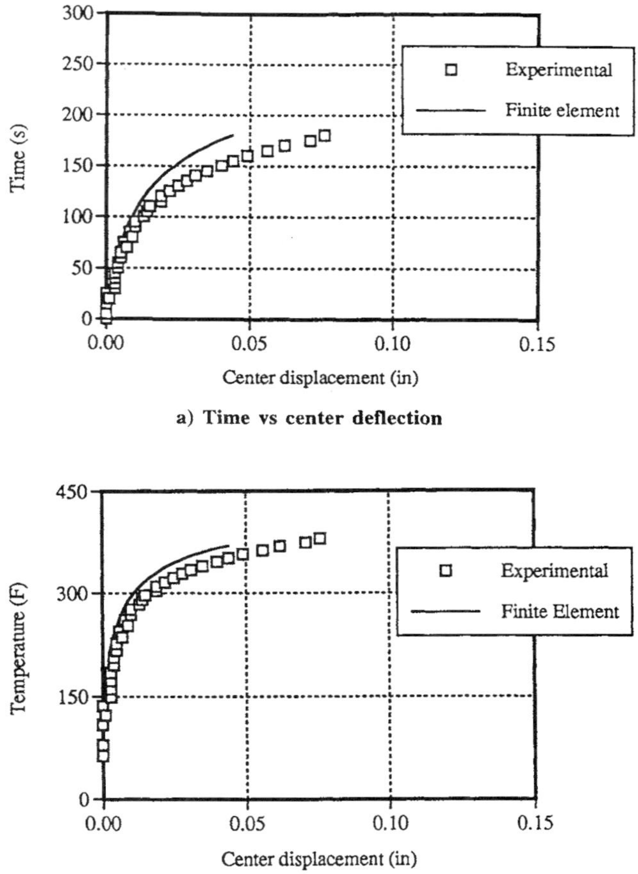

a) Time vs center deflection

b) Temperature vs center deflection

Fig. 10.19 Displacement response of a test plate.

Figure 10.18 shows the finite element and experimental temperature histories and profiles. Agreement between computations and experimental data is very good although the experimental temperatures are slightly higher than computed values at later times. Figure 10.19 compares the finite element and experimental displacement response at the plate center. Figure 10.19a compares the displacement histories, and Fig. 10.19b compares the temperature-displacement response at the plate center. Early in the response, agreement is good but, at later times, the analysis tends to underestimate the plate deflections. One point to note is that, during the nonlinear buckling response, a small change in temperature can cause a rather large change in displacement. This nonlinear relationship illustrates the importance of thermal-structural *interactions* in a nonlinear response.

10.3.5 Final Comments

The analytical and finite element solutions presented in the two preceding sections show the important effect that in-plane restraints have on thermal buckling of plates. We mentioned that a 15- × 10- × 0.125-in. Hastelloy-X plate with uniform temperature has a critical buckling temperature $\Delta T_{cr} = 18°F$ if it is fully restrained. Other calculations show that, with a parabolic temperature distribution, $(\Delta T_1)_{cr} = 33°F$. In contrast, if the plate can expand freely in the plane, it has a critical buckling temperature of $(\Delta T_1)_{cr} = 400°F$ for the tentlike temperature distribution and $(\Delta T_1)_{cr} = 312°F$ for the sinusoidal temperature distribution.

A closely related issue not described here concerns the plate's stress levels at buckling. For restrained plates, thermal buckling occurs for such small temperature changes that stresses are small and buckling is most likely to be elastic. For unrestrained plates, however, thermal stresses are much higher, and inelastic buckling is much more likely to occur.

10.4 Thermal Buckling of Shells

As we mentioned earlier, the author's 1993 article[12] gives a comprehensive review of thermal buckling of plates and shells. Here we highlight some conclusions reached for shell thermal buckling.

10.4.1 Historical Review

Buckling of shells has been the subject of extensive research in the aerospace industry because of the widespread use of shell structures in flight vehicles. Analytical, computational, and experimental studies have investigated the behavior of shell structures subjected to mechanical loads. Much less research has occurred for buckling and postbuckling of shells in a thermal environment. And yet, designers of airframes for future high-speed flight and spacecraft structures will have to consider carefully the effect of the thermal environment on structural and material behavior.

There are several excellent textbooks[20–29] on the theory of shell buckling, but most classic books do not include temperature in the formulation of the equations, nor do they address thermal buckling. Thus, with a few exceptions, solutions of shell thermal-buckling problems appear in papers and reports rather than in texts. Furthermore, in contrast to the study of shell buckling under mechanical loads, studies of thermal buckling are relatively more recent. For example, solutions for critical buckling loads of cylinders under axial loads were presented in the early 1900s, but studies of critical buckling temperatures for cylindrical shells were not undertaken until the 1950s.

In the 1950s and 1960s, there was significant research on thermal buckling of metallic shells as aerodynamic heating became important in the design of supersonic aircraft and missiles. Since then, research in thermal buckling of metallic shells has diminished. Beginning in the 1970s, there has been significant growth in the research efforts devoted to thermal buckling of composites. In the 1990s, shell thermal-buckling research, as indicated by published papers, has been devoted exclusively to composites.

Since the original investigations of Hoff,[30] numerous investigators have studied thermal buckling of shells. Many of the studies have been analytical particularly through the 1950s and 1960s but, since 1970, there has been increasing reliance on computational methods. Many of the computer programs have used finite difference methods, but analyses of composite shells are using the finite element method. An effort was also made from the 1950s through the 1970s to investigate thermal buckling of shells experimentally. The vast majority of the experimental studies considered monocoque shells. With the exception of the study by Anderson and

Card,[31] there have been no further investigations of stiffened shells. Experimental measurements were limited typically to a modest number of transducers. No data are available for initial imperfections, details of temperature distributions are lacking, and little documentation of postbuckling damage is available. Virtually no strain data were taken in the experiments because of strain-gauge limitations to low temperatures. Most experiments were conducted on steel or stainless steel. Data are limited for aluminum, and no experiments have studied the buckling of titanium shells. Experimental studies of composite shells have not yet been published.

Details of shell temperature distributions, particularly in stiffened shells, have an important role in the buckling response. The assumption of uniform temperature between stiffeners is usually not a good approximation. Because material properties such as thermal conductivity have significant effects on the shape of temperature distributions, thermal buckling tests should be conducted using materials that will be used in actual structures.

With one notable exception, there have been no validation studies of computer analyses of thermal buckling of shells. The exception is the very thorough analyses of several experiments by Bushnell[32] using the computer code BOSOR. Generally, agreement between analytical solutions, computer solutions, and experimental data has been only fair to good. A typical reason cited for discrepancies has been uncertainties in experimental boundary conditions. There have been relatively few tests with combined mechanical and thermal loads. For example, no tests of heated cylinders with internal pressures have been reported. There is a need for further experimental studies of thermally induced buckling of shells for both isotropic materials and laminated composites. Applications to aerospace structures strongly suggest the need for experimental programs for unstiffened and stiffened shells with thermal loads as well as combinations of thermal and mechanical loads.

10.4.2 Problem Statement

To illustrate thermal buckling of shells, we consider a circular cylindrical shell. We summarize the basic equations of classical shell buckling and present an example. The treatment serves as an introduction to shell buckling and, for further details, the reader is encouraged to examine the references at the end of this chapter. We follow here the approach of Brush and Almroth[25] with the addition of the thermal terms in appropriate equations. Here we present the equations without derivation; the development of shell equations is beyond the scope of our discussion.

We consider a circular cylindrical shell with the geometry shown in Fig. 10.20. Cylindrical coordinates x, θ locate a point on the middle surface of the shell. The displacements of a point on the shell's middle surface are represented by the components u, v, w. The components u and v are in-plane membrane displacements; the component w is an out-of-plane displacement. In classical thin-shell theory, we assume that the shell thickness h is much smaller than the shell radius a; that is, $h \ll a$. A typical shell segment has membrane forces, shear forces, and bending moments similar to those of a segment of a flat plate. We show the membrane forces in Fig. 10.21; a similar element (not shown) could be drawn to display the shear forces and bending moments.

Basic equations. The development of shell buckling theory is similar to the development of von Kármán's plate theory, but it is made more complicated by the shell geometry. For example, the constitutive laws for the membrane forces

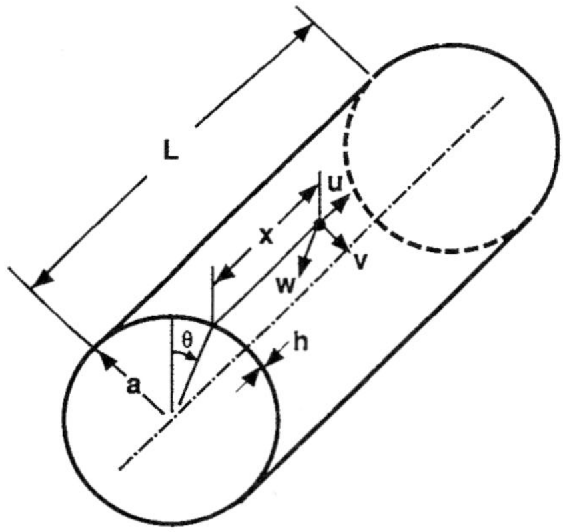

Fig. 10.20 Circular cylindrical shell.

shown in Fig. 10.21 are the same as for flat plates, namely,

$$N_x = \frac{Eh}{1-\nu^2}(\varepsilon_x + \nu\varepsilon_\theta) - N_T \tag{10.46a}$$

$$N_\theta = \frac{Eh}{1-\nu^2}(\varepsilon_\theta + \nu\varepsilon_x) - N_T \tag{10.46b}$$

$$N_{x\theta} = \frac{Eh}{2(1+\nu)}\varepsilon_{x\theta} \tag{10.46c}$$

but now the strain components are

$$\varepsilon_x = \frac{\partial u}{\partial x} + \frac{1}{2}\left(\frac{\partial w}{\partial x}\right)^2 \tag{10.47a}$$

$$\varepsilon_\theta = \frac{1}{a}\left(\frac{\partial v}{\partial \theta} + w\right) + \frac{1}{2}\left(\frac{1}{a}\frac{\partial w}{\partial \theta}\right)^2 \tag{10.47b}$$

$$\varepsilon_{x\theta} = \frac{\partial v}{\partial x} + \frac{1}{a}\frac{\partial u}{\partial \theta} + \frac{1}{a}\frac{\partial w}{\partial x}\frac{\partial w}{\partial \theta} \tag{10.47c}$$

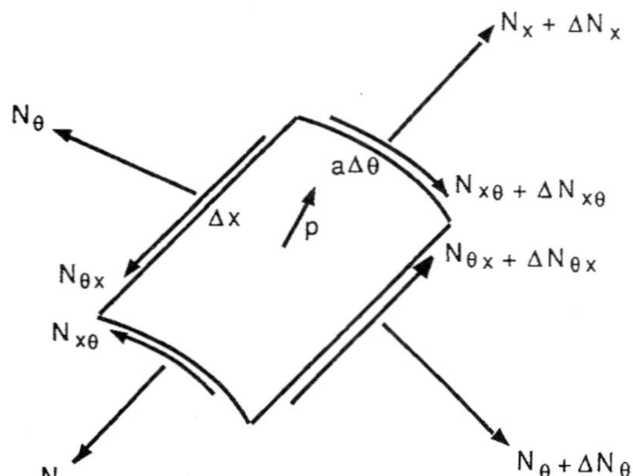

Fig. 10.21 Shell segment with membrane forces.

In the preceding equations, we use the displacement components u, v, w without subscripts to denote displacements of the middle surface whereas, for plates, we use a zero subscript to denote in-plane displacements. As with plates, the linear terms on the right-hand side of Eqs. (10.47) correspond to stretching of the middle surface due to small displacements. The linear terms are the same as those for plates except for the "hoop" strain component w/a in Eq. (10.47b). The nonlinear terms denote stretching of the middle surface due to bending rotation components $\partial w/\partial x$ and $\partial w/\partial(a\theta)$. These terms are the von Kármán strain components and are the same as those for plates. The thermal force N_T in Eqs. (10.46a) and (10.46b) is the same as that for a flat plate given in Eq. (7.8). We may write down a similar set of equations for the moment stress resultants but, for brevity, we will omit these relations.

The equilibrium equations for a cylindrical shell may be written as

$$\frac{\partial N_x}{\partial x} + \frac{1}{a}\frac{\partial N_{x\theta}}{\partial \theta} = 0 \tag{10.48a}$$

$$\frac{\partial N_{x\theta}}{\partial x} + \frac{1}{a}\frac{\partial N_{\theta}}{\partial \theta} = 0 \tag{10.48b}$$

$$D\nabla^4 w + \frac{1}{a}N_\theta - \left(N_x\frac{\partial^2 w}{\partial x^2} + \frac{2}{a}N_{x\theta}\frac{\partial^2 w}{\partial x\partial\theta} + \frac{1}{a^2}N_\theta\frac{\partial^2 w}{\partial\theta^2}\right) = p \tag{10.48c}$$

where

$$\nabla^4 w = \frac{\partial^4 w}{\partial x^4} + \frac{2}{a^2}\frac{\partial^4 w}{\partial x^2\partial\theta^2} + \frac{1}{a^4}\frac{\partial^4 w}{\partial\theta^4}$$

and p denotes the transverse surface pressure. Equations (10.48) are very similar to the corresponding equations we derive for large deflections of flat plates. The equations are, of course, nonlinear because of the nonlinear strain terms in Eqs. (10.47) and the products of the membrane forces and bending curvatures in Eq. (10.48c). The equations are referred to as the Donnell quasishallow shell equations. They have been widely used in the literature for analytical studies. In recent years, more sophisticated equations with fewer limitations are used for computational studies; see, for example, the recent book[29] by Palazotto and Dennis.

Linear stability equations. Equations (10.48) are the nonlinear equilibrium equations for circular cylindrical shells. For a prescribed temperature distribution, the solution of these equations furnishes the displacements describing the deformed shape of the shell. From the displacements, we can obtain the shell's stress components.

If our objective is to determine the bifurcation-buckling loads of the shell, an approach has been developed that permits the use of a linear set of differential equations. The approach involves the concept of an *adjacent equilibrium criterion*. Brush and Almroth[25] use the approach to study systematically the stability of beams, plates, and shells. The basic idea is to give small increments to the displacement components and determine if an adjacent equilibrium position exists under the same applied load. Existence of such an adjacent equilibrium position typically indicates a bifurcation point. To follow the approach here, we let

$$u \to u_0 + u_1 \qquad v \to v_0 + v_1 \qquad w \to w_0 + w_1 \tag{10.49}$$

where (u_0, v_0, w_0) is the initial equilibrium position, and (u, v, w) is an adjacent equilibrium position; the displacements (u_1, v_1, w_1) are small increments. The displacements with the zero subscripts are the prebuckling deformations, and the displacements with the 1 subscripts are the buckling deformations. When applied to the nonlinear equilibrium equations, two sets of equations are obtained. The first set of equations is the linear equilibrium equations for the prebuckled configuration, and the second set of equations is a linear set of stability equations for the incremental displacement components. Brush and Almroth[25] provide the details and give the final results as

$$\frac{\partial N_{x0}}{\partial x}+\frac{1}{a}\frac{\partial N_{x\theta 0}}{\partial \theta}=0 \tag{10.50a}$$

$$\frac{\partial N_{x\theta 0}}{\partial x}+\frac{1}{a}\frac{\partial N_{\theta 0}}{\partial \theta}=0 \tag{10.50b}$$

$$D\nabla^4 w_0+\frac{1}{a}N_{\theta 0}=p \tag{10.50c}$$

where the linear membrane forces are given by

$$N_{x0}=\frac{Eh}{1-\nu^2}\left[\frac{\partial u_0}{\partial x}+\frac{\nu}{a}\left(\frac{\partial v_0}{\partial \theta}+w_0\right)\right]-N_T \tag{10.51a}$$

$$N_{\theta 0}=\frac{Eh}{1-\nu^2}\left[\frac{1}{a}\left(\frac{\partial v_0}{\partial \theta}+w_0\right)+\nu\frac{\partial u_0}{\partial x}\right]-N_T \tag{10.51b}$$

$$N_{x\theta 0}=\frac{Eh}{2(1+\nu)}\left[\frac{\partial v_0}{\partial x}+\frac{1}{a}\frac{\partial u_0}{\partial \theta}\right] \tag{10.51c}$$

The expressions describing the stability equations may be written in two forms. The first form is similar to Eqs. (10.50),

$$\frac{\partial N_{x1}}{\partial x}+\frac{1}{a}\frac{\partial N_{x\theta 1}}{\partial \theta}=0 \tag{10.52a}$$

$$\frac{\partial N_{x\theta 1}}{\partial x}+\frac{1}{a}\frac{\partial N_{\theta 1}}{\partial \theta}=0 \tag{10.52b}$$

$$D\nabla^4 w_1+\frac{1}{a}N_{\theta 1}-\left(N_{x0}\frac{\partial^2 w_1}{\partial x^2}+\frac{2}{a}N_{x\theta 0}\frac{\partial^2 w_1}{\partial x\partial \theta}+\frac{1}{a^2}N_{\theta 0}\frac{\partial^2 w_1}{\partial \theta^2}\right)=0 \tag{10.52c}$$

Donnell showed that this last set of equations can be expressed in terms of the displacements in an uncoupled form,

$$\nabla^4 u_1=-\frac{\nu}{a}\frac{\partial^3 w_1}{\partial x^3}+\frac{1}{a^3}\frac{\partial w_1}{\partial x\partial \theta^2} \tag{10.53a}$$

$$\nabla^4 v_1=-\frac{2+\nu}{a^2}\frac{\partial^3 w_1}{\partial x^2\partial \theta}-\frac{1}{a^4}\frac{\partial^3 w_1}{\partial \theta^3} \tag{10.53b}$$

$$D\nabla^8 w_1+\frac{Eh}{a^2}\frac{\partial^4 w_1}{\partial x^4}-\nabla^4\left(N_{x0}\frac{\partial^2 w_1}{\partial x^2}+\frac{2}{a}N_{x\theta 0}\frac{\partial^2 w_1}{\partial x\partial \theta}+\frac{1}{a^2}N_{\theta 0}\frac{\partial^2 w_1}{\partial \theta^2}\right)=0 \tag{10.53c}$$

where $\nabla^8 w_1 = \nabla^4(\nabla^4 w_1)$. The equations are uncoupled because the last equation can be solved independently for w_1, and then the first two equations can be solved for u_1 and v_1, respectively. Equations (10.53) are called the *Donnell stability equations in uncoupled form.*

To determine the critical bifurcation-buckling load for a cylindrical shell, we first determine the membrane forces in the prebuckled configuration from the solution to the linear equilibrium problem. With these membrane forces, Eq. (10.53c) provides a linear, homogeneous equation for the buckling displacement w_1. The solution to this equation yields the bifurcation-buckling loads. Brush and Almroth[25] illustrate the approach for mechanical loads and, in the next section, we illustrate it for thermal buckling.

10.4.3 Cylindrical Shell with Uniform Temperature

As an example of shell thermal buckling, we consider a cylindrical shell subjected to a uniform temperature change ΔT. The shell is supported at each end by immovable simple supports. With uniform temperature, the prebuckling deformation of the shell is axisymmetric as shown in Fig. 10.22a. We seek the critical temperature change ΔT_{cr} corresponding to bifurcation buckling. We proceed in two steps; we first determine the prebuckling membrane forces and displacements and then determine the critical buckling temperature.

Prebuckling solution. For a uniform temperature change, we may assume that deformations are radially symmetric. Thus, we may take the displacements to be independent of the angle θ and a function of x alone. In addition, because the supports are immovable, the axial displacement component u_0 is identically zero. Because of the radial symmetry, the circumferential displacement v_0 is also zero. Thus, the prebuckling displacement components are

$$u_0 = 0 \qquad v_0 = 0 \qquad w_0 = w_0(x) \tag{10.54}$$

With these displacement components, the prebuckling membrane forces given in Eqs. (10.51) reduce to

$$N_{x0} = \frac{Eh}{1-\nu^2}\left(\frac{\nu}{a}w_0\right) - N_T \tag{10.55a}$$

$$N_{\theta 0} = \frac{Eh}{1-\nu^2}\left(\frac{w_0}{a}\right) - N_T \tag{10.55b}$$

$$N_{x\theta 0} = 0$$

where, since the temperature change is uniform,

$$N_T = \frac{E\alpha}{1-\nu}\int_{(-h/2)}^{h/2} \Delta T \,\mathrm{d}z = \frac{E\alpha h}{1-\upsilon}\Delta T \tag{10.55c}$$

Consider now the prebuckling equilibrium equations (10.50). The first of these equations shows that

$$\frac{\partial N_{x0}}{\partial x} = 0$$

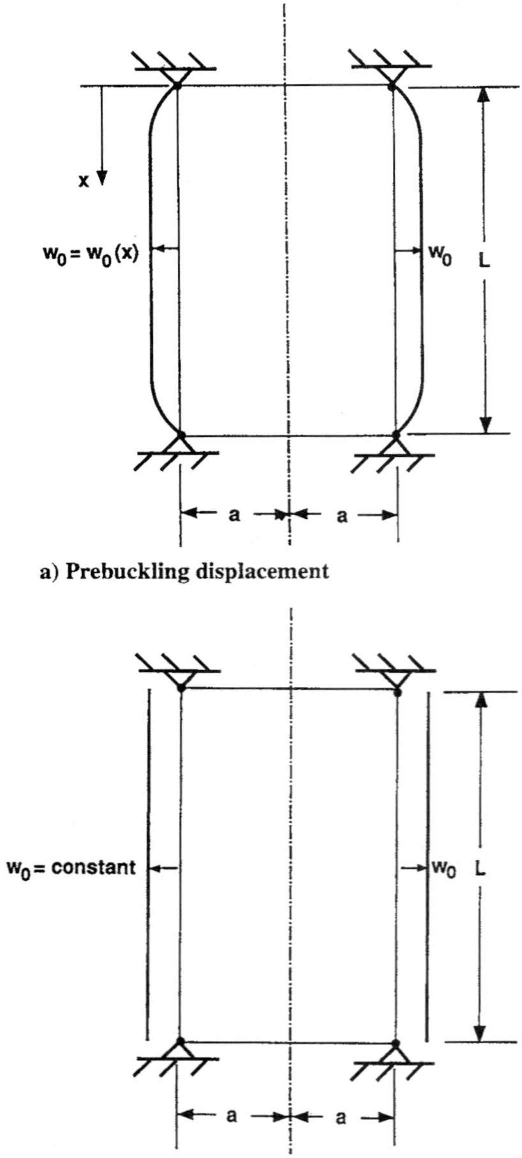

a) Prebuckling displacement

b) Assumed prebuckling displacement

Fig. 10.22 Restrained cylindrical shell.

Hence, the axial membrane force N_{x0} is constant. The second equilibrium equation is satisfied identically. The final equilibrium equation reduces to

$$D\frac{\mathrm{d}^4 w_0}{\mathrm{d}x^4} + \frac{N_{\theta 0}}{a} = 0 \tag{10.56}$$

and substituting for $N_{\theta 0}$ from Eq. (10.55b) yields

$$D\frac{\mathrm{d}^4 w_0}{\mathrm{d}x^4} + \frac{1}{a}\left[\frac{Eh}{1-\nu^2}\frac{w_0}{a} - \frac{E\alpha h \Delta T}{1-\nu}\right] = 0 \tag{10.57}$$

Solving the preceding equation provides the solution for the prebuckling radial displacement of the shell. The solution describes the deformation shown in Fig. 10.22a. Near the ends of the shell, the displacement reduces to zero at the supports, but some distance away from the ends, the displacement becomes nearly uniform. To obtain a relatively simple prebuckling solution for evaluating the critical buckling temperature, the uniform axisymmetric prebuckling displacement shown in Fig. 10.22b is assumed; that is, we take $w_0 = \text{const}$. With this assumption, Eqs. (10.56) and (10.57) show that

$$N_{\theta 0} = 0 \tag{10.58a}$$

and

$$w_0 = (1+\nu)\, a\alpha\Delta T \tag{10.58b}$$

With this result, we find the axial membrane force from Eqs. (10.55a) and (10.55c) to be given by

$$N_{x0} = -hE\alpha\Delta T \tag{10.58c}$$

In summary, the prebuckling solution consists of a uniform radial displacement w_0 and a uniform axial restraint force N_{x0}. All other displacements and membrane forces are zero.

Critical buckling temperature. We obtain the critical buckling temperature by solving the linear stability equation (10.53c) for the w_1 displacement. Substituting for the prebuckling forces, we obtain the stability equation

$$D\nabla^8 w_1 + \frac{Eh}{a^2}\frac{\partial^4 w_1}{\partial x^4} + hE\alpha\Delta T\nabla^4\left(\frac{\partial^2 w_1}{\partial x^2}\right) = 0 \tag{10.59}$$

For simply supported ends, the shell boundary conditions are similar to a beam or plate and require that

$$w_1(0) = w_1(L) = \frac{\partial^2 w_1}{\partial x^2}(0) = \frac{\partial^2 w_1}{\partial x^2}(L) = 0 \tag{10.60}$$

Unlike the prebuckling solution, the adjacent equilibrium position does not, in general, have an axisymmetric radial displacement. In fact, the shell may buckle into a mode shape with a significant number of circumferential waves. For this reason, we assume a solution to the stability equation (10.59) in the form

$$w_1(x,\theta) = c_1 \sin\frac{m\pi x}{L}\sin n\theta \tag{10.61}$$

where c_1 is a constant. The sine function of x on the right-hand side satisfies the simply supported boundary conditions specified in Eq. (10.60), and the sine function of θ ensures that the displacement is periodic in the circumferential direction. Because the stability equation (10.59) is homogeneous, we obtain a solution only when ΔT takes on critical buckling values. Substituting Eq. (10.59) into Eq. (10.61) produces the stability equation

$$\frac{D}{a^8}\left[\left(\frac{m\pi a}{L}\right)^2+n^2\right]^2+\frac{Eh}{a^6}\left(\frac{m\pi a}{L}\right)^4 - \frac{hE\alpha\Delta T_{cr}}{a^6}\left(\frac{m\pi a}{L}\right)^2\left[\left(\frac{m\pi a}{L}\right)^2+n^2\right]^2=0$$

Introducing $D = Eh^3/12(1-\nu^2)$ and the substitution

$$\bar{m} = m\pi a/L \tag{10.62a}$$

we solve for ΔT_{cr} to find

$$\Delta T_{cr} = \frac{1}{\alpha}\left[\frac{h^2}{12(1-\nu^2)a^2}\frac{(\bar{m}^2+n^2)^2}{\bar{m}^2}+\frac{\bar{m}^2}{(\bar{m}^2+\bar{n}^2)^2}\right] \tag{10.62b}$$

This last equation gives the critical uniform buckling temperature for an axially restrained, simply supported cylindrical shell. For each pair of values of m and n, there is a corresponding ΔT. The lowest value of ΔT is the critical buckling temperatureΔT_{cr}. It is important to note that the lowest value of ΔT does *not* occur for $m = n = 1$. Often, for example, the lowest value occurs for $m = 1$ and $n > 1$. The buckling shape then consists of one half-wave in the axial direction and several half-waves in the circumferential direction. We illustrate this behavior later in a numerical example.

Minimum buckling temperature. Brush and Almroth[25] show that the minimum value for ΔT may be estimated by analytically minimizing Eq. (10.62b). Let the continuous variable ϕ represent the quantity $(\bar{m}^2+n^2)^2/\bar{m}^2$; this is an approximation because m and n have only discrete integer values. With this substitution, the equation becomes

$$\Delta T_{cr} = \frac{1}{\alpha}\left[\frac{h^2}{12(1-\nu^2)a^2}\phi+\frac{1}{\phi}\right]$$

Differentiating with respect to ϕ, we find that the minimum value of ΔT occurs when

$$\phi_{min} = \sqrt{12(1-\nu^2)}(a/h)$$

Substituting this value into ΔT_{cr} produces the minimum or critical ΔT,

$$(\Delta T_{cr})_{min} = \frac{1}{\sqrt{3\left(1-\nu^2\right)}}\frac{h}{\alpha a} \tag{10.63a}$$

For $\nu = 0.3$,

$$(\Delta T_{cr})_{min} = 0.605(h/\alpha a) \tag{10.63b}$$

Note that the critical buckling temperature is independent of the length L of the shell.

Table 10.3 Cylindrical shell thermal-buckling modes

L, in.	m	n
1	1	260
2	1	130
5	1	52
10	1	26
20	1	13
40	2	13
50	5	26
60	3	13
100	5	13

Numerical example. As an example, we compute the critical buckling temperatures for an aluminum cylinder of radius $a = 20$ in., wall thickness $h = 0.2$ in., Poisson's ratio $\nu = 0.3$, and coefficient of thermal expansion $\alpha = 13 \times 10^{-6}$ 1/°F. The approximate value of the critical temperature change predicted by Eq. (10.63b) is

$$(\Delta T_{\mathrm{cr}})_{\mathrm{min}} = 465.6^{\circ}\mathrm{F}$$

Equation (10.62b) was also solved for ΔT_{cr} for various values of the shell length L; these results, with the corresponding values of m and n, are shown in Table 10.3. For the cases studied, the critical buckling temperatures predicted by Eq. (10.62b) were identical to four significant figures, with the results of Eq. (10.63) confirming that the buckling temperature is virtually independent of length. Table 10.3 shows, however, that the buckling mode changes with shell length. We should also note that, for a given length L, there are several combinations of m and n that yield the same eigenvalue. In other words, the buckling modes are not distinct. For example, for $L = 20$ in., the following pairs of m and n yield identical values for ΔT_{cr} in Eq. (10.62): $(m, n) = (1, 13), (2, 26), (3, 39)$, and so forth.

The numerical results demonstrate the earlier statement that the shell typically buckles with several half-waves in the circumferential direction.

10.5 Closing Remarks

In this chapter, we provide an introduction to thermal buckling for three basic structural components: beam columns, plates, and shells. There is a rich literature describing further developments in these and closely related problems. The author's survey paper[12] describes in more detail the literature for thermal buckling of plates and shells. This chapter limits discussion to isotropic metallic structures. In the years ahead, as advanced composites are developed for high temperature applications, we may expect that there will be significant analytical, computational, and experimental research for thermal buckling of composite structures. Indeed, in the late 1980s and early 1990s, thermal buckling analyses of plates and shells superseded research on metallic structures.

References

[1]Hoff, N. J., "Thin Shells in Aerospace Structures," *Astronautics and Aeronautics*, Vol. 5, No. 2, 1967, pp. 26–45.

[2]Hoff, N. J., "A Short History of the Development of Airplane Structures," *American Scientist*, Vol. 34, No. 2, 1946, pp. 212–225; No. 3, 1946, pp. 370–388.

[3]Thornton, E. A., "Thermal Structures: Four Decades of Progress," *Journal of Aircraft*, Vol. 29, No. 3, 1992, pp. 485–498.

[4]Poleski, S. P., Dechaumphai, P., Glass, C. E., and Pandey, A. K., "Three-Dimensional Thermal-Structural Analysis of a Swept Cowl Leading Edge Subjected to Skewed Shock-Shock Interference Heating," *Journal of Thermophysics and Heat Transfer*, Vol. 6, No. 1, 1992, pp. 48–54.

[5]van der Neut, A., "Buckling Caused by Thermal Stresses," *High Temperature Effects in Aircraft Structures*, edited by N. J. Hoff, Pergamon, New York, 1958, Chap. 11, pp. 215–266.

[6]Boley, B. A., and Weiner, J. H., *Theory of Thermal Stresses*, Wiley, New York, 1960.

[7]Johns, D. J., *Thermal Stress Analyses*, Pergamon, Oxford, England, UK, 1965.

[8]Parkus, H., *Thermoelasticity*, Blaisdell, Waltham, MA, 1968.

[9]Nowinski, J. L., *Theory of Thermoelasticity with Applications*, Sijhoff and Noordhoff, Alphen aan den Rijn, The Netherlands, 1978.

[10]Reynolds, J., "The Analysis of the Deployed Space Telescope Solar Array," British Aerospace, ESA Document TN-SA-B142, Bristol, England, UK, Jan. 1983.

[11]*STEM Design Characteristics and Parameters*, Astro Aerospace Corporation, TR AAC- B-006, Carpinteria, CA, Sept. 1985.

[12]Thornton, E. A., "Thermal Buckling of Plates and Shells," *Applied Mechanics Reviews*, Vol. 46, No. 10, 1993, pp. 485–506.

[13]Heldenfels R. R., and Roberts, W. M., "Experimental and Theoretical Determination of Thermal Stresses in a Flat Plate," NACA TN-2769, 1952.

[14]Gossard, M. L., Seide, P., and Roberts, W. M., "Thermal Buckling of Plates," NACA TN-2771, 1952.

[15]Klosner, J. M., and Forray, M. J., "Buckling of Simply Supported Plates under Arbitrary Symmetrical Temperature Distributions," *Journal of the Aeronautical Sciences*, March 1958, pp. 181–184.

[16]Miura, K., "Thermal Buckling of Rectangular Plates," *Journal of Aerospace Science*, Vol. 28, 1961, pp. 341–342; see also "Thermal Buckling of Rectangular Plates," Aeronautical Research Inst., Univ. of Tokyo, Rept. 53, Vol. 26, No. 4, May 1960, pp. 51–75.

[17]Wilcox, M. W., and Clemmer, L. E., "Large Deflection Analysis of Heated Plates," *Journal of Engineering, Mechanical Division*, ASCE, Dec. 1964, pp. 165–189.

[18]Thornton, E. A., Coyle, M. F., and McLeod, R. N., "Experimental Study of Plate Buckling Induced by Spatial Temperature Gradients," *Journal of Thermal Stresses*, Vol. 17, No. 2, 1994, pp. 191–212.

[19]Hernan, P. R., "Finite Element Analysis of Plate Thermal Buckling Tests—Comparison of Analyses and Experiments," M. S. Thesis, Univ. of Virginia, Charlottesville, May 1994.

[20]Flügge, W., *Stresses in Shells*, Springer–Verlag, Berlin, 1960.

[21]Timoshenko, S. P., and Gere, J. M., *Theory of Elastic Stability*, McGraw–Hill, New York, 1961.

[22]Gerard, G., *Introduction to Structural Stability Theory*, McGraw–Hill, New York, 1962.

[23]Kraus, H., *Thin Elastic Shells*, Wiley, New York, 1967.

[24]Croll, J. G., and Walker, A. C., *Elements of Structural Stability*, Wiley, New York, 1972.

[25]Brush, D. O., and Almroth, B. O., *Buckling of Bars, Plates and Shells*, McGraw–Hill, New York, 1975.

[26]Simitses, G. J., *An Introduction to the Elastic Stability of Structures*, Prentice–Hall, Englewood Cliffs, NJ, 1976.

[27]Elishakoff, I., Arbocz, J., Babcock, C. D., Jr., and Libai, A., *Buckling of Structures: Theory and Experiment*, Studies in Applied Mechanics 19, Elsevier, New York, 1988.

[28]Vinson, J. R., *The Behavior of Thin-Walled Structures: Beams, Plates, and Shells*, Klüwer Academic, Norwell, MA, 1989.

[29]Palazotto, A. N., and Dennis, S. T., *Nonlinear Analysis of Shell Structures*, AIAA Education Series, edited by J. S. Przemieniecki, AIAA, Washington, DC, 1992.

[30]Hoff, N. J., "Buckling at High Temperature," *Journal of the Royal Aeronautical Society*, Vol. 61, Nov. 1957, pp. 756–774.

[31]Anderson, M. S., and Card, M. F., "Buckling of Ring-Stiffened Cylinders under a Pure Bending Moment and a Nonuniform Temperature Distribution," NASA TN D-1513, Nov. 1962.

[32]Bushnell, D., *Computerized Buckling Analysis of Shells*, Klüwer Academic, Norwell, MA, 1989, pp. 168–179.

[33]Gellin, S., and Lee, G. C., "Finite Elements Available for the Analysis of Non-Curved, Thin-Walled Structures," *Finite Element Analysis of Thin-Walled Structures*, edited by S. W. Bull, Elsevier Applied Science, London, 1988, pp. 1–46.

Problems

10.1. Consider the simply supported, *unrestrained* beam column shown in Fig. P10.1. The beam has an immovable simple support at $x = 0$, but the simple support at $x = L$ is free to displace in the x direction. The beam supports an axial compressive force P and has a prescribed temperature distribution such that the thermal force $N_T = 0$ and the thermal moment $M_T = a_0 + a_1 x$, where a_0 and a_1 are known constants.

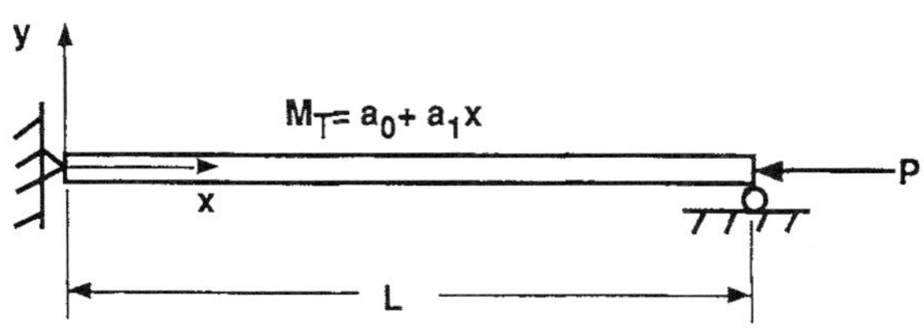

Fig. P10.1

(a) Determine the transverse deflection $v = v(x)$ in terms of P, a_0, a_1, and other beam parameters.

(b) Determine the critical buckling load P_{cr} for the beam column.

(c) Determine the relationship between the axial displacement $u_0(L)$ at the right-hand support and the force P.

(d) Discuss, using sketches, the behavior of $v(L/2)$ and $u_0(L)$ for increasing values of the compressive force P.

10.2. The behavior of a beam column may be formulated in a more general context using the nonlinear large-deflection equations for a beam. If we assume bending in the x–y plane, Eqs. (6.27) and (6.28) have the form

$$EA\frac{\partial^2 u}{\partial x^2} + \frac{1}{2}EA\frac{\partial}{\partial x}\left[\left(\frac{\partial v}{\partial x}\right)^2\right] = \frac{\partial N_T}{\partial x}$$

$$EI\frac{\partial^4 v}{\partial x^4} - \left[EA\frac{\partial u}{\partial x} + \frac{1}{2}EA\left(\frac{\partial v}{\partial x}\right)^2 - N_T\right]\frac{\partial^2 v}{\partial x^2} = -\frac{\partial^2 M_T}{\partial x^2}$$

where u, v are displacements of points on the beam's centroidal axis, and the thermal force N_T and the thermal moment M_T may vary with x. The beam's stress resultants are given by

$$N_x = EA\frac{\partial u}{\partial x} + \frac{1}{2}EA\left(\frac{\partial v}{\partial x}\right)^2 - N_T$$

$$M = -EI\frac{\partial^2 v}{\partial x^2} - M_T$$

$$V = -EI\frac{\partial^3 v}{\partial x^3} - \frac{\partial M_T}{\partial x}$$

The objective of this problem is to develop a finite element formulation for the large-deflection beam response that we may apply to beam columns. The primary variables in the formulation at nodes along the beam's axis are

$$u,\ v,\ \theta = \frac{\partial v}{\partial x}$$

where θ denotes the beam's slope. From Sec. 8.3.1, Eq. (8.32), the interpolation equation for the axial displacement $u(x)$ is

$$u(x) = N_1(x)u_1 + N_2(x)u_2$$

where $N_1(x) = 1 - x/L$, $N_2(x) = x/L$, and L is the element length. From Sec. 8.3.2, Eq. (8.49), the interpolation equation for the beam's bending displacement $v(x)$ is

$$v(x) = N_3(x)v_1 + N_4(x)\theta_1 + N_5(x)v_2 + N_6(x)\theta_2$$

where

$$N_3(x) = 1 - \frac{3x^2}{L^2} + \frac{2x^3}{L^3}$$

$$N_4(x) = x - \frac{2x^2}{L} + \frac{x^3}{L^2}$$

$$N_5(x) = \frac{3x^2}{L^2} - \frac{2x^3}{L^3}$$

$$N_6(x) = -\frac{x^2}{L} + \frac{x^3}{L^2}$$

are standard beam interpolation functions.

(a) Using the method of weighted residuals (MWR) described in Chapter 8, develop a nonlinear finite element formulation for the problem. Write an element's equilibrium equations in the form

$$\begin{bmatrix} [K_{11}] & [K_{12}] \\ [K_{21}] & [K_{22}] \end{bmatrix} \begin{Bmatrix} \{u\} \\ \{v\} \end{Bmatrix} = \begin{Bmatrix} \{F_1\} \\ \{F_2\} \end{Bmatrix}$$

where $\{u\}^T = [u_1 \, u_2]$ and $\{v\}^T = [v_1 \, \theta_1 \, v_2 \, \theta_2]$, and develop equations for an element's submatrices in terms of integrals of the interpolation functions.

(b) Evaluate all element integrals for the given interpolation functions. Assume that the thermal force and moment are constant within an element but that they may vary from element to element.

(c) Develop a Newton–Raphson iteration procedure for solving the nonlinear beam deflection equations.

10.3. Using the formulation developed in the preceding problem, write a finite element program for large-deflection beam problems, assuming that distributions of the thermal force and thermal moment are prescribed.

(a) Validate the computer program by solving the axially restrained beam with the thermal-bending problem presented in Sec. 10.2.3. Using the data for the restrained stainless steel plate in cylindrical bending, compare the finite element predictions with the analytical results presented in Fig. 10.3.

(b) Use the finite element computer program to investigate the response of a cantilever beam with a prescribed axial force as shown in Fig. P10.3. Assume that the beam has the properties of the spacecraft boom given in Table 10.1. The beam experiences steady heating on its top surface such that the beam temperature is uniform along its length and varies linearly through the beam depth. The temperature change at the beam centerline is 120 K, and a temperature gradient of 20 K develops through the beam depth. Determine the variation of the displacement components u and v at the end of the beam with increasing P. Plot load-deflection curves, and discuss the results.

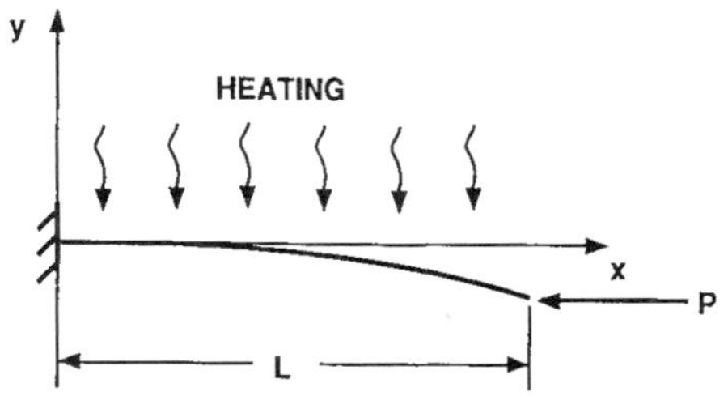

Fig. P10.3

10.4. A simply supported beam is restrained against thermal expansion by immovable supports (Fig. 10.1a). The beam is made of a thin-walled slit tube of radius R and wall thickness h. As part of Example 6.5, the properties of a slit tube's cross section are evaluated. Since the tube cross section is slit, the shear center does not coincide with the centroid; in fact, it is located outside the tube. Chapter 6 shows for such cross sections that bending and torsional deformations are coupled. If the restrained tube experiences a uniform temperature change ΔT, the tube may experience bending *and* torsional buckling.

(a) For this problem, show that the general bending and torsional equilibrium equations (6.70) from Chapter 6 reduce to

$$EI_z \frac{d^4 v}{dx^4} + P\frac{d^4 v}{dx^2} + Pe_z \frac{d^2\theta_x}{dx^2} = 0$$

$$EI_y \frac{d^4 v}{dx^4} + P\frac{d^2 w}{dx^2} = 0$$

$$E\Gamma \frac{d^4\theta_x}{dx^4} + \left(\frac{PI_E}{A} - GJ\right)\frac{d^2\theta_x}{dx^2} + Pe_z\frac{d^2 v}{dx^2} = 0$$

where $P = E\alpha A\Delta T$ is the axial restraint force. Define all terms, and evaluate the cross section's properties in terms of R and h.

(b) Write the six boundary conditions needed to complete the problem formulation. For torsion, assume that cross sections at each support cannot twist but are free to warp.

(c) By assuming deformations of the form,

$$v = A \sin m\pi x/L \qquad w = B \sin m\pi x/L \qquad \theta_x = C \sin m\pi x/L$$

determine the characteristic equation(s) that determines the critical buckling forces P_{cr}.

(d) Determine the critical buckling temperatures ΔT_{cr} for the lowest three buckling modes for a stainless steel slit tube with the following data:

$$L = 5.91 \text{ m} \qquad E = 1.93 \times 10^{12} \text{ N/m}^2$$

$$R = 1.09 \times 10^{-2} \text{ m} \qquad \alpha = 1.69 \times 10^{-5}\ 1/\text{m}$$

$$h = 2.35 \times 10^{-4} \text{ m} \qquad \nu = 0.33$$

(e) Sketch the mode shapes corresponding to the critical buckling temperatures in part d.

10.5. Read the chapter by Gellin and Lee, Ref. 33.

10.6. Develop a finite element approach for thermal buckling of a restrained thin-wall beam. The beam's axial force is prescribed as $N_x = -P = E\alpha A\Delta T$. The three coupled equations (6.70) are to be solved for the beam displacements v, w, and angle of twist θ_x. The beam's thermal stress σ_{xT} and thermal moments are assumed to be known in terms of ΔT.

(a) Using the method of weighted residuals, develop a finite element formulation for the problem. Select appropriate nodal unknowns and interpolation functions. Derive equations for the submatrices in an element's stiffness matrix and load vector. Give these equations in terms of integrals of the interpolation functions.

(b) Describe an approach for computing the thermal-buckling response for a restrained thin-walled beam. Discuss the roles of σ_{xT} and the thermal moments.

10.7. Consider a rectangular plate whose edges are simply supported for bending and are fully restrained against in-plane displacement. The plate experiences a temperature distribution $T = T(z)$ that is symmetrical about the middle surface. Derive an equation for the critical thermal-buckling force $(N_T)_{cr}$.

10.8. Consider an aluminum-alloy plate 0.25 in. thick with dimensions $a = 30$ in. and $b = 20$ in. The plate experiences a symmetrical temperature distribution through its thickness. The temperature change ΔT is given by

$$\Delta T(z) = \Delta T_m \left(1 - \frac{2z}{h}\right) \qquad 0 \leq z \leq h/2$$
$$= \Delta T_m \left(1 + \frac{2z}{h}\right) \qquad -h/2 \leq z \leq 0$$

Assuming that all plate edges are simply supported and are fully restrained against in-plane displacement,

(a) Determine the critical temperature change $(\Delta T_m)_{cr}$ for thermal buckling. Take $E = 10^7 \text{lb/in.}^2$ and $\alpha = 13 \times 10^{-7}$ 1/°F.

(b) Determine the plate stress at the critical buckling stress. Is the stress within the elastic range? Take $\sigma_y = 30 \times 10^3$ lb/in.^{2}

10.9. We wish to consider buckling of a cylindrical shell under the simultaneous loading of uniform lateral external pressure p and a uniform temperature change ΔT. The shell is supported at each end by immovable simple supports. In Ref. 25, pp. 161–166, Brush and Almroth present the solution for the uniform lateral pressure loading and, following the same approach, Sec. 10.4.3 presents the solution for uniform temperature.

(a) Derive the critical buckling equation for a shell under simultaneous pressure and temperature loading.

(b) For a cylindrical shell with radius $a = 20$ in., thickness $h = 0.2$, length $L = 20$ in. and Poisson's ratio $\nu = 0.3$, solve the critical buckling equation for various combinations of p and ΔT for the lowest buckling mode. Plot the interaction curve of p vs ΔT for the shell.

This page intentionally left blank

11
Thermoviscoplasticity

11.1 Introduction

In the 1970s and 1980s, modeling of inelastic material behavior at elevated temperatures was motivated by research in hypersonic flight and advanced engine structures. The commitment to develop the National Aerospace Plane generated new interest in the technology required to design structures for hypersonic flight. Such structures are exposed to aerodynamic heating of unprecedented magnitudes. As a vehicle accelerates or decelerates at hypersonic speeds in the atmosphere, shocks sweep across it and interact with local shocks and boundary layers (Fig. 11.1). These interactions introduce severe local pressures and heating rates. As the design requirements of gas turbine engines push operating temperatures to new limits, accurate modeling of thermoviscoplastic behavior of advanced alloys becomes increasingly important. Rate-dependent plasticity and creep are major concerns since engine components sustain extreme loads at elevated temperatures. Because of intense local heating and pressures, leading edges of engine structures present a significant design problem. Preliminary structural analyses[1,2] of an impingement-cooled leading edge at NASA Langley Research Center showed high local plasticity that seriously degraded the structure's load-carrying capacity at elevated temperatures. At NASA Lewis Research Center, a thermostructural analysis with experimental verification[3] of cowl lip designs confirmed that inelastic effects occur and can be significant.

In this chapter, we provide an introduction to thermoviscoplasticity. The subject is developing rapidly, and there are significant current research programs, particularly in modeling behavior of metal-matrix composites. Research literature on constitutive models can be highly specialized, requiring knowledge of advanced concepts from continuum mechanics and materials science.

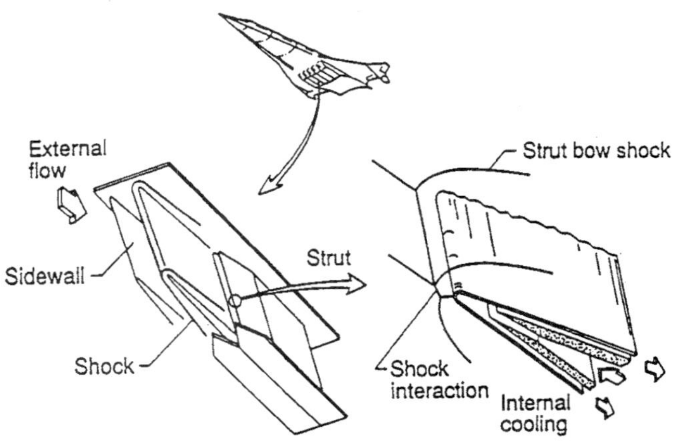

Fig. 11.1 Shock/shock interaction on the National Aerospace Plane scramjet engine leading edge.

We approach the subject from a more practical viewpoint. The chapter begins with a brief description of unified constitutive models. One constitutive model is described in some detail, and highlights of the experimental techniques used to determine the model's parameters are presented. Then a finite element formulation for thermoviscoplastic analysis is described, and a viscoplastic solution method is detailed. The chapter concludes with a description of computational studies of thermoviscoplastic behavior for thermal-structures applications.

11.2 Unified Constitutive Models

Unified viscoplastic constitutive models have evolved over the last 20 years to provide a means for analytically representing a material's response from the elastic through the plastic range, including strain-rate–dependent plastic flow, creep, and stress relaxation. The theories are guided by physical considerations, including dislocation dynamics, and are based on the principles of continuum mechanics. The first multidimensional formulations of elastic–viscoplastic constitutive equations are due to Bodner and Partom. Since their early work, a number of constitutive models have appeared; many of these theories are summarized in review articles that appear in Ref. 4. NASA Lewis Research Center sponsored a four-year research effort[5,6] by the Southwest Research Institute to develop unified constitutive models for isotropic materials and to demonstrate their usefulness for analysis of high-temperature gas turbine engines. One result of this study is material property data for high-temperature nickel-based alloys over a wide temperature range. One of the most widely used constitutive models is the Bodner–Partom model; we present some details of this model in the following section.

11.2.1 Bodner–Partom Model

Unified elastic-viscoplastic constitutive theories contain time- and temperature-dependent equations that describe elastic and inelastic behavior of materials subjected to thermal and mechanical loads. In these theories, mechanisms of inelastic deformation are represented with a kinetic equation that defines $\dot{\varepsilon}^p$, the inelastic strain rate.[7,8] The three basic elements that constitute the Bodner–Partom unified model are 1) a flow law relating inelastic strain rate to deviatoric stresses, 2) a kinetic equation that relates the inelastic strain rate to stress invariants through temperature and internal state variables, and 3) evolution equations defining the growth of internal state variables. The internal state variables represent resistance to inelastic deformation. The evolution equations for internal state variables follow the form of the Orowan–Bailey theory[7] incorporating both deformation and recovery terms. Bodner's anisotropic unified model contains evolutionary state variables for both isotropic and kinematic hardening.

In 1968, S. R. Bodner[9] published a unified elastic-viscoplastic constitutive theory based on continuum mechanics and guided by phenomenological considerations. This first representation of the Bodner–Partom unified constitutive model includes a flow law and a kinetic equation that together describe elastic and inelastic deformations. In 1975,[10] the model was reformulated to include evolutionary state variables to describe resistance to plastic flow. This strain-hardening law corresponds to isotropic hardening. The model was later expanded to include directional, or kinematic, hardening, with the introduction of the anisotropic form of the flow law.[11] Kinematic hardening is important for describing cyclic behavior. Inclusion of this hardening variable also improves the accuracy of the model for monotonic loading. The Bodner–Partom unified constitutive law assumes small deformations, less than 5% strain.[11]

The Bodner–Partom formulation begins by decomposing the total strain into elastic and inelastic components,

$$\varepsilon_{ij} = \varepsilon_{ij}^{e} + \varepsilon_{ij}^{p} \tag{11.1}$$

where ε_{ij} represents engineering strain, ε_{ij}^{e} are the elastic strain components, and ε_{ij}^{p} represents inelastic strain components. By taking the time derivative of Eq. (11.1), we can express the Bodner–Partom formulation in rate form as

$$\dot{\varepsilon}_{ij} = \dot{\varepsilon}_{ij}^{e} + \dot{\varepsilon}_{ij}^{p} \tag{11.2}$$

The elastic strain-rate components are provided by the time derivative of generalized Hooke's law (for isothermal conditions),

$$\dot{\varepsilon}_{ij}^{e} = \frac{1}{2\bar{\mu}}\dot{\sigma}_{ij} - \frac{\bar{\lambda}}{2\bar{\mu}(3\bar{\lambda} + 2\bar{\mu})}\dot{\sigma}_{kk}\delta_{ij} \tag{11.3}$$

where $\bar{\lambda}$ and $\bar{\mu}$ are the Lamé constants of the material. The Lamé constants for an elastic material are defined in Eq. (3.24). The inelastic strain-rate components $\dot{\varepsilon}_{ij}^{p}$ are represented by the isotropic form of the Prandtl–Reuss flow law,

$$\dot{\varepsilon}_{ij}^{p} = \lambda S_{ij} \tag{11.4}$$

where λ is a scalar representation of kinematic hardening defined by the equation

$$\lambda^2 = D_2^p / J_2 \tag{11.5}$$

and S_{ij}, the deviatoric stresses, are given by Eq. (3.5) as

$$S_{ij} = \sigma_{ij} - \tfrac{1}{3}\sigma_{kk}\delta_{ij} \tag{11.6}$$

Bodner relates the second invariant of plastic strain rate, D_2^p, to the invariant of the deviatoric stress, J_2, through an exponential function. The J_2 invariant is defined in Eq. (3.7). The kinetic equation is

$$D_2^p = D_0^2 \exp\left[-\left(\frac{Z^2}{3J_2}\right)^n\right] \tag{11.7}$$

where D_0 and n are material parameters, and Z is the internal state variable that represents resistance to plastic flow. The anisotropic form of Bodner's model[10] divides this internal state variable into components that correspond to isotropic and kinematic hardening,

$$Z = Z^I + Z^D \tag{11.8}$$

The internal state variables that correspond to hardening effects are described by a set of evolutionary equations. The Orowan–Bailey hardening/recovery format is used in formulating evolution equations for both isotropic and kinematic hardening variables. The isotropic hardening parameter is provided by

$$\dot{Z}^I(t) = m_1[Z_1 - Z^I(t)]\dot{W}_p(t) - A_1 Z_1 \left[\frac{Z^I(t) - Z_2}{Z_1}\right]^{r_1} \tag{11.9}$$

where m_1, Z_1, Z_2, A_1, and r_1 are material parameters, and $\dot{W}_p$ is the plastic work rate defined by

$$\dot{W}_p = \sigma_{ij}\dot{\varepsilon}_{ij}^p \tag{11.10}$$

The initial condition on $Z^I(t)$ is

$$Z^I(0) = Z_0 \tag{11.11}$$

where Z_0 is an experimentally determined parameter. Similarly, the kinematic, or directional, hardening variable is provided by the equation

$$Z^D(t) = \beta_{ij}(t)u_{ij}(t) \tag{11.12}$$

where $\beta_{ij}(t)$ is provided by the evolutionary equation

$$\dot{\beta}_{ij}(t) = m_2[Z_3\mu_{ij}(t) - \beta_{ij}(t)]\dot{W}_p(t) - A_2Z_1\left\{\frac{[\beta_{kl}(t)\beta_{kl}(t)]^{1/2}}{Z_1}\right\}^{r_2}\nu_{ij}(t) \tag{11.13}$$

where μ_{ij} is defined by

$$\mu_{ij}(t) = \frac{\sigma_{ij}(t)}{[\sigma_{kl}(t)\sigma_{kl}(t)]^{1/2}} \tag{11.14}$$

and ν_{ij} is

$$\nu_{ij}(t) = \frac{\beta_{ij}(t)}{[\beta_{kl}(t)\beta_{kl}(t)]^{1/2}} \tag{11.15}$$

In Eq. (11.12), u_{ij} are the direction cosines of the current stress state,

$$u_{ij} = \frac{\sigma_{ij}(t)}{[\sigma_{kl}(t)\sigma_{kl}(t)]^{1/2}} \tag{11.16}$$

The initial condition for the evolutionary hardening variable $\dot{\beta}_{ij}$ is

$$\beta_{ij}(0) = 0 \tag{11.17}$$

The material parameters m_2, Z_3, A_2, and r_2 found in Eq. (11.13) are similar to those found in Eq. (11.9). For the uniaxial case, the expressions for μ_{ij} and ν_{ij} reduce to 1, and the evolutionary equation for kinematic hardening reduces to

$$\dot{Z}^D(t) = m_2[Z_3 - Z^D(t)]\dot{W}_p - A_2Z_1\left|\frac{Z^D(t)}{Z_1}\right|^{r_2} \tag{11.18}$$

The initial condition for the kinematic hardening is

$$Z^D(0) = 0 \tag{11.19}$$

There are 11 parameters in the Bodner–Partom formulation. Among them are both temperature-dependent and temperature-independent parameters. All parameters are independent of strain rate. Each parameter has a physical or phenomenological significance.

Two of the material parameters, D_0 and n, are introduced in the kinetic equation (11.7). D_0, the limiting shear-strain rate, is an upper bound for D_2^p and is governed by principles of dislocation theory. For strain rates less than 10 s^{-1}, D_0 has been taken[12,13] to have a value of 10^4 1/s. The parameter n is a rate sensitivity parameter that is temperature-dependent; n influences the sharpness of yielding and the steepness of the stress-strain curve in the inelastic region. It also controls the overall level of the flow stress.

The remaining material parameters are found in the evolution equations that describe hardening, Eqs. (11.9) and (11.12). The parameters m_1 and m_2 are measures of the hardening rates. Both parameters are independent of temperature.

The evolution equations contain four variables that govern the upper and lower bounds of the internal state variables. These material parameters are Z_0, Z_1, Z_2, and Z_3. The initial value of isotropic hardening, Z_0 represents the internal state of the material at the onset of hardening and is phenomenologically related to yield stress. Z_1 is the limiting, or saturation, value of isotropic hardening and is independent of temperature. The evolutionary variable Z^I approaches this value asymptotically for large strains. Z_2 is the fully recovered or minimum value of isotropic hardening and is often taken to be equal to Z_0, the initial value of Z_1. Both Z_0 and Z_2 are temperature-dependent parameters and tend to decrease as temperature is increased. The limiting value, or upper bound, on kinematic hardening is provided by the parameter Z_3. This is a temperature-independent parameter.

The remaining parameters are found in the recovery terms of the evolution equations. These temperature-sensitive parameters tend to be small at relatively low temperatures but increase with increasing temperature. A_1 and A_2 are recovery coefficients and are often equated to each other and written as the single term A. Similarly, both recovery exponents, r_1 and r_2, that control growth of thermal recovery in the evolution equations are assumed equal and are written as r.

Parameter data for several materials have been determined and are available in the literature. However, the material parameters in the Bodner–Partom constitutive model must be determined from experimental data for new materials of interest. Tensile or compression tests performed over a range of temperatures and strain rates provide the data from which hardening parameters are derived. The thermal recovery terms in the evolution equations are often associated with creep behavior. Based on this phenomenological consideration, creep tests are used as the primary source from which the parameters A and r are determined.

11.2.2 Experimental Determination of Model Parameters

Chan et al.[12–14] developed a procedure through which isothermal, monotonic uniaxial tension tests and isothermal uniaxial creep tests are the only tests required for developing the model parameters. As will be illustrated later, creep tests are required only for developing certain model parameters that describe elevated temperature effects. These tests are not required if creep behavior of the material is negligible in the temperature range of interest. This method has been validated for several materials, including the super alloys B1900-Hf and MAR-M247.[13]

An interactive scheme involving linear and polynomial fits, interpretation of the physical meaning of the parameters, and calculations based on plasticity theory provides an effective method for obtaining the model parameters. To illustrate the procedure, we describe briefly the determination of the constitutive model parameters for Hastelloy-X, a nickel-based super alloy. This alloy has moderately good strength characteristics up to 1600°F (870°C) and oxidation resistance up

Table 11.1 Tensile test schedule for Hastelloy-X

	Temperature			
Strain rate	77°F, 300 K	400°F, 480 K	700°F, 640 K	1000°F, 800 K
2.63E-3 1/s	2 tests	2	2	2
2.63E-4 1/s	3	2	3	2
2.67E-5 1/s	0	1	0	1

to 2200°F (1200°C). Details of the test program and parameter determination procedure are given in Ref. 15.

Test program. Viscoplastic materials exhibit sensitivity to both temperature and strain-rate effects. To capture these effects, a series of tests were conducted on Hastelloy-X over a range of temperatures and strain rates. A minimum of three temperatures and three strain rates are required[9,10] to provide sufficient data for obtaining the model parameters. Strain rates are typically separated by an order of magnitude to distinguish hardening-rate effects.

Each tensile test was performed at a constant temperature and strain rate to separate the effects of these two test conditions. This was done to ensure that the full range of the two conditions was investigated independently. Strain measurements for all tests were accomplished with a high-temperature extensometer. Data were sampled from zero to 5% engineering strain.

The test schedule for Hastelloy-X is presented in Table 11.1. Tension tests were conducted at equal intervals over the temperature range of interest. These test temperatures ranged from 77 to 1000°F. This temperature range was chosen based on the operational limitations of the available extensometers. Three decades of strain rates at each temperature were available, allowing 12 independent test conditions. Multiple tests were conducted at each condition to improve results. Because creep behavior is not appreciable in the temperature range of interest for Hastelloy-X, creep tests were not performed. The recovery terms in the evolution equations would be expected to have little or no effect on the model, and the recovery coefficients A_1 and A_2 were set to zero.

Parameter determination. An outline of the general procedure for Bodner–Partom parameter determination is given in Fig. 11.2. In the parameter determination, published data for the temperature-dependent modulus of elasticity were used. From the tensile test data, the procedure of Chan et al. determines the temperature-independent hardening parameters m_1, m_2, Z_1, and Z_3, the temperature-dependent hardening parameters Z_0 and Z_2, and the n parameter. In general, data from creep tests are used to evaluate the A and r parameters but, as mentioned previously, in the Hastelloy-X tests these data were not taken.

Results. The model parameters for Hastelloy-X are presented in Tables 11.2. These parameters were interpreted from the stress–strain behavior of the experimental tensile data. Eulerian time marching was used to integrate numerically the Bodner–Partom constitutive model to generate data for a series of monotonic, isothermal uniaxial tests. Plots comparing the model to experimental data were generated showing temperature effects; a typical result is shown in Fig. 11.3. The magnitude of the stress decreases with increasing temperature. The model's curves

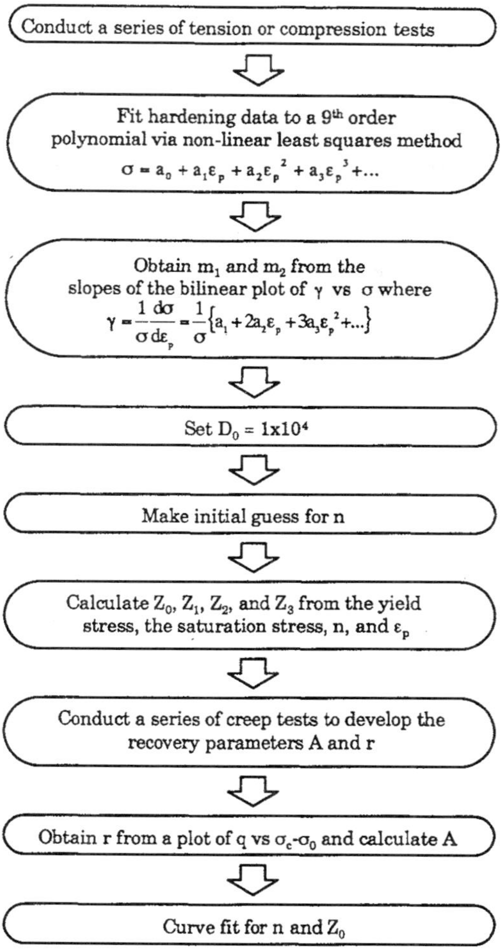

Fig. 11.2 Outline of the parameter determination procedure (Ref. 15).

exhibit the basic trends of the experimental data well and offer good predictive capabilities. The model's curves are more accurate at the middle of the temperature range and agree most closely with the experimental data at the middle of the three strain rates (not shown). This agreement is a direct result of the averaging scheme used in determining the model parameters.

Table 11.2a Material parameters for Hastelloy-X: temperature-independent

Parameter	Value
D_0, s^{-1}	10,000
Z_1, Mpa	2390
Z_3, Mpa	603
m_1, Mpa^{-1}	0.139
m_2, Mpa^{-1}	3.49

Table 11.2b Material parameters for Hastelloy-X: temperature-dependent

Parameter	Temperature 77°F, 300 K	400°F, 480 K	700°F, 640 K	1000°F, 800 K
E, Gpa	196.6	186.9	174.5	161.4
n	1.000	0.901	0.847	0.824
Z_0, Mpa	1860	1830	1790	1760
Z_2, Mpa	1860	1830	1790	1760
A, s^{-1}	0.0	0.0	0.0	0.0
r	2	2	2	2

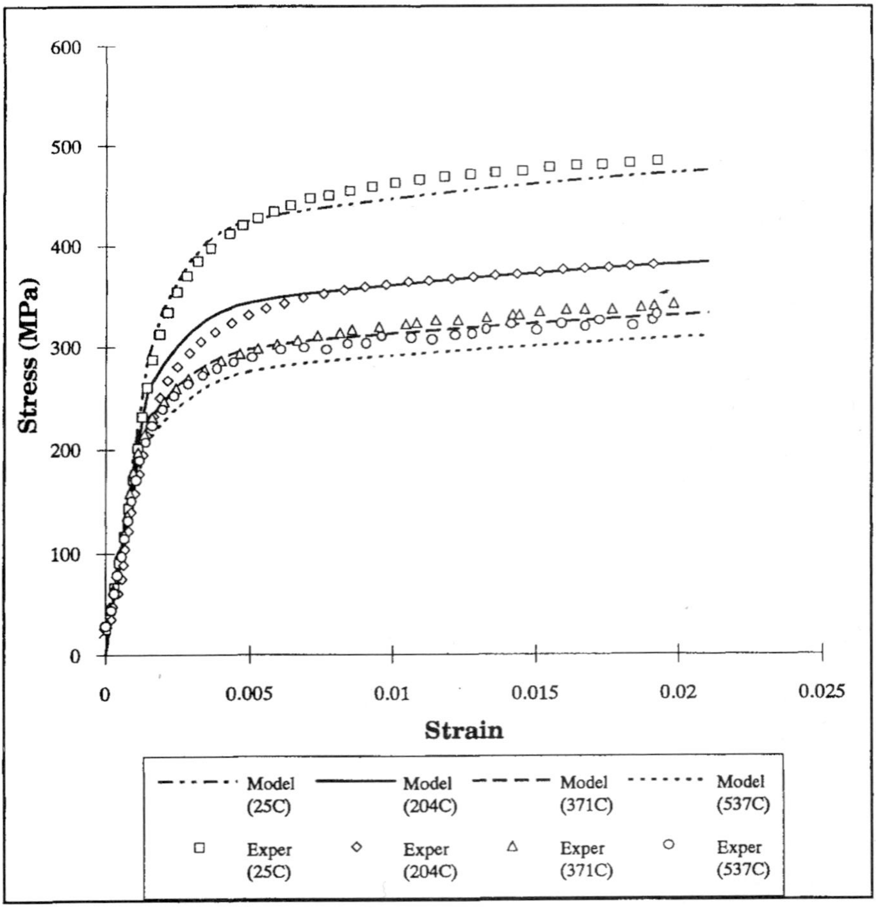

Fig. 11.3 Comparison plot for Hastelloy-X at four temperatures for $\dot{\varepsilon}$ = 2.63E-4 1/s (Ref. 15).

Temperature-dependent parameters follow expected theoretical trends. The kinetic parameter n (Table 11.2b) decreases with increasing temperature. This behavior of the kinetic parameter corresponds to an increase in the rate sensitivity of the material at elevated temperatures. Similarly, the parameters Z_0 and Z_2 decrease with increasing temperatures. The parameters Z_0 and Z_2 are directly related to the yield stress, and the decrease in the value of these parameters corresponds to the decrease in yield stress at elevated temperatures.

Assessment. Analysts may conclude from these results that the data for the Bodner–Partom constitutive model provide good representation for the Hastelloy-X stress–strain behavior, but the accuracy may vary with temperature and strain rate. A typical result illustrating this point is shown in Fig. 11.3, where the effect of temperature is demonstrated. Similar behavior for strain rates is also demonstrated in Ref. 15. An additional limitation of the model concerns the type of loading. The tests and parameter determination procedure described apply for monotonic loading, and applications of the constitutive model using these data are limited to these conditions. There is evidence to indicate that different model parameters are obtained for cyclic loadings.

Constitutive models have been implemented in finite element analysis by several analysts, and we illustrate such computations in the next section. In performing these computations and interpreting the predicted inelastic structural behavior, we should keep in mind that, whereas the unified constitutive models provide a powerful approach for modeling complex material behavior, there are a number of uncertainties inherent in the models themselves and in the values of the model parameters for a material.

11.3 Finite Element Analysis

Unified viscoplastic theories have been implemented by a number of finite element researchers. The first use of unified constitutive models in a finite element program was by Newman et al.[16] in 1976. Later, under the NASA HOST program, the Bodner–Partom and Walker models were implemented in finite element computer programs at General Electric and Pratt & Whitney. Kaufman et al.[17] describe these efforts as applied to gas turbine components. Moreno and Jordan,[18] as part of the NASA HOST program, developed and applied a unified constitutive model to gas turbine combustors using the MARC finite element program. Detailed studies of several rate-dependent plasticity models and their numerical implementation using adaptive finite element methods were conducted by Bass[19] and Bass and Oden.[20,21] In another finite element application,[22] the Bodner–Partom and Walker theories were compared for a thin, circular plate subject to highly localized, transient heating. Thornton et al.[23,24] developed a thermoviscoplastic computational method for hypersonic structures.

The behavior of a thermoviscoplastic structure is analyzed assuming that 1) thermomechanical coupling in the conservation of energy equation can be neglected, 2) the structural response is quasistatic, and 3) deformations are small. With these assumptions, an unsteady thermal analysis may be performed to determine the temperatures. Then, once these temperatures are known, the structure's viscoplastic response is determined. The solution is thus obtained by separately solving initial/boundary-value problems, first for the thermal and then for the structural response.

11.3.1 Initial-Value Viscoplasticity Problem

Consider a viscoplastic structure occupying a region Ω with boundary $\partial\Omega$. The behavior of the structure is described by the following system of differential equations.

Equilibrium in rate form:

$$\dot{\sigma}_{ij,j} + \dot{B}_i = 0 \tag{11.20}$$

where σ_{ij} denote components of the stress tensor, B_i are the body-force components per unit volume, and the summation convention is employed.

Kinematic relation for velocity gradients:

$$\dot{\varepsilon}_{ij} = \dot{\varepsilon}^e_{ij} + \dot{\varepsilon}^p_{ij} = \frac{(\dot{u}_{i,j} + \dot{u}_{j,i})}{2} \tag{11.21}$$

where ε_{ij} denotes the total strain components, and superscripts e and p denote elastic and inelastic strain components, respectively. The components of the displacement rates are $\dot{u}_i$.

Constitutive relations:

$$\dot{\sigma}_{ij} = E_{ijkl}\dot{\varepsilon}^e_{kl} - E_{ijkl}\alpha_{kl}\Delta\dot{T} \tag{11.22a}$$

$$\dot{\varepsilon}^p_{ij} = f_{ij}(\sigma_{ij}, Z_k) \qquad \text{(no sum)} \tag{11.22b}$$

$$\dot{Z}_i = g_i(\sigma_{ij}, Z_k) \qquad \text{(no sum)} \tag{11.22c}$$

where E_{ijkl} represents Hooke's tensor of elasticity parameters, α_{kl} are components of a tensor of thermal expansion parameters, and $\Delta\dot{T}$ represents the rate of the change in temperature. Both E_{ijkl} and α_{kl} are temperature-dependent. The constitutive functions are f_{ij} and g_i, where Z_k represents internal state variables. These functions and state variables characterize the viscoplastic response of the material.

The description of the problem is completed by prescribing the boundary and initial conditions,

$$\dot{u} = \dot{\bar{u}}_i \qquad \text{on } \partial\Omega_1 \tag{11.23a}$$

$$\dot{\sigma}_{ij}n_j = \dot{\bar{\sigma}} \qquad \text{on } \partial\Omega_2 \tag{11.23b}$$

where $\dot{\bar{u}}_i$ are prescribed surface displacement rates, n_j the components of a unit normal vector, and $\dot{\bar{\sigma}}_i$ the prescribed surface traction rates. The initial conditions include specifying the displacements $u_i(\boldsymbol{x}, 0)$, stresses $\sigma_{ij}(\boldsymbol{x}, 0)$, and internal state variables $Z_i(x, 0)$.

11.3.2 Finite Element Formulation

The finite element approach approximates displacement rates within an element by taking the displacement rates as

$$\{\dot{u}\} = [N]\{\dot{\delta}\} \tag{11.24}$$

where $[N]$ are the interpolation functions and $\{\dot{\delta}\}$ represents the nodal displacement rates. Using the strain-displacement equations in rate form, Eq. (11.21), an element's strain rates can be computed as

$$\{\dot{\varepsilon}\} = [B]\{\dot{\delta}\} \tag{11.25}$$

where $[B]$ is the strain-displacement matrix. Following usual finite element procedures, the finite element equations for a typical element are

$$[K(T)]\{\dot{\delta}\} = \{\dot{F}_p\} + \{\dot{F}_T\} + \{\dot{F}_\sigma\} + \{\dot{F}_B\} \tag{11.26}$$

where $[K(T)]$ is the element stiffness matrix, and the terms on the right side are element load vectors due to the rate of plastic strains, temperature, surface tractions, and body forces, respectively. These matrices are defined by

$$[K(T)] = \int_{\Omega_e} [B]^T [E(T)][B]\,\mathrm{d}\Omega \tag{11.27a}$$

$$\{\dot{F}_p\} = \int_{\Omega_e} [B]^T [E(T)]\{\dot{\varepsilon}^p\}\,\mathrm{d}\Omega \tag{11.27b}$$

$$\{\dot{F}_T\} = \int_{\Omega_e} [B]^T [E(T)]\{\alpha(T)\}\Delta\dot{T}\,\mathrm{d}\Omega \tag{11.27c}$$

$$\{\dot{F}_\sigma\} = \int_{\partial\Omega_e} [N]^T \{\dot{\bar{\sigma}}\}\,\mathrm{d}s \tag{11.27d}$$

$$\{\dot{F}_B\} = \int_{\Omega_e} [N]^T \{\dot{B}\}\,\mathrm{d}\Omega \tag{11.27e}$$

where $\partial\Omega_e$ denotes an element surface where tractions are defined.

The temperature affects the viscoplastic structural analysis directly in three ways:

1) The elasticity matrix $[E(T)]$ and the coefficients of thermal expansion $\{\alpha(T)\}$ depend on temperature.

2) Nodal loads $\{\dot{F}_T\}$ depend on local temperature rates.

3) Parameters in the Bodner–Partom constitutive model are temperature-dependent.

11.3.3 Viscoplastic Solution Method

Since the present approach uses an uncoupled (sometimes called "one-way coupled") formulation, the thermal problem is solved first, followed by the viscoplastic analysis. The transient thermal problem is solved by time-marching with a time step Δt_T, and the nodal temperatures at successive times $t_1, t_2, \ldots$ are obtained. These temperature vectors are used as input to the structural analysis.

The first computation in the structural analysis is to solve an initial statics problem if the initial temperature distribution $T(\boldsymbol{x}, 0)$ is not equal to a uniform reference temperature or if any initial static loads are present. The results of this analysis are the initial conditions (displacements and stresses) for the transient viscoplastic analysis.

The viscoplastic analysis time-marches with a time step Δt_s. Experience has shown that the time step required for the structural analysis is usually smaller than that for the thermal analysis, that is, $\Delta t_s < \Delta t_T$. At intermediate times in the structural analysis, the temperatures are linearly interpolated from the temperatures known at the beginning and end of the larger thermal time-step intervals.

The strategy employed in the viscoplastic algorithm is as follows. With the initial distribution of stress, temperature, and internal variables specified, use the equilibrium condition (11.26) to obtain the nodal displacement rates. Then, integrate the constitutive equations forward in time at the element Gauss integration points. With updated values of the stress, temperature, and internal variables at the new time, the equilibrium equation is solved again. This sequence of determining the nodal displacement rates and then advancing the constitutive equations in time is continued until the desired history of the initial/boundary-value problem has been obtained. Thus, the algorithm proceeds through the following steps:

1) At time t, initialize σ_{ij}, Z_i for each element.
2) Calculate $\dot{\varepsilon}^p_{ij} = f_{ij}(\sigma_{ij}, Z_k)$ for each element.
3) Assemble and solve $[K]\{\dot{\delta}\} = \{\dot{F}\}$.
4) Calculate $\dot{\varepsilon}_{ij}$ for each element; $\{\dot{\varepsilon}\} = [B]\{\dot{\delta}\}$.
5) Calculate $\dot{\sigma}_{ij}$ for each element, $\{\dot{\sigma}\} = [E]\{\dot{\varepsilon} - \dot{\varepsilon}^p\} - [E]\alpha\Delta\dot{T}$.
6) Calculate Z_i for each element, $Z_i = g_i\left(\sigma_{ij}, Z_k\right)$.
7) Integrate $\dot{\sigma}_{ij}$, $\dot{Z}$ forward for each element to get σ_{ij} and Z_i at $t + \Delta t_S$.
8) If $t + \Delta t_S < t_{\text{final}}$, go to 2; otherwise stop.

This computational method has been presented for a constant time step Δt_S. Computational experience by several investigators indicates that a very small time step can be required because of the stiff nature of the ordinary differential equations describing the internal state variables. To gain improved efficiency and reliability, a variable time-step algorithm may be implemented (see Ref. 23).

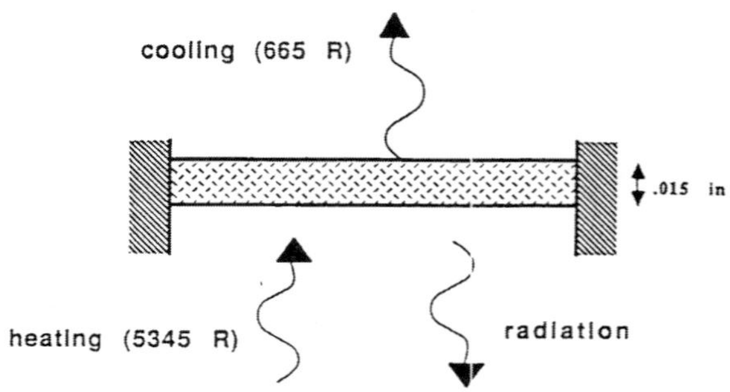

a) Thermal-structural model

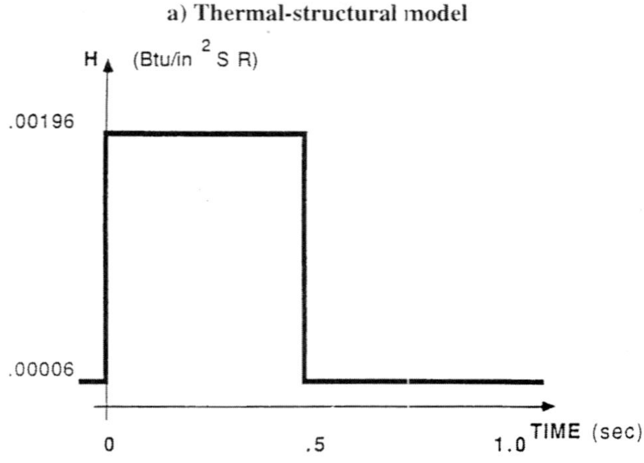

b) Convection coefficient history

Fig. 11.4 One-dimensional model of a convectively cooled structure.

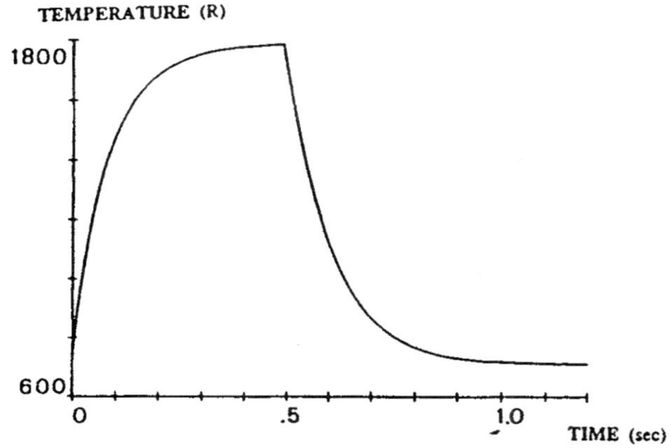

a) Temperature history

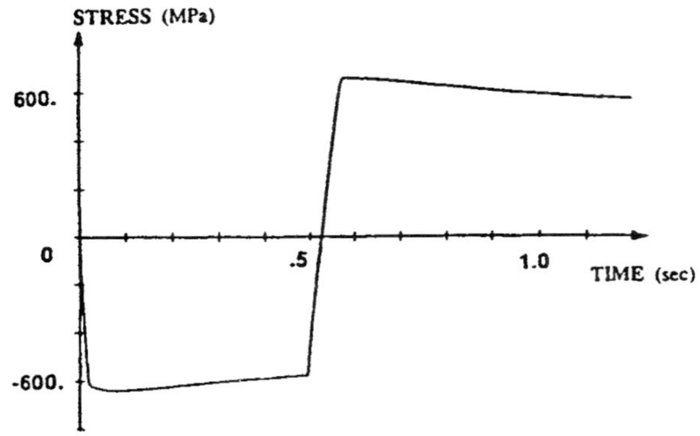

b) Stress history

Fig. 11.5 Thermoviscoplastic response of a one-dimensional model.

11.4 Applications

As mentioned in Sec. 11.1, leading edges of engine structures can experience intense local heating due to shock interactions. Two models of such structures are analyzed for heat loads representative of shock interactions. Material properties used represent a high-temperature, nickel-based super alloy. Computations are performed first for a simplified one-dimensional model to provide insight into the computational procedure. Then, a two-dimensional model of a realistic convectively cooled structure is analyzed in detail.

11.4.1 Simplified One-Dimensional Model

To gain a preliminary understanding of the behavior of a convectively cooled structure subjected to convective heating, a simple, one-dimensional bar model was investigated. A segment (Fig. 11.4a) of the aerodynamic skin is modeled assuming uniform temperature. The coolant is assumed to have a constant specified temperature, and the skin is heated convectively. A strong time variation of the convection coefficient (Fig. 11.4b) simulates the passage of a moving shock. The aerodynamic surface experiences an energy loss due to radiation, and we assume that the surface emissivity has a value of 0.8. The resulting transient temperature and the bar's compressive stress exhibit features to be expected in a more complex model.

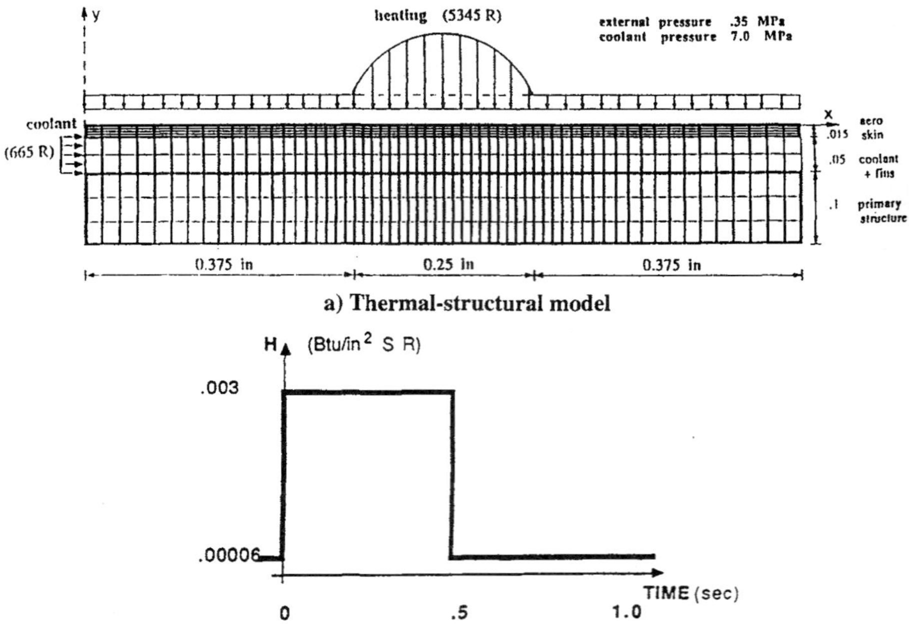

a) Thermal-structural model

b) Convection coefficient history

Fig. 11.6 Two-dimensional model of a convectively cooled structure.

The transient temperature history (Fig. 11.5a) shows a rapid rise following the sudden increase in the convective coefficient and then a smooth decay after the large convective heating is removed. The convective cooling causes the temperature to return quickly to equilibrium. The initial rapid increase in temperature causes the bar to yield in compression very early in the response (Fig. 11.5b). After the temperature begins to decay at $t = 0.5$ s, the stress rapidly changes to tension; hence, as the temperature returns to equilibrium, a large residual tensile stress remains in the bar. The tensile stress is induced because the initial high temperature causes compressive yielding. This yielding tends to shorten the bar permanently. However, the bar's fixed-end boundary conditions prohibit the bar from decreasing in length as the temperature decreases, so that a tensile stress develops as the material cools.

11.4.2 Convectively Cooled Structure

A more realistic two-dimensional model of a convectively cooled hypersonic structure Ref. 23, is shown in Fig. 11.6. This model represents a segment of a convectively cooled structure, such as a wall of a scramjet engine fuel-injection strut. The finite element thermal model (Fig. 11.6a) includes the following:

1) Conduction heat transfer in the aerodynamic skin, heat-exchanger fins, and primary structure.

2) Convective heat transfer between the walls of the coolant passage and coolant.

3) Mass-transport convection in the coolant, which has an unknown temperature.

4) Surface radiation on the aerodynamic skin.

The aerodynamic skin is convectively heated uniformly over its length, and superimposed on the uniform heating is a local, intense heating simulating a transient

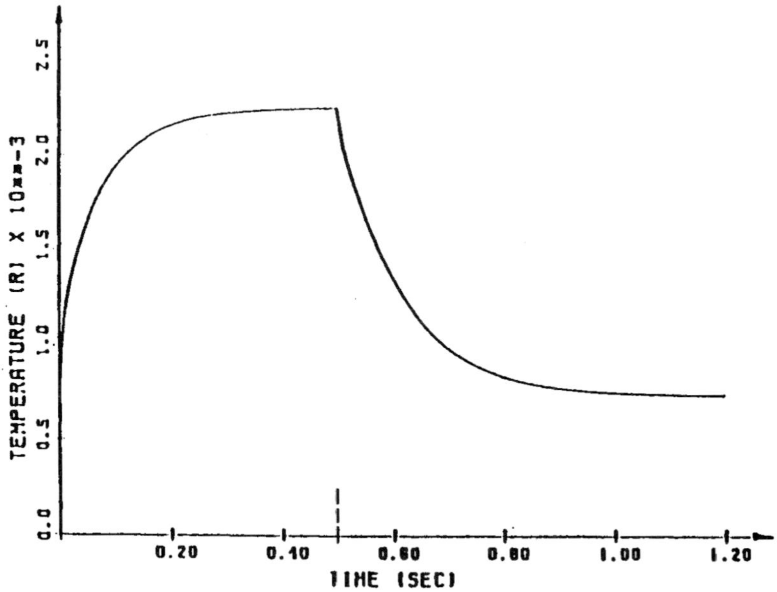

a) Temperature history at $x = 0.5$ in., $y = 0$

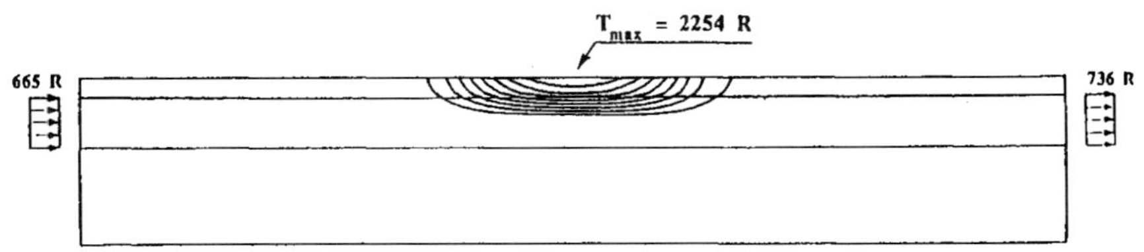

b) Temperature contours at $t = 0.5$ s

Fig. 11.7 Thermal response of a two-dimensional model.

shock. The transient heating is induced by the time-dependent convection coefficient shown in Fig. 11.6b.

In the plane strain structural finite element model, the primary structure and aerodynamic skin have unit thickness, but the heat-exchanger fins in the coolant passage are represented approximately by a single fin with a total thickness of 0.060 in. The wall segment has fixed displacement boundary conditions at the left and right ends. The aerodynamic skin, exchanger fins, and primary structure are also loaded by internal pressure (7.0 MPa = 1000 psi) in the coolant passage, which is relatively large compared to external aerodynamic pressure (0.35 MPa = 50 psi).

The model was analyzed first for steady aerodynamic heating and internal pressure operating conditions. Then the sudden localized aerodynamic heating was applied. Thus, the steady temperatures and stresses serve as initial conditions for the transient thermoviscoplastic analysis.

The thermal response of the wall segment is shown in Fig. 11.7. Figure 11.7a graphs the time history of the temperature at a point on the aerodynamic skin directly under the transient heating. Figure 11.7b plots contours of temperatures

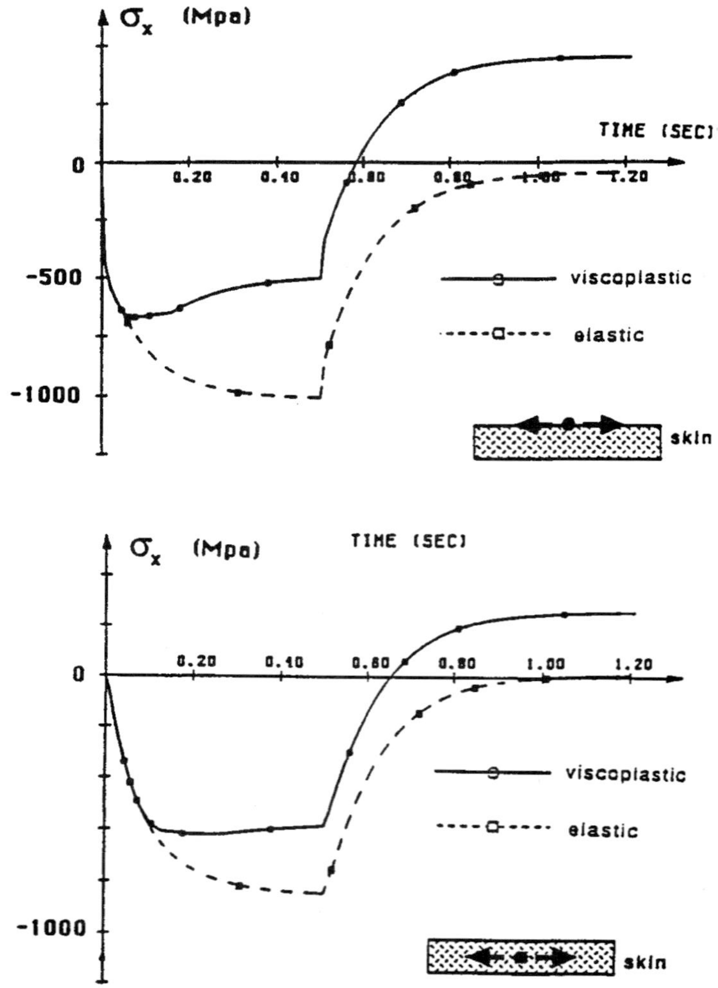

Fig. 11.8 Viscoplastic and elastic stress histories in aerodynamic skin for a two-dimensional model.

at $t = 0.5$ s, when temperatures are maximum. The temperature history is qualitatively similar to the results obtained in the one-dimensional model, indicating the rapid rise and fall of skin temperature with the simulated shock heating. The temperature contours trace relatively steep thermal gradients induced by the local heating and show that the coolant substantially limits the extent of the induced high temperatures. Thus, the high temperatures are confined to the aerodynamic skin, and the primary structure experiences only small temperature changes. The temperature gradients in the skin, particularly at the coolant/skin interface, are not predicted with high accuracy because of the engineering model of the coolant heat transfer. However, local temperature levels in the skin are reasonably accurate since net energy transfer to the coolant is modeled satisfactorily.

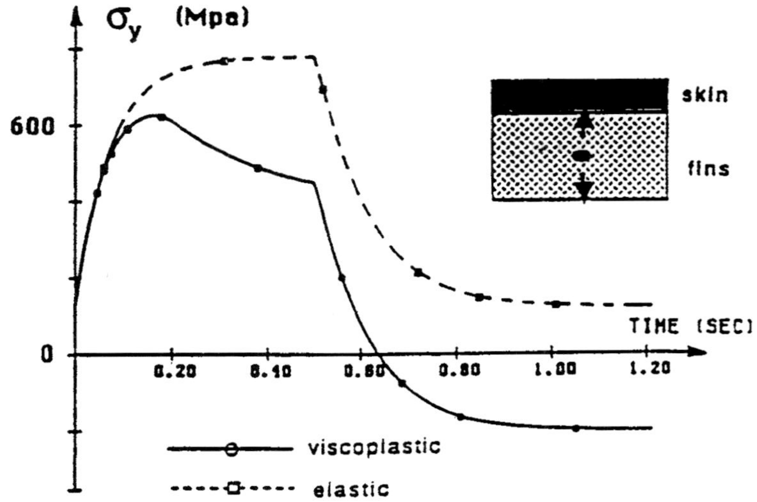

Fig. 11.9 Viscoplastic and elastic stress history in heat-exchanger fins for a two-dimensional model.

Viscoplastic histories of the horizontal stress component σ_x at two points through the skin thickness ($x = 0.4$, $y = 0$, and $y = -0.0075$) are shown and compared with an elastic solution in Fig. 11.8. The stress histories follow the temperature and, under the intense local heating stresses, are very similar to the results obtained from the one-dimensional model. At this location, the skin yields through most of its thickness. After the heating ceases, there is a rapid decay of stress and, under the intense local heating, there are residual tensile stresses. Figure 11.9 graphs the time history of the vertical component of stress σ_y in the heat-exchanger fins ($x = 0.5, y = -0.04$). High tensile stresses are induced, with significant local yielding, which are followed by residual compressive stress. The crude finite element model of the fins only approximates these stresses, but the high tensile stresses have the potential to cause a bond failure at the heat-exchanger/skin joint.

To illustrate the significance of viscoplastic phenomena, predictions of elastic theory (dashed line) are included for comparison in Figs. 11.8 and 11.9. The elastic computations were made with temperature-dependent elastic properties. The loss of stiffness due to the elevated temperatures for $0 < t < 0.5$ s accounts for the stress "rolling over" with an appearance of yielding. Generally, the elastic stresses are too high because yielding would actually occur. And, of course, because yielding does not occur, residual stresses are not predicted.

11.5 Closing Remarks

This chapter describes thermoviscoplasticity, which involves the use of unified constitutive models of sophisticated material behavior at elevated temperatures. One constitutive model—the Bodner–Partom model—is presented in some detail. The experiments needed to determine the material parameters are highlighted. A computational approach for implementing the Bodner–Partom constitutive models is presented. Two thermal-structural applications illustrate finite element computations using the Bodner–Partom model.

Data for unified constitutive models for several materials may be found in the literature, and commercial finite element programs provide thermoviscoplastic analysis capability. Material parameter determination techniques frequently use laboratory tests with one-dimensional states of stress. The concepts of continuum mechanics extend the models to two- and three-dimensional stress states in finite element applications. The performance of constitutive models for two- and three-dimensional stress states in complex structures with spatial temperature gradients has received limited evaluation. The difficulty and costs of performing thermal-structural experiments have precluded thorough evaluation of the models for complex structures. Thus, while thermoviscoplasticity provides formidable analysis capability, we must recognize that it remains under development. The concept is powerful and, as the subject matures with further research, we may expect that we will be able to apply the approach with confidence to model inelastic behavior of thermal structures under very severe operating conditions.

References

[1]Dechaumphai, P., Thornton, E. A., and Wieting, A. R., "Flow-Thermal-Structural Study of Aerodynamically Heated Leading Edges," *Journal of Spacecraft and Rockets*, Vol. 26, No. 4, 1989, pp. 201–209.

[2]Polesky, S., Dechaumphai, P., Glass, D. E., and Pandey, A. K., "Three-Dimensional Thermal-Structural Analysis of a Swept Cowl Leading Edge Subjected to Skew Shock-Shock Interference Heating," *Journal of Thermophysics and Heat Transfer*, Vol. 6, No. 1, 1992, pp. 48–54.

[3]Melis, M. W., and Gladden, H. J., "Thermostructural Analysis with Experimental Verification in a High Heat Flux Facility of a Simulated Cowl Lip," AIAA Paper 88-2222, April 1988.

[4]Miller, A. K., *Unified Constitutive Equations for Creep and Plasticity*, Elsevier Applied Science, New York, 1987.

[5]Chan, K. S., Lindholm, U. S., Bodner, S. R., Hill, J. R., Weber, R. M., and Meyer, T. G., "Constitutive Modeling for Isotropic Materials (HOST)," NASA CR-179522, Aug. 1986.

[6]Chan, K. S., Lindholm, U. S., and Bodner, S. R., "Constitutive Modeling for Isotropic Materials (HOST)," NASA CR-182132, June 1988.

[7]Chan, K.S., Bodner, S. R., Walker, K. P., and Lindholm, U. S., "A Survey of Unified Constitutive Theories," *Proceedings of the Second Symposium on Nonlinear Constitutive Relations for High Temperature Applications*, NASA-CP-2369, Cleveland, OH, 1984, pp. 1–23.

[8]James, G. H., Imbrie, P. K., and Hill, P. S., "An Experimental Comparison of Several Current Viscoplastic Constitutive Models at Elevated Temperatures," *Journal of Engineering Materials and Technology*, Vol. 109, April 1987, pp. 130–139.

[9]Bodner, S. R., and Partom, Y., "A Representation of Elastic-Viscoplastic Strain Hardening Behavior for Generalized Straining Histories," *Institute of Physics Conference Series No. 21*, 1975.

[10]Bodner, S. R., and Partom, Y., "Constitutive Equations for Elastic-Viscoplastic Strain-Hardening Materials," *Journal of Applied Mechanics*, Vol. 42, June 1975, pp. 385–389.

[11]Bodner, S. R., "Review of a Unified Elastic-Viscoplastic Theory," *Unified Constitutive Equations for Creep and Plasticity*, edited by A. K. Miller, Elsevier Applied Science, New York, 1987, pp. 273–301.

[12]Chan, K. S., Bodner, S. R., and Lindholm, U. S., *Constitutive Modeling for Isotropic Materials (HOST) 2nd Annual Report*, NASA-CR 174980, San Antonio, TX, 1985.

[13]Chan, K. S., Bodner, S. R., and Lindholm, U. S., *Constitutive Modeling for Isotropic Materials (HOST) Third Annual Report*, NASA-CR 174980, San Antonio, TX, 1986.

[14]Chan, K. S., Bodner, S. R., and Lindholm, U. S., "Phenomenological Modeling of Hardening and Thermal Recovery in Metals," *Journal of Engineering Materials and Technology*, Vol. 110, Jan. 1988, p. 108.

[15]Rowley, M. A., and Thornton, E. A, "Constitutive Modeling of the Visco-Plastic Response of Hastelloy-X and Aluminum Alloy 8009," *Journal of Engineering Materials and Technology*, Vol. 118, No. 1, 1996, pp. 19–27.

[16]Newman, M., Zaphir, Z., and Bodner, S. R., "Finite Element Analysis for Time-Dependent Inelastic Material Behavior," *Computers and Structures*, Vol. 6, 1976, pp. 157–162.

[17]Kaufman, A., Laflen, J. H., and Lindholm, U. S., "Unified Constitutive Material Models for Nonlinear Finite-Element Structural Analysis," AIAA Paper 85-1418, July 1985.

[18]Moreno, V., and Jordan, E. H., "Prediction of Material Thermomechanical Response with a Unified Viscoplastic Constitutive Model," *International Journal of Plasticity*, Vol. 2, March 1986, pp. 223–245.

[19]Bass, J. M., "Numerical Implementation of Constitutive Models in Rate-Dependent Plasticity," Ph.D. Dissertation, Univ. of Texas, Austin, 1985.

[20]Bass, J. M., and Oden, J. T., "Adaptive Finite Element Methods for a Class of Evolution Problems in Viscoplasticity," *International Journal of Engineering Science*, Vol. 25, No. 6, 1987, pp. 623–653.

[21]Bass, J. M., and Oden, J. T., "Numerical Solution of the Evolution Equations of Damage and Rate-Dependent Plasticity," *International Journal of Engineering Science*, Vol. 26, No. 7, 1988, pp. 713–740.

[22]Chang, H. T., and Allen, D. H., "A Finite Element Analysis of a Viscoplastic Plate Subjected to Rapid External Heating," *Mechanics and Structures and Machines, An International Journal*, Vol. 17, 1989, pp. 73–93.

[23]Thornton, E. A., Oden, J. T., Tworzydlo, W. W., and Youn, S. K., "Thermoviscoplastic Analysis of Hypersonic Structures Subjected to Severe Aerodynamic Heating," *Journal of Aircraft*, Vol. 27, No. 9, 1990, pp. 826–835.

[24]Thornton, E. A., and Kolenski, J. D., "Viscoplastic Response of Structures for Intense Local Heating," *Journal of Aerospace Engineering*, Vol. 7, No. 1, 1994, pp. 50–71.

Problems

11.1. Develop the equations for a Bodner–Partom model for a one-dimensional state of stress. Assume that directional hardening may be neglected and that the internal state variable $Z = Z^I$.

11.2. Use the one-dimensional equations formulated in the preceding problem to develop a computer simulation for a uniaxial tension test performed at a prescribed strain rate. To integrate the equations, use Euler integration (forward difference) with a time step Δt.

(a) Use the data for Hastelloy-X (Tables 11.2) to predict the stress–strain curves shown in Fig. 11.3.

(b) For Hastelloy-X, use the Bodner–Partom model to investigate the effect of the prescribed strain rate on the material's stress-strain behavior.

11.3. Use the computer simulation for a uniaxial test developed in the preceding problem to investigate the inelastic stress–strain behavior of the super alloy B1900 + Hf under cyclic loading. Subject the material to several cycles of alternating

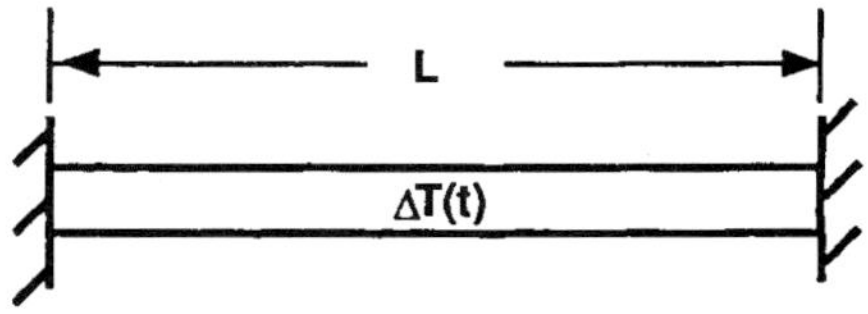

Fig. P11.4

tension and compression at a constant strain rate. Investigate the effects of temperature and strain rate on the material's cyclic behavior. Reference 24 gives the Bodner–Partom constants for B1900 + Hf.

11.4. A restrained bar (Fig. P11.4) is subjected to a prescribed uniform temperature rise $\Delta T(t)$, Ref. 24. Assume that buckling cannot occur and that the bar experiences uniaxial stress. The temperature varies with time according to $\Delta T(t) = T_r[1 - \exp(-t/t_T)]$, where T_r is the radiation equilibrium temperature, and t_T is the thermal rise time.

(a) Using a Bodner–Partom constitutive model, develop a computer simulation for the time history of the bar's axial stress.

(b) Use the simulation model to compute stress histories for a Hastelloy-X bar for $T_r = 400$, 700, and 1000°F. Use a rise time $t_T = 50$ s, and take the coefficient of thermal expansion as $\alpha_T = 15\text{E-6}$ 1/K.

(c) Plot temperature and stress histories, and discuss the results.

(d) Investigate the effect of the thermal rise time t_T on the stress response. What is the bar's strain rate? How does the strain rate compare with the experimental strain rates (Table 11.2) used in evaluation of the Hastelloy-X constitutive model's parameters?

11.5. In Ref. 23, a variable time-step algorithm is used to integrate the constitutive equations

(a) Explain how the variable time-step algorithm works.

(b) Implement the algorithm in the computer simulation developed in the preceding problem.

(c) Evaluate the benefits of the variable time-step scheme.

Appendix: Finite Element Development of von Kármán Plate Equations

This appendix presents the development of the weak form of the von Kármán large deflection plate equations. The results of the formulation presented here are used in Chapter 8 in Sec. 8.3.4 for a finite element formulation for large deflection of plates. The development starts by multiplying Eqs. (8.66) by a test function $v(x, y)$ and integrating over the plate area Ω.

$$\int_\Omega \left[\frac{\partial N_x}{\partial x} + \frac{\partial N_{xy}}{\partial y}\right] v \, \mathrm{d}\Omega = 0 \tag{A.1a}$$

$$\int_\Omega \left[\frac{\partial N_{xy}}{\partial x} + \frac{\partial N_y}{\partial y}\right] v \, \mathrm{d}\Omega = 0 \tag{A.1b}$$

$$\int_\Omega \left[\frac{\partial^2 M_x}{\partial x^2} + 2\frac{\partial^2 M_{xy}}{\partial x \partial y} + \frac{\partial^2 M_y}{\partial y^2}\right] v \, \mathrm{d}\Omega + \int_\Omega \left[\frac{\partial}{\partial x}\left(N_x \frac{\partial w}{\partial x}\right) + \frac{\partial}{\partial y}\left(N_{xy}\frac{\partial w}{\partial x}\right)\right.$$
$$\left. + \frac{\partial}{\partial x}\left(N_{xy}\frac{\partial w}{\partial y}\right) + \frac{\partial}{\partial y}\left(N_y \frac{\partial w}{\partial y}\right)\right] v \, \mathrm{d}\Omega + \int_\Omega q v \, \mathrm{d}\Omega = 0 \tag{A.1c}$$

These equations are then integrated by parts to introduce the boundary conditions on the plate edge. Equations (A.1a) and (A.1b) become

$$\int_\Omega \left[N_x \frac{\partial v}{\partial x} + N_{xy}\frac{\partial v}{\partial y}\right] \mathrm{d}\Omega = \int_S v(N_x n_x + N_{xy} n_y) \, \mathrm{d}s \tag{A.2a}$$

$$\int_\Omega \left[N_{xy} \frac{\partial v}{\partial x} + N_y \frac{\partial v}{\partial y}\right] \mathrm{d}\Omega = \int_S v(N_{xy} n_x + N_y n_y) \, \mathrm{d}s \tag{A.2b}$$

where n_x and n_y are components of a unit vector normal to the plate edge. Equation (A.1c) requires two integrations by parts to yield

$$\int_S \left[\left(\frac{\partial M_x}{\partial x} + \frac{\partial M_{xy}}{\partial y}\right) v n_x + \left(\frac{\partial M_{xy}}{\partial x} + \frac{\partial M_y}{\partial y}\right) v n_y\right] \mathrm{d}s - \int_S \left[(M_x n_x + M_{xy} n_y)\frac{\partial v}{\partial x}\right.$$
$$\left. + (M_y n_y + M_{xy} n_x)\frac{\partial v}{\partial y}\right] \mathrm{d}s + \int_\Omega \left[M_x \frac{\partial^2 v}{\partial x^2} + 2M_{xy}\frac{\partial^2 v}{\partial x \partial y} + M_y \frac{\partial^2 v}{\partial y^2}\right] \mathrm{d}\Omega$$
$$- \int_\Omega \left[\left(N_x \frac{\partial w}{\partial x} + N_{xy}\frac{\partial w}{\partial y}\right)\frac{\partial v}{\partial x} + \left(N_{xy}\frac{\partial w}{\partial x} + N_y \frac{\partial w}{\partial y}\right)\frac{\partial v}{\partial y}\right] \mathrm{d}\Omega + \int_S \left[\left(N_x \frac{\partial w}{\partial x}\right.\right.$$
$$\left.\left. + N_{xy}\frac{\partial w}{\partial y}\right) n_x + \left(N_{xy}\frac{\partial w}{\partial x} + N_y \frac{\partial w}{\partial y}\right) n_y\right] v \, \mathrm{d}s + \int_\Omega q v \, \mathrm{d}\Omega = 0 \tag{A.2c}$$

The surface integrals apply to a plate of arbitrary shape. To simplify the preceding equations, we now use the following derivative transformations in the boundary integral terms

$$\frac{\partial}{\partial x} = \frac{\partial}{\partial n}n_x - \frac{\partial}{\partial s}n_y \qquad \frac{\partial}{\partial y} = \frac{\partial}{\partial n}n_y + \frac{\partial}{\partial s}n_x$$

and the following relations/definitions:

$$Q_x = \frac{\partial M_{xy}}{\partial y} + \frac{\partial M_x}{\partial x}$$

$$Q_y = \frac{\partial M_y}{\partial y} + \frac{\partial M_{xy}}{\partial x}$$

$$V_n = \frac{\partial M_{ns}}{\partial s} + Q_n$$

$$\bar{N}_x = N_x n_x + N_{xy} n_y$$

$$\bar{N}_y = N_{xy} n_x + N_y n_y$$

$$N_n = N_x n_x^2 + N_{xy} n_x n_y + N_y n_y^2$$

$$N_{ns} = (N_y - N_x) n_x n_y + \left(n_x^2 - n_y^2\right) N_{xy}$$

These equations appear in many classical plate theory texts. The membrane forces $\bar{N}_x$ and $\bar{N}_y$ are surface tractions, and $\bar{N}_n$ and $\bar{N}_{ns}$ are the normal and tangential components of the edge membrane forces. With these relations, Eqs. (A.2) may be written as

$$\int_\Omega \left[N_x \frac{\partial v}{\partial x} + N_{xy} \frac{\partial v}{\partial y}\right] \mathrm{d}\Omega = \int_S v \bar{N}_x \, \mathrm{d}s \tag{A.3a}$$

$$\int_\Omega \left[N_{xy} \frac{\partial v}{\partial x} + N_y \frac{\partial v}{\partial y}\right] \mathrm{d}\Omega = \int_S v \bar{N}_y \, \mathrm{d}s \tag{A.3b}$$

$$\begin{aligned}
&\int_\Omega \left(M_x \frac{\partial^2 v}{\partial x^2} + 2M_{xy} \frac{\partial^2 v}{\partial x \partial y} + M_y \frac{\partial^2 v}{\partial y^2}\right) \mathrm{d}\Omega + \int_\Omega q v \, \mathrm{d}\Omega \\
&- \int_\Omega \left[N_x \frac{\partial w}{\partial x}\frac{\partial v}{\partial x} + N_{xy}\left(\frac{\partial w}{\partial x}\frac{\partial v}{\partial y} + \frac{\partial w}{\partial y}\frac{\partial v}{\partial x}\right) + N_y \frac{\partial w}{\partial y}\frac{\partial v}{\partial y}\right] \mathrm{d}\Omega \\
&+ \int_S \left[\left(N_x \frac{\partial w}{\partial x} + N_{xy}\frac{\partial w}{\partial y}\right) n_x + \left(N_{xy}\frac{\partial w}{\partial x} + N_y \frac{\partial w}{\partial y}\right) n_y\right] v \, \mathrm{d}s \\
&+ \int_S \left(V_n + N_n \frac{\partial w}{\partial n} + N_{ns}\frac{\partial w}{\partial s}\right) v \, \mathrm{d}s - \int_S M_n \frac{\partial v}{\partial n} \, \mathrm{d}s - [M_{ns} v]_1^2 = 0
\end{aligned} \tag{A.3c}$$

The formulation is completed by letting the test function v be a typical finite element interpolation function, where the interpolation function for the in-plane membrane displacements are N_i^m, and the interpolation functions for the bending displacement are N_i^b; the subscript i ranges over the nodal degrees of freedom. In Eqs. (A.3a) and (A.3b), the test function $v = N_i^m$ since these equations represent membrane equilibrium. In Eq. (A. 3c), the test function $v = N_i^b$ since this equation represents bending equilibrium. Then, writing the equations in matrix form yields

$$\int_\Omega \left[\frac{\partial N_i^m}{\partial x} + \frac{\partial N_i^m}{\partial y} \right] \begin{Bmatrix} N_x \\ N_{xy} \end{Bmatrix} \mathrm{d}\Omega = \int_S N_i^m \bar{N}_x \, \mathrm{d}s \tag{A.4a}$$

$$\int_\Omega \left[\frac{\partial N_i^m}{\partial x} + \frac{\partial N_i^m}{\partial y} \right] \begin{Bmatrix} N_{xy} \\ N_y \end{Bmatrix} \mathrm{d}\Omega = \int_S N_i^m \bar{N}_y \, \mathrm{d}s \tag{A.4b}$$

$$\begin{aligned} &\int_\Omega \left[\frac{\partial^2 N_i^b}{\partial x^2} \quad \frac{\partial^2 N_i^b}{\partial y^2} \quad 2\frac{\partial^2 N_i^2}{\partial x \partial y} \right] \begin{Bmatrix} M_x \\ M_y \\ M_{xy} \end{Bmatrix} \mathrm{d}\Omega + \int_\Omega q N_i^b \, \mathrm{d}\Omega \\ &- \int_\Omega \left[\frac{\partial N_i^b}{\partial x} \quad \frac{\partial N_i^b}{\partial y} \right] \begin{bmatrix} N_x & N_{xy} \\ N_{xy} & N_y \end{bmatrix} \begin{Bmatrix} \dfrac{\partial w}{\partial x} \\ \dfrac{\partial w}{\partial y} \end{Bmatrix} \mathrm{d}\Omega \\ &+ \int_S \left(V_n + N_n \frac{\partial w}{\partial n} + N_{ns} \frac{\partial w}{\partial s} \right) N_i^b \, \mathrm{d}s \\ &- \int_S M_n \frac{\partial N_i^b}{\partial n} \, \mathrm{d}s - \left[M_{ns} N_i^b \right]^2 = 0 \end{aligned} \tag{A.4c}$$

The terms involving stress resultants integrated over the plate area lead to element stiffness matrices. The surface integral terms and the $[M_{ns} N_i^b]_1^2$ term represent traction boundary conditions or load vectors.

The final matrix form in Eq. (8.75) is developed from Eqs. (A.4) by expressing force and moment resultants in terms of the displacement degrees of freedom with matrices of interpolation function and their derivatives, expressing the integrals as stiffness matrices and load vectors, and assembling the three equations into the single partitioned matrix equation.

This page intentionally left blank

Subject Index

This page intentionally left blank

Index of Authors Cited